PRACTICAL
OPTIMIZATION

PRACTICAL OPTIMIZATION

PHILIP E. GILL
WALTER MURRAY
MARGARET H. WRIGHT

Systems Optimization Laboratory
Department of Operations Research
Stanford University
California, USA

1981

 ACADEMIC PRESS, INC.

Harcourt Brace Jovanovich, Publishers

London   San Diego   New York   Berkeley
Boston   Sydney   Tokyo   Toronto

ACADEMIC PRESS LIMITED
24/28 Oval Road,
London NW1 7DX

United States Edition published by
ACADEMIC PRESS, INC.
San Diego, CA 92101

British Library Cataloguing in Publication Data
Gill, P.E.
   Practical optimization
   1. Mathematics optimization
   I. Title   II. Murray, W.   III. Wright, M.H.
   515   QA402.5

   ISBN 0-12-283950-1 (hardback)
   ISBN 0-12-283952-8 (paperback)

   LCCCN 81-66366

Printed in Great Britain by
Thomson Litho Ltd, East Kilbride, Scotland

PREFACE

As researchers at the National Physical Laboratory and Stanford University, and as contributors to the Numerical Algorithms Group (NAG) software library, we have been involved for many years in the development of numerical methods and software for the solution of optimization problems. Within the past twenty years, there has been a dramatic increase in the efficiency and reliability of optimization methods for almost all problem categories. However, this improved capability has been achieved by the use of more complicated ideas, particularly from the areas of numerical linear algebra and finite-precision calculation.

The best methods available today are extremely complex, and their manner of operation is far from obvious, especially to users from other disciplines. This book is intended as a treatment — necessarily broad — of the subject of *practical optimization*. The word "practical" is included in the title in order to convey our concern not only with the motivation for optimization methods, but also with details of implementation that affect the performance of a method in practice. In particular, we believe that some consideration of the effects of finite-precision computation is essential in order for any description of a method to be useful. We also believe that it is important to discuss the linear algebraic processes that are used to perform certain portions of all optimization methods.

This book is meant to be largely self-contained; we have therefore devoted one chapter to a description of the essential results from numerical linear algebra and the analysis of rounding errors in computation, and a second chapter to a treatment of optimality conditions.

Selected methods for unconstrained, linearly constrained, and nonlinearly constrained optimization are described in three chapters. This discussion is intended to present an overview of the methods, including the underlying motivation as well as particular theoretical and computational features. Illustrations have been used wherever possible in order to stress the geometric interpretation of the methods. The methods discussed are primarily those with which we have had extensive experience and success; other methods are described that provide special insights or background. References to methods not discussed and to further details are given in the extensive Notes and Bibliography at the end of each section. The methods have been presented in sufficient detail to allow this book to be used as a text for a university-level course in numerical optimization.

Two chapters are devoted to selected less formal, but nonetheless crucial, topics that might be viewed as "advice to users". For example, some suggestions concerning modelling are included because we have observed that an understanding of optimization methods can have a beneficial effect on the modelling of the activities to be optimized. In addition, we have presented an extensive discussion of topics that are crucial in using and understanding a numerical optimization method — such as selecting a method, interpreting the computed results, and diagnosing (and, if possible, curing) difficulties that may cause an algorithm to fail or perform poorly.

In writing this book, the authors have had the benefit of advice and help from many people. In particular, we offer special thanks to our friend and colleague Michael Saunders, not only for many helpful comments on various parts of the book, but also for all of his characteristic good humour and patience when the task of writing this book seemed to go on forever. He has played a major role in much of the recent work on algorithms for large-scale problems described in Chapters 5 and 6.

We gratefully acknowledge David Martin for his tireless help and support for the optimization group at the National Physical Laboratory, and George Dantzig for his efforts to assemble the algorithms group at the Systems Optimization Laboratory, Stanford University.

We thank Brian Hinde, Susan Hodson, Enid Long and David Rhead for their work in developing and testing many of the algorithms described in this book.

We are grateful to Greg Dobson, David Fuchs, Stefania Gai, Richard Stone and Wes Winkler, who have been helpful in many ways during the preparation of this manuscript. The clarity of certain parts of the text has been improved because of helpful comments from Paulo Benevides-Soares, Andy Conn, Laureano Escudero, Don Iglehart, James Lyness, Jorge Moré, Michael Overton and Danny Sorensen.

We thank Clive Hall for producing the computer-generated figures on the Laserscan plotter at the National Physical Laboratory. The remaining diagrams were expertly drawn by Nancy Cimina.

This book was typeset by the authors using the TₑX mathematical typesetting system of Don Knuth*. We are grateful to Don Knuth for his efforts in devising the TₑX system, and for making available to us various TₑX macros that improved the quality of the final text. We also thank him for kindly allowing us to use the Alphatype CRS typesetter, and Chris Tucci for his substantial help in producing the final copy.

Finally, we offer our deepest personal thanks to those closest to us, who have provided encouragement, support and humour during the time-consuming process of writing this book.

Stanford University
May, 1981

P. E. G.
W. M.
M. H. W.

---

*D. E. Knuth, *TEX and METAFONT, New Directions in Typesetting*, American Mathematical Society and Digital Press, Bedford, Massachusetts (1979).

CONTENTS

Those sections marked with "*" contain material of a rather specialized nature that may be omitted on first reading.

CHAPTER ONE

INTRODUCTION

Mankind always sets itself only such problems as it can solve....

—KARL MARX (1859)

1.1. DEFINITION OF OPTIMIZATION PROBLEMS

An optimization problem begins with a set of independent variables or parameters, and often includes conditions or restrictions that define acceptable values of the variables. Such restrictions are termed the *constraints* of the problem. The other essential component of an optimization problem is a single measure of "goodness", termed the *objective function*, which depends in some way on the variables. The solution of an optimization problem is a set of allowed values of the variables for which the objective function assumes an "optimal" value. In mathematical terms, optimization usually involves *maximizing* or *minimizing*; for example, we may wish to maximize profit or minimize weight.

Problems in all areas of mathematics, applied science, engineering, economics, medicine, and statistics can be posed in terms of optimization. In particular, mathematical models are often developed in order to analyze and understand complex phenomena. Optimization is used in this context to determine the form and characteristics of the model that corresponds most closely to reality. Furthermore, most decision-making procedures involve explicit solution of an optimization problem to make the "best" choice. In addition to their role *per se*, optimization problems often arise as critical sub-problems within other numerical processes. This situation is so common that the existence of the optimization problem may pass unremarked — for example, when an optimization problem must be solved to find points where a function reaches a certain critical value.

Throughout this book, it will be assumed that the ultimate objective is to *compute* the solution of an optimization problem. In order to devise solution techniques, it is helpful to assume that optimization problems can be posed in a standard form. It is clearly desirable to select a standard form that arises naturally from the nature of most optimization problems, in order to reduce the need for re-formulation. The general form of optimization problem to be considered may be expressed in mathematical terms as:

$$
\begin{array}{lll}
\text{NCP} & \underset{x \in \Re^n}{\text{minimize}} & F(x) \\
& \text{subject to} & c_i(x) = 0, \quad i = 1, 2, \ldots, m'; \\
& & c_i(x) \geq 0, \quad i = m' + 1, \ldots, m.
\end{array}
$$

The objective function F and constraint functions $\{c_i\}$ (which, taken together, are termed the *problem functions*) are real-valued scalar functions.

To illustrate some of the flavour and diversity of optimization problems, we consider two specific examples. Firstly, the layout of the text of this book was designed by the TEX computer

1

typesetting system using optimization techniques. The aim of the system is to produce a visually pleasing arrangement of text with justified margins. Two means are available to achieve this objective: the spaces between letters, words, lines and paragraphs can be adjusted, and words can be split between lines by hyphenation. An "ideal" spacing is specified for every situation in which a space may occur. These spaces may then be stretched or compressed within given limits of acceptability, subject to penalties for increasing the amount of deviation from the ideal. Varying penalties are also imposed to minimize undesirable features, such as two consecutive lines that end with hyphenated words or a page that begins with a displayed equation. The process of choosing a good text layout includes many of the elements of a general optimization problem: a single function that measures quality, parameters that can be adjusted in order to achieve the best objective, and restrictions on the form and extent of the allowed variation.

We next consider a simplified description of a real problem that illustrates the convenience of the standard form. The problem is to design the nose cone of a vehicle, such that, when it travels at hypersonic speeds, the air drag is minimized. Hence, the function to be optimized is the drag, and the parameters to be adjusted are the specifications of the structure of the nose cone. In order to be able to compute the objective function, it is necessary first to devise a model of the nose cone in terms of the chosen parameters. For this problem, the nose cone is represented as a series of conical sections, with a spherical section as the front piece and a fixed final radius R. Figure 1a illustrates the chosen model, and shows the parameters to be adjusted. Although the idealized model deviates from a real-world nose cone, the approximation should not noticeably impair the quality of the solution, provided that the number of conical sections is sufficiently large.

The next step is to formulate the drag as a scalar function of the eight parameters $\alpha_1, \ldots, \alpha_4$, $r_1, \ldots, r_4$. The function $D(\alpha_1, \ldots, \alpha_4, r_1, \ldots, r_4)$ will be assumed to compute an estimate of the drag on the nose cone for a set of particular values of the free variables, and thus D will be the objective function of the resulting optimization problem.

In order to complete the formulation of the problem, some restrictions must be imposed on the values of the design parameters in order for the mathematical model to be meaningful and for the optimal solution to be implementable. In particular, the radii $r_1, \ldots, r_4$ must not be negative, so that the constraints

$$r_i \geq 0, \quad i = 1, \ldots, 4,$$

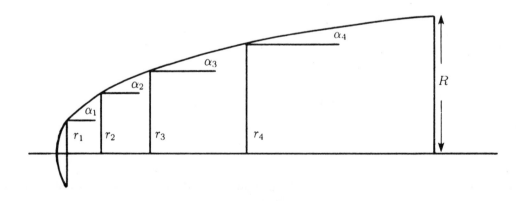

Figure 1a. Cross-section of a conical representation of a nose cone.

will be included in the problem specification. The nature of the design also means that the values of $\{\alpha_i\}$ must lie in the range $[0, \pi/2]$; and that the angles must be non-increasing. Such further restrictions add the constraints

$$0 \leq \alpha_i \leq \frac{\pi}{2}, \qquad i = 1, \ldots, 4;$$
$$\alpha_4 \leq \alpha_3 \leq \alpha_2 \leq \alpha_1.$$

In addition to these constraints, the nose cone may be required to perform some function — for example, its volume must be sufficient to carry objects up to a certain size. Furthermore, the overall length may be limited in order to ensure a sensible shape. Such requirements mean that additional functions of the design parameters will be required to satisfy constraints such as

$$\text{volume}(\alpha_1, \ldots, \alpha_4, r_1, \ldots, r_4) \geq V;$$
$$\text{length}(\alpha_1, \ldots, \alpha_4, r_1, \ldots, r_4) \leq L.$$

Further constraints may be required in order for the final nose cone to meet safety standards, and so on.

The final mathematical statement of the nose cone problem is then

$$\begin{aligned}
\underset{\alpha_1, \ldots, r_4}{\text{minimize}} \quad & D(\alpha_1, \ldots, r_4) \\
\text{subject to} \quad & 0 \leq r_i \leq R, \quad i = 1, \ldots, 4; \\
& 0 \leq \alpha_i \leq \frac{\pi}{2}, \quad i = 1, \ldots, 4; \\
& \alpha_4 \leq \alpha_3 \leq \alpha_2 \leq \alpha_1; \\
& 0 \leq \text{volume}(\alpha_1, \ldots, \alpha_4, r_1, \ldots, r_4) - V; \\
& 0 \leq L - \text{length}(\alpha_1, \ldots, \alpha_4, r_1, \ldots, r_4).
\end{aligned}$$

This example indicates the steps involved in posing a typical problem, as well as the way in which desired qualities of the solution can be transformed naturally into the chosen mathematical representation of the problem. Clearly, an algorithm capable of solving problems of the form NCP can be applied directly to the nose cone design problem.

1.2. CLASSIFICATION OF OPTIMIZATION PROBLEMS

The vast majority of optimization problems can be expressed in the form NCP. Even those that do not fit into this framework can often be posed as a sequence of standard problems. However, the existence of a standard, very general representation does not imply that all distinctions among problems should be ignored. When faced with *any* problem, it is usually advantageous to determine special characteristics that allow the problem to be solved more efficiently. For example, it may be possible to omit tests for situations that cannot occur, or to avoid re-computing quantities that do not vary. Therefore, we consider how to classify optimization problems in order to enhance the efficiency of solution methods.

The most extreme form of classification would be to assign every problem to a separate category. However, this approach is based on the false premise that *every* difference is significant with respect to solving the problem. For instance, with such a scheme it would be necessary to change methods if the number of variables in a problem changed from 3 to 4! Although no set of categories is ideal for every circumstance, a reasonable classification scheme can be developed based on balancing the improvements in efficiency that are possible by taking advantage of special properties against the additional complexity of providing a larger selection of methods and software.

The most obvious distinctions in problems involve variations in the mathematical characteristics of the objective and constraint functions. For example, the objective function may be very smooth in some cases, and discontinuous in others; the problem functions may be of a simple form with well understood properties, or their computation may require the solution of several complicated sub-problems.

The following table gives a typical classification scheme based on the nature of the problem functions, where significant algorithmic advantage can be taken of each characteristic:

Properties of $F(x)$	Properties of $\{c_i(x)\}$
Function of a single variable	No constraints
Linear function	Simple bounds
Sum of squares of linear functions	Linear functions
Quadratic function	Sparse linear functions
Sum of squares of nonlinear functions	Smooth nonlinear functions
Smooth nonlinear function	Sparse nonlinear functions
Sparse nonlinear function	Non-smooth nonlinear functions
Non-smooth nonlinear function	

For example, a particular problem might be categorized as the minimization of a smooth nonlinear function subject to upper and lower bounds on the variables.

Other features may also be used to distinguish amongst optimization problems. The *size* of a problem affects both the storage and the amount of computational effort required to obtain the solution, and hence techniques that are effective for a problem with a few variables are usually unsuitable when there are hundreds of thousands of variables. However, the definition of "size" is by no means absolute, and the tradeoffs are necessarily environment-dependent. A method that is impossible for a mini-computer may run comfortably on a large computer; a user with a limited computer budget has a different perspective on what constitutes acceptable computing time than a problem-solver with his own computer.

Another way in which problems vary involves the computable information that may be available to an algorithm during the solution process. For example, in one instance it may be possible to compute analytic first and second derivatives of the objective function, while in another case only the function values may be provided. A further refinement of such a classification would reflect the amount of computation associated with obtaining the information. The best strategy for a problem clearly depends on the relative effort required to compute the function value compared to the operations associated with the solution method.

Finally, applications of optimization may include special needs that reflect the source of the problem and the framework within which it is to be solved. Such "external" factors often dictate requirements in solving the problem that are not contained in the mathematical statement of the problem. For example, in some problems it may be highly desirable for certain constraints to be satisfied exactly at every iteration. The required accuracy also differs with the application; for example, it might be wasteful to solve an optimization problem with maximum accuracy when the results are used only in a minor way within some outer iteration.

The discussion of methods in this book will consider the influence of several problem characteristics on the efficiency of each algorithm. We shall also indicate how the choice of method and analysis of the computed results depend on the problem category.

1.3. OVERVIEW OF TOPICS

In earlier years, many optimization methods were simple enough so that it was appropriate for someone with a problem to consult a research paper, or even develop a special-purpose method, and then to write a computer program to execute the steps of the method. This approach is no longer feasible today, for several reasons.

Firstly, there has been tremendous progress in optimization methods in all problem categories, and many problems that were considered intractable even in recent years can now be successfully solved. The resulting wide variety in optimization techniques means that many users are unaware of the latest developments, and hence cannot make an adequately informed choice of method.

Furthermore, recently developed methods tend to be so complex that it is unlikely that a typical user will have the time or inclination to write his own computer program. A related phenomenon, which affects *all* numerical computation, is that modern numerical analysis has shown that even apparently simple computations must be implemented with great care, and that naïve implementations are subject to a high probability of serious error and numerical instability.

These developments mean that the typical person who wishes to solve an optimization problem would not (and, in our view, should not) start from scratch to devise his own optimization method or write his own implementation. Rather, he should be able to use selected routines from high-quality mathematical software libraries. However, this does not mean that the user should remain in complete ignorance of the nature of the algorithms that he will use, or the important features of optimization software.

This book is designed to help problem solvers make the best use of optimization software — i.e., to use existing methods most effectively whenever possible, and to adapt and modify techniques for particular problems if necessary. The contents of this book therefore include some topics that are essential for all those who wish to solve optimization problems. In addition, certain topics are treated that should be of special interest in most practical optimization problems. For example, advice is given to users who wish to take an active role in formulating their problems so as to enhance the chances of solving them successfully, and to users who need to understand why a certain method fails to solve their problem.

The second chapter of this book contains a review of selected aspects of numerical analysis, and may be skipped by readers who are already familiar with the subject matter. This introductory chapter contains a substantial amount of material concerning errors in numerical computation and certain topics in numerical linear algebra, since these areas are essential background for any understanding of numerical optimization.

The remaining chapters treat various aspects of numerical optimization, including methods, problem formulation, use of software, and analysis of the computed results. We emphasize that the reader should not expect to become an expert in the myriad details of each topic, but instead to understand the underlying principles.

CHAPTER TWO

FUNDAMENTALS

The only fence against the world is a thorough knowledge of it.

—JOHN LOCKE (1693)

2.1. INTRODUCTION TO ERRORS IN NUMERICAL COMPUTATION

2.1.1. Measurement of Error

We shall often need to measure how well an exact quantity is approximated by some computed value. Intuitively, a satisfactory measure of error would be zero if the approximation were exact, "small" if the two quantities were "close", and "large" if the approximation were "poor". The use of subjective terms like "small" and "poor" indicates the complications in defining an appropriate error criterion.

One obvious way to measure error is simply to compute the difference between the exact and approximate values. Let x be the exact quantity, and $\hat{x}$ the computed result. The error in $\hat{x}$ is then defined as $\hat{x} - x$, and the non-negative number $|\hat{x} - x|$ is termed the *absolute error* in $\hat{x}$.

Absolute error is not satisfactory for all situations. For example, if $x = 2$ and $\hat{x} = 1$, the absolute error in $\hat{x}$ is 1; $\hat{x}$ would probably be regarded as a poor approximation to x, and thus the absolute error would not be "small". However, if $x = 10^{10}$ and $\hat{x} = 10^{10} + 1$, the absolute error is again 1, yet $\hat{x}$ would probably be considered a "good" approximation because the difference between x and $\hat{x}$ is small in relation to x. The *relative error* is defined as

$$\frac{|\hat{x} - x|}{|x|}$$

if x is non-zero, and is undefined if x is zero; thus, a relative error measurement takes into account the size of the exact value. In the two examples above, the relative errors are 0.5 and 10^{-10} respectively, so that the relative error is indeed "small" in the second case.

Caution should be exercised in computing relative error when the exact values involved become close to zero. In practice, it is often convenient to use the following measure, which combines the features of absolute and relative error:

$$\frac{|\hat{x} - x|}{1 + |x|}.$$

This measure is similar to relative error when $|x| \gg 1$, and to absolute error when $|x| \ll 1$.

7

2.1.2. Number Representation on a Computer

Having defined what is meant by "error", we next consider how errors occur in numerical computation. The first source of error to be discussed arises from the very process of representing numbers on a computer. The details of representation vary from one machine range to another, but the same basic principles apply in all cases.

A standard way of representing information is as an ordered sequence of digits. The familiar decimal number system uses this principle, since a plus or minus sign, a string of the digits $0, 1, \ldots, 9$ and a decimal point (sometimes implicit) are interpreted as a real number. In interpreting the string of digits, the position of each digit with respect to the decimal point indicates the relevant power of ten.

Exactly these same ideas are applied when numbers are stored on a computer. In hardware, the simplest distinction is whether a switch is "on" or "off". These values can be considered to define a binary (two-valued) number that must be either zero or one, and is termed a *bit* (binary digit). Because of the use of binary logic circuits, the number bases used in computers are usually powers of 2; the three most common are 2 (binary arithmetic), 8 (octal, with digits $0, 1, 2, \ldots, 7$), and 16 (hexadecimal, with digits $0, 1, 2, \ldots, 9, A, \ldots, F$).

It is customary to allocate a fixed number of digits (usually termed a *word*) for storing a single number. We now consider two number formats, which are essentially rules for interpreting the digits within a word.

The first format is called *fixed-point* format because the "point" that divides the integer from the fractional part of a real number is assumed to be in a fixed position. With four decimal digits, for example, the numbers 0 through 9999 can be represented exactly as they would be written, except that leading zeros are not omitted — e.g., "0020" would represent 20.

To represent signed numbers in fixed point, a specified number (called the *bias* or *offset*) can be added to the desired number before it is stored — e.g., if the bias were 4999 in the previous example, "0000" would represent —4999. Alternatively, since the sign is a binary choice, a single bit may be designated as the "sign bit".

Fixed-point format is acceptable when all the numbers to be represented are known to lie in a certain range, but is too restrictive for most scientific computation, where numbers may vary widely in magnitude. The second format to be considered is *floating-point format*, which is analogous to the usual scientific notation where a number is written as a signed fraction times a power of 10.

In a floating-point format, a non-zero number x is written in the form

$$x = m\beta^e, \tag{2.1}$$

where β is the machine base, and e is a signed integer known as the *exponent* of x; the number m is known as the *mantissa* of x. Given a particular value of β, this representation can be made unique by requiring that the mantissa be *normalized*, i.e., m must satisfy

$$\frac{1}{\beta} \le |m| < 1.$$

Since the exponent e is a signed integer, it may be stored in some convenient fixed-point format; any binary-valued quantity can be used to store the sign of the mantissa. The magnitude of the fraction m is assumed to be stored as a string of τ digits $m_1, m_2, m_3, \ldots, m_\tau$, where $0 \le m_i \le \beta - 1$; this represents the fraction

$$\beta^{-\tau}(m_1\beta^{\tau-1} + m_2\beta^{\tau-2} + \cdots + m_\tau).$$

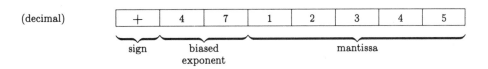

Figure 2a. Decimal word containing $+.12345 \times 10^{-3}$.

If m is normalized, $m_1 \neq 0$. With a normalized representation, the maximum value of $|m|$ is $1 - \beta^{-\tau}$, which corresponds to $m_i = \beta - 1, i = 1, \ldots, \tau$; the smallest value of $|m|$ is β^{-1}, which corresponds to $m_1 = 1, m_2 = \cdots = m_\tau = 0$.

Since zero is not a normalized number, any scheme for storing floating-point numbers must include a special form for zero.

These ideas will be illustrated with two examples. Consider first a hypothetical machine with decimal base arithmetic and eight-digit words. The leftmost ("first") digit contains a representation of the sign of m. The next two digits contain the exponent biased by 50, so that the exponent range is from -50 to $+49$. The last five digits represent the magnitude of the normalized mantissa. Figure 2a depicts a word of storage containing the number $+.12345 \times 10^{-3}$.

A more complicated example is based on the IBM 360/370 series of computers, which use hexadecimal arithmetic. A single-precision floating-point number is represented by a word composed of 32 bits, divided into four 8-bit *bytes*, or 8 hexadecimal digits. The first bit of the word contains the sign of the mantissa (0 for plus, 1 for minus). The next 7 bits (the remainder of the first byte) contain the exponent biased by 64. Bytes 2 through 4 contain the normalized mantissa.

Consider the representation of the number $-42/4096 = -16^{-1}(2/16 + 10/256)$. The true exponent is -1, so that the biased exponent is 63. The normalized mantissa is $.2A$ (base 16). Thus, the number is stored as depicted in Figure 2b.

2.1.3. Rounding Errors

Only a finite set of numbers can be represented by either fixed or floating-point format. If a word contains τ digits in base β, then at most β^τ distinct numbers can be represented, and they form the set of *representable numbers* for that machine. All other numbers cannot be represented exactly, and some error is incurred if an attempt is made to store such a number. This error is usually termed *rounding error*, or *error in representation*.

Some numbers cannot be represented because their magnitude lies outside the range of values that can be stored. If $e_{\max}$ is the maximum allowed exponent, the magnitude of the largest

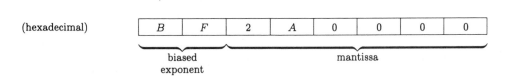

Figure 2b. Hexadecimal word containing $-42/4096$.

normalized floating-point number is $K = \beta^{e_{\max}}(1 - \beta^{-\tau})$. If $e_{\min}$ is the minimum allowed exponent, the smallest representable non-zero magnitude is $k = \beta^{(e_{\min}-1)}$. A number larger in magnitude than K is said to *overflow*, and a non-zero number of magnitude less than k is said to *underflow*.

Other numbers cannot be represented because their mantissa in the form (2.1) contains more than τ significant digits. For example, $\pi = 3.14159\ldots$ cannot be represented exactly with any finite number of digits. Note that whether or not a number is representable depends on β; the number $\frac{1}{10}$ is representable in one-digit decimal arithmetic, but not in any finite number of digits in bases 2, 8, or 16.

Given a non-representable number x in the form (2.1) that does not cause overflow or underflow, the question arises as to which representable number $\hat{x}$ should be selected to approximate x; $\hat{x}$ is usually denoted by $fl(x)$.

Since x lies between two representable numbers, a scheme that minimizes rounding error is one which chooses the nearest representable neighbour as $\hat{x}$. The following rule achieves this result when the nearest neighbour is unique: leave m_τ unchanged if the portion of m to be discarded is less than half a unit in the least significant digit to be retained ($\frac{1}{2}\beta^{-\tau}$); if the portion of m to be discarded is greater than $\frac{1}{2}\beta^{-\tau}$, add one unit to m_τ (and re-normalize if necessary). With this scheme on a decimal machine with a six-digit mantissa, the numbers $3.14159265\ldots$ and -20.98999 become 3.14159 and -20.9900, respectively.

The only remaining question is what to do when the exact number is halfway between two representable numbers (this situation is sometimes called "the tablemaker's dilemma"). There are several ways to resolve this ambiguity automatically. For example, the rule "round to the nearest even" is often used, in which a number is rounded to the nearest representable number whose last digit is even. With this rule, on a four-digit decimal machine, the numbers $.98435$ and $.98445$ would both be rounded to $.9844$.

Such schemes are termed "correct" rounding. Correct rounding produces an error in representation that is no greater than half a unit in the least significant digit of the mantissa. Let x be a non-zero exact number, and $\hat{x}$ its τ-digit rounded version. If the exponents of x and $\hat{x}$ in base β are the same, and m and $\hat{m}$ denote the mantissas of x and $\hat{x}$, respectively, then

$$|m - \hat{m}| \leq \frac{1}{2}\beta^{-\tau},$$

and the relative error in $\hat{x}$ is given by

$$\frac{|\hat{x} - x|}{|x|} = \frac{|m - \hat{m}|}{|m|},$$

so that

$$\frac{|fl(x) - x|}{|x|} \leq \frac{1}{2}\beta^{1-\tau}, \tag{2.2}$$

since $1/\beta \leq |m| < 1$. The number $\beta^{1-\tau}$ plays an important role in any discussion of floating-point computation, and is termed the *relative machine precision* (or simply the "machine precision"). Throughout this book, the relative machine precision will be denoted by ϵ_M.

Some computers use schemes other than correct rounding to choose a representation. With the rule of *truncation*, all digits of m beyond the last one to be retained are discarded. For example, in four-digit decimal arithmetic, all numbers between $.98340$ and $.983499\ldots9$ will be represented as $.9834$. The relative error bound analogous to (2.2) for truncation is:

$$\frac{|fl(x) - x|}{|x|} \leq \beta^{1-\tau},$$

so that there can be an error of a whole unit in the last place.

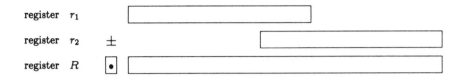

Figure 2c. Computer operands during addition and subtraction.

2.1.4. Errors Incurred During Arithmetic Operations

When arithmetic operations are performed on a computer, additional rounding errors occur because of the need to store non-representable results. Even if there is no error in representing x_1 or x_2, their exact sum or product is not necessarily representable. To illustrate the loss of accuracy through arithmetic operations, we discuss one method for carrying out a floating-point addition or subtraction.

Suppose we wish to add or subtract two floating-point numbers x_1 and x_2. The number of larger magnitude (say, x_1) is stored in a register r_1; the smaller number is stored in a register r_2 and right shifted until the exponents of r_1 and r_2 agree in value. The addition or subtraction then takes place with r_1 and r_2 as operands and the result is stored in the longer register R. Figure 2c shows a schematic view of the registers r_1, r_2 and R.

The register R has an extra overflow digit (the digit denoted by "•" in Figure 2c) to allow for the occurrence of an extra significant digit before the decimal point during addition or subtraction. If the calculation results in the overflow digit being set, then the register R must be re-normalized by performing a right shift. If the calculation results in leading zeros after the decimal point, then a suitable left shift must be made.

The error in addition and subtraction occurs because the number contained in the register R is generally longer than τ digits and must be rounded to τ digits. However, there may be another source of error during addition and subtraction. Some machines do not store all the digits of r_2 and R when digits of r_2 are moved beyond the least significant digit of r_1 during the right shift. In this case, there may be an additional loss of accuracy if insufficient digits are retained.

In practice, the stored result of a floating-point operation satisfies

$$fl(a \ op \ b) = (a \ op \ b)(1 + \eta),$$

where a and b are two representable numbers, "*op*" is one of the operations "$+$", "$-$", "$\times$", "$\div$", and η depends on a, b, the machine precision, and on the floating-point hardware.

The smallest possible bound on η is that associated with a single rounding error, and in practice η is usually bounded by a small multiple of the machine precision.

2.1.5. Cancellation Error

Certain computations carry the risk of introducing much greater relative error than a small multiple of the machine precision — in particular, the subtraction of nearly equal rounded numbers. The error associated with this procedure is termed *cancellation error*.

Consider two numbers x_1 and x_2, whose floating-point values are $\hat{x}_1 = x_1(1 + \epsilon_1)$ and $\hat{x}_2 = x_2(1 + \epsilon_2)$, where ϵ_1 and ϵ_2 are bounded in magnitude by the relative machine precision. The exact difference of $\hat{x}_1$ and $\hat{x}_2$ can be written as:

$$\Delta\hat{x} \equiv x_1(1 + \epsilon_1) - x_2(1 + \epsilon_2) = (x_1 - x_2)(1 + \eta), \tag{2.3}$$

so that η represents the relative error in $\Delta\hat{x}$ with respect to the exact difference of the original numbers.

If $x_1 = x_2$, we say that *complete cancellation* occurs. Otherwise, re-arranging (2.3) gives an expression for η:

$$\eta = \frac{\epsilon_1 x_1 - \epsilon_2 x_2}{x_1 - x_2}. \tag{2.4}$$

The relative error in $\Delta\hat{x}$ may therefore be bounded as follows:

$$\begin{aligned} |\eta| &= \frac{|\epsilon_1 x_1 - \epsilon_2 x_2|}{|x_1 - x_2|} \\ &= \frac{|\epsilon_2(x_1 - x_2) + x_1(\epsilon_1 - \epsilon_2)|}{|x_1 - x_2|} \\ &\leq \epsilon_M\left(1 + 2\frac{|x_1|}{|x_1 - x_2|}\right), \end{aligned} \tag{2.5}$$

assuming that $|x_1| \geq |x_2|$.

If $|x_1 - x_2|$ is small relative to $|x_1|$ (i.e., x_1 "nearly equals" x_2), (2.5) shows that the relative error in $\Delta\hat{x}$ is not restricted to be of order ϵ_M. The error may be large *not* because of errors in subtracting $\hat{x}_1$ and $\hat{x}_2$ (since $\Delta\hat{x}$ is their *exact* difference), but rather because of the initial errors incurred in rounding x_1 and x_2; note that if ϵ_1 and ϵ_2 are zero, $\eta = 0$. If x_1 nearly equals x_2, the original high-order significant digits cancel during subtraction, which means that low-order digits discarded in rounding are the most significant digits of the exact result — i.e., cancellation reveals the error of rounding. If x_1 and x_2 are not similar, the bound on the cancellation error becomes of the same order as the error resulting from any other floating-point operation, and is not of any special significance.

As an example, consider the subtraction of the numbers

$$x_1 = .2946796847 \quad \text{and} \quad x_2 = .2946782596 \tag{2.6}$$

on a machine with a mantissa of six decimal digits ($\epsilon_M = 10^{-5}$). If correct rounding is used, the values of $\hat{x}_1$ and $\hat{x}_2$ are .294680 and .294678, respectively, with the difference $\Delta\hat{x}$ (computed exactly) being $.2 \times 10^{-5}$. However, the difference between the *exact* values of x_1 and x_2 is $.14251 \times 10^{-5}$, which implies that $\Delta\hat{x}$ has a relative cancellation error of $\eta = .40341$, computed from (2.4). From (2.5) it can be seen that the bound on relative cancellation error decreases with ϵ_M; if eight figures are used to represent x_1 and x_2, the relative cancellation error is $.357 \times 10^{-2}$.

It is not possible to compute the exact value (2.4) of the relative cancellation error without utilizing the exact values of x_1 and x_2, and exact arithmetic. Therefore, we can only estimate the bound (2.5) on the cancellation error. For convenience, we shall usually refer to the estimate of a bound on the cancellation error as simply the "cancellation error", and shall be concerned with *computable* estimates of such a bound. Methods for estimating the cancellation error will be discussed further in Chapter 8.

2.1.6. Accuracy in a Sequence of Calculations

We shall often be concerned with computed values that are the result of a complex sequence of calculations involving other computed or measured quantities. In this section, we define terminology that will be used in subsequent discussions of accuracy.

Let f denote the *exact* value of the quantity of interest; thus, f is the value that would be obtained if all intermediate calculations were performed with exact arithmetic and with the exact values of all associated quantities. Let $fl(f)$ denote the final computed result. If

$$fl(f) = f + \sigma,$$

the quantity $|\sigma|$ is the *absolute error* in $fl(f)$ (see Section 2.1.1). We shall use the term *absolute precision* (or *noise level*) to denote a positive scalar ϵ_A that is an upper bound on the absolute error, i.e. $|\sigma| \le \epsilon_A$. When f is non-zero, the error in $fl(f)$ may sometimes be expressed in terms of *relative accuracy*. Relative errors are important because they are inherent in the nature of floating-point arithmetic and in the usual methods for computing standard functions. For example, on most machines the computed value of $\sqrt{x}$ contains an error of no more than one unit in the last place of the mantissa. When using relative errors, we write $fl(f)$ as

$$fl(f) = f(1 + \delta).$$

The *relative precision* ϵ_R is a positive scalar such that $|\delta| \le \epsilon_R$.

Algebraically, the errors δ and σ satisfy $|\delta f| = |\sigma|$. Unfortunately, there is no fixed relationship between the associated bounds ϵ_A and ϵ_R. In many instances, the relative and absolute precisions satisfy the approximate relationship $\epsilon_A \sim \epsilon_R |f|$. For example, when f is a standard function, $\epsilon_R = \epsilon_M$ and $\epsilon_A = \epsilon_M |f|$. However, the connection between ϵ_A and ϵ_R is generally much more complicated, particularly when $|f|$ is small.

2.1.7. Error Analysis of Algorithms

The final result of an algorithm to solve a problem on a computer is a set of representable numbers, which, in general, have been produced by a sequence of floating-point operations. In this section, we consider how to assess the merits of an algorithm in terms of producing an "acceptable" computed solution.

Ideally, one would like to state that the computed solution is "close" to the exact solution. The process of *forward error analysis* is directed to finding a relationship of the form:

$$\|s - \bar{s}\| \le \delta, \tag{2.7}$$

where s is the exact solution, $\bar{s}$ is the computed solution, and $\|\cdot\|$ denotes some reasonable measure of the difference. Unfortunately, forward error analysis is not useful for most algorithms, because the value of δ in (2.7) will not be small for all problems.

To see why this is so, we note that certain problems are inherently *ill-conditioned*, i.e., a small perturbation in the data of the problem can lead to an enormous change in the exact solution (such problems will be illustrated in Section 2.2.4). Ill-conditioning is *not* related to floating-point computation, but is a property of the problem itself.

Suppose that an algorithm were developed in which all computations are performed exactly except for a single rounding error. By any reasonable standard, this should qualify as a "good" algorithm. However, if this algorithm were applied to a highly ill-conditioned problem, the computed solution could differ enormously from the exact solution, and the bound in (2.7) would not be small. Hence, forward error analysis would imply that the algorithm was unsatisfactory, even though the computation deviated from perfection only by one small error.

By contrast, the form of error analysis that has been most useful is termed *backward error analysis*. Rather than considering the computed solution as a perturbation of the exact solution, backward error analysis usually considers the computed solution to be the exact solution of a perturbation of the original problem. In essence, the perturbations introduced by floating-point computation are reflected "backward" to the problem rather than "forward" to the solution.

Let P be the problem to be solved, and assume that the computed solution is the *exact* solution of some other problem $\bar{P}$. Backward error analysis typically provides a bound of the form:

$$\|P - \bar{P}\| \leq \Delta, \tag{2.8}$$

where Δ depends on the machine precision and on P.

Unlike the situation with forward error analysis, Δ can be shown to be "small" for most good algorithms. An algorithm for which a satisfactory result of the form (2.8) can be derived is said to be *numerically stable* because the errors introduced by computing the solution have a small effect on the deviation from the original problem. Note that merely representing P on the machine may imply that Δ is non-zero.

Notes and Selected Bibliography for Section 2.1

The pioneering work on rounding errors in algebraic processes was performed by J. H. Wilkinson — see Wilkinson (1963) for a discussion of the essential principles involved. A full discussion of how floating-point arithmetic is implemented on modern computers is beyond the scope of this introductory text. Interested readers are referred to Kahan (1973) for further information. Dahlquist and Björck (1974) is a good, general-purpose numerical analysis textbook.

2.2. INTRODUCTION TO NUMERICAL LINEAR ALGEBRA

This section will be concerned with aspects of numerical linear algebra that are most important in optimization. No attempt will be made to give a complete discussion of any of the topics; the reader who wishes to pursue an area in more detail should consult one of the cited references.

2.2.1. Preliminaries

2.2.1.1. Scalars. A single quantity or measurement is typically stated in terms of familiar entities called "scalars" or "real numbers". Real numbers display many remarkable properties in connection with the operations of addition and multiplication, which are crucial in subsequent definitions of operations on vectors and matrices.

Scalars will usually be denoted by lower-case Greek letters, e.g., α, β, δ.

2.2.1.2. Vectors. A vector can be formally defined as an *ordered collection of scalars*, where both the ordering and the numbers in the collection are crucial. Just as a real number serves to measure a single aspect of something — e.g., the height of a desk is 3 feet — a vector is used to characterize an ordered set of quantities or attributes. For example, suppose that the height of a desk were defined as property 1, its width as property 2, and its depth as property 3. Given this ordering, a desk that is 3 feet high, 3.5 feet wide, and 2 feet deep, could be described as the vector

$$\text{desk} = \begin{pmatrix} 3.0 \\ 3.5 \\ 2.0 \end{pmatrix}.$$

The question immediately arises of representing a vector in written symbols. The standard convention for this representation contains an implicit statement of the order that is based on the pattern in which we read, and thus a vector is indicated by a vertical column of numbers, with the first component at the top.

The number of scalars in a vector is termed the *dimension* of the vector, and the numbers themselves are called the *components* of the vector. The components are usually numbered for convenience from 1 to n, where n is the dimension of the vector.

Vectors will usually be named and denoted by lower case Roman letters — e.g., a, b, x. A particular component of a vector may be denoted by the subscripted name of the vector, so that x_2 may refer to the second component of the vector x.

Two vectors are said to be *equal* if and only if their components are equal, taken in the proper order. If two vectors are equal, they must be of the same dimension.

The representation of a vector as a *row*, or horizontal list of numbers, is sometimes useful. If the column vector x is given by:

$$x = \begin{pmatrix} x_1 \\ x_2 \\ \vdots \\ x_n \end{pmatrix},$$

then x^T will denote the corresponding row vector, with the obvious convention of numbering from left to right:

$$x^T = (x_1, x_2, \ldots, x_n).$$

2.2.1.3. Matrices. A "vector" generalizes the notion of a single scalar to an ordered set. The idea of a "matrix" carries this process one step further by associating two orderings with a set of scalars, and a matrix can be viewed as a two-dimensional collection of scalars.

As in the vector case, the standard convention for representing a matrix is based on the directions in which we read, with the two orderings indicated by horizontal and vertical displacement from a starting point. The "first" element in a matrix is in the upper left-hand corner of a rectangle. The "first" ordering runs from top to bottom, and defines the number of "rows" of scalars (the *row dimension*). The "second" ordering runs across the horizontal, from left to right, and defines the number of *columns* of scalars (the *column dimension*). If the number of rows is equal to the number of columns, the matrix is said to be *square*.

Matrices are usually denoted by capital Roman letters (e.g., A, W). A particular element in the matrix is referenced by the name of the matrix (upper or lower case), with two subscripts that give, respectively, the row and column indices of the element; thus, A_{ij} or a_{ij} refers to the element in row i and column j of the matrix A.

A matrix B with three rows and two columns would therefore be written as:

$$B = \begin{pmatrix} b_{11} & b_{12} \\ b_{21} & b_{22} \\ b_{31} & b_{32} \end{pmatrix}.$$

A column vector can be regarded as a special case of a matrix with only one column, so that the column index can be suppressed.

Equality between matrices is defined as element-by-element equality with the same two orderings; this implies that the row and column dimensions of equal matrices must match.

The *transpose* of a matrix is the matrix that results from interchanging the roles of the row and column indices. The transpose of A is denoted by A^T, and is defined by:

$$(A^T)_{ij} = (A)_{ji}.$$

For example, if

$$A = \begin{pmatrix} 1 & 2 & 3 \\ 4 & 5 & 6 \end{pmatrix}, \quad \text{then} \quad A^T = \begin{pmatrix} 1 & 4 \\ 2 & 5 \\ 3 & 6 \end{pmatrix}.$$

A matrix A is said to be *symmetric* if $A = A^T$, which implies that $a_{ij} = a_{ji}$ for all i and j, and also that A is square.

The set of elements a_{ii} in the matrix A are termed the *diagonal* elements, while all other elements are *off-diagonal*. The elements a_{ij} for which $j > i$ are said to be *super-diagonal*; those for which $j < i$ are said to be *sub-diagonal*.

2.2.1.4. Operations with vectors and matrices. The operations to be defined among vectors and matrices are an ordered collection of scalar operations, just as vectors and matrices are ordered sets of scalars.

The simplest operation is that of multiplication of a matrix (or vector) by a scalar. The result is a matrix (vector) with each element multiplied by the scalar. Thus,

$$\alpha \begin{pmatrix} x_1 \\ x_2 \\ x_3 \end{pmatrix} = \begin{pmatrix} \alpha x_1 \\ \alpha x_2 \\ \alpha x_3 \end{pmatrix};$$

$$-4 \begin{pmatrix} 1 & 0 \\ 3 & -4 \end{pmatrix} = \begin{pmatrix} -4 & 0 \\ -12 & 16 \end{pmatrix}.$$

The operation of *addition* of two matrices or vectors is based directly on scalar addition. The elements in the sum of two matrices (vectors) are simply the sums of the corresponding elements. Implicit in this definition is the requirement that the dimensions of the two matrices (vectors) must match. Consider the vector case:

$$\begin{pmatrix} a_1 \\ a_2 \\ a_3 \end{pmatrix} + \begin{pmatrix} b_1 \\ b_2 \\ b_3 \end{pmatrix} = \begin{pmatrix} a_1 + b_1 \\ a_2 + b_2 \\ a_3 + b_3 \end{pmatrix}.$$

The matrix or vector whose elements are all zero plays the same role here as zero does in scalar addition — i.e., addition of the zero matrix (vector) to any other leaves the original unaltered. The zero vector and zero matrix are denoted simply by the standard symbol "0", and the dimensions should always be clear from the context.

It is easy to see that vector/matrix addition satisfies the same properties as scalar addition, i.e.,

associativity: $A + (B + C) = (A + B) + C;$
commutativity: $A + B = B + A.$

The *scalar product* (*inner product, dot product*) of two n-vectors, say a and b, is the value γ defined by

$$\gamma = a_1 b_1 + a_2 b_2 + \cdots + a_n b_n = \sum_{i=1}^{n} a_i b_i.$$

The scalar product of a and b will be denoted by $a^T b$. Note that the order of the components is retained in forming the pairwise products, and that the dimensions of the vectors must match. The scalar product operation satisfies the following two properties of ordinary multiplication:

commutativity: $\qquad\qquad\qquad\qquad a^T b = b^T a;$

distributivity over vector addition: $\qquad a^T(b + c) = a^T b + a^T c.$

Although the scalar product might seem to be analogous to the simple product of two scalars, one critical difference is worth a special mention. It is possible for the scalar product of two non-zero vectors to be zero, which cannot happen with the product of two non-zero scalars. For example, if

$$a = \begin{pmatrix} 1 \\ -1 \\ 2 \end{pmatrix} \quad \text{and} \quad b = \begin{pmatrix} 1 \\ -1 \\ -1 \end{pmatrix},$$

then

$$a^T b = 1 + 1 - 2 = 0.$$

If $a^T b = 0$, the vectors a and b are said to be *orthogonal* to one another. Note also that if $a^T c = b^T c$, it is *incorrect* (in general) to conclude that a is equal to b. For example, consider

$$a = \begin{pmatrix} 4 \\ 1 \end{pmatrix}, \quad b = \begin{pmatrix} 2 \\ 3 \end{pmatrix}, \quad c = \begin{pmatrix} 1 \\ 1 \end{pmatrix}.$$

By the distributive property, however, it is correct to conclude that $(a - b)^T c = 0$, and hence that $a - b$ is orthogonal to c. Only the zero vector is orthogonal to *all* vectors.

The product C of two matrices A and B is a matrix defined as follows: the (i, j)-th element of C is the scalar product of the i-th *row* of A and the j-th *column* of B. This definition implies that the column dimension of A must be equal to the row dimension of B, in order for the scalar products to be properly defined. The row dimension of C is the row dimension of A, and the column dimension of C is the column dimension of B.

If a_i^T denotes the i-th row of A, and b_j denotes the j-th column of B, then the product of A and B can be written as:

$$C = AB = \begin{pmatrix} a_1^T \\ a_2^T \\ \cdot \\ \cdot \\ \cdot \end{pmatrix} (b_1 \ b_2 \ \dots) = \begin{pmatrix} a_1^T b_1 & a_1^T b_2 & \cdot & & \cdot \\ a_2^T b_1 & a_2^T b_2 & \cdot & & \cdot \\ \cdot & \cdot & \cdot & & \cdot \\ \cdot & \cdot & \cdot & \cdot & \cdot \end{pmatrix}.$$

For example, if

$$A = \begin{pmatrix} 1 & 1 \\ 2 & -3 \end{pmatrix}, \quad B = \begin{pmatrix} 4 & 1 \\ 0 & 2 \end{pmatrix},$$

then

$$AB = \begin{pmatrix} 4 & 3 \\ 8 & -4 \end{pmatrix}.$$

It is essential to note that there is a clear distinction between the treatment of rows and columns in the two matrices of a matrix product, and hence the order of the two matrices is of critical importance.

In the matrix product AB, the leftmost or "first" matrix A is said to *multiply B on the left*, or to *pre-multiply B*. Pre-multiplication by A affects each column of B separately, so that, for example, the first column of B affects only the first column of the product. Similarly, the "second" matrix B is said to *multiply A on the right*, or to *post-multiply A*. Post-multiplication of A by B acts on each *row* of A separately, and the i-th row of A affects only the i-th row of the product.

Matrix multiplication satisfies the following properties:

associativity: $\qquad\qquad\qquad\qquad\qquad\qquad$ $(AB)C = A(BC);$
distributivity over matrix addition: $\qquad\quad$ $A(B+C) = AB + AC.$

However, matrix multiplication is *not* commutative in general because of the non-interchangeable roles of rows and columns. Even if the matrices are square (the only case in which the dimensions of AB would match those of BA), in general $AB \neq BA$. An additional property of matrix multiplication is that $(AB)^T = B^T A^T$.

The scalar product already defined is equivalent to a special case of matrix multiplication, between a matrix with only one row and a matrix of only one column. A matrix-vector product is also a special case, where the second matrix has only one column. In order for the product of the matrix A and the vector x to be defined, the dimension of x must equal the column dimension of A.

The *identity matrix* of order n, usually denoted by I_n, is a square matrix of the form:

$$I_n = \begin{pmatrix} 1 & 0 & 0 & \cdot & 0 & 0 \\ 0 & 1 & 0 & \cdot & 0 & 0 \\ 0 & 0 & 1 & \cdot & 0 & 0 \\ \cdot & \cdot & \cdot & \cdot & \cdot & \cdot \\ 0 & 0 & \cdot & \cdot & 1 & 0 \\ 0 & 0 & \cdot & \cdot & 0 & 1 \end{pmatrix}.$$

The role of the identity matrix in matrix multiplication is analogous to that of the scalar "1" in ordinary multiplication. If A is $m \times n$, I_n satisfies $AI_n = A$, and the m-dimensional identity matrix I_m satisfies $I_m A = A$. The subscript indicating the dimension of an identity matrix is usually suppressed when the dimension is clear from the context. The i-th column of the identity matrix is often denoted by e_i.

2.2.1.5. Matrices with special structure. We note here certain matrices that display important special patterns in the placement of zeros, or in the relationships among their elements.

A matrix is *diagonal* if all its off-diagonal elements are zero. A diagonal matrix is often denoted by the letter D — for example

$$D = \begin{pmatrix} -2 & 0 & 0 \\ 0 & \frac{1}{2} & 0 \\ 0 & 0 & 1 \end{pmatrix}.$$

An n-dimensional diagonal matrix may be conveniently specified in terms of its diagonal elements as $D = \text{diag}(d_1, d_2, \ldots, d_n)$.

A matrix is *upper* (or *right*) *triangular* if all its sub-diagonal elements are zero, i.e.,

$$a_{ij} = 0, \quad \text{if} \quad i > j. \tag{2.9}$$

An upper-triangular matrix is often denoted by the letter R or U — for example,

$$R = \begin{pmatrix} 1 & 2 & 4 \\ 0 & -7 & 1 \\ 0 & 0 & 3 \end{pmatrix}.$$

When the property (2.9) applies to a matrix with more columns than rows, the matrix is said to be *upper trapezoidal*, e.g.:

$$T = \begin{pmatrix} 1 & 2 & 4 & 6 \\ 0 & -3 & 5 & 2 \\ 0 & 0 & 1 & 1 \end{pmatrix}.$$

Exactly analogous definitions apply for a *lower-triangular matrix*, commonly denoted by L, which satisfies

$$l_{ij} = 0, \quad \text{if} \quad i < j,$$

and for a lower-trapezoidal matrix. A square triangular matrix (either upper or lower) is said to be a *unit triangular matrix* if all of its diagonal elements are unity.

A set of vectors $\{q_1, q_2, \ldots, q_k\}$ is said to be *orthogonal* if

$$q_i^T q_j = 0, \quad i \neq j.$$

If the additional property holds that $q_i^T q_i = 1$, the vectors are said to be *orthonormal*. A square matrix is said to be an *orthogonal (orthonormal) matrix* if its columns are orthogonal (orthonormal). If $Q = (q_1, q_2, \ldots, q_n)$ and Q is orthonormal, then $Q^T Q = I_n$.

A matrix of the form uv^T, where u and v are vectors, is termed a *matrix of rank one*. Every column of uv^T is a multiple of the vector u, and every row is a multiple of the vector v^T. Certain choices of the vectors u and v lead to special rank-one matrices. For example, if $v = e_i$ (the i-th column of the identity matrix), the matrix uv^T is zero except for its i-th column, which is the vector u. A matrix of the form $I + \alpha uv^T$, where α is a scalar, is termed an *elementary matrix*.

2.2.2. Vector Spaces

In this subsection, a subscript on a vector refers to a particular element of a set of vectors.

2.2.2.1. Linear combinations.
Given a set of k vectors $\{a_1, a_2, \ldots, a_k\}$, and a set of k scalars $\{\alpha_1, \alpha_2 \ldots, \alpha_k\}$, we form a *linear combination* of the vectors by multiplying the i-th vector by the i-th scalar, and then adding up the resulting vectors. Thus, the vector b,

$$b = \alpha_1 a_1 + \alpha_2 a_2 + \cdots + \alpha_k a_k, \tag{2.10}$$

is said to be a linear combination of the vectors $\{a_i\}$.

The process of forming a linear combination involves ordered sets of vectors and scalars, which suggests that the result could be written as a matrix-vector product. If A is the matrix whose i-th column is a_i, then (2.10) can be written as

$$b = \begin{pmatrix} a_1 & a_2 & \cdots & a_k \end{pmatrix} \begin{pmatrix} \alpha_1 \\ \alpha_2 \\ \vdots \\ \alpha_k \end{pmatrix}.$$

Any matrix-vector product yields a linear combination of the columns of the matrix, where the coefficients in the linear combination are the components of the vector.

A linear combination where all of the coefficients are zero is called a *trivial* linear combination; if at least one of the coefficients is non-zero, the result is called a *non-trivial* linear combination.

2.2.2.2. Linear dependence and independence. If a vector a_{k+1} can be written as a linear combination of the set of vectors $\{a_1, a_2, \ldots, a_k\}$, a_{k+1} is said to be *linearly dependent* on the set of vectors. For example, the vector

$$a_3 = \begin{pmatrix} 3 \\ 2 \\ 3 \end{pmatrix} \tag{2.11a}$$

is a linear combination of the vectors

$$a_1 = \begin{pmatrix} 1 \\ 0 \\ 1 \end{pmatrix}, \quad a_2 = \begin{pmatrix} 0 \\ \frac{1}{2} \\ 0 \end{pmatrix}, \tag{2.11b}$$

since $a_3 = 3a_1 + 4a_2$, and so a_3 is linearly dependent on the set $\{a_1, a_2\}$.

Linear dependence can alternatively be defined as a property of a set of vectors — i.e., the set $\{a_1, a_2, \ldots, a_k, a_{k+1}\}$ is linearly dependent if the zero vector can be written as a non-trivial linear combination of all the vectors. In (2.11)

$$3a_1 + 4a_2 - a_3 = 0,$$

and thus the set of vectors $\{a_1, a_2, a_3\}$ is linearly dependent.

An analogous definition concerns the opposite situation. If a vector a_{k+1} *cannot* be written as a linear combination of the set of vectors $\{a_1, a_2, \ldots, a_k\}$, a_{k+1} is said to be *linearly independent* of $\{a_1, a_2, \ldots, a_k\}$. Similarly, the set of vectors $\{a_1, a_2, \ldots, a_k, a_{k+1}\}$, is said to be linearly independent if only a trivial linear combination of the vectors can yield the zero vector. For example, with a_1 and a_2 as in (2.11b), the vector

$$a_4 = \begin{pmatrix} 1 \\ 1 \\ -1 \end{pmatrix}$$

cannot be written as a linear combination of a_1 and a_2, and the set $\{a_1, a_2, a_4\}$ is therefore linearly independent.

If the columns of the matrix A are linearly independent, A is said to have *full column rank*; A is said to have *full row rank* if its rows are linearly independent. Note that if the columns of A are linearly independent, the relationship $Ax = 0$ necessarily implies that $x = 0$, since, by definition, a non-trivial linear combination of the columns may not be zero.

2.2.2.3. Vector spaces; subspaces; basis. Consider any vector x from the collection of all possible vectors of dimension n. For any scalar α, the vector αx is also a vector of dimension n; for any other n-vector, y, the vector $x + y$ is also a vector of dimension n. Consequently, the set of all n-vectors is said to be *closed* with respect to these operations and, by extension, to linear combination. Because of these properties, the set of all n-dimensional vectors is said to form a *linear vector space*, denoted by $\Re^n$ (or E^n).

Given a set of k vectors from $\Re^n$, say $\{a_1, a_2, \ldots, a_k\}$, consider the set L of all vectors that can be expressed as linear combinations of $\{a_1, a_2, \ldots, a_k\}$. Let A be the matrix whose i-th column is a_i. The vector b is in the set L (which we write as "$b \in L$") if $b = Ax$ for some k-vector x.

Any linear combination of vectors from L is also in L. To verify this, observe that if $b \in L$, then

$$\alpha b = \alpha Ax = A(\alpha x), \tag{2.12}$$

so that $\alpha b \in L$. Furthermore, if $b \in L$ and $c \in L$ (where $b = Ax$ and $c = Ay$), then

$$b + c = Ax + Ay = A(x + y), \tag{2.13}$$

using the distributive property of matrix multiplication. Thus, $(b + c) \in L$ also. From (2.12) and (2.13) we see that the set L is closed under linear combination, and is therefore said to be a *subspace* of $\Re^n$.

The subspace of all possible linear combinations of the vectors $\{a_1, a_2, \ldots, a_k\}$ is said to be the *range space* generated or *spanned* by the set of vectors. If $\{a_i\}$ are the columns of a matrix A, this subspace is called the *column space* of A. The vectors in the subspace are often said to *lie* in the subspace.

For any non-trivial subspace L, it can be shown that there is a unique smallest integer r — termed the *dimension* or *rank* of the subspace — such that the entire subspace can be generated by a set of r vectors. Such a set of vectors must be linearly independent, and is said to form a *basis* for the subspace. In fact, any set of k linearly independent vectors is a basis for a k-dimensional subspace. However, the vectors in the basis for a particular subspace are not unique — for example, the two-dimensional subspace generated by

$$a_1 = \begin{pmatrix} 1 \\ 0 \\ 1 \end{pmatrix} \quad \text{and} \quad a_2 = \begin{pmatrix} 0 \\ 2 \\ 0 \end{pmatrix} \tag{2.14}$$

can be summarized as the set of all vectors b of the form

$$b = \begin{pmatrix} \alpha \\ \beta \\ \alpha \end{pmatrix},$$

for some scalars α and β. Clearly, the vectors

$$a_1 = \begin{pmatrix} 1 \\ 2 \\ 1 \end{pmatrix} \quad \text{and} \quad a_2 = \begin{pmatrix} 0 \\ -1 \\ 0 \end{pmatrix}$$

are linearly independent, and also form a basis for the subspace.

If the vectors $\{a_1, a_2, \ldots, a_k\}$ form a basis for a subspace L, every vector in that subspace can be *uniquely* represented as a linear combination of the basis vectors. To see this, suppose that the vector $b \in L$ can be written as two different linear combinations of $\{a_i\}$:

$$b = \alpha_1 a_1 + \alpha_2 a_2 + \cdots + \alpha_k a_k = \beta_1 a_1 + \beta_2 a_2 + \cdots + \beta_k a_k.$$

Re-arranging, we obtain

$$(\alpha_1 - \beta_1)a_1 + (\alpha_2 - \beta_2)a_2 + \cdots + (\alpha_k - \beta_k)a_k = 0. \qquad (2.15)$$

Since the $\{a_i\}$ are linearly independent, (2.15) can be true only if $\alpha_i = \beta_i$ for all i, and so the coefficients are unique.

Several computational advantages result when the vectors in a basis are orthogonal. In particular, given any vector b in L, the coefficients in the linear combination that defines b are easily determined. If

$$b = \alpha_1 a_1 + \alpha_2 a_2 + \cdots + \alpha_k a_k,$$

then

$$a_i^T b = \alpha_1 a_i^T a_1 + \alpha_2 a_i^T a_2 + \cdots + \alpha_k a_i^T a_k$$
$$= \alpha_i a_i^T a_i,$$

since all other terms vanish by orthogonality. Because a_i is in the basis, it must be non-zero, and hence the coefficient α_i is given by $a_i^T b / a_i^T a_i$.

For example, the vectors a_1 and a_2 in (2.14) are orthogonal. The coefficients in the expansion of

$$b = \begin{pmatrix} 3 \\ -1 \\ 3 \end{pmatrix}$$

in terms of a_1 and a_2 are given by

$$\alpha_1 = \frac{a_1^T b}{a_1^T a_1} = \frac{3+3}{2} = 3$$

$$\alpha_2 = \frac{a_2^T b}{a_2^T a_2} = -\frac{2}{4} = -\frac{1}{2}$$

so that

$$\begin{pmatrix} 3 \\ -1 \\ 3 \end{pmatrix} = 3 \begin{pmatrix} 1 \\ 0 \\ 1 \end{pmatrix} - \frac{1}{2} \begin{pmatrix} 0 \\ 2 \\ 0 \end{pmatrix}.$$

2.2.2.4. The null space. For every subspace L in $\Re^n$, there is a complementary set $\bar{L}$ of $\Re^n$ whose members are defined as follows: y is in $\bar{L}$ if, for every x in L,

$$x^T y = 0,$$

i.e., y is orthogonal to every vector in L. The set $\bar{L}$ is also a subspace, since any linear combination of vectors in $\bar{L}$ is orthogonal to every vector in L. The subspace $\bar{L}$ is termed the *orthogonal complement* of L, and the two subspaces are completely disjoint. If the dimension of L is k, the dimension of $\tilde{L}$ is $n - k$.

Any vector in $\Re^n$ can be written as a linear combination of vectors from L and $\tilde{L}$, so that the combined bases for these two complementary subspaces span all of $\Re^n$. When L is defined by the k basis vectors $\{a_1, a_2, \ldots, a_k\}$, the subspace $\tilde{L}$ is termed the *null space* corresponding to $\{a_1, a_2, \ldots, a_k\}$.

As an illustration, suppose that L is defined by a_1 and a_2 from (2.14). The vector

$$a_3 = \begin{pmatrix} 1 \\ 0 \\ -1 \end{pmatrix}$$

is orthogonal to a_1 and a_2, and thus lies in their null space (it is, in fact, a basis for the null space, which is of dimension unity).

2.2.3. Linear Transformations

2.2.3.1. Matrices as transformations. We have seen that a general real matrix can be regarded as a two-dimensional set of data, and as a collection of column (or row) vectors. A matrix can also be interpreted as a means for *transforming* one vector into another, since multiplying a vector on the left by a matrix yields a second vector. In this context, the matrix is said to be *applied* to the vector. For simplicity, we shall ignore the fact that a matrix merely represents a transformation, and shall refer to the matrix and transformation interchangeably.

The transformation of a vector by a matrix is a *linear* transformation because of the familiar properties already observed concerning matrix multiplication. For any matrix A, vectors x and y, and scalars α and β, it holds that

$$A(\alpha x + \beta y) = A(\alpha x) + A(\beta y) = \alpha(Ax) + \beta(Ay),$$

so that the transformation of a linear combination of vectors is the same linear combination of the transformed vectors.

2.2.3.2. Properties of linear transformations. Given a matrix (transformation) A, one property of interest is whether any non-zero vector exists that is transformed by A into the zero vector, i.e., whether for some non-zero x,

$$Ax = 0. \tag{2.16}$$

It has already been observed that (2.16) can be satisfied only if the columns of A are linearly dependent. A square matrix whose columns are linearly dependent is said to be *singular*; a square matrix whose columns are linearly independent is said to be *non-singular*.

If (2.16) holds, then for any vector y and scalar α

$$A(y + \alpha x) = Ay + \alpha Ax = Ay, \tag{2.17}$$

so that the result of transforming $y + \alpha x$ by A is identical to that of transforming y alone. From (2.17), it is impossible to determine the original vector from the transformed vector.

On the other hand, if (2.16) holds only when x is the zero vector, the columns of A must be linearly independent. In this case, two distinct vectors cannot be transformed to the same vector. If the columns of A are linearly independent and $Ax = Ay$, then $Ax - Ay = A(x - y) = 0$, which implies that $x = y$. Thus, it is legitimate to "cancel" a full-rank matrix A that appears on the left of both sides of a vector or matrix equation.

2.2.3.3. Inverses. Suppose that some n-vector x has been transformed by the non-singular matrix A to yield the vector b, so that $b = Ax$. We seek a transformation that can be applied to b to "undo" the effect of the transformation A, and return the original vector x. (As noted above, this is possible only if A is non-singular.) The matrix that transforms Ax back into x is termed the *inverse* transformation, and is denoted by A^{-1} ("A inverse").

For all x, A^{-1} should satisfy

$$A^{-1}(Ax) = x,$$

or

$$(A^{-1}A - I)x = 0,$$

which implies that $A^{-1}A = I$ (the identity matrix). If A^{-1} exists, it is unique, non-singular, and satisfies $AA^{-1} = I$. If A and B are non-singular, then $(AB)^{-1} = B^{-1}A^{-1}$.

The inverses of certain matrices have a special form. When Q is an orthonormal matrix, $Q^{-1} = Q^T$. The inverse of a lower (upper) triangular matrix is lower (upper) triangular. The inverse of a non-singular elementary matrix $I + \alpha uv^T$ is also an elementary matrix that involves the same two vectors: if $\alpha u^T v \neq -1$, then

$$(I + \alpha uv^T)^{-1} = I + \beta uv^T,$$

where $\beta = -\alpha/(1 + \alpha u^T v)$.

2.2.3.4. Eigenvalues; eigenvectors. For any square matrix, there is at least one special scalar λ and a corresponding non-zero vector u such that:

$$Au = \lambda u, \qquad (2.18)$$

i.e., the transformed vector is simply a multiple of the original vector. When (2.18) holds, the scalar λ is said to be an *eigenvalue* of A; the vector u is termed an *eigenvector* corresponding to the eigenvalue λ. For example, $\lambda = 2$ is an eigenvalue of the matrix

$$A = \begin{pmatrix} 1 & 2 \\ 0 & 2 \end{pmatrix},$$

since

$$A \begin{pmatrix} 2 \\ 1 \end{pmatrix} = \begin{pmatrix} 4 \\ 2 \end{pmatrix} = 2 \begin{pmatrix} 2 \\ 1 \end{pmatrix}.$$

The eigenvector u is uniquely determined in direction only, since any non-zero multiple of u will also satisfy (2.18). The relationship (2.18) can also be written as:

$$(A - \lambda I)u = 0,$$

which implies that the matrix $A - \lambda I$ is singular if λ is an eigenvalue of A. The set of eigenvalues of A is often written as $\{\lambda_i[A]\}$.

The function $\Pi(A)$ is defined as the product of the eigenvalues of A. If A and B are arbitrary square matrices, Π satisfies the useful property that

$$\Pi(AB) = \Pi(A)\Pi(B).$$

Two matrices are said to be *similar* if they have the same eigenvalues. If W is a non-singular matrix, the matrix WAW^{-1} is similar to A since, if λ is an eigenvalue of A,

$$Ax = \lambda x \quad \text{and} \quad WAW^{-1}(Wx) = \lambda(Wx).$$

A real $n \times n$ matrix has n eigenvalues (not necessarily distinct), and at most n linearly independent eigenvectors. In general, the eigenvalues of a real matrix are complex numbers; however, we shall be concerned almost exclusively with the eigenvalues of symmetric matrices. The following two properties hold if $A = A^T$ (A is symmetric):

(i) all eigenvalues of A are real numbers;

(ii) the matrix A has n distinct eigenvectors.

The eigenvectors can be made to form an orthonormal set, i.e., if $\{u_i\}$, $i = 1, 2, \ldots, n$ are the eigenvectors, then

$$u_i^T u_j = 0, \quad i \neq j; \qquad u_i^T u_i = 1.$$

Consequently, the eigenvectors of a symmetric matrix form an orthonormal basis for $\Re^n$.

If A is non-singular, all its eigenvalues are non-zero, and the eigenvalues of A^{-1} are the reciprocals of the eigenvalues of A.

The maximum and minimum eigenvalues of A satisfy

$$\lambda_{\max}[A] = \max_{x \neq 0} \frac{x^T A x}{x^T x} \quad \text{and} \quad \lambda_{\min}[A] = \min_{x \neq 0} \frac{x^T A x}{x^T x}.$$

The *spectral radius* of A is defined as $\rho(A) = \max|\lambda_i[A]|$.

2.2.3.5. Definiteness. If all the eigenvalues of a symmetric matrix A are strictly positive, the matrix is said to be *positive definite*. If A is positive definite, then for any non-zero vector x

$$x^T A x > 0.$$

Furthermore, all the diagonal elements of a positive-definite matrix are strictly positive. There is a corresponding definition of *negative definite*, when all the eigenvalues are negative. If all the eigenvalues of a symmetric matrix A are non-negative, A is said to be *positive semi-definite*. If a symmetric matrix A has both positive and negative eigenvalues, A is said to be *indefinite*.

2.2.4. Linear Equations

2.2.4.1. Properties of linear equations. One of the most fundamental problems in numerical linear algebra is that of solving a system of linear equations: given an $m \times n$ matrix A, and an m-vector b, find an n-vector x such that

$$Ax = b. \tag{2.19}$$

In (2.19), the vector b is often called (for obvious reasons) the *right-hand side*, and x is called the *vector of unknowns*. As discussed previously, the vector x in (2.19) can be interpreted as the set of coefficients in a linear combination of the columns of A, or as a vector that is transformed by A into b.

In order for (2.19) to have a solution, the vector b must lie in the subspace spanned by the columns of A. If b is in the range of A, the system (2.19) is said to be *compatible*. For example, the system

$$\begin{pmatrix} 3 & 1 \\ 2 & 4 \end{pmatrix} \begin{pmatrix} x_1 \\ x_2 \end{pmatrix} = \begin{pmatrix} b_1 \\ b_2 \end{pmatrix}$$

is compatible for every b, since the columns of A are linearly independent and hence span $\Re^2$.

The system

$$\begin{pmatrix} 1 & 0 \\ 0 & 1 \\ 1 & 0 \end{pmatrix} \begin{pmatrix} x_1 \\ x_2 \end{pmatrix} = \begin{pmatrix} 2 \\ 3 \\ 2 \end{pmatrix} \tag{2.20}$$

is also compatible, since b is a linear combination of the columns of A. In fact, (2.20) has a unique solution, even though A is not square.

If b does *not* lie in the column space of A, then (2.19) is said to be *incompatible* or *inconsistent*, and no vector x can satisfy (2.19). For example, if the right-hand side of (2.20) is altered as follows:

$$\begin{pmatrix} 1 & 0 \\ 0 & 1 \\ 1 & 0 \end{pmatrix} \begin{pmatrix} x_1 \\ x_2 \end{pmatrix} = \begin{pmatrix} 1 \\ 0 \\ -1 \end{pmatrix}, \tag{2.21}$$

then (2.21) is an *incompatible* system, since no linear combination of the columns of A can produce b.

If b is compatible with A, the system (2.19) has a *unique* solution if and only if the columns of A are linearly independent. If the columns of A are linearly dependent, there are an infinite number of solutions to (2.19). To see why, note that by definition of linear dependence there must exist a non-zero z such that $Az = 0$. If x solves (2.19), then for any scalar δ,

$$A(x + \delta z) = Ax + \delta Az = Ax = b,$$

so that $x + \delta z$ also solves (2.19). For example, the vector $x = (3, 0)^T$ solves the compatible system

$$\begin{pmatrix} 1 & -2 \\ 0 & 0 \\ 1 & -2 \end{pmatrix} \begin{pmatrix} x_1 \\ x_2 \end{pmatrix} = \begin{pmatrix} 3 \\ 0 \\ 3 \end{pmatrix}. \tag{2.22}$$

Since

$$\begin{pmatrix} 1 & -2 \\ 0 & 0 \\ 1 & -2 \end{pmatrix} \begin{pmatrix} 2 \\ 1 \end{pmatrix} = \begin{pmatrix} 0 \\ 0 \\ 0 \end{pmatrix},$$

it follows that (2.22) has an infinite number of solutions of the form

$$\begin{pmatrix} 3 \\ 0 \end{pmatrix} + \delta \begin{pmatrix} 2 \\ 1 \end{pmatrix}.$$

2.2.4.2. Vector and matrix norms. In subsequent discussion, it will be necessary to have some means of "measuring" vectors and matrices, in order to say that one vector or matrix is "larger" or "smaller" than another. The definition of a "norm" gives a rule for associating a non-negative scalar with a vector or matrix.

A *vector norm*, which will be denoted by $\|\cdot\|$, must satisfy the following three properties:

(i) $\|x\| \geq 0$ for all vectors x, and $\|x\| = 0$ if and only if x is the zero vector;

(ii) for any real number δ, $\|\delta x\| = |\delta|\,\|x\|$;

(iii) for any two vectors x and y, $\|x + y\| \leq \|x\| + \|y\|$ (*the triangle inequality*).

The *p-norm* of an *n*-vector x is denoted by $\|x\|_p$, and is defined as

$$\|x\|_p = \left(\sum_{i=1}^{n} |x_i|^p \right)^{1/p}.$$

The *p*-norm satisfies properties (i) through (iii).

The three most common values of p are $p = 1, 2$, and ∞; the corresponding norms are termed the *one-*, *two-* and *infinity-norms*. The two-norm is sometimes termed the *Euclidean* norm; note that

$$\|x\|_2^2 = x_1^2 + \cdots + x_n^2 = x^T x.$$

When $p = \infty$, $\|x\|_\infty = \max_i |x_i|$. For example, if $x^T = (1, -2, -3)$ then $\|x\|_1 = 6$, $\|x\|_2 = \sqrt{14}$ and $\|x\|_\infty = 3$.

There are several useful inequalities that relate the inner product of two vectors with their respective vector norms. If we consider the vectors x, y and $y - x$ as defining a triangle in $\Re^n$, the cosine formula can be used to relate the angle θ between x and y and the lengths of the vectors x, y and $y - x$:

$$\|y - x\|_2^2 = \|y\|_2^2 + \|x\|_2^2 - 2\|y\|_2\,\|x\|_2 \cos\theta.$$

Expanding the expression $\|y - x\|_2^2$ as $(y - x)^T(y - x)$, we obtain

$$\cos\theta = \frac{y^T x}{\|x\|_2 \|y\|_2}.$$

Since $\cos\theta$ lies between -1 and $+1$, it follows that

$$|y^T x| \leq \|x\|_2 \|y\|_2,$$

which is known as the *Schwartz inequality*.

It is also desirable to be able to assign norms to matrices. A *matrix norm*, denoted by $\|\cdot\|$ (the same notation as for the vector case), is a non-negative scalar that satisfies three analogous properties:

(i) $\|A\| \geq 0$ for all A; $\|A\| = 0$ if and only if A is the zero matrix;

(ii) $\|\delta A\| = |\delta|\,\|A\|$;

(iii) $\|A + B\| \leq \|A\| + \|B\|$.

Because matrices can be multiplied together to form other matrices, an additional property is desirable for a matrix norm, namely:

(iv) $\|AB\| \leq \|A\|\,\|B\|$.

A matrix norm can be conveniently defined in terms of a vector norm, as follows. Given a vector norm $\|\cdot\|$ and a matrix A, consider $\|Ax\|$ for all vectors such that $\|x\| = 1$. The matrix norm *induced by*, or *subordinate to*, the vector norm, is given by:

$$\|A\| = \max_{\|x\|=1} \|Ax\|. \tag{2.23}$$

It is important to note that the norms on the right of (2.23) are vector norms, and that the maximum is attained for some vector (or vectors) x.

The three vector norms defined earlier induce three corresponding matrix norms for the $m \times n$ matrix A:

$\|A\|_1 = \max_{1 \le j \le n}(\sum_{i=1}^{m} |a_{ij}|)$, the maximum absolute column sum;

$\|A\|_2 = (\lambda_{\max}[A^T A])^{\frac{1}{2}}$, the square root of the largest eigenvalue of $A^T A$;

$\|A\|_\infty = \max_{1 \le i \le m}(\sum_{j=1}^{n} |a_{ij}|)$, the maximum absolute row sum.

The matrix two-norm is sometimes known as the *spectral norm*.

An important matrix norm that is not subordinate to a vector norm is the *Frobenius norm*, denoted by $\|\cdot\|_F$. This norm arises from regarding an $m \times n$ matrix A as a vector with mn elements, and then computing the Euclidean norm of that vector:

$$\|A\|_F = \left(\sum_{i=1}^{m} \sum_{j=1}^{n} a_{ij}^2 \right)^{\frac{1}{2}}.$$

A vector norm $\|\cdot\|$ and a matrix norm $\|\cdot\|'$ are said to be *compatible* if, for every A and x,

$$\|Ax\| \le \|A\|' \, \|x\|. \tag{2.24}$$

By definition, (2.24) always holds for a vector norm and its subordinate matrix norm, with equality possible for some vector (or vectors) x. The Euclidean vector norm and Frobenius matrix norm are also compatible.

2.2.4.3. Perturbation theory; condition number. The linear system

$$Ax = b \tag{2.25}$$

has a unique solution for every right-hand side only if A is square and non-singular. Before considering how to solve (2.25), it is of interest to see how the solution is affected by small changes (perturbations) in the right-hand side and in the elements of the matrix.

The exact solution of (2.25) is given by

$$x = A^{-1}b.$$

Suppose that the right-hand side of (2.25) is perturbed to $b + \delta b$, and that the exact solution of the perturbed system is $x + \delta x$, i.e.,

$$A(x + \delta x) = b + \delta b,$$

where "δ" denotes a small change in a vector or matrix. Therefore,

$$x + \delta x = A^{-1}(b + \delta b),$$

and since $x = A^{-1}b$,

$$\delta x = A^{-1}\delta b.$$

To measure δx, we invoke the properties of compatible vector and matrix norms:

$$\|\delta x\| \le \|A^{-1}\|\,\|\delta b\|, \tag{2.26}$$

with equality possible for some vector δb. The perturbation in the exact solution is thus bounded above by $\|A^{-1}\|$ times the perturbation in the right-hand side.

To determine the *relative* effect of this same perturbation, note that

$$\|b\| \le \|A\|\,\|x\|. \tag{2.27}$$

Combining the inequalities (2.26) and (2.27) and re-arranging yields

$$\frac{\|\delta x\|}{\|x\|} \le \|A\|\,\|A^{-1}\|\,\frac{\|\delta b\|}{\|b\|}. \tag{2.28}$$

If the matrix in (2.25) is perturbed by δA, a similar procedure gives $(A + \delta A)(x + \delta x) = b$. This equation can be rewritten as

$$\delta x = -A^{-1}\delta A(x + \delta x),$$

so that

$$\|\delta x\| \le \|A^{-1}\|\,\|\delta A\|\,\|x + \delta x\|$$

or

$$\frac{\|\delta x\|}{\|x + \delta x\|} \le \|A^{-1}\|\,\|\delta A\|.$$

When the change $\|\delta A\|$ is considered relative to $\|A\|$, this becomes

$$\frac{\|\delta x\|}{\|x + \delta x\|} \le \|A\|\,\|A^{-1}\|\,\frac{\|\delta A\|}{\|A\|}. \tag{2.29}$$

In both (2.28) and (2.29), the relative change in the exact solution is bounded by the factor $\|A\|\,\|A^{-1}\|$ multiplied by the relative perturbation in the data (right-hand side or matrix). The number $\|A\|\,\|A^{-1}\|$ is defined as the *condition number of A* with respect to solving a linear system (the last phrase is usually omitted) and is denoted by cond(A).

Since $\|I\| = 1$ for any subordinate norm, and $I = AA^{-1}$, it follows that

$$1 = \|I\| \le \|A\|\,\|A^{-1}\|,$$

so that

$$\text{cond}(A) \ge 1$$

for every matrix.

The condition number of A indicates the *maximum* effect of perturbations in b or A on the exact solution of (2.25). The results (2.28) and (2.29) indicate that if cond(A) is "large", the exact solution may be changed substantially by even small changes in the data. The matrix A is said to be *ill-conditioned* if cond(A) is "large", and *well-conditioned* if cond(A) is "small".

To illustrate these ideas, consider the matrix and vector

$$A = \begin{pmatrix} .550 & .423 \\ .484 & .372 \end{pmatrix}, \quad b = \begin{pmatrix} .127 \\ .112 \end{pmatrix}. \tag{2.30}$$

The exact solution of $Ax = b$ is

$$x = \begin{pmatrix} 1 \\ -1 \end{pmatrix}.$$

If b is perturbed to be

$$b + \delta b = \begin{pmatrix} .12707 \\ .11228 \end{pmatrix}, \quad \text{so that} \quad \delta b = \begin{pmatrix} .00007 \\ .00028 \end{pmatrix},$$

the exact solution becomes

$$x + \delta x = \begin{pmatrix} 1.7 \\ -1.91 \end{pmatrix}, \quad \text{with} \quad \delta x = \begin{pmatrix} .7 \\ -.91 \end{pmatrix}.$$

In this case, $\|\delta b\|/\|b\| = .0022$, whereas $\|\delta x\|/\|x\| = .91$, using $\|\cdot\|_\infty$. Clearly, the relative change in the solution is much larger than the relative change in the right-hand side.

Similarly, if the $(2,1)$-th element of A is perturbed in the third place, so that

$$A + \delta A = \begin{pmatrix} .550 & .423 \\ .483 & .372 \end{pmatrix},$$

then the solution to the perturbed system

$$(A + \delta A)(x + \delta x) = b$$

is (rounded to four figures)

$$x + \delta x = \begin{pmatrix} -.4535 \\ .8899 \end{pmatrix}.$$

Once again, the relative change in the solution is much larger than the relative change in A. For the matrix A of (2.30)

$$A^{-1} = \begin{pmatrix} -2818 & 3205 \\ 3667 & -4167 \end{pmatrix},$$

rounded to four figures. Using $\|\cdot\|_\infty$, $\|A\| = .973$, $\|A^{-1}\| = 7834$, and $\text{cond}(A) = 7622$, so that the two seemingly large perturbations above are not even close to the worst case for the given matrix.

It should be emphasized that this perturbation analysis concerns the *exact solution* of a linear system, and is therefore an inherent characteristic of the mathematical problem. The ill-conditioning of a matrix does not involve computation of the solution in finite precision, but simply reflects the different ways in which the transformations A and A^{-1} affect different vectors.

2.2.4.4. Triangular linear systems. Suppose that we wish to compute the solution of the non-singular linear system:

$$Ax = b. \tag{2.31}$$

Under what conditions will this problem be easy to solve? Suppose that A is triangular (say, lower triangular), so that (2.31) looks like:

$$
\begin{pmatrix}
l_{11} & & & & \\
l_{21} & l_{22} & & & \\
l_{31} & l_{32} & l_{33} & & \\
 & \cdot & \cdot & \cdot & \\
 & \cdot & \cdot & \cdot & \cdot \\
l_{n1} & l_{n2} & l_{n3} & \cdot & \cdot & l_{nn}
\end{pmatrix}
\begin{pmatrix}
x_1 \\ x_2 \\ x_3 \\ \cdot \\ \cdot \\ x_n
\end{pmatrix}
=
\begin{pmatrix}
b_1 \\ b_2 \\ b_3 \\ \cdot \\ \cdot \\ b_n
\end{pmatrix}.
$$

The first equation involves only one unknown, x_1, which can be obtained immediately by division as

$$
x_1 = \frac{b_1}{l_{11}}.
$$

The second equation involves the two components x_1 and x_2; however, x_1 is no longer unknown, and thus the second equation is sufficient to solve for x_2, which is given by

$$
x_2 = \frac{b_2 - l_{21}x_1}{l_{22}}.
$$

Continuing in this manner, each successive equation involves only one additional unknown, and thus a triangular system can be solved easily if we solve for the unknowns *in a specified order*. The non-singularity of the matrix is crucial to this solution procedure, since the process would break down if $l_{ii} = 0$ (in which case x_i would not appear in the i-th equation).

The above procedure can also be applied to an upper-triangular matrix, except that x_n is the first unknown to be solved for in the upper-triangular case. The process of solving a lower-triangular system is termed *forward solution*, while that of solving an upper-triangular system is termed *backward* solution, or *back substitution*.

The number of arithmetic operations required to solve a general triangular system in this manner can be estimated as follows: one division is required for the first unknown; computing the second unknown requires one multiplication, one addition, and one division; and so on, until finally $n - 1$ additions, $n - 1$ multiplications, and one division are required to solve for the n-th unknown. Summing these up, the total number of additions (or multiplications) is a quadratic polynomial in n, with leading coefficient $\frac{1}{2}$; hence, we say that solving a triangular system requires about $\frac{1}{2}n^2$ additions and multiplications, and n divisions.

An even simpler problem occurs when A in (2.31) is *diagonal* (a special case of triangular). In this case, each equation involves exactly one unknown, whose value does not depend on any of the others. Only n divisions are required to solve a diagonal linear system.

2.2.4.5. Error analysis. In any computational process, the idea of a *backward error analysis* is to show how the *computed* solution of the original problem is related to the *exact* solution of the perturbed problem (see Section 2.1.7). Error analysis is essential to understanding numerical algorithms, but unfortunately the details are extremely tedious. Therefore, the standard techniques will be illustrated through an abbreviated version of the backward error analysis of solving a lower-triangular linear system. For this purpose, we rely on the axiomatization of floating-point arithmetic described in Section 2.1.4.

To solve

$$Lx = b,$$

where L is lower triangular, the first step involves computing x_1, as:

$$x_1 = fl\left(\frac{b_1}{l_{11}}\right) = \frac{b_1}{l_{11}(1 + \delta_1)}, \tag{2.32}$$

where $|\delta_1|$ is bounded by a small number ϵ_M that indicates the machine precision. The expression on the right-hand side of (2.32) involves *exact* arithmetic, and (2.32) can be re-written as

$$l_{11}(1 + \delta_1)x_1 = b_1.$$

To obtain x_2, the calculation is:

$$\begin{aligned}
x_2 &= fl\left(\frac{-l_{21}x_1 + b_2}{l_{22}}\right) \\
&= \frac{-l_{21}x_1(1 + \xi_{21})(1 + \eta_{22}) + b_2(1 + \eta_{22})}{l_{22}(1 + \delta_2)},
\end{aligned} \tag{2.33}$$

where ξ_{21} arises from multiplication, η_{22} from addition, and δ_2 from division. The magnitudes of ξ_{21}, η_{22}, and δ_2 are all bounded above by ϵ_M. After combining all terms that reflect the computational error, (2.33) becomes

$$l_{21}(1 + \beta_{21})x_1 + l_{22}(1 + \beta_{22})x_2 = b_2,$$

where $|\beta_{21}|$ and $|\beta_{22}|$ are bounded above by a constant multiple of ϵ_M.

Continuing in this fashion, at the r-th step the result is:

$$l_{r1}(1 + \beta_{r1})x_1 + l_{r2}(1 + \beta_{r2})x_2 + \cdots + l_{rr}(1 + \beta_{rr})x_r = b_r,$$

where each $|\beta_{rj}|$ is bounded above by ϵ_M times a linear function of r. After n steps, this analysis reveals that the computed vector x is the *exact solution of a perturbed lower-triangular system*

$$(L + \delta L)x = b, \tag{2.34}$$

where each element in the r-th row of δL is bounded in magnitude by an expression involving machine precision, r, and the corresponding element of the original L. Hence, although the δL associated with any particular b depends on the computed solution, the bound on $|(\delta L)_{ij}|$ is independent of b, x, and the condition number of L. The relationship (2.34) and the bound on $\|\delta L\|$ together imply that *the computed solution is the exact solution of a nearby problem*.

2.2.5. Matrix Factorizations

For many problems in computational linear algebra, it is useful to represent a matrix as the product or sum of matrices of special form. Such a representation is termed a *factorization* or *decomposition* of the original matrix.

In this section, several factorizations that are frequently used in optimization problems will be described in varying degrees of detail.

2.2.5.1. The *LU* factorization; Gaussian elimination. To solve a general non-singular linear system

$$Ax = b, \tag{2.35}$$

a direct approach is to devise an equivalent linear system in which the matrix is triangular. This can be achieved by generating a non-singular matrix M such that MA is triangular, since the solution of the transformed problem

$$MAx = Mb$$

is the same as the solution of (2.35). It is interesting that the formal process by which M is most frequently formed is, in large part, equivalent to the way in which people tend to solve a small linear system by hand.

Consider the linear system

$$
\begin{aligned}
2x_1 + 2x_2 + x_3 &= 5 \\
x_1 + 3x_2 + 2x_3 &= 6 \\
4x_1 - 2x_2 + 3x_3 &= 5
\end{aligned}
\tag{2.36}
$$

where

$$
A = \begin{pmatrix} 2 & 2 & 1 \\ 1 & 3 & 2 \\ 4 & -2 & 3 \end{pmatrix} \quad \text{and} \quad b = \begin{pmatrix} 5 \\ 6 \\ 5 \end{pmatrix}.
\tag{2.37}
$$

The first step would be to "eliminate" x_1 from the second and third equations by subtracting half the first row from the second row, and twice the first row from the third row; these operations introduce zeros into the $(2,1)$-th and $(3,1)$-th positions. The modified matrix would then look like

$$
A^{(2)} = \begin{pmatrix} 2 & 2 & 2 \\ 0 & 2 & \frac{3}{2} \\ 0 & -6 & 1 \end{pmatrix},
$$

which has been partially reduced to upper-triangular form. To complete the process, one would add three times the modified second row to the third row, yielding the desired upper-triangular matrix

$$
A^{(3)} = \begin{pmatrix} 2 & 2 & 1 \\ 0 & 2 & \frac{3}{2} \\ 0 & 0 & \frac{11}{2} \end{pmatrix}.
$$

This systematic way of triangularizing A by introducing zeros below the diagonal is termed *Gaussian elimination*.

At the k-th stage of the process, the first $k-1$ columns have been transformed to upper-triangular form. To carry out the k-th step of elimination, the (j, k)-th element will be annihilated by subtracting a multiple m_{jk} of row k from row j, where

$$
m_{jk} = \frac{a_{jk}^{(k)}}{a_{kk}^{(k)}}, \quad j = k+1, \dots, n.
$$

The (k, k)-th element $a_{kk}^{(k)}$ of the partially reduced matrix is termed the *pivot* element. It has a special role because the process will break down if the pivot is zero at any step. The $n - k$ numbers $\{m_{jk}\}$, $j = k+1, \dots, n$, are referred to as the *multipliers* at the k-th stage.

After the matrix has been reduced to upper-triangular form, the same transformations must be applied to the right-hand side. In example (2.36), b is transformed as follows:

$$b = \begin{pmatrix} 5 \\ 6 \\ 5 \end{pmatrix} \rightarrow b^{(2)} = \begin{pmatrix} 5 \\ \frac{7}{2} \\ -5 \end{pmatrix} \rightarrow b^{(3)} = \begin{pmatrix} 5 \\ \frac{7}{2} \\ \frac{11}{2} \end{pmatrix}.$$

The solution is then obtained by back-substitution as $x_3 = 1$, $x_2 = 1$ and $x_1 = 1$.

The steps of elimination can be described formally as multiplication of the original matrix by a sequence of special elementary matrices (see Section 2.2.1.5), whose effect is to subtract the appropriate multiple of the pivot row from all lower rows. For example, the matrix that performs the first step of elimination in (2.36) is

$$M_1 = I - \begin{pmatrix} 0 \\ \frac{1}{2} \\ 2 \end{pmatrix} (\,1 \quad 0 \quad 0\,) = \begin{pmatrix} 1 & 0 & 0 \\ -\frac{1}{2} & 1 & 0 \\ -2 & 0 & 1 \end{pmatrix}.$$

Note that M_1 is unit lower triangular, and differs from the identity only in the first column, where it contains the negative multipliers.

When all pivots are non-zero, Gaussian elimination is equivalent to applying a special sequence $\{M_i\}$ of unit lower-triangular matrices to A on the left, to obtain an upper-triangular matrix U:

$$MA = M_{n-1} \cdots M_2 M_1 A = U. \tag{2.38}$$

Each M_i reduces the i-th column of the partially triangularized matrix to the desired form. We emphasize that the formalism of introducing the matrices $\{M_i\}$ does *not* mean that these transformations are stored as matrices; rather, only $n-i$ multipliers need to be stored to represent M_i, so that the full sequence of transformations can be stored in a strict lower triangle.

After A has been triangularized, x is the solution of the upper-triangular system

$$MAx = Ux = Mb, \tag{2.39}$$

in which the right-hand side Mb is computed by applying to b the transformations that reduced A.

Since M is non-singular, (2.38) can be written as

$$A = M^{-1}U = M_1^{-1} M_2^{-1} \cdots M_{n-1}^{-1} U. \tag{2.40}$$

The special structure of the matrices M_i means that M^{-1} is a unit lower-triangular matrix whose j-th column simply contains the multipliers used at the j-th stage of the elimination. If we define M^{-1} as L, (2.40) gives a representation of A as the product of a unit lower-triangular matrix L and an upper-triangular matrix U:

$$A = LU.$$

This form is termed the *LU factorization* of A.

In summary, to solve $Ax = b$ by Gaussian elimination, the steps are:

(i) compute L and U such that $A = LU$ (assuming non-zero pivots);
(ii) solve $Ly = b$ (this is equivalent to forming $y = Mb$);
(iii) solve $Ux = y$.

Except for small n, the main effort in Gaussian elimination arises from the $\frac{1}{3}n^3$ operations of forming L and U, since the two triangular systems in (ii) and (iii) can be solved using n^2 operations.

Unfortunately, the preceding discussion has ignored a crucial aspect of Gaussian elimination. Recall that the process will break down at the k-th step if the pivot element is zero, which can happen even if the matrix is non-singular. For example, consider the matrix

$$A = \begin{pmatrix} 0 & 1 & 1 \\ 2 & 3 & 4 \\ 1 & 0 & 7 \end{pmatrix}.$$

To reduce A to upper-triangular form, one cannot subtract a multiple of the first row from rows 2 and 3. However, the difficulty is resolved by *interchanging* rows 1 and 2 (or rows 1 and 3), so that the $(1,1)$-th element is non-zero.

Gaussian elimination will be well-defined in theory if all the pivots are non-zero. In practice, matters are more complicated. It is (almost) an axiom of numerical computation that if a process is undefined when a certain quantity is zero, the process will be numerically unstable (i.e., may lead to large errors) when that quantity is "small". In the case of Gaussian elimination, this observation can be verified rigorously, and the general rule is to *avoid small pivots*. Gaussian elimination can be made numerically stable only by introducing a "pivoting strategy" to select the next pivot element.

The most common pivoting strategy is called *partial pivoting*. It is based on the idea of interchanging rows so that the element of largest magnitude in the part of the column to be reduced will serve as the pivot element. For example, if partial pivoting were used in triangularizing (2.37), the third row would be the pivot row, and rows 1 and 3 would be interchanged, giving the matrix

$$\begin{pmatrix} 4 & -2 & 3 \\ 1 & 3 & 2 \\ 2 & 2 & 1 \end{pmatrix}.$$

Such a row interchange may be carried out at every step, and can be described formally in terms of applying a *permutation matrix* — one whose elements are either "0" or "1", with exactly one "1" in each row and column. When applied to a matrix on the left, a permutation matrix interchanges the rows; when applied on the right, it interchanges the columns. For example, to interchange rows 1 and 3 of A in (2.37), the matrix

$$P_1 = \begin{pmatrix} 0 & 0 & 1 \\ 0 & 1 & 0 \\ 1 & 0 & 0 \end{pmatrix}$$

is applied to A on the left.

When a partial pivoting strategy is added to Gaussian elimination as described earlier, a permutation matrix is applied on the left of each partially triangularized matrix if an interchange is made. The final result is that the computed LU factorization corresponds to a permutation of the rows of the original matrix. Thus, for some permutation matrix P,

$$PA = LU.$$

When partial pivoting is used in forming LU, the sub-diagonal elements of L are all bounded above in magnitude by unity, since at each step of the reduction the pivot element has the largest magnitude in the relevant column, and hence the multipliers are less than or equal to unity.

The backward error analysis of Gaussian elimination with partial pivoting is rather complicated, and can be summarized as follows: *the computed L and U are the exact triangular factors of a matrix $P(A + E)$.* Let

$$g = \frac{\max_k |a_{ij}^{(k)}|}{\max |a_{ij}|};$$

g is termed the *growth factor*, and indicates the "growth" in elements of the intermediate partially reduced matrices compared to those of the original matrix. Then

$$\|E\|_\infty \leq n^2 \gamma \epsilon_M g \|A\|_\infty,$$

where γ is a constant of order unity, independent of A and n, and ϵ_M is the machine precision.

This bound indicates that the main source of instability in the algorithm is growth in the elements of the intermediate matrices. Without an interchange strategy, growth could be unbounded, and the computed factors might bear no relation to any matrix close to the original matrix.

The ideas introduced in this section — reduction to a convenient form, strategies to ensure numerical stability, and the backward error analysis bound — will recur in all subsequent treatments of matrix factorizations.

2.2.5.2. The LDL^T and Cholesky factorizations. A positive-definite symmetric matrix A can be written as:

$$A = LDL^T, \tag{2.41}$$

where L is a *unit* lower-triangular matrix, and D is a diagonal matrix with strictly positive diagonal elements. This representation of a positive-definite matrix will be termed its LDL^T *factorization*.

Since the diagonal of D is strictly positive, (2.41) can also be written as:

$$A = LD^{\frac{1}{2}}D^{\frac{1}{2}}L^T = \bar{L}\bar{L}^T = R^T R, \tag{2.42}$$

where $\bar{L}$ is a general lower-triangular matrix, and R is a general upper-triangular matrix. This factorization is known as the *Cholesky factorization* and the matrix R is called the Cholesky factor, or "square root" of A, by analogy with the scalar case. For simplicity, we shall refer to *either* of the forms (2.41) or (2.42) as the Cholesky factorization.

The Cholesky factors can be computed directly from element-by-element equality. For the $R^T R$ representation, we see that:

$$
\begin{pmatrix}
a_{11} & a_{12} & a_{13} & . & . & a_{1n} \\
a_{12} & a_{22} & a_{23} & . & . & a_{2n} \\
a_{13} & a_{23} & a_{33} & . & . & a_{3n} \\
. & . & . & . & . & . \\
. & . & . & . & . & . \\
a_{1n} & a_{2n} & a_{3n} & . & . & a_{nn}
\end{pmatrix}
=
\begin{pmatrix}
r_{11} & & & & & \\
r_{12} & r_{22} & & & & \\
r_{13} & r_{23} & r_{33} & & & \\
. & . & . & . & & \\
. & . & . & . & . & \\
r_{1n} & r_{2n} & r_{3n} & . & . & r_{nn}
\end{pmatrix}
\begin{pmatrix}
r_{11} & r_{12} & r_{13} & . & . & r_{1n} \\
& r_{22} & r_{23} & . & . & r_{2n} \\
& & r_{33} & . & . & r_{3n} \\
& & & . & . & . \\
& & & & . & . \\
& & & & & r_{nn}
\end{pmatrix}. \tag{2.43}
$$

Equating the $(1,1)$-th elements gives:

$$a_{11} = r_{11}^2, \quad r_{11} = \sqrt{a_{11}}.$$

Equating the first rows of both sides of (2.43) yields:

$$a_{12} = r_{11}r_{12}, \quad a_{13} = r_{11}r_{13}, \quad \cdots \quad , a_{1n} = r_{11}r_{1n},$$

so that all elements in the first row of R can be computed after r_{11} is known.

For the $(2, 2)$-th element

$$a_{22} = r_{12}^2 + r_{22}^2.$$

Since r_{12} is known, r_{22} can be computed, and then the second row of R, and so on. This algorithm is sometimes referred to as the *row-wise* Cholesky factorization. However, the elements of R can equally well be computed in the order r_{11}, r_{12}, r_{22}, r_{13}, r_{23}, r_{33}, ..., etc. For obvious reasons, this latter procedure is known as the *column-wise* Cholesky factorization.

The Cholesky factorization of (2.42) can be computed in about $\frac{1}{6}n^3$ multiplications (and additions), and n square roots (which can be avoided if the form (2.41) is used).

A remarkable feature of the Cholesky algorithm is that, in contrast to Gaussian elimination, *no interchanges are necessary* for numerical stability. This property holds because of the following relationship between the elements of A and R:

$$r_{1k}^2 + r_{2k}^2 + \cdots + r_{kk}^2 = a_{kk}, \quad k = 1, \dots, n.$$

This expression provides an *a priori* bound on the elements of R in terms of the diagonal elements of A and thus there can be *no growth* in the elements of R, even if a small pivot occurs during the factorization. The backward error analysis for the Cholesky algorithm yields:

$$R^T R = A + E,$$

where the bound on $\|E\|$ includes no growth factor.

It is tempting to think that every symmetric matrix can be written in the form (2.41), but with no restriction in sign on the diagonal elements of D. However, it should be strongly emphasized that the LDL^T factorization need not exist for a general symmetric indefinite matrix — for example, consider the matrix

$$A = \begin{pmatrix} 0 & 1 \\ 1 & 0 \end{pmatrix}.$$

Even if the factorization exists for an indefinite matrix, the algorithm given above cannot be guaranteed to be numerically stable, because there is no *a priori* bound on the sub-diagonal elements of the triangular factor. For example, if ϵ is small relative to unity:

$$\begin{pmatrix} \epsilon & 1 \\ 1 & \epsilon \end{pmatrix} = \begin{pmatrix} 1 & 0 \\ \frac{1}{\epsilon} & 1 \end{pmatrix} \begin{pmatrix} \epsilon & 0 \\ 0 & \epsilon - \frac{1}{\epsilon} \end{pmatrix} \begin{pmatrix} 1 & \frac{1}{\epsilon} \\ 0 & 1 \end{pmatrix},$$

and the elements of the factors are extremely large.

2.2.5.3. The QR factorization. In Section 2.2.5.1, a square matrix was reduced to upper-triangular form by applying a sequence of lower-triangular elementary matrices. A similar reduction can be performed using a sequence of *orthogonal* elementary matrices.

For any non-zero vector w, we define the corresponding *Householder transformation* as a symmetric elementary matrix of the form

$$H = I - \frac{1}{\beta}ww^T,$$

where $\beta = \frac{1}{2}\|w\|_2^2$. A Householder matrix is orthogonal, and hence preserves Euclidean length when applied to a vector. For any two distinct vectors a and b of equal Euclidean length, there exists a Householder matrix H that will transform one into the other:

$$Ha = (I - \frac{1}{\beta}ww^T)a = a - w(\frac{w^Ta}{\beta}) = b. \tag{2.44}$$

Re-arranging (2.44), we see that w must be a vector in the direction $(a - b)$.

Note also that the transformed vector Ha is simply the difference between the original vector and a multiple of the Householder vector w. Consequently, the transformed vector is identical to the original vector in all components where the Householder vector is zero. Furthermore, if $w^Ta = 0$, from (2.44) we see that the vector a is unaltered by the Householder transformation.

A *plane rotation* is a special orthogonal transformation, and is most commonly used to annihilate a single element of a vector. A plane rotation is equivalent to a Householder matrix where the Householder vector has non-zero elements only in the two positions i and j. Thus, a plane rotation differs from the identity matrix only in rows and columns i and j. Assuming that i is less than j, the (i, i), (i, j), (j, i) and (j, j) elements have the configuration

$$\begin{pmatrix} c & s \\ s & -c \end{pmatrix},$$

where $c^2 + s^2 = 1$ (hence, $c = \cos\theta$ and $s = \sin\theta$ for some θ).

When a vector is pre-multiplied by a plane rotation, only the i-th and j-th elements are affected. Usually, θ is chosen so that application of the rotation will introduce a zero into the j-th position of a vector (say, x). This can be achieved by choosing

$$s = \pm\frac{x_j}{r} \quad \text{and} \quad c = \pm\frac{x_i}{r},$$

where $r = (x_i^2 + x_j^2)^{\frac{1}{2}}$ and it is customary to take the positive sign. A plane rotation can be represented compactly with various combinations of information — e.g., i, j, s and c.

The properties of Householder matrices imply that a sequence of n matrices $\{H_i\}$, $i = 1, \ldots, n$, may be applied on the left to reduce an $m \times n$ matrix A of rank n to upper-triangular form. The transformation H_i is designed to annihilate elements $i + 1$ through m of the i-th column of the partially triangularized matrix, without altering the first $i - 1$ columns. (If $m = n$, only $n - 1$ transformations are required.)

In particular, H_1 is constructed to transform the first column a_1 of A to a multiple of e_1. The transformed first column will be $(r_{11}, 0, \ldots, 0)^T$, where $|r_{11}| = \|a_1\|_2$. Thus, the vector corresponding to H_1 is $(a_{11} - r_{11}, a_{21}, \ldots, a_{m1})^T$. To avoid cancellation error in computing the first component of this vector, the sign of r_{11} is chosen to be opposite to that of a_{11}.

After n Householder transformations, we obtain

$$H_n \cdots H_2 H_1 A = QA = \begin{pmatrix} R \\ 0 \end{pmatrix}, \tag{2.45}$$

where R is an $n \times n$ non-singular upper-triangular matrix, and the orthogonal matrix Q is the product $H_n \cdots H_1$. The form (2.45) is termed the *QR factorization* of A.

The assumption of full column rank is essential in the derivation of (2.45) in order to ensure that the next column to be reduced is not identically zero. If the rank of A is r, where r is less than n, column interchanges can be carried out to ensure that r linearly independent columns are processed first. The result of r transformations from the left is then

$$QAP = \begin{pmatrix} T \\ 0 \end{pmatrix},$$

where P is a permutation matrix and T is an $r \times n$ upper-trapezoidal matrix. A further sequence of r Householder transformations $\bar{H}_i$ may then be applied on the right to annihilate the last $n - r$ columns of T, while preserving the triangular structure (but not the elements!) of the first r columns. Thus we obtain

$$QAP\bar{H}_r \cdots \bar{H}_1 = QAV = \begin{pmatrix} R & 0 \\ 0 & 0 \end{pmatrix},$$

where R is a $r \times r$ non-singular upper-triangular matrix and V is an $n \times n$ orthogonal matrix. This form is termed the *complete orthogonal factorization* of A.

The orthogonal factors of A provide information about its column and row spaces. Let Q^T be partitioned as

$$Q^T = (Q_1 \ Q_2),$$

where Q_1 and Q_2 are $m \times r$ and $m \times (m - r)$ matrices respectively. The columns of Q_1 form an orthonormal basis for the column space of A, and the columns of Q_2 form an orthonormal basis for the corresponding null space of vectors orthogonal to the columns of A. In addition, if $r < n$, the last $n - r$ rows of V form an orthonormal basis for the set of vectors orthogonal to the rows of A.

The number of operations required to compute Q, R, and V in the rectangular case depends on the relative values of m, n, and r. If $m \geq n \geq r$, the number of additions or multiplications is approximately $mn^2 - \frac{1}{3}n^3 + (n - r)r^2$.

When a similar factorization is required for a matrix A with full *row* rank (i.e., whose rows are linearly independent), it is often more convenient to represent the factorization as

$$AQ = (L \ \ 0),$$

which is known as the *LQ factorization*.

The most common use of the orthogonal factors is in solving the *linear least-squares problem*: find x such that

$$\|Ax - b\|_2^2$$

is a minimum. Because Euclidean length is preserved by orthogonal matrices, the residual of a problem transformed by Q is equal in length to the residual of the original problem. Thus

$$\|Ax - b\|_2 = \|Q(Ax - b)\|_2 \equiv \|\hat{R}x - Qb\|_2,$$

and consequently

$$\|Ax - b\|_2 = \left\|\begin{pmatrix} R \\ 0 \end{pmatrix} x - Qb\right\|_2.$$

From the last expression, we see that the residual vector of the transformed problem will be minimized when the first n components of $\hat{R}x$ equal the first n components of Qb. When $n = r$, the minimum residual is therefore achieved for the unique solution of the linear system

$$Rx = Q_1^T b.$$

2.2.5.4. The spectral decomposition of a symmetric matrix.

When a matrix A is symmetric, its eigenvalues are all real, and its eigenvectors form an orthonormal basis for $\Re^n$. Let the eigenvalues of A be $\{\lambda_1, \ldots, \lambda_n\}$, and the eigenvectors be $\{u_1, \ldots, u_n\}$. By definition,

$$Au_i = \lambda_i u_i, \quad i = 1, \ldots, n. \tag{2.46}$$

If we define U as the matrix whose i-th column is u_i, the set of relations in (2.46) can be written in matrix form as

$$AU = U\Lambda,$$

where $\Lambda = \operatorname{diag}(\lambda_1, \ldots, \lambda_n)$. Since U is orthonormal, the result is:

$$A = U\Lambda U^T,$$

or explicitly

$$A = \sum_{i=1}^n \lambda_i u_i u_i^T, \tag{2.47}$$

the *spectral decomposition* of A.

An iterative algorithm is necessary to compute the eigensystem of a matrix, requiring approximately $3n^3$ additions and multiplications.

2.2.5.5. The singular-value decomposition.

Any real $m \times n$ matrix can be written as

$$A = U\Sigma V^T,$$

where U is an $m \times m$ orthonormal matrix, V is an $n \times n$ orthonormal matrix, and Σ is an $m \times n$ diagonal matrix, $\Sigma = \operatorname{diag}(\sigma_1, \ldots, \sigma_n)$, with $\sigma_i \geq 0$ for all i.

The numbers $\{\sigma_i\}$ are termed the *singular values* of A, and are generally assumed to be ordered so that $\sigma_1 \geq \sigma_2 \geq \cdots \geq 0$. If A is of rank r, then $\sigma_r > 0$, $\sigma_{r+1} = 0$, so that a matrix of rank r has r non-zero singular values. The singular values of A satisfy:

    (i) $\sigma_i^2(A) = \lambda_i[A^T A]$;

    (ii) $\sigma_1(A) = \|A\|_2$ (the two-norm of A is its largest singular value).

For a square, non-singular A, σ_n gives the "distance" (measured in the matrix two-norm) between A and the nearest singular matrix, i.e., $\sigma_n = \min\|A - S\|$, with S a singular matrix. For a symmetric matrix, the singular-value decomposition can be obtained from the spectral decomposition, and $\sigma_i = |\lambda_i|$.

2.2.5.6. The pseudo-inverse. When A is non-singular, A^{-1} is the unique matrix such that the vector $x = A^{-1}b$ solves the linear system $Ax = b$. If A is singular or rectangular, the inverse of A does not exist; furthermore, the system $Ax = b$ may be incompatible. However, a generalization of the concept of the inverse transformation can be formulated in terms of a related least-squares problem.

The *pseudo-inverse* of the $m \times n$ matrix A, denoted by A^+, is the unique $n \times m$ matrix such that $x = A^+b$ is the vector of minimum Euclidean length that minimizes $\|b - Ax\|_2$.

There are several mathematically equivalent expressions for the pseudo-inverse of A. When A is non-singular, $A^+ = A^{-1}$. When A has full column rank, the pseudo-inverse may be written as

$$A^+ = (A^TA)^{-1}A^T;$$

however, it should not be computed in this form; if the QR factorization (2.45) is available, A^+ is given by

$$A^+ = R^{-1}Q_1^T.$$

When A is rank-deficient, the most convenient form of the pseudo-inverse is based on the singular-value decomposition. If $A = U\Sigma V^T$, with r non-zero singular values, then

$$A^+ = V\Omega U^T,$$

where $\Omega = \text{diag}(\omega_i)$, with $\omega_i = 1/\sigma_i$, $i = 1,\ldots,r$, and $\omega_i = 0$, $i > r+1$.

2.2.5.7. Updating matrix factorizations. In many instances, we wish to factorize a matrix $\bar{A}$ that is closely related to another matrix A for which the same factorization has already been computed. Naturally, one would hope to be able to use the known factorization to reduce the amount of computation required to form the new factorization.

The most effective method of achieving this economy is to *modify* or *update* the factorization of the original matrix. In most cases this modification process requires significantly less work than computing the factorization *ab initio*. Techniques have been developed for updating all the factorizations discussed thus far, and a detailed description of these methods is beyond the scope of this text. Therefore, we shall outline only two particular updates corresponding to the most frequent case in optimization, when the matrices A and $\bar{A}$ differ by a matrix of rank one (see Section 2.2.1.5).

Since the process of adding a column to a matrix can be described formally as a rank-one modification, some factorizations that proceed column by column can usefully be viewed in terms of updating. For example, consider computing the QR factorization of a matrix $\bar{A}$ that is generated by adding a new column a to the end of the $m \times n$ matrix A. Then

$$Q\bar{A} = (\, QA \quad Qa \,)$$
$$= \begin{pmatrix} R & v_1 \\ 0 & v_2 \end{pmatrix},$$

where v_1 and v_2 are the relevant partitions of the m-vector $v = Qa$. Let H_{n+1} denote the Householder matrix that annihilates elements $n+2$ through m of v, and leaves the first n elements of v unchanged, so that

$$H_{n+1}v = \begin{pmatrix} v_1 \\ \gamma \\ 0 \end{pmatrix},$$

where $|\gamma| = \|v_2\|_2$. When H_{n+1} is applied to $Q\bar{A}$, we obtain

$$H_{n+1}Q\bar{A} = \begin{pmatrix} R & v_1 \\ 0 & \gamma \\ 0 & 0 \end{pmatrix},$$

or

$$\bar{Q}\bar{A} \equiv \begin{pmatrix} \bar{R} \\ 0 \end{pmatrix},$$

which defines the QR factorization of $\bar{A}$. Note that this factorization has been obtained by computing just one Householder matrix.

Other factorizations can also be interpreted in terms of updating the factorization of an augmented matrix. Matrix modification schemes are a natural extension of such techniques, and allow not only a more general change between stages, but also the possibility of reversing the process.

The other common modification that we shall consider arises when we require the Cholesky factorization of a positive-definite matrix $\bar{B}$ that is a rank-one modification of a positive-definite matrix B:

$$\bar{B} = B \pm vv^T.$$

In the case of a positive correction and the form (2.41), we have

$$\bar{B} = LDL^T + vv^T$$
$$= L(D + pp^T)L^T,$$

where p is the solution of the triangular system $Lp = v$. The very special nature of the matrix $D + pp^T$ means that its Cholesky factors (which will be denoted by $\hat{L}$ and $\hat{D}$) can be computed directly. Thus,

$$\bar{B} = L\hat{L}\hat{D}\hat{L}^TL^T$$
$$= \bar{L}\bar{D}\bar{L}^T,$$

where $\bar{L}$ is the matrix $L\hat{L}$ and $\bar{D}$ is $\hat{D}$. The product of two unit lower-triangular matrices is itself a unit lower-triangular matrix, and consequently $\bar{L}$ and $\bar{D}$ are precisely the triangular factors required. The vector p can be computed during the multiplication of L and $\hat{L}$; moreover, multiplication of these matrices requires only $O(n^2)$ operations.

The elements of $\bar{L}$ and $\bar{D}$ can be computed using the following recurrence relations:

(i) define $t_0 = 1$, $v^{(1)} = v$;

(ii) for $j = 1, 2, \ldots, n$, compute:

$$p_j = v_j^{(j)}, \qquad t_j = t_{j-1} + p_j^2/d_j,$$
$$\bar{d}_j = d_j t_j/t_{j-1}, \quad \beta_j = p_j/(d_j t_j),$$
$$\left.\begin{array}{l} v_r^{(j+1)} = v_r^{(j)} - p_j l_{rj} \\ \bar{l}_{rj} = l_{rj} + \beta_j v_r^{(j+1)} \end{array}\right\} r = j+1, \ldots, n.$$

When $\bar{B}$ results from a negative correction, extreme care must be taken in order to prevent a loss of positive definiteness when computing the updated factors — for example, rounding errors may cause elements of $\bar{D}$ to become zero or negative. The recurrence relations for a negative update are as follows:

(i) solve the equations $Lp = v$ and define $t_{n+1} = 1 - p^T D^{-1} p$; if $t_{n+1} \leq \epsilon_M$, set $t_{n+1} = \epsilon_M$, where ϵ_M is the relative machine precision;

(ii) for $j = n, n-1, \ldots, 1$ compute:

$$t_j = t_{j+1} + p_j^2/d_j \qquad \bar{d}_j = d_j t_{j+1}/t_j,$$

$$\beta_j = -p_j/(d_j t_{j+1}) \quad v_j^{(j)} = p_j,$$

$$\left. \begin{array}{l} \bar{l}_{rj} = l_{rj} + \beta_j v_r^{(j+1)} \\ v_r^{(j)} = v_r^{(j+1)} + p_j l_{rj} \end{array} \right\} r = j+1, \ldots, n.$$

If the modified matrix is theoretically guaranteed to be positive definite, the *exact* value of t_{n+1} must be positive. The error introduced by replacing a too-small computed value of t_{n+1} by ϵ_M is comparable with the error that would be made in explicitly computing the elements of $\bar{B} = B - vv^T$. With the procedure given above, it can be easily verified that all values $\bar{d}_j$, $j = 1, 2, \ldots, n$, are guaranteed to be positive, regardless of any rounding errors made during the computation.

Similar techniques can be applied to update the Cholesky factorization in the form (2.42).

2.2.6. Multi-Dimensional Geometry

Let $a = (a_1, \ldots, a_n)^T$ be a non-zero vector. In this section, we consider a geometric characterization of the vectors x that satisfy

$$a^T x = \beta \tag{2.48}$$

for some scalar β.

In the special case of $\beta = 0$, the set of vectors z that satisfy

$$a^T z = 0$$

form a vector space of dimension $n - 1$ (the null space of the vector a), for which there is an orthonormal basis $Z = \{z_1, z_2, \ldots, z_{n-1}\}$. Since a is linearly independent of $\{z_1, \ldots, z_{n-1}\}$, the set composed of Z and the vector a forms a basis for $\Re^n$. Every vector x that satisfies (2.48) can be written as:

$$x = \alpha_1 z_1 + \alpha_2 z_2 + \cdots + \alpha_{n-1} z_{n-1} + \alpha_n a. \tag{2.49}$$

Pre-multiplying (2.49) by the transpose of a, we obtain

$$a^T x = \alpha_n a^T a = \beta,$$

so that

$$\alpha_n = \frac{\beta}{a^T a} = \frac{\beta}{\|a\|_2^2}.$$

The representation (2.49) is said to be the set Z *translated* by the vector a, and the set of all such vectors defines a *hyperplane*. The vector a is termed the *normal* to the hyperplane, and the normalized vector $a/\|a\|_2$, which has Euclidean length unity, is said to be the *unit normal* to the hyperplane.

One can think of a hyperplane as a shift from the origin of the $(n-1)$-dimensional subspace orthogonal to a. Note that if $\beta = 0$, the hyperplane passes through the origin. Otherwise, the

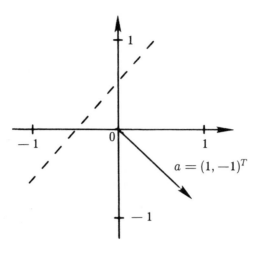

Figure 2d. Two-dimensional example of a hyperplane.

squared distance to the origin from any point on the hyperplane is

$$x^T x = \alpha_1^2 + \alpha_2^2 + \cdots + \alpha_{n-1}^2 + \alpha_n^2 (a^T a).$$

Since α_n is fixed, the minimum distance occurs at the point where $\alpha_1 = \cdots = \alpha_{n-1} = 0$ (i.e., x is a multiple of a), and is $|\beta|/\|a\|_2$.

As a graphical illustration, consider the set of two-dimensional vectors that satisfy

$$a^T x = a_1 x_1 + a_2 x_2 = \beta$$

for $a_1 = 1, a_2 = -1, \beta = -\frac{1}{2}$. The dashed line in Figure 2d illustrates this set of values. Note that the vector $a = (1, -1)^T$ is orthogonal to the hyperplane (which is a straight line in this case), and that the shortest distance from any point on the hyperplane to the origin is

$$\frac{|\beta|}{\|a\|_2} = \frac{|\frac{1}{2}|}{\sqrt{2}} = \frac{\sqrt{2}}{4}.$$

Figure 2e illustrates a hyperplane in the three-dimensional case. Notice that the vectors $z_1, z_2, \ldots, z_{n-1}$ may be regarded as the axes of a perpendicular co-ordinate system, lying in the hyperplane, whose center is at the point on the hyperplane nearest to the origin. As in the

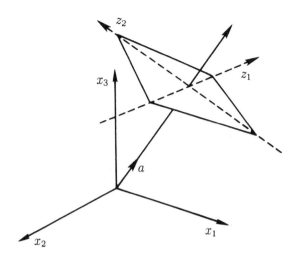

Figure 2e. Three-dimensional example of a hyperplane.

two-dimensional case, a different value of β yields a parallel (translated) hyperplane.

We can translate a more general subspace L by considering the set of all vectors of the form $x_0 + y$, where $y \in L$ and x_0 is a given vector; this set is termed a *linear manifold*. Note that a hyperplane is an example of a manifold where L is of dimension $n - 1$.

Notes and Selected Bibliography for Section 2.2

There are many introductory texts on numerical linear algebra; see, for example, Stewart (1973), Strang (1976), and the relevant chapters of Dahlquist and Björck (1974). More specialized or advanced topics are treated in Forsythe and Moler (1967) and Lawson and Hanson (1974). Details of methods for modifying matrix factorizations have not yet filtered through to linear algebra textbooks. However, the interested reader may refer to the research papers of Gill, Golub, Murray and Saunders (1974) and Daniel, Gragg, Kaufman and Stewart (1976).

A complete and thorough background for the mature scholar is contained in Wilkinson (1965).

Much effort has been devoted to the production of high-quality software for linear algebraic problems. Excellent sources of software are Wilkinson and Reinsch (1971), Smith *et al.* (1974), and Dongarra, Bunch, Moler and Stewart (1979).

2.3. ELEMENTS OF MULTIVARIATE ANALYSIS

2.3.1. Functions of Many Variables; Contour Plots

In this section, we shall be concerned with the basic properties of scalar functions of n variables. A typical function will be denoted by $F(x)$, where x is the real n-vector $(x_1, x_2, \ldots, x_n)^T$. If $n = 1$, $F(x)$ is a function of a single variable, or a *univariate function*; in this case, the function is denoted by $f(x)$ and the subscript on x is usually omitted. We shall occasionally be concerned with an ordered set of multivariate functions $(f_1(x), f_2(x), \ldots, f_m(x))^T$.

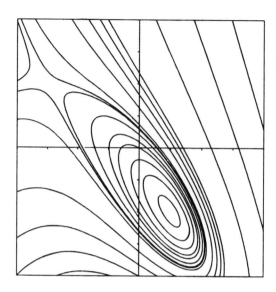

Figure 2f. Contour map of the function $F(x) = e^{x_1}(4x_1^2 + 2x_2^2 + 4x_1x_2 + 2x_2 + 1)$.

Throughout all subsequent discussion of multivariate functions, much insight can be gained from examples of low dimensionality — in particular, contour plots will be used for illustration.

For any multivariate function, the equation $z = F(x)$ defines a surface in the $(n + 1)$-dimensional space $\Re^{n+1}$. In the case of $n = 2$, the points $z = F(x_1, x_2)$ represent a surface in the three-dimensional space (z, x_1, x_2). If we let c be a particular value of $F(x_1, x_2)$, the equation $F(x_1, x_2) = c$ can be regarded as defining a curve in x_1 and x_2 on the plane $z = c$. If the plane curves are drawn on a single plane for a selection of values of c, we obtain a figure which is equivalent to a contour map of the function. Figure 2f shows a contour map of the function $F(x) = e^{x_1}(4x_1^2 + 2x_2^2 + 4x_1x_2 + 2x_2 + 1)$. The contours shown correspond to the c-values 0.2, 0.4, 0.7, 1, 1.7, 1.8, 2, 3, 4, 5, 6 and 20.

2.3.2. Continuous Functions and their Derivatives

Methods for solving optimization problems depend on the ability to use information about a function at a particular point to analyze its general behaviour. We now consider some properties of functions that allow such conclusions to be drawn. Central among such properties is *continuity*.

In the univariate case, a function f is said to be *continuous* at the point x, if, given any ϵ $(\epsilon > 0)$, there exists a δ $(\delta > 0)$ such that $|y - x| < \delta$ implies that $|f(y) - f(x)| < \epsilon$. Pictorially, continuity of f at x means that the graph does not contain a "break" at x. If f is not continuous at x, f is said to be *discontinuous*. In this case, the function value at x need not be close to the value of f at any nearby point, as shown in Figure 2g.

In order to define continuity of multivariate functions, we introduce the concept of a *neighbourhood* in n-space. Let δ be any positive number. A δ-*neighbourhood* of the point x is defined as the set of all points y such that $\|y - x\| \leq \delta$. The choice of norm is not usually significant,

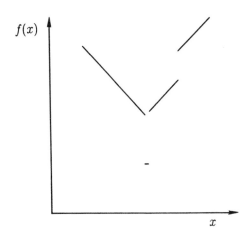

Figure 2g. A discontinuous function.

although it does affect the "shape" of the neighbourhood; the Euclidean norm is the most frequently used. With the Euclidean norm in three-space, a δ-neighbourhood of x is the sphere of radius δ with center at x. The term "neighbourhood" alone simply implies a δ-neighbourhood for some specified value of δ.

The definition of continuity extends directly to the multivariate case by considering the value of the function at all points in a neighbourhood. A multivariate function F is *continuous* at x if, given any $\epsilon > 0$, there exists $\delta > 0$ such that if $\|y - x\| \leq \delta$ (i.e., y is in a δ-neighbourhood of x), then $|F(y) - F(x)| < \epsilon$. A vector-valued function is said to be continuous if each of its component functions is continuous.

Note that the continuity of F at x does not depend on the choice of norm to measure $\|y - x\|$, but the value of δ will, in general, be norm-dependent.

In the case $n = 2$, a discontinuity can be observed in a contour plot of F when many contours "merge" — for example, at a vertical "cliff" on the surface.

An additional important property of functions is *differentiability*. In the univariate case, consider the graph of f in an interval containing a particular point $\bar{x}$. The chord joining $f(\bar{x})$ and $f(y)$ for any y ($y \neq \bar{x}$) in the interval has slope $m(y)$ given by

$$m(y) = \frac{f(y) - f(\bar{x})}{y - \bar{x}}.$$

If the value of $m(y)$ approaches a limit as y approaches x, the limiting value is the slope of the tangent to f at $\bar{x}$, and is defined to be the *first derivative*, or *gradient*, of f at $\bar{x}$. The first derivative is denoted by $f'(\bar{x})$, $f^{(1)}(\bar{x})$, or by $df/dx \mid_{\bar{x}}$. Formally,

$$f'(\bar{x}) = \lim_{y \to \bar{x}} m(y) = \lim_{h \to 0} \frac{f(x+h) - f(x)}{h}. \tag{2.50}$$

The function f is said to be *differentiable* at $\bar{x}$ if $f'(\bar{x})$ exists. Figure 2h illustrates the function

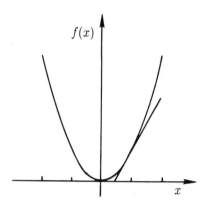

Figure 2h. The function $f(x) = x^2$.

$f(x) = x^2$, with $\bar{x} = 1$, and the interval $[-2, 2]$. In general,

$$\frac{f(\bar{x} + h) - f(\bar{x})}{h} = \frac{\bar{x}^2 + 2h\bar{x} + h^2 - \bar{x}}{h} = 2\bar{x} + h. \tag{2.51}$$

The limit of (2.51) as $h \to 0$ exists for every $\bar{x}$, and is given by

$$f'(\bar{x}) = 2\bar{x};$$

hence, $f'(1) = 2$.

In order for the derivative defined by (2.50) to exist, the limit must exist as y approaches $\bar{x}$ from either the left or the right, i.e. for positive or negative values of h. However, it may happen that the limit exists as y approaches $\bar{x}$ from one side only, or the value of the limit may differ as $\bar{x}$ is approached from either side. For example, consider the continuous function $f(x) = |x|$, shown in Figure 2i in the interval $[-1, 1]$. The slope of f is clearly -1 for $x < 0$, and $+1$ for $x > 0$. At the point $\bar{x} = 0$, $m(y) = -1$ if $y < \bar{x}$, and $m(y) = +1$ for $y > \bar{x}$, so that the limit (2.50) does not exist; hence, $f(x)$ is not differentiable at the origin.

This situation suggests extending the definition (2.50) to a *one-sided derivative*, as follows. If the limit

$$\lim_{\substack{h \to 0 \\ h > 0}} \frac{f(\bar{x} + h) - f(\bar{x})}{h} \tag{2.52}$$

exists, (2.52) defines the *right-hand (right-sided)* derivative at $\bar{x}$, which will be denoted by $f'_+(\bar{x})$.

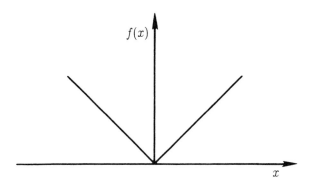

Figure 2i. The function $f(x) = |x|$.

Similarly, a left-hand derivative can be defined if a limit of the form (2.52) exists, restricted to $h < 0$. In the previous example of $f(x) = |x|$, the right-hand and left-hand derivatives exist everywhere, and are equal at every point except the origin.

If f is differentiable in an interval, then the derivative f' can itself be treated as a univariate function of x, which may be written as $f'(x)$; the notation "(x)" in this context indicates that x is the independent variable rather than a particular point. If the derivative function $f'(x)$ is in its turn also differentiable at a point $\bar{x}$, its first derivative is termed the *second derivative* of f at $\bar{x}$, and is denoted by $f''(\bar{x})$, $f^{(2)}(\bar{x})$, or $d^2 f/dx^2 \mid_{\bar{x}}$. Again using the example of $f(x) = x^2$, it is easy to see that f'' exists for all x, and is a constant:

$$\frac{f'(x+h) - f'(x)}{h} = \frac{2\bar{x} + 2h - 2\bar{x}}{h} = 2.$$

The process of differentiating each new derivative can be continued (if all the required limits exist), and we can thereby define the n-th derivative of f at $\bar{x}$, which is usually written as $f^{(n)}(\bar{x})$, or $d^n f/dx^n \mid_{\bar{x}}$.

The reader should consult the references cited in the Notes for details concerning the rules that apply for differentiating sums, products and quotients of differentiable functions. However, we shall state without proof a general rule for differentiating a composite function of the form $g(f(x))$. The *chain rule* states that, under certain mild conditions on f, g and x,

$$\frac{d}{dx} g(f(x)) = g'(f(x)) f'(x).$$

To generalize the ideas of differentiability to multivariate functions, the variation in the function at $\bar{x}$ is observed with respect to each variable separately. For example, at the point $\bar{x} = (\bar{x}_1, \bar{x}_2, \ldots, \bar{x}_n)^T$, we consider the charge in F resulting from a change in the first variable only, while the remaining $n - 1$ variables remain constant. The *partial derivative of F with respect to x_1 at $\bar{x}$* is denoted by $\partial F/\partial x_1 \mid_{\bar{x}}$, and is defined as the following limit (when it exists):

$$\left. \frac{\partial F}{\partial x_1} \right|_{\bar{x}} = \lim_{h \to 0} \frac{F(\bar{x}_1 + h, \bar{x}_2, \ldots, \bar{x}_n) - F(\bar{x})}{h}.$$

This number gives the slope of the tangent to F at $\bar{x}$ along the x_1 direction.

In general, the slope of $F(x)$ at $\bar{x}$ in the direction of the i-th co-ordinate axis is denoted by $\partial F/\partial x_i|_{\bar{x}}$. When all n partial derivatives of F are continuous at $\bar{x}$, F is said to be *differentiable* at $\bar{x}$.

Under conditions analogous to those in the univariate case, each partial derivative may be regarded as a multivariate function of x. The n-vector of these partial derivatives is then a vector function of x, termed the *gradient vector* of F, and will be denoted by $\nabla F(x)$, or $g(x)$:

$$\nabla F(x) \equiv g(x) \equiv \begin{pmatrix} \dfrac{\partial F}{\partial x_1} \\ \vdots \\ \dfrac{\partial F}{\partial x_n} \end{pmatrix}.$$

For example, if $F(x)$ is defined by

$$F(x) = x_1 x_2^2 + x_2 \cos x_1,$$

then

$$g(x) = \begin{pmatrix} \dfrac{\partial F}{\partial x_1} \\ \dfrac{\partial F}{\partial x_2} \end{pmatrix} = \begin{pmatrix} x_2^2 - x_2 \sin x_1 \\ 2x_1 x_2 + \cos x_1 \end{pmatrix}.$$

If the gradient of F is a constant vector, F is said to be a *linear function* of x; in this case, F is of the form:

$$F(x) = c^T x + \alpha,$$

for some fixed vector c and scalar α, and $\nabla F(x) = c$.

We recall that a one-dimensional derivative defines the slope of the tangent line to the curve defined by $F(x)$. Similarly, for a differentiable multivariate function the *tangent hyperplane* at the point $\bar{x}$ is defined by the gradient vector (the normal to the hyperplane) and the scalar value $F(\bar{x})$ (the distance of the hyperplane from the origin).

As in the univariate case, there are multivariate functions such that the condition of differentiability is not satisfied. For example, Figure 2j shows a contour plot of the non-differentiable function $F(x) = \max\{|x_1|, |x_2|\}$. At those points where $|x_1| = |x_2|$, the partial derivatives do not exist; these points are the "corners" in the contour plot. It can also happen that F is differentiable with respect to some of the variables, but not all.

Higher derivatives of a multivariate function are defined as in the univariate case, and the number of associated quantities increases by a factor of n with each level of differentiation. Thus, the "first derivative" of an n-variable function is an n-vector; the "second derivative" of an n-variable function is defined by the n^2 partial derivatives of the n first partial derivatives with respect to the n variables:

$$\frac{\partial}{\partial x_i}\left(\frac{\partial F}{\partial x_j}\right) \quad i = 1,\ldots,n; \quad j = 1,\ldots,n. \tag{2.53}$$

The quantity (2.53) is usually written as

$$\frac{\partial^2 F}{\partial x_i \, \partial x_j}, \quad i \neq j; \qquad \frac{\partial^2 F}{\partial x_i^2}, \quad i = j.$$

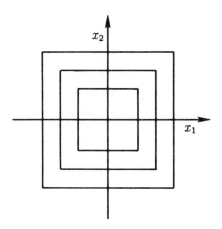

Figure 2j. Contour plot of $F(x) = \max\{|x_1|, |x_2|\}$.

If the partial derivatives $\partial F/\partial x_i$, $\partial F/\partial x_j$ and $\partial^2 F/\partial x_i\,\partial x_j$ are continuous, then $\partial^2 F/\partial x_j\,\partial x_i$ exists and $\partial^2 F/\partial x_i\,\partial x_j = \partial^2 F/\partial x_j\,\partial x_i$. These n^2 "second partial derivatives" are usually represented by a square, symmetric matrix, termed the *Hessian matrix* of $F(x)$, which will be denoted by $\nabla^2 F(x)$, or $G(x)$:

$$\nabla^2 F(x) \equiv G(x) \equiv \begin{pmatrix} \dfrac{\partial^2 F}{\partial x_1^2} & \cdots & \dfrac{\partial^2 F}{\partial x_1\,\partial x_n} \\ \vdots & & \vdots \\ \dfrac{\partial^2 F}{\partial x_1\,\partial x_n} & \cdots & \dfrac{\partial^2 F}{\partial x_n^2} \end{pmatrix}.$$

If the Hessian matrix of F is constant, F is said to be a *quadratic function*. In this case, F can be expressed as:

$$F(x) = \frac{1}{2}x^T G x + c^T x + \alpha, \tag{2.54}$$

for a constant matrix G, vector c and scalar α (multiplication by $\frac{1}{2}$ is included in the quadratic term to avoid the appearance of a factor of two in the derivatives). Note that the derivatives of (2.54) are:

$$\nabla F(x) = G x + c \qquad \text{and} \qquad \nabla^2 F(x) = G.$$

Further differentiation of a general multivariate function leads to higher order derivatives. The r-th derivative of F will be denoted by

$$\frac{\partial^r F}{\partial x_{i_1}\,\partial x_{i_2}\ldots\partial x_{i_r}}, \quad i_j = 1,\ldots,n,$$

or by $\nabla^r F(x)$. However, derivatives of higher than second order are seldom useful in optimization.

For a vector-valued function, derivatives can be defined in a consistent manner by simply differentiating each component function separately. The *Jacobian matrix* of a vector function

$f(x) = (f_1(x), \ldots, f_m(x))^T$ is defined as the $m \times n$ matrix whose (i,j)-th element is the derivative of f_i with respect to x_j, i.e. whose i-th row is the transposed gradient vector of f_i. Note that the Hessian matrix of the scalar function $F(x)$ is the Jacobian matrix of the vector function $g(x)$.

The class of functions with continuous derivatives of order 1 through k is denoted by C^k. The class C^2 is usually called the set of twice-continuously differentiable functions. Functions with high degrees of differentiability are referred to as "smooth" functions, where the terminology is derived from the observed interpretation of a discontinuity in the derivative as a "corner".

Let g be a multivariate function of n variables and let f be a vector function such that $f = (f_1(\theta), f_2(\theta), \ldots, f_n(\theta))^T$ with each f_i a univariate function of the scalar θ. The extension of the chain rule to multivariate functions gives the derivative of g with respect to θ as:

$$\frac{d}{d\theta}[g(f(\theta))] = (f_1'(\theta) \; f_2'(\theta) \cdots f_n'(\theta)) \nabla g(f).$$

2.3.3. Order Notation

Let $f(h)$ be a univariate function of h. Then the function $f(h)$ is said to be *of order h^p* (written as $f(h) = O(h^p)$) if there exists a finite number M, $M > 0$, independent of h, such that as $|h|$ approaches zero

$$|f(h)| \leq M|h|^p. \tag{2.55}$$

The importance of the relationship (2.55) is that for sufficiently small $|h|$, the rate at which an $O(h^p)$ term goes to zero will increase as p increases. Therefore, if $p > q$, a term that is $O(h^p)$ is of "higher order" than a term of $O(h^q)$, and its magnitude will decrease to zero more rapidly for sufficiently small h. For this reason, it is often desirable to find the largest value of p for which (2.55) holds.

When considering approximations and their associated errors, it is usually considered worthwhile to achieve an error of maximum possible order. However, some caution must be exercised before drawing unwarranted conclusions from the order alone, since the actual size of an $O(h^p)$ term depends on the associated constant M and on the value of h. For example, if $f_1(h) = 10^{-4}h$ and $f_2(h) = 10^6 h^2$, $|f_2(h)|$ is larger than $|f_1(h)|$ for $h > 10^{-10}$ even though $f_2(h)$ is of higher order.

2.3.4. Taylor's Theorem

The results from analysis that are most frequently used in optimization come from the group of "Taylor" or "mean-value" theorems. These theorems are of fundamental importance because they show that if the function and its derivatives are known at a single point, then we can compute approximations to the function at all points in the immediate neighbourhood of that point.

In the univariate case, we have the following theorem:

Taylor's theorem. *If $f(x) \in C^r$ then there exists a scalar θ $(0 \leq \theta \leq 1)$, such that*

$$f(x + h) = f(x) + hf'(x) + \frac{1}{2}h^2 f''(x) + \cdots$$
$$+ \frac{1}{(r-1)!}h^{r-1}f^{(r-1)}(x) + \frac{1}{r!}h^r f^{(r)}(x + \theta h), \tag{2.56}$$

where $f^{(r)}(x)$ denotes the r-th derivative of f evaluated at x.

Using order notation, the Taylor-series expansion of f about x can be written as:

$$f(x + h) = f(x) + hf'(x) + \frac{1}{2}h^2 f''(x) + \cdots$$
$$+ \frac{1}{(r-1)!} h^{r-1} f^{(r-1)}(x) + O(h^r),$$

assuming that $|f^{(r)}|$ is finite in the interval $[x, x + h]$.

A Taylor theorem also applies for sufficiently smooth multivariate functions. Let x be a given point, p a vector of unit length, and h a scalar. The function $F(x + hp)$ can be regarded as a univariate function of h, and the univariate expansion (2.56) can be applied directly:

$$F(x + hp) = F(x) + hg(x)^T p + \frac{1}{2}h^2 p^T G(x) p + \cdots$$
$$+ \frac{1}{(r-1)!} h^{r-1} D^{r-1} F(x) + \frac{1}{r!} h^r D^r F(x + \theta hp),$$

for some θ ($0 \le \theta \le 1$), and where

$$D^s F(x) = \sum_{i_1=1}^{n} \sum_{i_2=1}^{n} \cdots \sum_{i_s=1}^{n} \{ p_{i_1} p_{i_2} \cdots p_{i_s} \frac{\partial^s F(x)}{\partial x_{i_1} \partial x_{i_2} \cdots \partial x_{i_s}} \}.$$

For practical computation, we shall be interested only in the first three terms of this expansion:

$$F(x + hp) = F(x) + hg(x)^T p + \frac{1}{2}h^2 p^T G(x) p + O(h^3).$$

Note that the rate of change of F at the point x along the direction p is given by the quantity $g(x)^T p$, which is termed the *directional derivative* (or the *first derivative along p*). Similarly, the scalar $p^T G(x) p$ can be interpreted as the second derivative of F along p, and is commonly known as the *curvature of F along p*. A direction p such that $p^T G(x) p > 0$ (< 0) is termed a *direction of positive curvature (negative curvature)*.

The Taylor-series expansion of a general function F about a point $\hat{x}$ allows us to construct simple approximations to the function in a neighbourhood of $\hat{x}$. For example, ignoring all but the linear term of the Taylor series gives:

$$F(\hat{x} + p) \approx F(\hat{x}) + g(\hat{x})^T p.$$

The expression $F(\hat{x}) + g(\hat{x})^T p$ defines a linear function of the n-vector p, the displacement from $\hat{x}$, and will approximate F with error of order $\|p\|^2$. Similarly, including one additional term from the Taylor series produces a quadratic approximation:

$$F(\hat{x} + p) \approx F(\hat{x}) + g(\hat{x})^T p + \frac{1}{2}p^T G(\hat{x}) p,$$

with error of order $\|p\|^3$.

2.3.5. Finite-Difference Approximations to Derivatives

The ability to expand smooth functions in Taylor series allows the values of derivatives to be approximated, rather than calculated analytically. In the univariate case, the Taylor-series expansion (2.56) of a twice-continuously differentiable function $f(x)$ can be re-arranged to give:

$$\frac{f(x+h)-f(x)}{h} = f'(x) + \frac{1}{2}hf''(x+\theta_1 h), \qquad (2.57)$$

where $0 \le \theta_1 \le 1$. The relationship (2.57) can also be written as

$$\frac{f(x+h)-f(x)}{h} = f'(x) + O(h). \qquad (2.58)$$

The error in approximating $f'(x)$ by the expression on the left is due to neglecting the term $\frac{1}{2}hf''(x+\theta_1 h)$, which is termed the *truncation error*, since it arises from truncating the Taylor series. In general, θ_1 is unknown, and it is possible only to compute or estimate an upper bound on the truncation error.

The quantity h in (2.58) is known as the *finite-difference interval*. If $|h|$ is small relative to $|f'(x)|$ and $|f''(x+\theta_1 h)|$, the truncation error will be "small". The left-hand side of (2.58), which gives the slope of the chord joining $f(x)$ and $f(x+h)$, is called a *forward-difference approximation* to $f'(x)$.

Similarly, f can be expanded about the point $x-h$, and we obtain:

$$f(x-h) = f(x) - hf'(x) + \frac{1}{2}h^2 f''(x-\theta_2 h), \qquad (2.59)$$

where $0 \le \theta_2 \le 1$. This expansion leads to a *backward-difference approximation*

$$\frac{f(x)-f(x-h)}{h} = f'(x) + O(h), \qquad (2.60)$$

with associated truncation error $\frac{1}{2}hf''(x-\theta_2 h)$.

If the Taylor expansions (2.57) and (2.59) are carried one term further, the result is:

$$f(x+h) = f(x) + hf'(x) + \frac{1}{2}h^2 f''(x) + O(h^3) \qquad (2.61)$$

and

$$f(x-h) = f(x) - hf'(x) + \frac{1}{2}h^2 f''(x) + O(h^3). \qquad (2.62)$$

Subtracting and dividing by h, we obtain:

$$\frac{f(x+h)-f(x-h)}{2h} = f'(x) + O(h^2), \qquad (2.63)$$

since the terms involving $f''(x)$ cancel. The expression on the left-hand side of (2.63) is called a *central-difference approximation* to $f'(x)$; the associated truncation error is of second order, and involves values of $f^{(3)}$ in $[x-h, x+h]$. Note that two function values (in addition to $f(x)$) are required to approximate $f'(x)$ using (2.63), as compared to one evaluation for (2.58) or (2.60). In general, additional information is needed to obtain an approximation with higher-order truncation error.

To illustrate the use of finite-difference formulae, consider $f(x) = x^3$, for which $f'(x) = 3x^2$. The forward-difference formula (2.58) yields:

$$\frac{f(x+h) - f(x)}{h} = 3x^2 + 3xh + h^2,$$

so that the truncation error is $3xh + h^2$. The central-difference formula (2.63) gives:

$$\frac{f(x+h) - f(x-h)}{2h} = \frac{6x^2h + 2h^3}{2h} = 3x^2 + h^2,$$

and the truncation error is simply h^2.

The Taylor expansion can also be used to approximate higher derivatives of f by finite differences. Forming a linear combination of (2.61) and (2.62) so as to eliminate the terms involving $f'(x)$, we obtain:

$$\frac{f(x+h) - 2f(x) + f(x-h)}{h^2} = f''(x) + O(h),$$

where the $O(h)$ term involves $f^{(3)}$. Similarly, an $O(h^2)$ approximation to $f''(x)$ is given by:

$$\frac{1}{4h^2}(f(x+2h) - f(x+h) - f(x-h) + f(x+2h)) = f''(x) + O(h^2).$$

As we might expect, finite-difference formula for derivatives of arbitrary order are defined from high-order differences of f. To illustrate how these formulae work, we shall consider the simplest formula for a derivative of order k. We define the *forward-difference operator* Δ by the relation

$$\Delta f(x) = f(x+h) - f(x),$$

where the dependence of Δ upon h is suppressed. The second-order forward-difference operator Δ^2 is defined as

$$\Delta^2 f(x) = \Delta(\Delta f(x)) = \Delta f(x+h) - \Delta f(x)$$
$$= f(x+2h) - 2f(x+h) + f(x),$$

and higher-order differences can be defined in a similar way. Let f_j denote $f(x+jh)$. The numbers $\Delta^i f_j$ can be arranged into the following *difference table*:

$$
\begin{array}{ccccccccc}
f_0 & & & & & & & & \\
& \Delta f_0 & & & & & & & \\
f_1 & & \Delta^2 f_0 & & & & & & \\
& \Delta f_1 & & \Delta^3 f_0 & & & & & \\
f_2 & & \Delta^2 f_1 & & \Delta^4 f_0 & & & & \\
& \Delta f_2 & & \Delta^3 f_1 & & & & & \\
f_3 & & \Delta^2 f_2 & & & & & & \\
& \Delta f_3 & & & & & & & \\
f_4 & & & & & & & &
\end{array}
$$

where each difference is computed by subtraction of two entries in the previous column.

It can be shown that $\Delta^k f_0 = h^k f^{(k)}(x) + O(h^{k+1})$. Given a difference table constructed from the function values $f(x + jh)$, $j = 0, 1, \ldots, k$, we thus obtain

$$f^{(k)}(x) \approx \frac{1}{h^k} \Delta^k f_0.$$

Finite-difference formulae may be defined for multivariate functions. Let h_j be the finite-difference interval associated with the j-th component of x, and let e_j be the vector with unity in the j-th position and zeros elsewhere. Then:

$$F(x + h_j e_j) = F(x) + h_j g(x)^T e_j + \frac{1}{2} h_j^2 e_j^T G(x + \theta_1 h_j e_j) e_j,$$

where $0 \leq \theta_1 \leq 1$. The terms $g(x)^T e_j$ and $e_j^T G(x + \theta_1 h_j e_j) e_j$ are just the j-th element of $g(x)$ and the j-th diagonal element of $G(x + \theta_1 h_j e_j)$ respectively, giving:

$$g_j(x) = \frac{1}{h_j} \big(F(x + h_j e_j) - F(x) \big) + O(h_j),$$

which is a forward-difference expression analogous to (2.58). The backward- and central-difference formulae corresponding to (2.60) and (2.63) are:

$$\frac{1}{h_j} \big(F(x) - F(x - h_j e_j) \big) \quad \text{and} \quad \frac{1}{2h_j} \big(F(x + h_j e_j) - F(x - h_j e_j) \big).$$

As in the univariate case, we can compute finite-difference approximations to second derivatives. A first-order approximation to $G_{ij}(x)$ is given by:

$$\frac{1}{h_i h_j} \big(F(x + h_j e_j + h_i e_i) - F(x + h_j e_j) - F(x + h_i e_i) + F(x) \big).$$

This formula may be derived by writing down the first-order approximation to the j-th column of $G(x)$,

$$\frac{1}{h_j} \big(g(x + h_j e_j) - g(x) \big)$$

and approximating both $g(x + h_j e_j)$ and $g(x)$ by first-order differences of $F(x)$.

2.3.6. Rates of Convergence of Iterative Sequences

The majority of methods for optimization are *iterative*, in that an infinite sequence $\{x_k\}$ is generated of estimates of the optimal x^*. Even if it can be proved theoretically that this sequence will converge in the limit to the required point, a method will be practicable only if convergence occurs with some rapidity. In this section we briefly discuss means of characterizing the *rate* at which such sequences converge. A completely rigorous analysis will not be given, but the most important and relevant concepts from convergence theory will be introduced (a more comprehensive treatment may be found in the suggested reference). The reader should be forewarned that only selected results from convergence theory have any relevance to practical computation. Although the effectiveness of an algorithm is determined to some extent by the associated rate of convergence, the conditions under which a theoretical convergence rate is actually achieved may be rare (for example, an "infinite sequence" does not exist on a computer). Furthermore, the absence of a theorem on the rate of convergence of a method may be just as much a measure of the difficulty of the proof as of the inadequacy of the method.

In all that follows, it is assumed that the sequence $\{x_k\}$ converges to x^*. To simplify the discussion, we shall also assume that the elements of $\{x_k\}$ are distinct, and that x_k does not equal x^* for any value of k.

The most effective technique for judging the rate of convergence is to compare the improvement at each step to the improvement at the previous step, i.e., to measure the closeness of x_{k+1} to x^* relative to the closeness of x_k to x^*.

A sequence $\{x_k\}$ is said to converge with *order* r when r is the largest number such that

$$0 \le \lim_{k \to \infty} \frac{\|x_{k+1} - x^*\|}{\|x_k - x^*\|^r} < \infty.$$

Since we are interested in the value of r that occurs in the limit, r is sometimes known as the *asymptotic convergence rate*. If $r = 1$, the sequence is said to display *linear* convergence; if $r = 2$, the sequence is said to have *quadratic* convergence.

If the sequence $\{x_k\}$ has order of convergence r, the *asymptotic error constant* is the value γ that satisfies

$$\gamma = \lim_{k \to \infty} \frac{\|x_{k+1} - x^*\|}{\|x_k - x^*\|^r}. \tag{2.64}$$

When $r = 1$, γ must be strictly less than unity in order for convergence to occur.

To illustrate these ideas, we mention two examples of infinite sequences that may be readily computed on a scientific calculator. First, consider the scalar sequence

$$x_k = c^{2^k}, \tag{2.65}$$

where c is a constant that satisfies $0 \le c < 1$. Each member of this sequence is the square of the previous element, and the limiting value is zero. Furthermore,

$$\frac{|x_{k+1} - 0|}{|x_k - 0|^2} = c,$$

so that the convergence rate r is equal to two, and the sequence converges quadratically. Quadratic convergence generally means that, roughly speaking, eventually the number of correct figures in x_k doubles at each step. The second column of Table 2a contains the first fourteen iterates in the sequence (2.65), with $c = 0.99$. Note that the number of correct figures does not begin to double at each step until $k \ge 7$.

The second sequence is given by

$$y_k = c^{2^{-k}}, \tag{2.66}$$

where $c \ge 0$. Each member of this sequence is the square root of the previous element. For any positive value of c, $\lim_{k \to \infty} y_k = 1$. Thus,

$$\lim_{k \to \infty} \frac{c^{2^{-(k+1)}} - 1}{c^{2^{-k}} - 1} = \lim_{k \to \infty} \frac{1}{c^{2^{-(k+1)}} + 1} = \frac{1}{2},$$

which implies that the sequence (2.66) has *linear* convergence. Some values of this sequence with $c = 2.2$ are given in the third column of Table 2a. Note that an additional correct figure is obtained approximately every third iteration.

If a sequence has linear convergence, the step-wise decrease in $\|x_k - x^*\|$ will vary substantially with the value of the asymptotic error constant. If the limit (2.64) is zero when r is taken as unity, the associated type of convergence is given the special name of *superlinear* convergence. Note that a convergence rate greater than unity implies superlinear convergence.

Table 2a

Examples of quadratic, linear and superlinear convergence.

k	x_k	y_k	z_k
0	.99	2.2	—
1	.9801	1.4832397	1.0
2	.96059601	1.2178833	.25
3	.92274469	1.1035775	.03703704
4	.85145777	1.0505130	.00390625
5	.72498033	1.0249453	.00032
6	.52559649	1.0123958	.00002143
7	.27625167	1.0061788	.00000121
8	.07631498	1.0030847	.0000000596
9	.00582398	1.0015411	.0000000026
10	.00003392	1.0007703	
11	$.11515 \times 10^{-8}$	1.0003851	
12	$.13236 \times 10^{-17}$	1.0001925	
13	$.17519 \times 10^{-36}$	1.0000963	
14	$.30692 \times 10^{-72}$	1.0000481	

A scalar sequence that displays superlinear convergence is

$$z_k = \frac{1}{k^k}.$$ (2.67)

The limit of (2.67) is zero, and

$$\lim_{k \to \infty} \frac{z_{k+1}}{z_k} = \lim_{k \to \infty} \frac{1}{k(1 + 1/k)^{k+1}} = 0.$$

The first nine members of the sequence (2.67) are listed in the fourth column of Table 2a.

Notes and Selected Bibliography for Section 2.3

Much of the material contained in this section can be found in standard texts on advanced calculus; see, for example, Courant (1936) and Apostol (1957). A detailed treatment of the rates of convergence of iterative sequences is given by Ortega and Rheinboldt (1973).

CHAPTER THREE

OPTIMALITY CONDITIONS

*The statement of the cause is incomplete, unless
in some shape or other we introduce all the conditions.*

—JOHN STUART MILL (1846)

3.1. CHARACTERIZATION OF A MINIMUM

As discussed in Chapter 1, optimization problems involve minimizing an objective function, subject to a set of constraints imposed on the variables. In essentially all problems of concern, the constraints will be expressible in terms of relationships involving continuous functions of the variables; other constraint types will be discussed in Chapter 7.

The general problem class to be considered is known as *nonlinearly constrained optimization*, and may be expressed in mathematical terms as:

$$
\begin{array}{lll}
\text{NCP} & \underset{x \in \Re^n}{\text{minimize}} & F(x) \\
& \text{subject to} & c_i(x) = 0, \quad i = 1, 2, \ldots, m'; \\
& & c_i(x) \geq 0, \quad i = m' + 1, \ldots, m.
\end{array}
$$

Any point $\hat{x}$ that satisfies all the constraints of NCP is said to be *feasible*. The set of all feasible points is termed the *feasible region*. For example, in a two-dimensional problem with the single constraint $x_1 + x_2 = 0$, the feasible region would include all points on the dashed line in Figure 3a; if the constraint were instead $x_1^2 + x_2^2 \leq 1$, the feasible region would include the interior and boundary of the unit circle, as shown in Figure 3a. (This figure also shows a convention that will be observed throughout this book: hatching on one side of an inequality constraint indicates the *infeasible* side.) A problem for which there are no feasible points is termed an *infeasible problem*.

In order to select an efficient method for solving a particular problem of the form NCP, it is necessary to analyze and classify the problem in various ways, which will be discussed in detail in Chapters 4, 5 and 6. Before considering methods for solving problems, however, we must be able to define a "solution" of NCP. Firstly, we note that only feasible points may be optimal. Secondly, optimality of a point x^* is defined by its *relationship with neighbouring points* — in contrast, say, to seeking a point $\hat{x}$ where $F(\hat{x}) = 0$. Formally, the set of relevant points is defined as follows. Let x^* denote a feasible point for problem NCP, and define $N(x^*, \delta)$ as the set of feasible points contained in a δ-neighbourhood of x^*.

Definition A. *The point x^* is a strong local minimum of NCP if there exists $\delta > 0$ such that*

A1. *$F(x)$ is defined on $N(x^*, \delta)$; and*

A2. *$F(x^*) < F(y)$ for all $y \in N(x^*, \delta)$, $y \neq x^*$.*

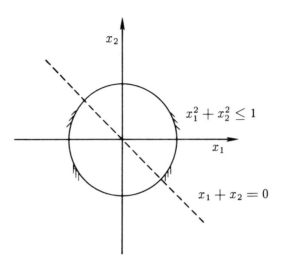

Figure 3a. The constraints $x_1 + x_2 = 0$ and $x_1^2 + x_2^2 \leq 1$.

In the special circumstance when x^* is the *only* feasible point in $N(x^*, \delta)$, x^* is also considered to be a strong local minimum.

Definition B.   *The point x^* is a weak local minimum of NCP if there exists $\delta > 0$ such that*

B1. *$F(x)$ is defined on $N(x^*, \delta)$;*

B2. *$F(x^*) \leq F(y)$ for all $y \in N(x^*, \delta)$; and*

B3. *x^* is not a strong local minimum.*

These definitions imply that x^* is *not* a local minimum if every neighbourhood of x^* contains at least one feasible point with a strictly lower function value.

In some applications, it is important to find the feasible point at which $F(x)$ assumes its least value. Such a point is termed the *global minimum*. It is usually not possible to find a global minimum except in special cases; however, this is rarely an impediment to the satisfactory solution of practical problems. Figure 3b displays some of the types of minima discussed above.

For the most general case of NCP, there may exist none of these types of minima. In particular, $F(x)$ may be unbounded below in the feasible region — e.g., the unconstrained function $F(x) = x_1 + x_2^3$. Even if the function is bounded below, its least value may occur at a limit as $\|x\|$ approaches infinity — e.g., $f(x) = e^{-x}$.

The remainder of this chapter will be concerned with optimality conditions. Our interest in optimality conditions arises in two ways: firstly, we wish to be able to verify whether or not a given point is optimal; secondly, the properties of an optimal point may suggest algorithms for finding a solution.

Unfortunately, definitions A and B of strong and weak local minima are not satisfactory for these purposes. In order to use them to verify optimality, it would be necessary to evaluate F at the (generally infinite) set of feasible points in a δ-neighbourhood of any proposed solution. Even if this process were carried out on a digital computer, where F would be evaluated only at

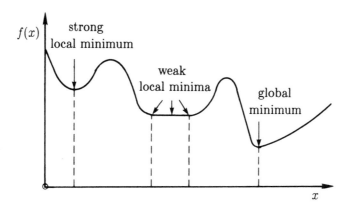

Figure 3b. Examples of minima in the univariate case.

the finite number of representable feasible points in the neighbourhood, an unacceptably large amount of computation would generally be required.

Fortunately, if $F(x)$ and $\{c_i(x)\}$ have some smoothness properties, it is possible to state other, more practical, conditions that characterize a minimum. Unless otherwise stated, henceforth it will be assumed that the objective function and all constraint functions are twice-continuously differentiable. In the remainder of this chapter, we present optimality conditions for successively more complicated problem categories. Two themes appear in all the derivations: analysis of the behaviour of the objective function at a possibly optimal point (often using a Taylor-series expansion) and, in the constrained case, a characterization of neighbouring feasible points.

3.2. UNCONSTRAINED OPTIMIZATION

3.2.1. The Univariate Case

The simplest problem that we shall consider is the unconstrained minimization of the univariate function f:

$$\underset{x \in \Re^1}{\text{minimize}} \ f(x).$$

If f is everywhere twice-continuously differentiable, and a local minimum of f exists at a finite point x^*, the following two conditions must hold at x^*.

Necessary conditions for an unconstrained univariate minimum.

A1. $f'(x^*) = 0$; and

A2. $f''(x^*) \geq 0$.

Condition A1 is proved by contradiction; we shall show that if $f'(x^*)$ is non-zero, every neighbourhood of x^* contains points with a strictly lower function value than $f(x^*)$. Since there are no constraints, all points are feasible, and hence we need to be concerned only with the value of f at neighbouring points. Because f is smooth, it can be expanded in its Taylor series about

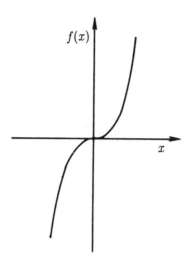

Figure 3c. The function $f(x) = x^3$.

x^*, which gives:

$$f(x^* + \epsilon) = f(x^*) + \epsilon f'(x^*) + \frac{1}{2}\epsilon^2 f''(x^* + \theta\epsilon), \qquad (3.1)$$

for some θ ($0 \leq \theta \leq 1$). Suppose that x^* is a local minimum, but that $f'(x^*)$ is strictly negative; then there must exist $\bar{\epsilon}$ ($\bar{\epsilon} > 0$) such that $\epsilon f'(x^*) + \frac{1}{2}\epsilon^2 f''(x^* + \theta\epsilon) < 0$ for all $0 < \epsilon \leq \bar{\epsilon}$. The relationship (3.1) then implies that $f(x^* + \epsilon) < f(x^*)$ for all such ϵ, and hence every neighbourhood of x^* contains points with a strictly lower function value. This contradicts the assumed optimality of x^*. Similarly, it can be shown that x^* is non-optimal if $f'(x^*)$ is positive. Therefore, $f'(x^*)$ must be zero in order for x^* to be a minimum.

Any point $\hat{x}$ such that $f'(\hat{x})$ is zero is termed a *stationary point* of f. We have just shown that any local minimum of a smooth function must be a stationary point; however, the first derivative must also be zero at a local *maximum* of f. In addition, the first derivative can vanish at a point that is neither a local minimum nor a local maximum; such a point is said to be a *point of inflection*. For example, if $f(x) = x^3$, the origin is not a local maximum or minimum, yet $f'(0) = 0$, as seen in Figure 3c.

The requirement that $f'(x^*)$ must vanish is termed a *first-order* condition for optimality, since it involves the first derivative of f. The necessary condition A2, which involves the second derivative of f, is termed a *second-order* condition, and is also proved by contradiction. Assume that x^* is a local minimum; from condition A1, $f'(x^*)$ vanishes, and so (3.1) becomes

$$f(x^* + \epsilon) = f(x) + \frac{1}{2}\epsilon^2 f''(x^* + \theta\epsilon), \qquad (3.2)$$

for some θ, $0 \leq \theta \leq 1$. If $f''(x^*)$ is strictly negative, by continuity, f'' will remain negative within some neighbourhood of x^*. If $|\epsilon|$ is small enough so that $x^* + \epsilon$ is in that neighbourhood, (3.2) shows that $f(x^* + \epsilon) < f(x)$. Consequently, x^* cannot be a local minimum if $f''(x^*) < 0$.

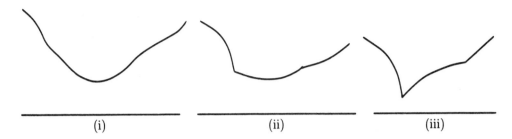

Figure 3d. Three types of minimum in the univariate case.

Necessary conditions are often invoked to show that a given point is not optimal; however, it is also helpful to determine *sufficient* conditions whose satisfaction guarantees that a certain point *is* optimal. The following two conditions are sufficient to guarantee that the point x^* is a strong local minimum.

Sufficient conditions for an unconstrained univariate minimum.

B1. $f'(x^*) = 0$; *and*

B2. $f''(x^*) > 0$.

Condition B1 has already been shown to be necessary for optimality. If B2 holds, then by continuity, $f''(x^*+\epsilon)$ will be strictly positive for all sufficiently small $|\epsilon|$. Hence, by choosing $|\epsilon|$ in (3.2) to be small enough, $f(x^*+\epsilon) > f(x^*)$. This implies that $f(x^*)$ is strictly less than the value of f at any other point in some neighbourhood, and thus x^* must be a strong local minimum.

If $f(x)$ or $f'(x)$ is discontinuous, few conditions of computational value can be added to the basic definition of a strong local minimum. If x^* is not at a point of discontinuity, and f is twice-continuously differentiable in the neighbourhood of x^*, the conditions just presented apply at x^*. If $f(x)$ is continuous, but x^* is at a point of discontinuity in $f'(x)$, sufficient conditions for x^* to be a strong local minimum are that $f'_+(x^*) > 0$ and $f'_-(x^*) < 0$.

Figure 3d illustrates three possible situations in the univariate case: (i) f is everywhere twice-continuously differentiable; (ii) f' is discontinuous, but x^* is not at a point of discontinuity; and (iii) x^* is at a point of discontinuity of f'.

3.2.2. The Multivariate Case

In this section, we consider the unconstrained minimization problem in n dimensions:

UCP	$\displaystyle \operatorname*{minimize}_{x \in \Re^n} F(x).$

As in the univariate case, a set of necessary conditions for optimality will be derived first. The necessary conditions for x^* to be a local minimum of UCP are as follows.

Necessary conditions for a minimum of UCP.

C1. $\|g(x^*)\| = 0$, *i.e.* x^* *is a stationary point; and*

C2. $G(x^*)$ *is positive semi-definite.*

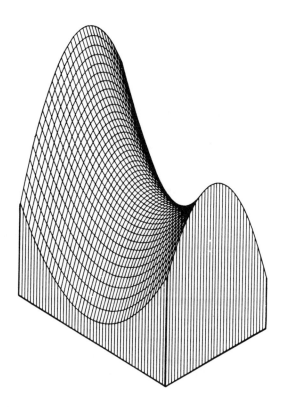

Figure 3e. A saddle point in two dimensions.

As in the univariate case, the optimality conditions can be derived from the Taylor-series expansion of F about x^*:

$$F(x^* + \epsilon p) = F(x^*) + \epsilon p^T g(x^*) + \frac{1}{2}\epsilon^2 p^T G(x^* + \epsilon\theta p)p, \qquad (3.3)$$

where θ satisfies $0 \leq \theta \leq 1$, ϵ is a scalar, and p is an n-vector. We can assume without loss of generality that the scalar ϵ in (3.3) is positive.

The necessity of C1 is proved by contradiction. Assume that x^* is a local minimum, but not a stationary point. If $g(x^*)$ is non-zero, then there must exist a vector p for which

$$p^T g(x^*) < 0; \qquad (3.4)$$

for example, p can be taken as $-g(x^*)$. Any vector p that satisfies (3.4) is termed a *descent direction* at x^*. Given any descent direction p, there exists a positive scalar $\bar{\epsilon}$ such that for all positive ϵ satisfying $\epsilon \leq \bar{\epsilon}$, it holds that $\epsilon p^T g(x^*) + \frac{1}{2}\epsilon^2 p^T G(x^* + \epsilon\theta p)p < 0$. From (3.3), it follows that $F(x^* + \epsilon p) < F(x^*)$ for all such ϵ. Hence, unless $g(x^*)$ is zero, every neighbourhood of x^* contains points with a strictly lower function value than $F(x^*)$; this proves that every local minimum must be a stationary point.

As in the one-dimensional case, the gradient vector can vanish at a point that is not a local minimum. If the gradient is zero at a point $\hat{x}$ that is neither a minimum nor maximum, $\hat{x}$ is known as a *saddle point*. Figure 3e depicts a saddle point in two dimensions.

The necessity of the second-order condition C2 is also proved by contradiction. Note that from (3.3) and C1,

$$F(x^* + \epsilon p) = F(x^*) + \frac{1}{2}\epsilon^2 p^T G(x^* + \epsilon \theta p)p. \tag{3.5}$$

If $G(x^*)$ is indefinite, by continuity G will be indefinite for *all* points in some neighbourhood of x^*, and we can choose $|\epsilon|$ in (3.5) to be small enough so that $x^* + \epsilon p$ is inside that neighbourhood. By definition of an indefinite matrix, p can then be chosen so that $p^T G(x^* + \epsilon \theta p)p < 0$. Therefore, from (3.5), every neighbourhood of x^* contains points where the value of F is strictly lower than at x^*, which contradicts the optimality of x^*.

The following conditions are *sufficient* for x^* to be a strong local minimum of UCP.

Sufficient conditions for a minimum of UCP.

D1. $\|g(x^*)\| = 0$; *and*

D2. $G(x^*)$ *is positive definite.*

To verify that these conditions are sufficient, observe that condition D1 has already been shown to be necessary, and the expansion of F about x^* is therefore given by (3.5). If $G(x^*)$ is positive definite, by continuity G is positive definite for all points in some neighbourhood of x^*. If $|\epsilon|$ is small enough, then $x^* + \epsilon p$ will be inside that neighbourhood. Hence, for all such ϵ and *every* direction p, it holds that $p^T G(x^* + \epsilon \theta p)p > 0$. From (3.5), this implies that $F(x^*)$ is strictly less than the value of F for all points in some neighbourhood of x^*, and thus x^* is a strong local minimum.

3.2.3. Properties of Quadratic Functions

The Taylor-series expansion (3.3) of a smooth function F indicates that F can be closely approximated by a quadratic function in a sufficiently small neighbourhood of a given point. Many algorithms are based on the properties of quadratic functions, and hence it is useful to study them in some detail.

Consider the quadratic function $\Phi(x)$ given by

$$\Phi(x) = c^T x + \frac{1}{2}x^T G x, \tag{3.6}$$

for some constant vector c and constant symmetric matrix G (the Hessian matrix of Φ). The definition of Φ implies the following relationship between $\Phi(\hat{x})$ and $\Phi(\hat{x} + \alpha p)$ for any vectors $\hat{x}$ and p, and any scalar α:

$$\Phi(\hat{x} + \alpha p) = \Phi(\hat{x}) + \alpha p^T (G\hat{x} + c) + \frac{1}{2}\alpha^2 p^T G p. \tag{3.7}$$

The function Φ has a stationary point only if there exists a point x^* where the gradient vector vanishes, i.e. it must hold that $\nabla \Phi(x^*) = Gx^* + c = 0$. Consequently, a stationary point x^* must satisfy the following system of linear equations:

$$Gx^* = -c. \tag{3.8}$$

If the system (3.8) is incompatible — i.e., the vector c cannot be expressed as a linear combination of the columns of G — then Φ has no stationary point, and is consequently unbounded above and below. If (3.8) is a compatible system, there is at least one stationary point, which is unique if G is non-singular.

If x^* is a stationary point, it follows from (3.7) and (3.8) that

$$\Phi(x^* + \alpha p) = \Phi(x^*) + \frac{1}{2}\alpha^2 p^T G p. \tag{3.9}$$

Hence, the behaviour of Φ in a neighbourhood of x^* is determined by the matrix G. Let λ_j and u_j denote the j-th eigenvalue and eigenvector of G, respectively. By definition,

$$G u_j = \lambda_j u_j.$$

The symmetry of G implies that the set of vectors $\{u_j\}$, $j = 1, \ldots, n$, are orthonormal (see Section 2.2.3.4). When p is equal to u_j, (3.9) becomes

$$\Phi(x^* + \alpha u_j) = \Phi(x^*) + \frac{1}{2}\alpha^2 \lambda_j.$$

Thus, the change in Φ when moving away from x^* along the direction u_j depends on the sign of λ_j. If λ_j is positive, Φ will strictly increase as $|\alpha|$ increases. If λ_j is negative, Φ is monotonically decreasing as $|\alpha|$ increases. If λ_j is zero, the value of Φ remains constant when moving along any direction parallel to u_j, since $G u_j = 0$; furthermore, Φ reduces to a *linear* function along any such direction, since the quadratic term in (3.7) vanishes.

When all eigenvalues of G are positive, x^* is the unique global minimum of Φ. In this case, the contours of Φ are ellipsoids whose principal axes are in the directions of the eigenvectors of G, with lengths proportional to the reciprocals of the square roots of the corresponding eigenvalues. If G is positive semi-definite, a stationary point (if it exists) is a weak local minimum. If G is indefinite and non-singular, x^* is a saddle point, and Φ is unbounded above and below.

In Figure 3f the contours are depicted of the following three quadratic functions, to illustrate differing combinations of sign in the eigenvalues of G:

(i) two positive eigenvalues

$$G = \begin{pmatrix} 5 & 3 \\ 3 & 2 \end{pmatrix}, \qquad c = \begin{pmatrix} -5.5 \\ -3.5 \end{pmatrix};$$

(ii) one positive eigenvalue, one zero eigenvalue

$$G = \begin{pmatrix} 4 & 2 \\ 2 & 1 \end{pmatrix}, \qquad c = \begin{pmatrix} -4 \\ -2 \end{pmatrix};$$

(iii) one positive eigenvalue, one negative eigenvalue

$$G = \begin{pmatrix} 3 & -1 \\ -1 & -8 \end{pmatrix}, \qquad c = \begin{pmatrix} -0.5 \\ 8.5 \end{pmatrix}.$$

The foregoing analysis can be applied to predict the local behaviour of a general nonlinear function $F(x)$ in a small neighbourhood of a stationary point, based on the eigenvalues of the Hessian matrix $G(x^*)$. For example, if any eigenvalue of $G(x^*)$ is close to zero, F can be expected to change very little when moving away from x^* along the corresponding eigenvector.

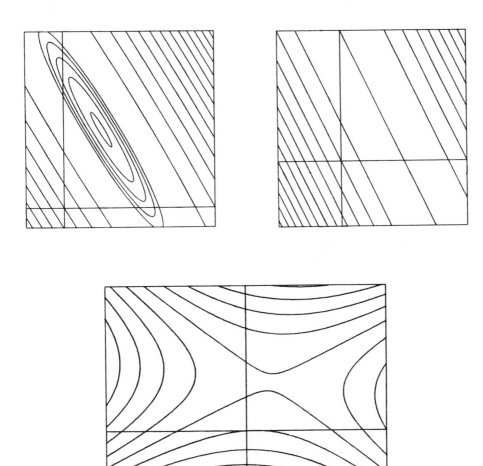

Figure 3f. Contours of: (i) a positive-definite quadratic function; (ii) a positive semi-definite quadratic function; and (iii) an indefinite quadratic function.

3.3. LINEARLY CONSTRAINED OPTIMIZATION

In most practical problems, not all possible values of the variables are acceptable, and it is often necessary or desirable to impose constraints. A frequent form of constraint involves specifying that a certain *linear* function of the variables must be exactly zero, non-negative, or non-positive. The general form of a linear function is $\ell(x) = a^T x - \beta$, for some row vector a^T and scalar β. By linearity, the column vector a is the (constant) gradient of $\ell(x)$. The types of linear constraints to be considered are:

(i) $a^T x - \beta = 0$ (equality constraint);

(ii) $a^T x - \beta \geq 0$ (inequality constraint).

Clearly, constraints of the form $a^T x - \beta \leq 0$ can equivalently be stated as $-a^T x + \beta \geq 0$. By convention, the linear constraints (i) and (ii) will be written in the forms $a^T x = \beta$ and $a^T x \geq \beta$.

A particularly simple form of linear constraint occurs when $\ell(x)$ involves only one variable. In this case, if the relevant variable is x_i, the possible constraint forms are:

(iii) $x_i = \beta$ (x_i is fixed at β);

(iv) $x_i \geq \beta$ (β is a lower bound for x_i);

(v) $x_i \leq \beta$ (β is an upper bound for x_i).

The constraint forms (iv) and (v) are termed *simple bounds* on the variable x_i.

3.3.1. Linear Equality Constraints

In this section, we consider optimality conditions for a problem that contains only linear equality constraints, i.e.

$$
\begin{array}{ll}
\text{LEP} & \underset{x \in \Re^n}{\text{minimize}} \quad F(x) \\[2ex]
& \text{subject to} \quad \hat{A}x = \hat{b}.
\end{array}
$$

The i-th row of the $m \times n$ matrix $\hat{A}$ will be denoted by $\hat{a}_i^T$, and contains the coefficients of the i-th linear constraint:

$$\hat{a}_i^T x = \hat{a}_{i1}x_1 + \cdots + \hat{a}_{in}x_n = \hat{b}_i.$$

From the discussion in Section 3.1, the feasible point x^* is a local minimum of LEP only if $F(x^*) \leq F(x)$ for all feasible x in some neighbourhood of x^*. In order to derive optimality conditions for LEP, we first consider means of characterizing the set of feasible points in a neighbourhood of a feasible point.

There is no feasible point if the constraints are inconsistent, and thus we assume that $\hat{b}$ lies in the range of $\hat{A}$. If t rows of $\hat{A}$ are linearly independent, the constraints remove t degrees of freedom from the choice of x^*. In two dimensions, for example, with the constraint $x_1 + x_2 = 0$, any solution must lie on the dashed line in Figure 3g.

It should be emphasized that the *rank* of the set of constraints, rather than the number of constraints, is the significant value. If the constraint $2x_1 + 2x_2 = 0$ (which is linearly dependent on the first) were added to the example of Figure 3g, no further degrees of freedom would be removed. On the other hand, if the additional constraint were $x_1 - x_2 = 1$ (the dotted line in Figure 3g), the constraints remove two degrees of freedom, and in this case completely determine the solution.

Because the constraints are a linear system, the properties of linear subspaces make it possible to state a simple characterization of *all* feasible moves from a feasible point. Consider the step between two feasible points $\bar{x}$ and $\hat{x}$; by linearity $\hat{A}(\bar{x} - \hat{x}) = 0$, since $\hat{A}\bar{x} = \hat{b}$ and $\hat{A}\hat{x} = \hat{b}$. Similar reasoning shows that the step p from any feasible point to any other feasible point must be orthogonal to the rows of $\hat{A}$, i.e. must satisfy

$$\hat{A}p = 0. \tag{3.10}$$

Any vector p for which (3.10) holds is termed a *feasible direction* with respect to the equality constraints of LEP. Any step from a feasible point along such a direction does not violate the constraints, since $\hat{A}(\hat{x} + \alpha p) = \hat{A}\hat{x} = \hat{b}$. The relationship (3.10) completely characterizes feasible perturbations, since even an infinitesimal step along a vector p for which $\hat{A}p \neq 0$ causes the

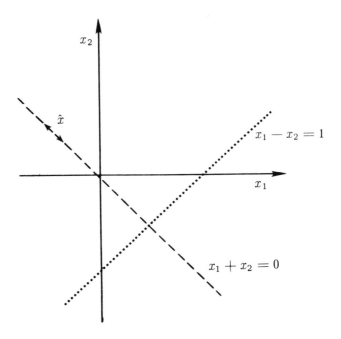

Figure 3g. The constraints $x_1 + x_2 = 0$ and $x_1 - x_2 = 1$.

perturbed point to become infeasible. In Figure 3g, the arrows represent feasible directions for the equality constraint $x_1 + x_2 = 0$ at the feasible point $\hat{x}$.

From Section 2.2.2.4, there must exist a basis for the subspace of vectors that satisfy (3.10). Let the columns of the matrix Z form such a basis; then $\hat{A}Z = 0$, and *every* feasible direction can be written as a linear combination of the columns of Z. Therefore, if p satisfies (3.10), p can be written as Zp_z for some vector p_z.

In order to determine the optimality of a given feasible point x^*, we examine the Taylor-series expansion of F about x^* along a feasible direction p ($p = Zp_z$):

$$F(x^* + \epsilon Zp_z) = F(x^*) + \epsilon p_z^T Z^T g(x^*) + \frac{1}{2}\epsilon^2 p_z^T Z^T G(x^* + \epsilon\theta p)Zp_z, \tag{3.11}$$

where θ satisfies $0 \le \theta \le 1$, and ϵ is taken without loss of generality as a positive scalar. Using an argument similar to that in the unconstrained case, (3.11) shows that if $p_z^T Z^T g(x^*)$ is negative, then every neighbourhood of x^* will contain feasible points with a strictly lower function value. Thus, a *necessary* condition for x^* to be a local minimum of LEP is that $p_z^T Z^T g(x^*)$ must vanish for *every* p_z, which implies that

$$Z^T g(x^*) = 0. \tag{3.12}$$

The vector $Z^T g(x^*)$ is termed the *projected gradient* of F at x^*. Any point at which the projected gradient vanishes is termed a *constrained stationary point*.

The result (3.12) implies that $g(x^*)$ must be a *linear combination of the rows of* $\hat{A}$, i.e.

$$g(x^*) = \sum_{i=1}^{m} \hat{a}_i \lambda_i^* = \hat{A}^T \lambda^*, \tag{3.13}$$

for some vector λ^*, which is termed the vector of *Lagrange multipliers*. *The Lagrange multipliers are unique only if the rows of $\hat{A}$ are linearly independent.*

Condition (3.13) is equivalent to (3.12) because every n-vector can be expressed as a linear combination of the rows of $\hat{A}$ and the columns of Z (see Section 2.2.2.4), and hence $g(x^*)$ can be written as $g(x^*) = \hat{A}^T\lambda^* + Zg_z$ for some vectors λ^* and g_z. Pre-multiplying $g(x^*)$ by Z^T and using (3.12), it follows that $Z^TZg_z = 0$; since Z^TZ is non-singular by definition of a basis, this will be true only if $g_z = 0$.

As in the unconstrained case, it is possible to derive higher-order necessary conditions for a local minimum. Since $Z^Tg(x^*) = 0$, the Taylor-series expansion (3.11) becomes

$$F(x^* + \epsilon Zp_z) = F(x^*) + \frac{1}{2}\epsilon^2 p_z^T Z^T G(x^* + \epsilon\theta p)Zp_z. \tag{3.14}$$

By an argument similar to that in the unconstrained case, (3.14) indicates that if the matrix $Z^TG(x^*)Z$ is indefinite, every neighbourhood of x^* contains feasible points with a strictly lower value of F. Therefore, a second-order necessary condition for x^* to be optimal for LEP is that the matrix $Z^TG(x^*)Z$, which is termed the *projected Hessian matrix*, must be positive semi-definite. It is important to note that $G(x^*)$, the Hessian matrix itself, is not required to be positive semi-definite.

As an example of the need for a second-order condition, consider the two-variable problem of minimizing the function $-x_1^2 + x_2^2$, subject to $x_2 = 1$. In this case, $g(x) = (-2x_1, 2x_2)^T$, $\hat{A} = (0, 1)$, and

$$G(x) = \begin{pmatrix} -2 & 0 \\ 0 & 2 \end{pmatrix}.$$

At $\hat{x} = (0, 1)^T$, condition (3.13) is satisfied, with $\lambda = \frac{1}{2}$. However, any vector p of the form $(\delta, 0)^T$, for any δ, satisfies $\hat{A}p = 0$, and thus $p^TG(\hat{x})p = -2\delta^2 < 0$ if δ is non-zero. Hence, $\hat{x}$ is not a local optimum.

In summary, the necessary conditions for x^* to be a local minimum of LEP are the following.

Necessary conditions for a minimum of LEP.

E1. $\hat{A}x^* = \hat{b}$;

E2. $Z^Tg(x^*) = 0$; *or, equivalently,* $g(x^*) = \hat{A}^T\lambda^*$; *and*

E3. $Z^TG(x^*)Z$ *is positive semi-definite.*

Sufficient conditions for optimality can be derived that are directly analogous to the sufficient conditions in the unconstrained case, except that they involve the projected gradient and the projected Hessian matrix.

Sufficient conditions for a minimum of LEP.

F1. $\hat{A}x^* = \hat{b}$;

F2. $Z^Tg(x^*) = 0$; *or, equivalently,* $g(x^*) = \hat{A}^T\lambda^*$; *and*

F3. $Z^TG(x^*)Z$ *is positive definite.*

3.3.2. Linear Inequality Constraints

3.3.2.1. General optimality conditions.
Consider the problem in which the constraints are a set of linear inequalities:

> **LIP**
>
> $$\underset{x \in \Re^n}{\text{minimize}} \quad F(x)$$
> $$\text{subject to} \quad Ax \geq b.$$

As in the case of linear equality constraints, we shall derive a characterization of the feasible points in the neighbourhood of a possible solution. In so doing, it will be important to distinguish between the constraints that hold exactly and those that do not. At the feasible point $\hat{x}$, the constraint $a_i^T x \geq b_i$ is said to be *active* (or *binding*) if $a_i^T \hat{x} = b_i$, and *inactive* if $a_i^T \hat{x} > b_i$. The constraint is said to be *satisfied* if it is active or inactive. If $a_i^T \bar{x} < b_i$, the constraint is said to be *violated* at $\bar{x}$.

The active constraints have a special significance because they restrict feasible perturbations about a feasible point. If the j-th constraint is inactive at the feasible point $\hat{x}$, it is possible to move a non-zero distance from $\hat{x}$ in *any* direction without violating that constraint; i.e., for any vector p, $\hat{x} + \epsilon p$ will be feasible with respect to an inactive constraint if $|\epsilon|$ is small enough.

On the other hand, an active constraint restricts feasible perturbations in *every* neighbourhood of a feasible point. Suppose that the i-th constraint is active at $\hat{x}$, so that $a_i^T \hat{x} = b_i$. There are two categories of feasible directions with respect to an active inequality constraint. Firstly, if p satisfies

$$a_i^T p = 0,$$

the direction p is termed a *binding perturbation* with respect to the i-th constraint, since the i-th constraint remains active at all points $\hat{x} + \alpha p$ for any α. A move along a binding perturbation is said to remain "on" the constraint.

Secondly, if p satisfies

$$a_i^T p > 0,$$

p is termed a *non-binding perturbation* with respect to the i-th constraint. Since it holds that $a_i^T(\hat{x} + \alpha p) = b_i + \alpha a_i^T p > b_i$ if $\alpha > 0$, the i-th constraint becomes *inactive* at the perturbed point $\hat{x} + \alpha p$. A positive step along a non-binding perturbation is said to move "off" the constraint.

Figure 3h displays some feasible directions for the constraint $x_1 + x_2 \geq 1$, from the feasible point $\hat{x} = (\frac{1}{2}, \frac{1}{2})^T$.

In order to determine whether the feasible point x^* is optimal for LIP, it is necessary to identify the active constraints. Let the t rows of the matrix $\hat{A}$ contain the coefficients of the constraints active at x^*, with a similar convention for the vector $\hat{b}$, so that $\hat{A}x^* = \hat{b}$. The numbering of the rows of $\hat{A}$ corresponds to the order of the active constraints, so that $\hat{a}_1^T$ contains the coefficients of the "first" active constraint. For simplicity in the proof, we assume that the rows of $\hat{A}$ are linearly independent; however, the derived conditions hold even when $\hat{A}$ does not have full row rank. Let Z be a matrix whose columns form a basis for the set of vectors orthogonal to the rows of $\hat{A}$. Every vector p satisfying $\hat{A}p = 0$ can therefore be written as a linear combination of the columns of Z.

Consider the Taylor-series expansion of F about x^* along a binding perturbation p $(p = Zp_z)$:

$$F(x^* + \epsilon Zp_z) = F(x^*) + \epsilon p_z^T Z^T g(x^*) + \frac{1}{2}\epsilon^2 p_z^T Z^T G(x^* + \epsilon\theta p)Zp_z, \tag{3.15}$$

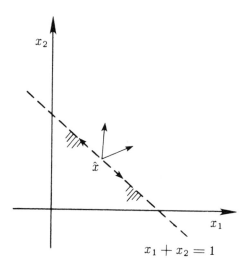

Figure 3h. Feasible directions with respect to the linear constraint $x_1 + x_2 \geq 1$.

where θ satisfies $0 \leq \theta \leq 1$, and ϵ is taken without loss of generality to be positive. As in the equality-constraint case, (3.15) reveals that if $p_z^T Z^T g(x^*)$ is non-zero for *any* p_z, then x^* cannot be a local minimum. Thus, a necessary condition for optimality of x^* is that $Z^T g(x^*) = 0$, or, equivalently, that

$$g(x^*) = \hat{A}^T \lambda^*. \tag{3.16}$$

The convention is sometimes adopted of associating a zero Lagrange multiplier with each inactive constraint. However, we shall use the convention that Lagrange multipliers correspond only to the active constraints.

The condition (3.16) ensures that F is stationary along all binding perturbations from x^*. However, since non-binding perturbations are also feasible directions with respect to the active inequality constraints, the point x^* will not be optimal if there exists any non-binding perturbation p that is a descent direction for F. If such a direction were to exist, a sufficiently small positive step along it would remain feasible and produce a strict decrease in F. To avoid this possibility, we seek a condition to ensure that for all p satisfying $\hat{A}p \geq 0$, it holds that $g(x^*)^T p \geq 0$. Since we know already from (3.16) that $g(x^*)$ is a linear combination of the rows of $\hat{A}$, the desired condition is that

$$g(x^*)^T p = \lambda_1^* \hat{a}_1^T p + \cdots + \lambda_t^* \hat{a}_t^T p \geq 0, \tag{3.17}$$

where $\hat{a}_i^T p \geq 0$, $i = 1, \ldots, t$.

The condition (3.17) will hold only if $\lambda_i^* \geq 0$, $i = 1, \ldots, t$, i.e. x^* will not be optimal if there are any negative Lagrange multipliers. To see why, assume that x^* is a local minimum (so that (3.16) must hold), but that $\lambda_j^* < 0$ for some j. Because the rows of $\hat{A}$ are linearly independent, corresponding to such a value of j there must exist a non-binding perturbation p such that

$$\hat{a}_j^T p = 1; \quad \hat{a}_i^T p = 0, \quad i \neq j. \tag{3.18}$$

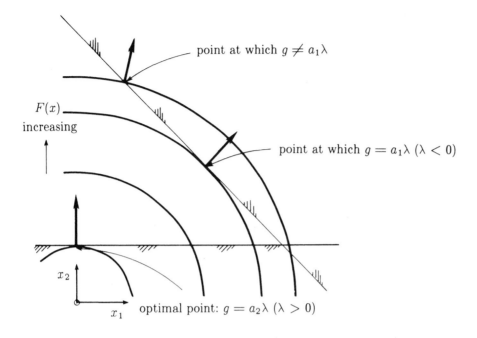

Figure 3i. The effect of first-order optimality conditions.

For such a p,

$$g(x^*)^T p = \lambda_j^* \hat{a}_j^T p = \lambda_j^* < 0;$$

hence, p is a feasible descent direction, which contradicts the optimality of x^*. Thus, a necessary condition for a solution of LIP is that *all the Lagrange multipliers must be non-negative*. Figure 3i displays the effect of the sign of the Lagrange multipliers.

By considering the Taylor-series expansion of F about x^* along binding perturbations, we can derive the second-order necessary condition that the projected Hessian matrix $Z^T G(x^*) Z$ must be positive semi-definite. This condition is precisely analogous to the second-order necessary condition for the equality-constrained problem.

In summary, the necessary conditions for x^* to be a local minimum of LIP are the following.

Necessary conditions for a minimum of LIP.

G1. $Ax^* \geq b$, with $\hat{A}x^* = \hat{b}$;

G2. $Z^T g(x^*) = 0$; or, equivalently, $g(x^*) = \hat{A}^T \lambda^*$;

G3. $\lambda_i^* \geq 0$, $i = 1, \ldots, t$; and

G4. $Z^T G(x^*) Z$ is positive semi-definite.

Sufficient conditions can also be derived for LIP. The only complication over the equality-constrained case arises because of the possibility of a zero Lagrange multiplier corresponding to an active constraint. The Lagrange multiplier corresponding to the j-th active constraint is a first-order indication of the change in the objective function that would result from a positive step along a perturbation such as (3.18) that is non-binding with respect to the j-th constraint and binding with respect to all other constraints. For example, a positive Lagrange multiplier shows that such a move would cause the objective function to *increase*. A zero Lagrange multiplier means that the effect on F of such a feasible perturbation cannot be deduced from first-order information. This implies that any set of sufficient conditions must include extra restrictions on non-binding perturbations with respect to constraints with zero Lagrange multipliers.

To illustrate that positive-definiteness of the projected Hessian does not suffice to ensure optimality when there are zero Lagrange multipliers, consider the two-dimensional example of minimizing $x_1^2 - x_2^2$, subject to $x_2 \geq 0$. At the point $x^* = (0,0)^T$, we have that $g(x^*) = (0,0)^T$, $\hat{A} = (0,1)$, and $Z^T G(x^*) Z = 2$. Therefore, $g(x^*) = \hat{A}^T \lambda^*$, with $\lambda^* = 0$. Although the projected Hessian matrix is positive definite, x^* is *not* a local minimum, since any positive step along the vector $(0, \delta)^T$, where δ is positive, is a feasible perturbation that strictly reduces F. The origin is non-optimal because, although F is stationary along binding perturbations, it displays negative curvature along a non-binding perturbation.

One means of avoiding such complications is to include the condition that all the Lagrange multipliers be strictly positive, which guarantees that F will sustain a strict increase for any non-binding perturbation. With this approach, the following conditions are sufficient for x^* to be a strong local minimum of LIP.

Sufficient conditions for a minimum of LIP.

H1. $Ax^* \geq b$, with $\hat{A}x^* = \hat{b}$;

H2. $Z^T g(x^*) = 0$; or, equivalently, $g(x^*) = \hat{A}^T \lambda^*$;

H3. $\lambda_i^* > 0$, $i = 1, \ldots, t$; and

H4. $Z^T G(x^*) Z$ is positive definite.

It is possible to state sufficient conditions even when zero Lagrange multipliers are present, by including extra restrictions on the Hessian matrix to ensure that F displays positive curvature along any perturbation that is binding for all constraints with positive Lagrange multipliers, but may be binding or non-binding for constraints with zero Lagrange multipliers. Let $\hat{A}_+$ contain the coefficients of the active constraints with *positive* Lagrange multipliers, and let Z_+ be a matrix whose columns span the null space of $\hat{A}_+$. In this case, sufficient conditions for x^* to be a strong local minimum of LIP are as follows.

Sufficient conditions for a minimum of LIP (with zero Lagrange multipliers).

I1. $Ax^* \geq b$, with $\hat{A}x^* = \hat{b}$;

I2. $Z^T g(x^*) = 0$; or, equivalently, $g(x^*) = \hat{A}^T \lambda^*$;

I3. $\lambda_i^* \geq 0$, $i = 1, \ldots, t$; and

I4. $Z_+^T G(x^*) Z_+$ is positive definite.

Note that these conditions do not hold in the example cited earlier. However, if the objective function in that example is changed to $x_1^2 + x_2^2$, the origin *is* a strong local minimum, and satisfies the second set of sufficient conditions.

When a linearly constrained problem includes both equalities and inequalities, the optimality conditions are a combination of those given for the separate cases. In a mixed problem, the "active" constraints include all the equalities as well as the binding inequalities, and there is no sign restriction on the Lagrange multipliers corresponding to equality constraints.

3.3.2.2. Linear programming. A *linear program* (LP) is an optimization problem in which the objective function and all the constraint functions are linear. The linear objective function will be denoted by

$$F(x) = c^T x, \tag{3.19}$$

for some constant n-vector c (the optimal x is not affected by any constant term in the objective, so such a term is usually omitted). The gradient of the function (3.19) is the constant vector c, and its Hessian matrix is identically zero.

When c is a non-zero vector, the unconstrained function (3.19) is unbounded below, since F can be made arbitrarily large and negative by moving along any non-zero $\hat{x}$ such that $c^T \hat{x} < 0$. Therefore, constraints are necessary if the non-trivial LP problem is to have a finite solution.

Since c is an arbitrary vector, the necessary condition $c = \hat{A}^T \lambda^*$ will hold only if $\hat{A}$ is non-singular, where $\hat{A}$ includes all equality constraints and the binding inequalities. In this case, x^* is the solution of a non-singular system of linear equations. When $c = \hat{A}^T \lambda^*$, but the row rank of $\hat{A}$ is less than n, the value of $c^T x$ is uniquely determined, but x^* is not. For example, if $F(x) = 3x_1 + 3x_2$, with the single constraint $x_1 + x_2 = 1$, the optimal objective value is 3, which is obtained for any vector of the form $(\frac{1}{2} - \delta, \frac{1}{2} + \delta)^T$.

When considering methods for solving linear programming problems in Chapter 5, we shall refer to the corresponding *dual* problem. Given an optimization problem P (the *primal* problem), we can sometimes define a related problem D of the same type (the *dual* problem) such that the Lagrange multipliers of P are part of the solution of D, and the Lagrange multipliers of D are contained in the solution of P.

Consider the following primal linear programming problem:

$$\begin{array}{ll} \underset{x \in \Re^n}{\text{maximize}} & c^T x \\ \text{subject to } Ax \geq b, \end{array}$$

where A is $m \times n$. The corresponding dual problem is

$$\begin{array}{ll} \underset{y \in \Re^m}{\text{maximize}} & b^T y \\ \text{subject to } A^T y = c, \\ \qquad\qquad y \geq 0. \end{array}$$

Although we have adopted the convention of associating Lagrange multipliers only with the active constraints, when discussing duality it is convenient to associate a zero Lagrange multiplier with an inactive constraint. We assume without loss of generality that the active constraints of LP comprise the first t rows of A. Let y^* denote the extended vector $(\lambda^*, 0)^T$ of Lagrange multipliers for the primal problem, where $\lambda^* > 0$.

The proof that y^* solves the dual problem involves showing that y^* is feasible for the dual problem, and satisfies the sufficient conditions for optimality.

Let $\hat{A}$ and $\bar{A}$ denote the active and inactive constraint matrices at the solution of the primal. By the optimality conditions of the primal,

$$c = \hat{A}^T \lambda^* = \left(\hat{A}^T \quad \bar{A}^T \right) \begin{pmatrix} \lambda^* \\ 0 \end{pmatrix} = A^T y^*,$$

so that y^* satisfies the equality constraints of the dual. Since $y^* \geq 0$ by the optimality conditions of the primal, y^* is feasible for the dual.

We must now show that y^* is the optimal solution of the dual. Each positive component of y^* (i.e., each element of λ^*) corresponds to an inactive non-negativity constraint of the dual; each zero component of y^* corresponds to an active non-negativity constraint of the dual. Hence, the active dual constraints at y^* are given by

$$\begin{pmatrix} \hat{A}^T & \bar{A}^T \\ 0 & I \end{pmatrix}\begin{pmatrix} \lambda^* \\ 0 \end{pmatrix} = \begin{pmatrix} c \\ 0 \end{pmatrix}.$$

Since maximizing $b^T y$ is equivalent to minimizing $-b^T y$, the Lagrange multipliers of the dual must satisfy

$$\begin{pmatrix} \hat{A} & 0 \\ \bar{A} & I \end{pmatrix}\begin{pmatrix} \hat{\sigma} \\ \bar{\sigma} \end{pmatrix} = -b = -\begin{pmatrix} \hat{b} \\ \bar{b} \end{pmatrix},$$

where $\hat{\sigma}$ are the dual multipliers corresponding to the equality constraints, and $\bar{\sigma}$ are the multipliers corresponding to the active lower bounds. The first set of equations gives $\hat{A}\hat{\sigma} = -\hat{b}$; since $\hat{A}$ is non-singular, $\hat{\sigma}$ must be equal to $-x^*$. Substituting this expression in the second set of equations, we obtain $\bar{\sigma} = \bar{A}x^* - \bar{b}$. By definition of the inactive constraints of the primal, it follows that $\bar{\sigma} > 0$. Thus, y^* is the solution of the dual linear program, and the solution of the primal can be determined from the solution of the dual.

3.3.2.3. Quadratic programming. A *quadratic program* is the special case of linearly constrained optimization that occurs when the objective function is the quadratic function

$$F(x) = c^T x + \frac{1}{2}x^T G x, \tag{3.20}$$

for some constant vector c and constant symmetric matrix G. The gradient of the function (3.20) is $Gx + c$, and the Hessian is the constant matrix G.

The concept of duality also applies to quadratic programs. Let the original (primal) quadratic program be given by

$$\begin{array}{ll} \underset{x \in \Re^n}{\text{minimize}} & c^T x + \frac{1}{2}x^T G x \\ \text{subject to} & Ax \geq b. \end{array}$$

The corresponding dual problem is then

$$\begin{array}{ll} \underset{y \in \Re^m, w \in \Re^n}{\text{maximize}} & \Psi(y, w) = b^T y - \frac{1}{2}w^T G w \\ \text{subject to} & A^T y = Gw + c, \\ & y \geq 0. \end{array}$$

Let y^* be the extended Lagrange multiplier vector for the primal QP problem, and let $w^* = x^*$. Then (y^*, w^*) is the solution of the dual. The feasibility and optimality of this vector are verified exactly as in the LP case.

In order to ensure that a solution of the primal can be recovered from a solution of the dual, it is necessary for G to satisfy two additional conditions, which are more restrictive than those necessary for x^* to be a strong local minimum of the primal problem: G must be non-singular, and $\hat{A}G^{-1}\hat{A}^T$ must be positive definite.

3.3.2.4. Optimization subject to bounds. In many problems, all the constraints of LIP are simple bounds on the variables:

$$l_i \le x_i \le u_i, \quad i = 1, 2, \ldots, n, \tag{3.21}$$

where either bound may be omitted.

The special form of bound constraints leads to considerable simplifications in the optimality conditions. Since an active constraint indicates that a variable lies on one of its bounds, the matrix $\hat{A}$ will contain only signed columns of the identity matrix; the corresponding components of x are termed the *fixed* variables. Let x_L denote the vector of variables fixed on their lower bounds, and x_U the set of variables on their upper bounds. Because of the special form of $\hat{A}$, the Lagrange multiplier for an active bound on x_i is given by $g_i(x^*)$ if $x_i^* = l_i$, and by $-g_i(x^*)$ if $x_i^* = u_i$ (since the constraint $x_i \le u_i$ can be written as $-x_i \ge -u_i$).

Furthermore, the matrix Z can be taken as a set of the columns of the identity matrix corresponding to the variables that are not on their bounds (the *free* variables, denoted by x_{FR}). Thus, the projected gradient $Z^T g(x^*)$ is simply the sub-vector of $g(x^*)$ corresponding to free variables, which we denote by g_{FR}^*; similarly, the rows and columns of the Hessian matrix $G(x^*)$ corresponding to the free variables comprise the projected Hessian, which we shall denote by G_{FR}^*.

Thus, sufficient conditions for x^* to be a strong local minimum of $F(x)$ subject to the bounds (3.21) are as follows.

Sufficient conditions for a bound-constrained minimum.

J1. $l \le x^* \le u$, with $l < x_{\mathrm{FR}}^* < u$;

J2. $g_{\mathrm{FR}}^* = 0$;

J3. $g_L^* > 0$ and $g_U^* < 0$; and

J4. G_{FR}^* is positive definite.

The simplest application of these conditions occurs when minimizing the univariate function f over the bounded interval $[l, u]$. A sufficient condition that $x^* = l$ is that $f'(l) > 0$ (note that the set of free variables is null).

3.4. NONLINEARLY CONSTRAINED OPTIMIZATION

The constraints imposed upon the variables do not always involve only linear functions. A *nonlinear constraint* involves the specification that a certain nonlinear function, say $c_i(x)$, must be exactly zero or non-negative, so that the constraint forms to be considered are:

> (i) $c_i(x) = 0$ (equality constraint);

> (ii) $c_i(x) \ge 0$ (inequality constraint).

Obviously, any constraint of the form $c_i(x) \le 0$ can be represented with the inequality reversed as $-c_i(x) \ge 0$.

Throughout this section, we shall discuss the case when *all* constraints are nonlinear, since the optimality conditions for a problem with a mixture of constraint types can be deduced from the results for the separate cases.

3.4.1. Nonlinear Equality Constraints

In this section, we consider optimality conditions for a problem that contains only nonlinear *equality* constraints, i.e.

$$
\begin{array}{ll}
\text{NEP} & \underset{x \in \Re^n}{\text{minimize}} \quad F(x) \\
& \text{subject to} \quad \hat{c}_i(x) = 0, \quad i = 1, \ldots, t.
\end{array}
$$

The gradient vector of the function $\hat{c}_i(x)$ will be denoted by the n-vector $\hat{a}_i(x)$, and its Hessian will be denoted by $\hat{G}_i(x)$. Given a set of constraint functions $\{\hat{c}_i(x)\}$, $i = 1, \ldots, t$, the $t \times n$ matrix $\hat{A}(x)$ whose i-th row is $\hat{a}_i(x)^T$ is termed the *Jacobian matrix* of the constraints.

To determine whether a feasible point x^* is optimal for NEP, it is necessary to characterize feasible perturbations in order to analyze the behaviour of F along them. With linear equality constraints, we have seen that all feasible perturbations can be defined in terms of a linear subspace (see Section 3.3.1). However, it is more complicated to characterize feasible perturbations with respect to *nonlinear* equality constraints. If $\hat{c}_i$ is a nonlinear function and $\hat{c}_i(x^*) = 0$, in general there is *no* feasible direction p such that $\hat{c}_i(x^* + \alpha p) = 0$ for all sufficiently small $|\alpha|$.

Therefore, in order to retain feasibility with respect to $\hat{c}_i$, it will be necessary to move along a *feasible arc* emanating from x^* (an arc is a directed curve in $\Re^n$ parameterized by a single variable θ). For example, when there is only a single constraint, the contour line corresponding to $\hat{c} = 0$ defines a feasible arc. Let $\alpha(\theta)$ denote the arc, where $\alpha(0) = x^*$, and let p denote the tangent to the arc at x^*.

In order to remain feasible, the function $\hat{c}_i$ must remain identically zero for all points on the arc. This implies that the rate of change, or derivative of $\hat{c}_i$ along the arc, must be zero at x^*. Applying the chain rule (see Section 2.3.2) to obtain this derivative gives

$$
\left. \frac{d}{d\theta} \hat{c}_i\big(\alpha(\theta)\big) \right|_{\theta=0} = \nabla \hat{c}_i\big(\alpha(0)\big)^T p = \hat{a}_i(x^*)^T p = 0.
$$

Hence, the tangent to a feasible arc for all constraints of NEP must satisfy

$$
\hat{A}(x^*)p = 0. \tag{3.22}
$$

In the case of linear equality constraints, the analogous relationship (3.10) *completely* characterized feasible perturbations, in that any feasible direction was required to satisfy (3.10), and all directions satisfying (3.10) were feasible. With nonlinear constraints, the first condition still applies — namely, that the tangent to any feasible arc must satisfy (3.22). This can be seen from the Taylor-series expansion of $\hat{c}_i$ about x^* along the direction p:

$$
\hat{c}_i(x^* + \epsilon p) = \hat{c}_i(x^*) + \epsilon \hat{a}_i(\hat{x})^T p + \frac{1}{2} \epsilon p^T \hat{G}_i(x^*) p + \cdots
$$

If $\hat{c}_i(x^*) = 0$ and $\hat{a}_i(x^*)^T p \neq 0$, an arbitrarily small move along the arc will cause the constraint to be violated. However, it is *not* true for nonlinear constraints that every vector satisfying (3.22) is tangent to a feasible arc.

To demonstrate this possibility, consider the two constraints $\hat{c}_1(x) = (x_1 - 1)^2 + x_2^2 - 1$ and $\hat{c}_2(x) = (x_1 + 1)^2 + x_2^2 - 1$. At the origin, any vector p of the form $(0, \delta)^T$ satisfies (3.22); however, since the origin is the only feasible point, obviously no feasible arc exists.

In order for (3.22) to be a complete characterization of the tangent p to a feasible arc, it is necessary to assume that the constraint functions satisfy certain conditions at x^*. These conditions are usually termed *constraint qualifications*. There are several forms of constraint qualification, many of which are of theoretical interest only. A practical constraint qualification for nonlinear constraints is the condition that *the constraint gradients at x^* are linearly independent*, i.e. that the matrix $\hat{A}(x^*)$ has full row rank. This ensures that every vector p satisfying (3.22) is tangent to a feasible arc. If this constraint qualification is satisfied, it is possible to state necessary conditions for optimality with respect to NEP. Therefore, it will henceforth be assumed that $\hat{A}(x^*)$ has full row rank. Note that the constraint qualification is irrelevant for linear constraints, since (3.22) completely characterizes feasible perturbations.

Firstly, we consider first-order necessary conditions. In order for x^* to be optimal, F must be stationary at x^* along any feasible arc, i.e. $\nabla F\big(\alpha(\theta)\big)\,|_{\theta=0} = 0$. Using the chain rule, the requisite condition is that

$$g(x^*)^T p = 0, \tag{3.23}$$

where p satisfies (3.22).

Let $Z(x^*)$ denote a matrix whose columns form a basis for the set of vectors orthogonal to the rows of $\hat{A}(x^*)$. In order for (3.23) to hold for every p that satisfies (3.22), it must be true that

$$Z(x^*)^T g(x^*) = 0. \tag{3.24}$$

This condition is analogous to the necessary condition (3.12) in the linearly constrained case, except that the matrix Z is no longer constant. The vector $Z(x^*)^T g(x^*)$ is termed the *projected gradient* of F at x^*.

As before, (3.24) is equivalent to the condition that $g(x^*)$ must be a linear combination of the rows of $\hat{A}(x^*)$:

$$g(x^*) = \hat{A}(x^*)^T \lambda^*, \tag{3.25}$$

for some t-vector λ^* of Lagrange multipliers.

Define the *Lagrangian function* as

$$L(x, \lambda) = F(x) - \lambda^T \hat{c}(x), \tag{3.26}$$

with x and the t-vector λ as independent variables. Condition (3.25) is a statement that x^* is a *stationary point* (with respect to x) of the Lagrangian function when $\lambda = \lambda^*$.

We shall give only a brief overview of the derivation of the second-order necessary condition, which is rather complicated to explain. As in all the previous discussion of second-order necessary conditions, it is clear that x^* will be optimal only if F has non-negative curvature at x^* along any feasible arc, i.e.

$$\left.\frac{d^2}{d\theta^2} F\big(\alpha(\theta)\big)\right|_{\theta=0} \geq 0. \tag{3.27}$$

We wish to derive an expression for $d^2 F/d\theta^2$ in terms of derivatives of the problem functions. From the chain rule, $d^2 F/d\theta^2$ involves the Hessian of F and $d^2\alpha/d\theta^2$ (the second derivative of the arc). It is possible to obtain an expression for this latter quantity as a combination of λ_i^* and $\hat{G}_i(x^*)$, $i = 1, \ldots, t$ because of two special properties of x^*: firstly, (3.25) holds; secondly, the curvature of $\hat{c}_i$ along the feasible arc must be zero at x^* in order for $\hat{c}_i$ to remain identically zero. Thus, the second-order necessary condition (3.27) can be expressed as follows: for all p satisfying

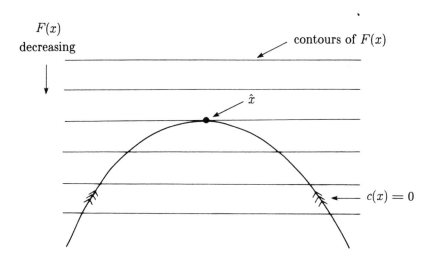

Figure 3j. The role of constraint curvature in optimality conditions.

(3.22), it must hold that

$$p^T(G(x^*) - \sum_{i=1}^{t} \lambda_i^* \hat{G}_i(x^*))p \geq 0. \tag{3.28}$$

Note that the significant matrix in the necessary conditions for a nonlinearly constrained problem is the *Hessian matrix of the Lagrangian function*, which will be denoted by

$$W(x, \lambda) \equiv G(x) - \sum_{i=1}^{t} \lambda_i \hat{G}_i(x).$$

The condition (3.28) is equivalent to the requirement that the matrix

$$Z(x^*)^T(G(x^*) - \sum_{i=1}^{t} \lambda_i^* \hat{G}_i(x^*))Z(x^*) \tag{3.29}$$

be positive semi-definite. The matrix (3.29) is termed the *projected Hessian of the Lagrangian function*.

We emphasize that the second-order optimality conditions for a nonlinearly constrained problem depend on a special combination of the Hessian of the objective function *and* the Hessian matrices of the constraints. Figure 3j illustrates the key role of constraint curvature by displaying an example with a linear objective function, which by definition has no curvature. Note that the point $\hat{x}$ satisfies (3.24), but is non-optimal because the projected Hessian of the Lagrangian function is indefinite.

In summary, if a constraint qualification holds at x^*, the following conditions are necessary for x^* to be optimal for NEP.

Necessary conditions for a minimum of NEP.

K1. $\hat{c}(x^*) = 0$;

K2. $Z(x^*)^T g(x^*) = 0$, or, equivalently, $g(x^*) = \hat{A}(x^*)^T \lambda^*$; and

K3. $Z(x^*)^T W(x^*, \lambda^*) Z(x^*)$ is positive semi-definite.

Using a similar derivation, it can be shown that the following conditions are *sufficient* for x^* to be a strong local minimum of NEP.

Sufficient conditions for a minimum of NEP.

L1. $\hat{c}(x^*) = 0$;

L2. $Z(x^*)^T g(x^*) = 0$, or, equivalently, $g(x^*) = \hat{A}(x^*)^T \lambda^*$; and

L3. $Z(x^*)^T W(x^*, \lambda^*) Z(x^*)$ is positive definite.

3.4.2. Nonlinear Inequality Constraints

Finally, we consider the problem in which all the constraints are nonlinear inequalities:

$$
\boxed{
\begin{array}{ll}
\text{NIP} & \displaystyle \operatorname*{minimize}_{x \in \Re^n} \quad F(x) \\[1ex]
& \text{subject to} \quad c_i(x) \geq 0, \quad i = 1, \ldots, m.
\end{array}
}
$$

As in the case of linear inequalities, it is necessary to identify the set of nonlinear inequality constraints that are exactly zero at an optimal point x^*. The constraint $c_i(x) \geq 0$ is said to be *active* at x^* if $c_i(x^*) = 0$, and *inactive* if $c_i(x^*) > 0$. Only the active constraints restrict feasible perturbations at x^*, since an inactive constraint will remain strictly satisfied within a sufficiently small neighbourhood of x^*.

The derivation of optimality conditions for NIP is based on combining the concepts of binding and non-binding perturbations for an active linear inequality (Section 3.3.2.1) and of a feasible arc with respect to a nonlinear constraint (Section 3.4.1) to define binding and non-binding feasible arcs with respect to nonlinear inequality constraints.

In order to derive necessary conditions for optimality with respect to NIP, we again need to assume that the constraint functions satisfy a constraint qualification at x^*. Let the vector $\hat{c}(x^*)$ denote the subset of t constraint functions that are active at x^*, and let $\hat{A}(x^*)$ be the matrix whose rows are the transposed gradient vectors of the active constraints. If $\hat{A}(x^*)$ has full row rank, the constraint qualification holds at x^*. Again we emphasize that a constraint qualification is relevant only to nonlinear constraints.

Under this assumption, a combination of the arguments from Section 3.3.2.1 and 3.4.1 shows that the following conditions are necessary for x^* to be a local minimum of NIP.

Necessary conditions for a minimum of NIP.

M1. $c(x) \geq 0$, with $\hat{c}(x^*) = 0$;

M2. $Z(x^*)^T g(x^*) = 0$, or, equivalently, $g(x^*) = \hat{A}(x^*)^T \lambda^*$;

M3. $\lambda_i^* \geq 0$, $i = 1, \ldots, t$; and

M4. $Z(x^*)^T W(x^*, \lambda^*) Z(x^*)$ is positive semi-definite.

Zero Lagrange multipliers cause complications in stating sufficient optimality conditions for nonlinear constraints, just as in the linearly constrained case. We shall therefore state two sets of sufficient conditions, the first of which avoids the difficulties by assuming that all Lagrange multipliers are positive.

Sufficient conditions for x^* to be a strong local minimum of NIP are as follows.

Sufficient conditions for a minimum of NIP.

N1. $c(x^*) \geq 0$, *with* $\hat{c}(x^*) = 0$;

N2. $Z(x^*)^T g(x^*) = 0$; *or, equivalently,* $g(x^*) = \hat{A}(x^*)^T \lambda^*$;

N3. $\lambda_i^* > 0$, $i = 1, \ldots, t$; *and*

N4. $Z(x^*)^T W(x^*, \lambda^*) Z(x^*)$ *is positive definite.*

When zero Lagrange multipliers are present, the sufficient conditions include extra restrictions on the Hessian matrix of the Lagrangian function to ensure that F displays positive curvature along any feasible arc that is binding for all constraints with positive Lagrange multipliers, but may be binding or non-binding for constraints with zero Lagrange multipliers. Let $\hat{A}_+(x^*)$ contain the coefficients of the active constraints with *positive* Lagrange multipliers, and let $Z_+(x^*)$ be a matrix whose columns span the null space of $\hat{A}_+(x^*)$. In this case, sufficient conditions for x^* to be a strong local minimum of NIP are as follows.

Sufficient conditions for a minimum of NIP (with zero Lagrange multipliers).

O1. $c(x^*) \geq 0$, *with* $\hat{c}(x^*) = 0$;

O2. $Z(x^*)^T g(x^*) = 0$; *or, equivalently,* $g(x^*) = \hat{A}(x^*)^T \lambda^*$;

O3. $\lambda_i^* \geq 0$, $i = 1, \ldots, t$; *and*

O4. $Z_+(x^*)^T W(x^*, \lambda^*) Z_+(x^*)$ *is positive definite.*

Notes and Selected Bibliography for Section 3.4

The optimality conditions for constrained optimization described in this chapter are often called the *Kuhn-Tucker conditions*; see Kuhn and Tucker (1951) and Kuhn (1976). Some good general discussions of optimality conditions are given in Fiacco and McCormick (1968), Powell (1974) and Avriel (1976).

We have been concerned in this chapter only with optimality conditions that are necessary *or* sufficient. Sets of conditions that are simultaneously necessary *and* sufficient are surprisingly complicated — even in the unconstrained case. These conditions have not been included because they can seldom be verified in practice. An interesting discussion of the unconstrained case is given by Gue and Thomas (1968).

CHAPTER FOUR

UNCONSTRAINED METHODS

Now, if a Muse cannot run when she is unfettered,
it is a sign she has but little speed.

—JOHN DRYDEN (1697)

In Chapters 4, 5 and 6, we shall describe a selection of the many methods available for various problem categories. The descriptions are intended to present an overview of certain methods, including the underlying motivation as well as selected theoretical and computational features.

The discussion has been restricted to certain methods for two reasons. Firstly, the limitations of space make it impossible to be comprehensive; a complete discussion of unconstrained optimization alone would fill an entire volume. Secondly, in Chapters 7 and 8 we shall present techniques and suggestions for formulating non-standard problems, for analyzing why a method may fail or be inefficient, and for obtaining the best performance from software. Although a few general observations of this nature are valid for all numerical methods, such suggestions are most useful when they refer to *specific algorithms*.

We shall describe primarily methods with which the authors have had extensive experience and success. Other methods will also be described that provide special insights or background. References to methods not discussed and to further details are given in the Notes at the end of each section. The authors of methods and results will be mentioned only in the Notes.

4.1. METHODS FOR UNIVARIATE FUNCTIONS

4.1.1. Finding the Zero of a Univariate Function

We have shown in Section 3.2.1 that a necessary condition for x^* to be an unconstrained minimum of a twice-continuously differentiable univariate function $f(x)$ is that $f'(x^*) = 0$. Since x^* is a zero of the derivative function, it can be seen that the problems of univariate minimization and zero-finding are very closely related (although not equivalent). Therefore, we shall first consider the problem of finding a point x^* in a bounded interval such that $f(x^*) = 0$, where f is a nonlinear function. We restrict ourselves to the case when f changes sign at x^*.

The first question that should be asked is: what is meant by "finding" a zero? We shall generally be working in finite precision on some computer, with only a finite number of representable values of x, and a computed value of f. Hence it is unrealistic to expect to find a machine-representable $\bar{x}$ such that $fl(f(\bar{x}))$ is exactly zero, where $fl(f)$ is the computed value of f. We shall be satisfied if an algorithm provides an interval $[a, b]$ such that

$$f(a)f(b) < 0 \quad \text{and} \quad |a - b| < \delta,$$

where δ is some "small" tolerance. Since a zero of f must lie in $[a, b]$, any point in the interval can be taken as an estimate of the zero.

In order to describe algorithms for zero-finding, we require two preliminary definitions. The smallest interval in which x^* is known to lie is called the *interval of uncertainty*. The zero is said to be *bracketed* in an interval if f changes sign in the interval.

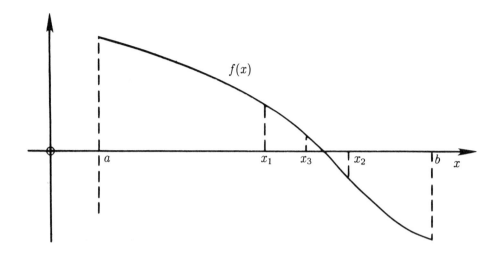

Figure 4a. The points x_1, x_2 and x_3 denote the first three members of the sequence generated by the bisection algorithm applied to $f(x)$ on the interval $[a, b]$.

4.1.1.1. The method of bisection. This method is based on *systematically reducing the interval of uncertainty by function comparison*. Suppose that an initial interval $[a, b]$ has been specified in which $f(a)f(b) < 0$. We evaluate f at the midpoint of the interval and test its sign. If the function value is zero, the algorithm terminates; otherwise, a new interval of uncertainty is produced by discarding the value of a or b, depending on whether $f(a)$ or $f(b)$ agrees in sign with f at the midpoint. This process is illustrated in Figure 4a.

Bisection is *guaranteed* to find x^* to within any specified tolerance δ *if f can be computed to sufficient accuracy so that its sign is correct*. A satisfactory interval may always be found by bisection under these conditions in about $\log_2((b-a)/\delta)$ evaluations of f. Bisection can be shown to be the "optimal" algorithm for the class of functions that change sign in $[a, b]$, in the sense that it yields the smallest interval of uncertainty for a *specified* number of function evaluations.

Some indication of the speed of a zero-finding algorithm can be obtained by finding the rate at which the interval of uncertainty converges to zero. For the bisection algorithm, the length of this interval converges to zero linearly, with asymptotic error constant $\frac{1}{2}$.

4.1.1.2. Newton's method. The defect of the bisection algorithm is that no account is taken of the relative magnitudes of the values of f at the various points. If f is known to be *well behaved*, it seems reasonable to use the values of f at the endpoints to determine the next estimate of x^*. A means of utilizing the magnitude of f during the search for a zero is to *approximate* or *model* f by a function $\hat{f}$ whose zero can be easily calculated. An iterative procedure can then be devised in which the zero of $\hat{f}$ is taken as a new estimate of the zero of f itself.

If f is differentiable, an obvious candidate for $\hat{f}$ is the *tangent line* at x_k, the current estimate of the zero. To compute the zero of the tangent line, we evaluate $f(x_k)$ and $f'(x_k)$; if $f'(x_k)$ is non-zero, the zero of the tangent line is given by

$$x_{k+1} = x_k - \frac{f(x_k)}{f'(x_k)}. \tag{4.1}$$

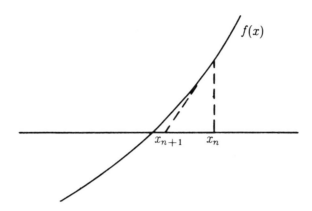

Figure 4b. The point x_{n+1} denotes the Newton approximation to the zero of $f(x)$. The new point is defined as the zero of the tangent line of $f(x)$ at x_n.

The iterative procedure defined by (4.1) is known as *Newton's method for finding a zero*, and is illustrated in Figure 4b. It has a different flavour from bisection in that there is no "interval of uncertainty" *per se*; rather, a new estimate of the zero is computed at each iteration.

Some well-known and seemingly unrelated procedures are actually equivalent to Newton's method. For example, with the "divide and average" technique taught to children for computing $\sqrt{a}$, the new estimate of the square root is given by

$$x_{k+1} = \frac{1}{2}(x_k + \frac{a}{x_k});$$

this is precisely the next iterate of Newton's method applied to the function $f(x) = x^2 - a$.

The underlying assumption of Newton's method is that, for the purposes of zero-finding, f is "like" a straight line. One therefore expects Newton's method to work well when f satisfies this assumption. In particular, if $f'(x^*)$ is non-zero, and the starting point x_0 is "close enough" to x^*, Newton's method converges *quadratically* to x^* (see Section 2.3.6).

Example 4.1. When $\sqrt{4}$ is computed by Newton's method with starting point $x_0 = 2.5$, the errors $\{x_k^2 - 4\}$, $k = 0, \ldots, 4$, are 2.25, $.2025$, 2.439×10^{-3}, 3.717×10^{-7}, and 8.0×10^{-16} (computed with approximately sixteen decimal digits of precision).

The major difficulty with Newton's method is that its spectacular convergence rate is only *local*. If x_0 is not close enough to x^*, the Newton iteration may diverge hopelessly. For example, the Newton approximation depicted in Figure 4c lies at a completely unreasonable value. In addition, since the Newton iteration is undefined when $f'(x_k)$ is zero, numerical difficulties occur when $f'(x_k)$ is "small". Newton's method must therefore be used with care, and it is not an all-purpose algorithm for zero-finding.

4.1.1.3. Secant and regula falsi methods. A further objection to Newton's method is that f' is required at every iterate. In practical problems, f' may be very expensive, troublesome, or even

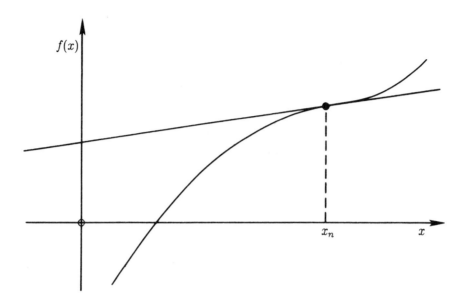

Figure 4c. Divergent case for Newton's method. The zero of the tangent line at x_n lies at a completely unreasonable value.

impossible to compute. A different method is suggested by using the same idea of approximating f by a straight line $\hat{f}$, but choosing as $\hat{f}$ the straight line that passes through the values of f at the two most recent iterates; in essence, $f'(x_k)$ is replaced in the Newton formula by the finite-difference approximation $(f_k - f_{k-1})/(x_k - x_{k-1})$, where f_k denotes $f(x_k)$. The iterates are then defined by:

$$x_{k+1} = x_k - \left(\frac{x_k - x_{k-1}}{f_k - f_{k-1}}\right)f_k,$$

and the method is called the *secant method*, or the *method of linear interpolation*. As with Newton's method, we compute a new estimate of the zero rather than an interval of uncertainty. The values of f at two points are required to initiate the secant method.

If $f'(x^*)$ is non-zero and x_0 and x_1 are sufficiently close to x^*, the secant method can be shown to have a superlinear convergence rate $r \approx 1.6180$, and thus it can be a rapidly convergent method. For example, when the secant method is applied to Example 4.1 with $x_0 = 1$, $x_1 = 2.5$, the sequence of errors $\{x_k^2 - 4\}$ for $k = 2, \ldots, 7$, is -5.51×10^{-1}, -6.53×10^{-2}, 2.44×10^{-3}, -1.00×10^{-5}, -1.53×10^{-9}, and 8.88×10^{-16} (computed with approximately sixteen decimal digits of precision).

One difficulty with the secant method is that the iterates may diverge if the straight line approximation is an *extrapolation* (i.e., when f_k and f_{k-1} have the same sign). If f_k and f_{k-1} were always of opposite sign, the predicted zero would necessarily lie strictly between x_{k-1} and x_k. This suggests a modification of the secant method — the *method of false position*, or *regula falsi*, in which x_{k+1} replaces either x_k or x_{k-1}, depending on which corresponding function value agrees in sign with f_{k+1}. The two initial points x_0 and x_1 must be chosen such that $f_0 f_1 < 0$. In

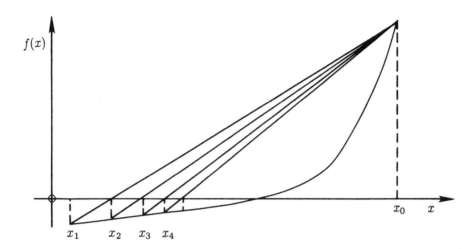

Figure 4d. Example of slow convergence for *regula falsi*. The initial interval of uncertainty is $[x_1, x_0]$. Since $f_i < 0$ for $i \geq 1$, the point x_0 is never discarded and the convergence is extremely slow.

this way the zero remains bracketed, and there is no danger from extrapolation. Unfortunately, although this modification of the secant method guarantees convergence, its rate of convergence is only linear, and the asymptotic error constant can be arbitrarily close to unity. In Figure 4d we show an example where x_0 will never be discarded, since $f_i < 0$ for $i \geq 1$; hence the convergence is extremely slow. In general the method of false position is much less efficient than bisection. This illustrates that care is necessary when attempting to ensure convergence, in order not to destroy the rapid convergence of the original method.

*4.1.1.4. Rational interpolation and higher-order methods. Other procedures for well-behaved functions can be developed by constructing approximations based on the values of f and its derivatives at any number of known points. However, higher-order schemes based upon polynomial interpolation have the disadvantage that one particular zero of the polynomial must be selected for the next iterate.

Methods without this drawback can be devised using a rational interpolation function of the form

$$\hat{f} = \frac{x - c}{d_0 + d_1 x + d_2 x^2}.$$

The values of c, d_0, d_1, and d_2 are chosen so that the function value and derivatives of $\hat{f}$ agree with those of f at two points. If derivatives are not available, a rational interpolation function $\hat{f}$ without the x^2 term can be computed from the values of f at three points.

4.1.1.5. Safeguarded zero-finding algorithms. The bisection algorithm can be viewed as a technique for constructing a set of intervals $\{I_j\}$, each containing x^*, such that the interval I_j lies wholly within I_{j-1}, and the length of each interval is strictly smaller than that of its predecessor.

Mathematically this process can be expressed as: *given I_0 such that $x^* \in I_0$, for $j = 1, \ldots,$ find $\{I_j\}$ such that $I_j \subset I_{j-1}$ and $x^* \in I_j$.* Note that, given an interval I_j, knowledge of the sign of f at a single interior point (say, u) in I_{j-1} enables a new interval I_j to be found, assuming that the sign of f is known at the endpoints of I_{j-1}. Methods that generate a set of nested intervals in this manner are known as *bracketing methods*.

With the bisection algorithm, u is obtained simply by bisecting I_{j-1}. However, we may just as easily compute u using Newton's method, linear approximation or rational approximation. Furthermore, the approximating function need not be based on the values of f at the end points of the interval of uncertainty, but rather could use the "best" points found so far. For example, in Figure 4d the points x_1 and x_2 could be used for linear interpolation, even though the interval of uncertainty is $[x_2, x_0]$.

The best methods available for zero-finding are the so-called *safeguarded* procedures. The idea is to combine a guaranteed, reliable method (such as bisection) with a fast-convergent method (such as linear or rational interpolation), to yield an algorithm that will converge rapidly if f is well-behaved, but is not much less efficient than the guaranteed method in the worst case.

A safeguarded linear interpolation method might include the following logic at each iteration. An interval of uncertainty is known, say $[a, b]$. The two "best" points found so far are used to obtain a point u by linear interpolation. Without safeguards, the next step would involve evaluating $f(u)$, and discarding one of the old points in order to form a new set of points for the linear fit. However, a safeguarded procedure ensures that u is a "reasonable" point before evaluating f at u.

The definition of "reasonable" includes several requirements. Firstly, u must lie in $[a, b]$. Secondly, u is replaced by the bisection step if the step to u from the best point exceeds half the interval of uncertainty, since the bisection step is then likely to yield a greater reduction in the interval of uncertainty. In this way the number of function evaluations required in the worst possible case should not be significantly more than that required by bisection.

Finally, safeguards are necessary to ensure that successive iterates are not "too close". It can happen that u is numerically indistinguishable from the previous best point, even when u is far from optimal; if u were accepted, no further progress would occur. If extrapolation is performed with nearly equal iterates, rounding error may cause the predicted point to be poorly defined, thereby impairing the rate of convergence. The most common technique used to prevent this phenomenon is to specify a "small" distance δ, and to "take a step of δ" whenever a newly generated point is too close to the current best point. The step of δ is taken in the direction that will result in the largest reduction of the interval of uncertainty. The use of this technique will usually cause the final two steps of any safeguarded algorithm to be of length δ, and x^* will therefore lie in an interval of length 2δ.

4.1.2. Univariate Minimization

The techniques used to solve the problem of univariate minimization in a bounded interval are analogous to those for zero-finding. It is generally advisable to use techniques designed specifically for minimization, although in some special circumstances it may be possible to apply bisection directly to locate a zero of f'.

To develop an interval-reduction procedure for minimization when only function values are available, we need to define a condition that ensures that there is a proper minimum in a given interval. For this purpose, we introduce the concept of *unimodality*, of which there are several definitions in the literature. One practical definition is: $f(x)$ is unimodal in $[a, b]$ if there exists a

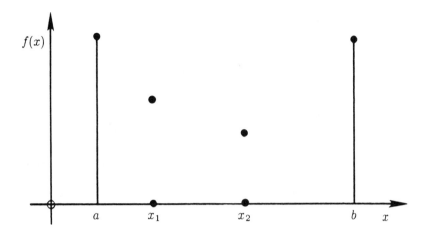

Figure 4e. Reduction of the interval of uncertainty for a unimodal function. The minimum must lie in the reduced interval $[x_1, b]$ or $[a, x_2]$, depending on the relative values of $f(x_1)$ and $f(x_2)$.

unique $x^* \in [a, b]$ such that, given any $x_1, x_2 \in [a, b]$ for which $x_1 < x_2$:

$$\text{if } x_2 < x^* \text{ then } f(x_1) > f(x_2); \quad \text{if } x_1 > x^* \text{ then } f(x_1) < f(x_2).$$

If f is known to be unimodal in $[a, b]$, it is possible to reduce the interval of uncertainty by comparing the values of f at *two* interior points. For example, in Figure 4e, the minimum must lie in the reduced interval $[x_1, b]$ or $[a, x_2]$, depending on the relative values of $f(x_1)$ and $f(x_2)$.

4.1.2.1. Fibonacci search. An algorithm for finding a univariate minimum by interval reduction must specify how to choose the two interior points at each step. Clearly it would be most efficient in terms of the number of evaluations of f to choose the points so that one of them could be re-used during the next iteration. In this way, only one new evaluation of f would be required at each iteration.

The "optimum" strategy (i.e., the strategy which yields the maximum reduction in the interval of uncertainty for a given number of function evaluations) is termed *Fibonacci search*. It is based upon the Fibonacci numbers $\{F_i\}$, which satisfy

$$F_k = F_{k-1} + F_{k-2}, \qquad F_0 = F_1 = 1.$$

The first few values of the sequence are $1, 1, 2, 3, 5, 8, 13, \dots$. Given a number N of function evaluations, the Fibonacci numbers F_{N-1}, $F_{N-2}, \dots$ can be used to define the points at which the function should be evaluated within a sequence of shrinking intervals of uncertainty, beginning with an original interval taken as $[0, F_N]$. We illustrate this process by considering the case where just two function evaluations are performed. If the original interval is $[0, F_2]$ $(F_2 = 2)$, f is evaluated at two points corresponding to F_0 and F_1, i.e. at $x_1 = 1$ and $x_2 = 1 + \delta$, where δ is "negligible". Regardless of the function values $f(x_1)$ and $f(x_2)$, the new interval of uncertainty

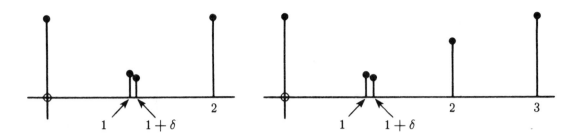

Figure 4f. The first figure depicts the position of points for Fibonacci search for $N = 2$. The original interval is $[0, 2]$; f is evaluated at $x_1 = 1$ and $x_2 = 1 + \delta$. The second figure gives the points for $N = 3$. Here, the original interval of uncertainty is considered to be $[0, 3]$, and the first two points are placed at $x_1 = 1$ and $x_2 = 2$. Depending on the values of $f(x_1)$ and $f(x_2)$, the reduced interval will either be $[0, 2]$ or $[1, 3]$.

is thereby reduced to arbitrarily close to half its original length (see the first diagram in Figure 4f).

Figure 4f depicts the case when three function evaluations are allowed. Here, the original interval of uncertainty is considered to be $[0, F_3]$ ($F_3 = 3$), and the first two points are placed at F_1 and F_2 (i.e., $x_1 = 1$ and $x_2 = 2$). Depending on the values of $f(x_1)$ and $f(x_2)$, the reduced interval will either be $[0, 2]$, or $[1, 3]$. Since the function value at the midpoint of the reduced interval is already available, only one additional evaluation is required to obtain a final interval of length $1 + \delta$.

When N function evaluations are allowed, a Fibonacci search procedure will essentially produce a final interval of uncertainty of length $1/F_N$ times the length of the original interval.

4.1.2.2. Golden section search. A disadvantage of Fibonacci search is that the desired accuracy in most problems will not usually be stated in terms of a specified number of function evaluations. Therefore, in order to determine how many evaluations are required to reduce the original interval by a given factor, it is necessary to store or generate the Fibonacci numbers. Furthermore, a Fibonacci search strategy cannot easily be adapted to the case when the termination criterion requires that the function values in the final interval of uncertainty differ by less than a certain amount.

A procedure that does not require the *a priori* selection of the final interval of uncertainty and is almost as efficient as Fibonacci search is *golden section search*. It can be shown that

$$\lim_{k \to \infty} \frac{F_{k-1}}{F_k} = \frac{2}{1 + \sqrt{5}} \equiv \tau \approx .6180,$$

where τ satisfies the quadratic equation $\tau^2 + \tau - 1 = 0$. With a golden section search procedure, if the initial interval is considered to be $[0, 1]$, the two points are placed at τ and $1 - \tau$ (i.e., at approximately .6180 and .3820). No matter how the interval is reduced, one of the old points will then be in the correct position with respect to the new interval; golden section search may

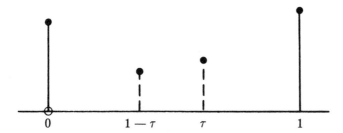

Figure 4g. Position of points for golden section search. If the initial interval is considered to be $[0, 1]$, two points are placed at τ and $1 - \tau$, where $\tau = .6180$. No matter how the interval is reduced, one of these points will be in the correct position with respect to the new interval.

thus be viewed as the limiting case of Fibonacci search. Figure 4g illustrates the configuration of these points.

With golden section search, there is a constant reduction of the interval of uncertainty by the factor τ at every step, and the length of the interval of uncertainty converges linearly to zero.

4.1.2.3. Polynomial interpolation. Golden section search and Fibonacci search are similar in the sense that each permits only two possible choices for the next interval of uncertainty. Moreover, the interval chosen is based solely on a comparative test on the last computed value of f. As in the zero-finding case, more efficient procedures can be developed for smooth functions by utilizing known values of f to define the next iterate. In particular, f can be approximated by a simple function $\hat{f}$ whose minimum is easy to compute, and the minimum of $\hat{f}$ can be used iteratively as an estimate of the minimum of f. Since a general straight line has no minimum, we consider approximating f by a parabola (quadratic), of the form:

$$\hat{f} = \frac{1}{2}ax^2 + bx + c.$$

If $a > 0$, $\hat{f}$ has a minimum at x^* such that $ax^* + b = 0$.

Three independent pieces of information are required in order to construct the quadratic approximation $\hat{f}$. For example, at a particular point x_k, given $f(x_k)$, $f'(x_k)$ and $f''(x_k)$, $\hat{f}$ can be defined by the first three terms of the Taylor-series expansion, i.e.,

$$\hat{f}(x) = f(x_k) + f'(x_k)(x - x_k) + \frac{1}{2}f''(x_k)(x - x_k)^2.$$

If $f''(x_k)$ is non-zero, $\hat{f}$ has a stationary point at x_{k+1} such that

$$f'(x_k) + f''(x_k)(x_{k+1} - x_k) = 0,$$

or equivalently

$$x_{k+1} = x_k - \frac{f'(x_k)}{f''(x_k)}. \tag{4.2}$$

The formula (4.2) is equivalent to Newton's method (4.1) applied to finding a zero of $f'(x)$.

A quadratic approximation $\hat{f}$ can also be fitted to three function values. When first, but not second, derivatives are known, the values of f and f' at two points provide four independent pieces of data, and $\hat{f}$ may be taken as a cubic polynomial.

Methods based upon polynomial fitting can be shown to have superlinear convergence under suitable conditions; for example, the rate of convergence for parabolic fitting with three function values is approximately 1.324, and the cubic fitting technique has quadratic convergence. However, because these methods are based on a well-behaved simple model, they are subject to the same difficulties as the rapidly converging zero-finding techniques. If $\hat{f}$ is not an accurate representation of the behaviour of f, the minimum of $\hat{f}$ may be a poor estimate — for example, the predicted minimum may lie outside the initial interval of uncertainty, or $\hat{f}$ may be unbounded below. In order to avoid such difficulties, polynomial fitting strategies can be modified so that the minimum is always bracketed. For example, in cubic fitting, the gradient can be required to be of opposite sign at the two points, in which case the minimum of the fitted cubic will lie in the original interval. Unfortunately, as with the similarly motivated modification of the secant method, this attempt to improve robustness destroys the superlinear convergence rate. With such a method, the convergence rate is linear, and can be arbitrarily slow.

4.1.2.4. Safeguarded polynomial interpolation. As in zero-finding, the best general methods are the so-called *safeguarded* procedures. In this case, polynomial interpolation can be combined with bisection (when derivatives are known) or golden section search (when only function values are available). A safeguarded method based upon parabolic interpolation with three function values requires an interval of uncertainty, say $[a, b]$, and the three "best" points found so far. Suppose that u is the step, computed by parabolic interpolation, from the best point x. Let $\bar{u}$ denote the "approximate" golden section step

$$\bar{u} = \begin{cases} \beta(a - x), & \text{if } x \geq \frac{1}{2}(a + b); \\ \beta(b - x), & \text{if } x < \frac{1}{2}(a + b), \end{cases}$$

where $\beta = 1 - \frac{1}{2}(\sqrt{5} - 1)$. If $u > \bar{u}$, $\bar{u}$ will be used during the next iteration instead of u. Safeguards are also included to prevent iterates from being too close (see Section 4.1.1.5).

Notes and Selected Bibliography for Section 4.1

For an excellent discussion on many aspects of univariate zero-finding and minimization, see Brent (1973a). A safeguarded linear interpolation algorithm was first suggested by Dekker (1969). Algorithms based upon linear and rational interpolation have been suggested by Jarratt and Nudds (1965), Anderson and Björck (1973), Kahan (1973), and Bus and Dekker (1975), amongst others.

The use of bisection and golden section within a safeguarded interpolation scheme can lead to serious inefficiencies if the current "best" point lies close to an end point (a situation that arises often when univariate minimization is performed in the context of multivariate optimization). For details of a scheme that biases the new point toward the best point and avoids this problem see Gill and Murray (1974e).

When the function f is known to have special properties, univariate minimization techniques can be designed specifically for f by choosing an approximation function $\hat{f}$ that reflects these properties. For example, f may be a sum of the squares of a set of functions $\{\phi_i(x)\}$, i.e.

$$f(x) = \sum_{i=1}^{m} \phi_i(x)^2.$$

In this case, each individual function ϕ_i may be approximated by a quadratic or cubic polynomial. The model function $\hat{f}$ is then a polynomial of order four or six, and an appropriate method can be used to find the smallest positive zero of its derivative.

When f has a singularity of known form, standard techniques for univariate minimization tend to be relatively inefficient. In such a circumstance, $\hat{f}$ can be constructed as a simple function with the correct type of singularity, and special iterative methods can be used to locate its minimum (see, for example, Murray and Wright, 1976).

4.2. METHODS FOR MULTIVARIATE NON-SMOOTH FUNCTIONS

In this section, we consider methods for unconstrained minimization of the multivariate function $F(x)$, where x is an n-vector and F is not a smooth function. Although non-differentiable functions are in general more difficult to minimize than smooth functions, a distinction must be made between a problem with random discontinuities in the function or its derivatives, and one in which a great deal of information is known about the nature of any discontinuities. We shall divide algorithms for non-smooth functions into two categories. Firstly, function comparison methods for *general* non-smooth functions are discussed in Sections 4.2.1 and 4.2.2. Methods for certain problems with highly structured discontinuities in the gradient will be discussed in Section 4.2.3. *If a function has just a "few" discontinuities in its first derivative, and these discontinuities do not occur in the neighbourhood of the solution, then the methods described in later sections for smooth problems are likely to be more efficient.*

4.2.1. The Use of Function Comparison Methods

Methods based on function comparison are intended only for problems in which $F(x)$ is discontinuous, the gradient vector $g(x)$ is discontinuous at the solution, or $g(x)$ has so many discontinuities that a method assuming continuity of $g(x)$ will fail. Furthermore, the discontinuities are presumed to have no special structure. A general method for non-smooth functions will, in general, be *inefficient* when F *does* have additional smoothness properties. *A method using function comparison should be used only when there is no other suitable alternative method available.* If a user decides to use a function comparison method only because of its simplicity and seeming generality, he may pay a severe price in speed and reliability. It is often mistakenly thought that, because of their simplicity, function comparison methods are very robust. They do not generally suffer from rounding errors that arise from complicated numerical processes; however, proper implementations of more sophisticated methods are also robust in this sense. The substantial disadvantage of function comparison methods is that few (if any) guarantees can be made concerning convergence.

In the univariate case, search procedures exist for systematically shrinking an interval in which the minimum is known to lie, using only function comparison (see Section 4.1.2). Unfortunately, such algorithms tend to be extremely inefficient in higher dimensions (if they work at all). For example, we might consider generating a grid of points in n-space and using function comparisons to reduce the region, but the required number of function evaluations increases *exponentially* with the problem dimension (this phenomenon is sometimes termed the "curse of dimensionality").

Methods based on function comparison are often called *direct search* methods. Many direct search methods were developed to solve specialized problems, and may be of very limited general usefulness. The heuristic nature of these methods is reflected by a large number of parameters to be selected; the success of the method often crucially depends on the choice of these parameters.

4.2.2. The Polytope Algorithm

In order to give the flavour of a typical direct search method, we shall consider briefly the *polytope method*, which is usually termed the "simplex" method. However, we prefer not to use the latter name, in order to avoid confusion with the better-known simplex method for linear programming.

At each stage of the algorithm, $n+1$ points $x_1, x_2, \ldots, x_{n+1}$ are retained, together with the function values at these points, which are ordered so that $F_{n+1} \geq F_n \geq \cdots \geq F_2 \geq F_1$. The method derives its name because these points can be considered to be the vertices of a polytope in n-space. At each iteration, a new polytope will be generated by producing a new point to replace the "worst" point x_{n+1} (i.e., the point with the highest function value).

Let c denote the centroid of the best n vertices $x_1, x_2, \ldots, x_n$, given by

$$c = \frac{1}{n} \sum_{j=1}^{n} x_j.$$

At the beginning of each iteration, a trial point is generated by a single *reflection step* in which we construct the point $x_r = c + \alpha(c - x_{n+1})$, where α ($\alpha > 0$) is the *reflection coefficient*. The function is evaluated at x_r, yielding F_r.

There are three cases to consider.

1. If $F_1 \leq F_r \leq F_n$ (i.e., x_r is not either a new best point or a new worst point), x_r replaces x_{n+1} and the next iteration is begun.

2. If $F_r < F_1$, so that x_r is the new best point, we assume that the direction of reflection is a "good" direction, and attempt to *expand* the polytope in this direction by defining an expanded point x_e such that $x_e = c + \beta(x_r - c)$, where β ($\beta > 1$) is the *expansion coefficient*. If $F_e < F_r$, the expansion is successful, and x_e replaces x_{n+1}. Otherwise, the expansion has failed, and x_{n+1} is replaced by x_r.

3. If $F_r > F_n$, the polytope is assumed to be too large, and should be *contracted*. A contraction step is carried out by defining

$$x_c = \begin{cases} c + \gamma(x_{n+1} - c), & \text{if } F_r \geq F_{n+1}; \\ c + \gamma(x_r - c), & \text{if } F_r < F_{n+1}, \end{cases}$$

where γ ($0 < \gamma < 1$) is the *contraction coefficient*. If $F_c < \min\{F_r, F_{n+1}\}$, the contraction step has succeeded, and x_c replaces x_{n+1}. Otherwise, a further contraction is carried out.

Figure 4h illustrates the position of the reflected and expansion steps for a polytope in two dimensions.

Occasionally, the most recent polytope is discarded and replaced by a regular polytope. This procedure is known as *restarting*. Restarting can be used to prevent the polytope from becoming unbalanced after several successive expansions are made. In this case the best two points are retained and their vector-difference determines the length of the side of the new regular polytope. A restart may also be made to check the validity of a solution. In this case, the regular polytope may be given the centre x_1 and side $\|x_1 - x_{n+1}\|_2$. If the algorithm re-converges to the same

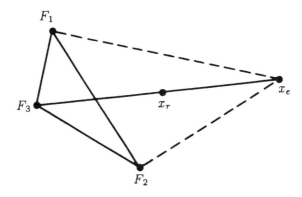

Figure 4h. This figure depicts the position of the reflected point x_r and the expanded point x_e for a polytope in two dimensions. The vertex of the polytope corresponding to the highest function value is marked on the figure as F_3.

point, or $2n$ iterations are made without finding a lower point, x_1 is regarded as an adequate solution.

Modifications can be made to the polytope method that significantly improve its performance. For example, during the contraction step described above, a new point is found on a line joining what could be two poor points (the worst point and the reflected point). In this case it is better to compute a point biased toward the best point of the polytope. A modified contraction step that achieves this bias is given in the following variation of Step 3 of the original algorithm.

$3'$. If $F_r > F_n$, a contraction step is carried out by defining

$$x_c = \begin{cases} x_1 + \gamma(x_{n+1} - x_1), & \text{if } F_r \geq F_{n+1}; \\ x_1 + \gamma(x_r - x_1), & \text{if } F_r < F_{n+1}. \end{cases}$$

Another modification involves *shrinking* the polytope if the contraction step is unsuccessful, or if the best point remains unchanged for many iterations. The polytope is shrunk by moving the vertices half-way toward the current best point in the order $x_2, x_3, \ldots$. Note that the best point may change during the shrinking process.

Example 4.2. Consider the two-dimensional function

$$F(x) = 100(x_2 - x_1^2)^2 + (1 - x_1)^2.$$

This function, commonly known as *Rosenbrock's function*, has a unique minimum at the point $(1, 1)^T$, which lies at the base of a banana-shaped valley. Figure 4i depicts the solution path of a polytope method when applied to Rosenbrock's function. The curved lines correspond to lines of equal function value; the linear segments correspond to the movement of the "best" vertex x_1 as the solution process progresses. The algorithm was started at the point $(-1.2, 1.0)^T$. Although the polytope method is not intended for problems with continuous derivatives, the figure illustrates that many evaluations of the objective function are required to locate the minimum. However, we must stress that the performance of the polytope method depicted in Figure 4i is much better than we would normally expect on a typical non-smooth problem.

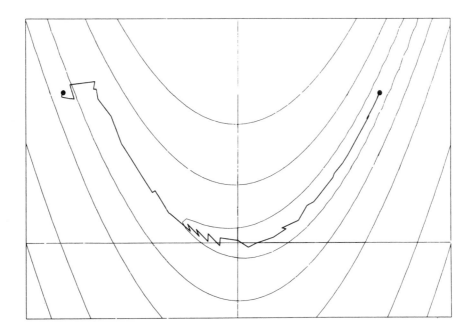

Figure 4i. Solution path of a polytope algorithm on Rosenbrock's function. The curved lines correspond to lines of equal function value; the linear segments correspond to the movement of the "best" vertex x_1 as the solution process progresses. The algorithm was started at the point $(-1.2, 1.0)^T$. Rosenbrock's function has a unique minimum at the point $(1, 1)^T$, which lies at the base of the banana-shaped valley.

*4.2.3. Composite Non-Differentiable Functions

In some well-known instances, the problem functions themselves are not smooth, but rather are *composites* of smooth functions. In this case, one method of solution is to transform the *unconstrained* composite non-differentiable problem into a smooth, but *nonlinearly constrained*, problem. This problem, although more complex than the original, can then be solved using a method for nonlinear constraints that takes special advantage of the structure of the problem. Methods for this class of nonlinearly constrained problem will be discussed in Section 6.8. It should be emphasized that the transformation would normally be performed *implicitly* by software designed to solve the original non-smooth problem.

To illustrate the type of transformation used, we shall consider three common composite problems.

Example 4.3. The ℓ_∞ *(minimax) problem* is defined as:

$$\underset{x \in \Re^n}{\text{minimize}} \ \underset{\{f_i\}}{\max} \ \{f_1(x), f_2(x), \ldots, f_m(x)\}, \tag{4.3}$$

where $\{f_i(x)\}$ are smooth functions.

This problem can be transformed into a smooth problem by introducing a new variable x_{n+1}, which is an upper bound on all the functions $\{f_i(x)\}$. A problem equivalent to (4.3) is then

$$\begin{array}{ll} \underset{x \in \Re^{n+1}}{\text{minimize}} & x_{n+1} \\ \text{subject to} & f_i(x) \le x_{n+1}, \quad i = 1, 2, \ldots, m. \end{array}$$

Note that the original unconstrained problem has been transformed into a nonlinearly constrained problem; this increase in complexity is typical of the transformations associated with non-differentiable composite functions.

Example 4.4. The ℓ_1 *problem* is defined as:

$$\underset{x \in \Re^n}{\text{minimize}} \sum_{i=1}^{m} |f_i(x)|. \tag{4.4}$$

The name arises because (4.4) is the one-norm of the vector $(f_1(x), f_2(x), \ldots, f_m(x))$ (see Section 2.2.4.2).

To transform (4.4), we note that a typical function $f_i(x)$ may be split into positive and negative parts by writing $f_i(x) = r_i - s_i$, where both r_i and s_i are non-negative. The relationship $|f_i(x)| \le r_i + s_i$ leads to the following transformation of (4.4):

$$\begin{array}{ll} \underset{x \in \Re^n; r, s \in \Re^m}{\text{minimize}} & \sum_{i=1}^{m} (r_i + s_i) \\ \text{subject to} & f_i(x) = r_i - s_i, \quad i = 1, 2, \ldots, m; \\ & r_i \ge 0; \quad s_i \ge 0, \quad i = 1, 2, \ldots, m. \end{array}$$

Example 4.5.

$$\underset{x \in \Re^n}{\text{minimize}} \sum_{i=1}^{m} \max \{f_i(x), 0\}.$$

A similar argument to that used in transforming (4.4) gives

$$\begin{array}{ll} \underset{x \in \Re^n; r, s \in \Re^m}{\text{minimize}} & \sum_{i=1}^{m} r_i \\ \text{subject to} & f_i(x) = r_i - s_i, \quad i = 1, 2, \ldots, m; \\ & r_i \ge 0; \quad s_i \ge 0, \quad i = 1, 2, \ldots, m. \end{array}$$

Although there is a significant increase in the number of variables with the latter two approaches, this need not be a serious obstacle if the transformed problems are solved by an algorithm designed to take advantage of the simple form of the bound constraints (see Section 5.5.1).

The ℓ_∞ and ℓ_1 problems often arise in the context of data fitting, in which case the optimal value of the objective function in (4.3) or (4.4) is expected to be small. If the optimal objective value is actually zero, then the solution of (4.3) or (4.4) is also the solution of the smooth nonlinear least-squares problem

$$\underset{x \in \Re^n}{\text{minimize}} \sum_{i=1}^{m} f_i^2(x).$$

If the optimal objective value is "small" (although not zero), an alternative to solving a ℓ_∞ or ℓ_1 problem is to solve a sequence of weighted unconstrained least-squares problems of the form

$$\underset{x\in\Re^n}{\text{minimize}} \sum_{i=1}^{m} w_i^{(k)} f_i(x)^2, \qquad k=1,2,3,\ldots. \tag{4.5}$$

(The problem (4.5) can be solved by specialized methods designed to minimize a weighted sum of squares of nonlinear functions, to be discussed in Section 4.7.)

Let $\hat{x}^{(k)}$ denote the solution of (4.5), and let $w_i^{(1)} = 1/m$, $i = 1,\ldots,m$. To solve the ℓ_∞ problem, the weights in the sequence of subproblems are chosen to be

$$w_i^{(k+1)} = \frac{1}{S} f_i^2(\hat{x}^{(k)}) w_i^{(k)},$$

where $S = \sum_i f_i^2(\hat{x}^{(k)}) w_i^{(k)}$, so that $\sum_i w_i^{(k+1)} = 1$. For the ℓ_1 problem, the sequence of weights is defined as

$$w_i^{(k+1)} = \frac{S}{|f_i(\hat{x}^{(k)})|},$$

where $S = \sum_i 1 / |f_i(\hat{x}^{(k)})|$ (assuming that $f_i(\hat{x}^{(k)})$ is non-zero for all i). Usually, only two or three least-squares problems of the form (4.5) must be solved to obtain convergence.

In either the ℓ_∞ or ℓ_1 case, it is not necessary to compute $\hat{x}^{(k)}$ to high accuracy. Since $\hat{x}^{(k)}$ is used as the initial point when solving for $\hat{x}^{(k+1)}$, only a few iterations are typically required to find the solution of (4.5) for $k > 1$.

This technique can encounter difficulties for the ℓ_1 problem if any $f_i(\hat{x}^{(k)})$ is exactly zero. Under such circumstances, the weights are not well-defined, and an improved set cannot be predicted. Several strategies may overcome the problem — for example, the least-squares problem can be re-solved with a different initial set of weights, or with the offending components of f omitted (possibly to be re-introduced at a later stage). However, there is no guarantee that these approaches will resolve the difficulty. Fortunately, the likelihood that this situation will occur in practice is small.

The success of these schemes depends critically on whether the model gives a close fit to the data. If so, the least-squares solution is close to the optimal solution in the other norms, and only a small additional effort is required to find the ℓ_1 or ℓ_∞ solution. Since each least-squares problem in the sequence does not need to be solved accurately, the total effort may be comparable to solving a single least-squares problem with high accuracy.

Notes and Selected Bibliography for Section 4.2

The polytope algorithm described is that of Spendley, Hext and Himsworth (1962), as modified by Nelder and Mead (1965). Modifications have been suggested by Parkinson and Hutchinson (1972), amongst others. A survey of direct search methods is given by Swann (1972).

Methods for minimization problems with composite functions have been the subject of active research in recent years. For further discussion and more references, see Section 6.8. The solution of linear ℓ_1 and ℓ_∞ problems by iterative linear least-squares was suggested by Lawson (see Lawson and Hanson, 1974).

4.3. METHODS FOR MULTIVARIATE SMOOTH FUNCTIONS

4.3.1. A Model Algorithm for Smooth Functions

In this section, we shall be concerned with methods for the unconstrained minimization of a twice-continuously differentiable function $F(x)$.

An essential concept in algorithm design is that of a *measure of progress*. For an iterative method, it is important to have some reasonable way of deciding whether a new point is "better" than the old point. This idea will recur throughout the discussion of optimization algorithms. In the case of unconstrained minimization, a natural measure of progress is provided by the value of the objective function. It seems reasonable to require a decrease in F at every iteration, and to impose the *descent condition* that $F_{k+1} < F_k$ for all $k \geq 0$. A method that imposes this requirement is termed a *descent method*. We shall discuss only descent methods for unconstrained minimization.

All methods to be considered are of the form of the following *model algorithm*.

Algorithm U (*Model algorithm for n-dimensional unconstrained minimization*). Let x_k be the current estimate of x^*.

U1. [Test for convergence.] If the conditions for convergence are satisfied, the algorithm terminates with x_k as the solution.

U2. [Compute a search direction.] Compute a non-zero n-vector p_k, the *direction of search*.

U3. [Compute a step length.] Compute a positive scalar α_k, the *step length*, for which it holds that $F(x_k + \alpha_k p_k) < F(x_k)$.

U4. [Update the estimate of the minimum.] Set $x_{k+1} \leftarrow x_k + \alpha_k p_k$, $k \leftarrow k+1$, and go back to step U1. ∎

We shall use the notation F_k for $F(x_k)$, g_k for $g(x_k)$, and G_k for $G(x_k)$. It is important to note that step U3 of the model algorithm involves the solution of a univariate problem (finding the scalar α_k).

In order to satisfy the descent condition, p_k and α_k must have certain properties. A standard way of guaranteeing that F can be reduced at the k-th iteration is to require that p_k should be a *descent direction* at x_k, i.e. a vector satisfying

$$g_k^T p_k < 0.$$

If p_k is a descent direction, there must exist a positive α such that $F(x_k + \alpha p_k) < F(x_k)$ (see Section 3.2.2).

4.3.2. Convergence of the Model Algorithm

Before considering specific algorithms, we consider the conditions necessary in order for an algorithm of the model form to converge to a point x^* where $g(x^*)$ vanishes. It is beyond the scope of this text to present a rigorous analysis of convergence. For more details, the reader should refer to the references cited in the Notes at the end of this section.

The requirement that $F_{k+1} < F_k$ is not sufficient in itself to ensure that the sequence $\{x_k\}$ converges to a minimum of F. For example, consider the univariate function x^2 at the points defined by the sequence $(-1)^k(\frac{1}{2} + 2^{-k})$.

If $F_{k+1} < F_k$ and $F(x)$ is bounded below, then clearly the sequence $\{F_k\}$ converges. However, we must be sure that the sequence converges to $F(x^*)$. There are two obvious situations in which the strictly decreasing sequence $\{F_k\}$ generated by the model algorithm might converge to a non-optimal point. Firstly, whatever the choice of search direction p_k, the sequence of step lengths $\{\alpha_k\}$ could be chosen so that F is reduced by an ever-smaller amount at each iteration. Secondly, $\{p_k\}$ could be chosen so that, to first order, F is almost constant along p_k. This will occur only if p_k is almost parallel to the first-order approximation to the contour line $F(x) = F_k$, so that g_k and p_k are almost orthogonal (i.e., $-g_k^T p_k / (\|g_k\|_2 \|p_k\|_2)$ is close to zero). To avoid these two situations, α_k must be chosen so that F is "sufficiently decreased" at each iteration, and there must be a limit on the closeness of p_k to orthogonality to the gradient.

Additional mild restrictions on F are also required in order to prove convergence of the model algorithm. For a given function F and a specified scalar β, the corresponding *level set* $L(\beta)$ is the set of points x such that $F(x) \leq \beta$. We shall always assume that the level set $L(F(x_0))$ is closed and bounded; this assumption excludes functions such as e^x that are bounded below but strictly decreasing as $\|x\|$ becomes infinite.

If the following conditions hold: (i) F is twice-continuously differentiable; (ii) the level set $L(F(x_0))$ is closed and bounded; (iii) F sustains a "sufficient decrease" at each iteration; and (iv) p_k remains bounded away from orthogonality to the gradient, then the iterates from Algorithm U satisfy

$$\lim_{k \to \infty} \|g_k\| = 0.$$

This type of convergence result is sometimes termed *global* convergence, since there is no restriction on the closeness of x_0 to a stationary point.

4.3.2.1. Computing the step length.

A fundamental requirement for a step-length algorithm associated with a descent method involves the change in F at each iteration. If convergence is to be assured, the step length must produce a "sufficient decrease" in $F(x)$. The "sufficient decrease" requirement can be satisfied by several alternative sets of conditions on α_k. For example, a sufficient decrease is achieved when α_k satisfies the *Goldstein-Armijo principle*:

$$0 < -\mu_1 \alpha_k g_k^T p_k \leq F_k - F_{k+1} \leq -\mu_2 \alpha_k g_k^T p_k, \qquad (4.6)$$

where μ_1 and μ_2 are scalars satisfying $0 < \mu_1 \leq \mu_2 < 1$. The upper and lower bounds of (4.6) ensure that α_k is neither "too large" nor "too small".

In typical algorithms based on (4.6), the trial values of the step length are defined in terms of an initial step $\alpha^{(0)}$ and a scalar w. The value of α_k is taken as the first member of the sequence $\{w^j \alpha^{(0)}\}$, $j = 0, \ldots$, for which (4.6) is satisfied for some μ_1 and μ_2.

The performance of these algorithms depends critically on the choice of $\alpha^{(0)}$, rather than on any particular merits of condition (4.6). The usual convention is to take $\alpha^{(0)}$ as unity — not in order to maximize the probability of satisfying (4.6), but to achieve the best convergence with Newton-type and quasi-Newton methods (see Sections 4.4 and 4.5). Step-length algorithms of this type thus perform well only for descent methods in which an *a priori* value of $\alpha^{(0)}$ tends to be a good step, so that only the first member of the sequence $\{w^j \alpha^{(0)}\}$ usually needs to be computed.

We emphasize that condition (4.6) alone does not guarantee a good value of α_k. Note that for almost all functions encountered in practice, choosing $\alpha^{(0)}$ as 10^{-5} would satisfy (4.6) for appropriate small values of μ_1 and μ_2. Although this strategy would be "efficient" in that a suitable α_k would be found with only a single function evaluation per iteration, any descent method that included such a step-length algorithm would be extremely *inefficient*. It is essential to consider the performance of a step-length algorithm not merely in terms of the number of function evaluations per iteration, but in terms of the overall reduction in F achieved at each iteration. This point is sometimes overlooked when considering only convergence theorems for algorithms.

Whatever the criteria specified for α_k, there is usually a *range* of suitable values, some of which are "better" than others. An intuitively satisfying rule of thumb is that the "better" the step, the greater the reduction in F. Whether a step-length algorithm can efficiently produce a "good" α_k in this sense depends on the method by which estimates of α_k are generated, as well as on the criteria imposed on α_k.

Obviously, there is a tradeoff between the effort expended to determine a good α_k at each iteration and the resulting benefits for the descent method. The balance varies with the type of algorithm as well as with the problem to be solved, and hence some flexibility is desirable in specifying the conditions to be satisfied by α_k. An alternative condition on α_k can be derived from an interpretation of a step-length procedure in terms of univariate minimization. It is sometimes desirable (or even essential) for α_k to be a close approximation to a minimum of F along p_k. A step to the first minimum of F along p_k would intuitively appear to yield a "significant" decrease in F, and this choice of α_k is important in numerous theoretical convergence results. A practical criterion based on interpreting α_k in terms of univariate minimization requires that the *magnitude* of the directional derivative at $x_k + \alpha p_k$ be sufficiently reduced from that at x_k:

$$|g(x_k + \alpha p_k)^T p_k| \leq -\eta g_k^T p_k, \tag{4.7}$$

where $0 \leq \eta < 1$. The value of η determines the accuracy with which α_k approximates a stationary point of F along p_k, and consequently provides a means of controlling the balance of effort to be expended in computing α_k. Condition (4.7) ensures that α_k is not too small.

When η is "small", the step-length procedure based on (4.7) will be termed an "accurate line search"; when $\eta = 0$, the step-length procedure will be described as an "exact line search". Note also that the terms "line search" and "linear search" will be used interchangeably.

Condition (4.7) does not involve the change in F, and hence is not adequate to ensure a sufficient decrease. To guarantee a suitable reduction in F, many step-length algorithms include the following condition:

$$F(x_k) - F(x_k + \alpha p_k) \geq -\mu \alpha g_k^T p_k, \tag{4.8}$$

where $0 < \mu \leq \frac{1}{2}$. Note that condition (4.8) requires $F(x_k + \alpha_k p_k)$ to lie on or below the line $l(\alpha) = F(x_k) + \mu \alpha g_k^T p_k$.

The set of points satisfying (4.7) and (4.8) will be denoted by $\Gamma(\eta, \mu)$. If $\mu \leq \eta$, it can be shown that $\Gamma(\eta, \mu)$ must contain at least one point.

An advantage of (4.7) as an acceptance criterion is that its interpretation in terms of a local minimum suggests efficient methods for computing a good value of α_k. In particular, safeguarded polynomial fitting techniques for univariate minimization converge very rapidly on well-behaved functions. Furthermore, if μ is chosen as a small value (say, $\mu = 10^{-4}$), any α that satisfies (4.7) almost always satisfies (4.8).

An effective class of methods for computing α_k are based upon computing a sequence of intervals $\{I_j\}$, each containing points of $\Gamma(\eta, \mu)$, such that the interval I_j lies wholly within I_{j-1}. The sequence of intervals can be computed using the methods of safeguarded polynomial interpolation described in Section 4.1.2. The first point generated by the polynomial interpolation that lies in $\Gamma(\eta, \mu)$ qualifies for α_k.

To initiate computation of the nested set of intervals, the required step length is assumed to lie in an interval of uncertainty such that

$$0 < \delta \leq \|\alpha p_k\| \leq \Delta, \tag{4.9}$$

where δ defines the *minimum distance* allowed between x_{k+1} and x_k, and Δ is an upper bound on the change in x; in general, δ and Δ depend on x_k and on F. The scalar δ is analogous to the minimum separation between iterates in a safeguarded zero-finding method (see Section 4.1.1.5), and its value should reflect the accuracy to which F can be computed (see Section 8.5· for further details). In practice, Δ is a bound upon the step length enforced by the presence of linear inequality constraints (see Section 5.2.1.2), or a user-provided estimate of the distance from the starting point to the minimum (see Section 8.1.4).

Like algorithms based upon the Goldstein-Armijo principle, safeguarded step-length algorithms require a step $\alpha^{(0)}$ to initiate the interpolation, and fewer function evaluations are usually required if a good *a priori* value for $\alpha^{(0)}$ is known. We shall show in Section 4.4.2.1 that there are occasions on which it is *impossible* to provide a meaningful estimate of $\alpha^{(0)}$. In this case, the ability to impose the upper bound in (4.9) on $\|\alpha p_k\|$ becomes essential if unnecessary evaluations of F during the line search are to be avoided.

In the preceding discussion, it has been assumed that the gradient of F will be evaluated at every trial point during the calculation of α_k. However, if $g(x)$ is relatively expensive to compute, it is more efficient to use a criterion different from (4.7) for accepting the step length, and to choose a step-length algorithm that requires only function values at the trial points. The following criterion can be used instead of (4.7) to ensure that α_k is not too small:

$$\frac{|F(x_k + \alpha p_k) - F(x_k + \nu p_k)|}{\alpha - \nu} \leq -\eta g_k^T p_k, \tag{4.10}$$

where ν is any scalar such that $0 \leq \nu < \alpha$. Criterion (4.10) can be combined with (4.8) to guarantee that F sustains a sufficient decrease at every iteration.

4.3.2.2. Computing the direction of search. The remaining question in the model algorithm is how to choose a "good" direction p_k. In contrast to the univariate case, where the only possible "moves" are in the positive or negative directions, even in two dimensions there are an infinite number of possible choices. The remainder of this chapter will be concerned with different methods (and the corresponding motivations) for specifying p_k. It is customary to name an algorithm of the model form according to the procedure by which it computes the search direction, since the determination of the step length is usually viewed as a separate procedure.

Consider the linear approximation to F based on the Taylor-series expansion about x_k:

$$F(x_k + p) \approx F_k + g_k^T p.$$

Assuming that a step of unity will be taken along p, it would appear that a good way to achieve a large reduction in F is to choose p so that $g_k^T p$ is large and negative. Obviously some normalization must be imposed on p; otherwise, for any $\bar{p}$ such that $g_k^T \bar{p} < 0$, one could simply choose p as an arbitrarily large positive multiple of $\bar{p}$. The aim is to choose p so that, amongst all suitably normalized vectors, $g_k^T p$ is a minimum. Given some norm $\|\cdot\|$, p_k is thus the solution of the minimization problem

$$\underset{p \in \Re^n}{\text{minimize}} \ \frac{g_k^T p}{\|p\|}. \tag{4.11}$$

The solution of the problem (4.11) depends on the choice of norm. When the norm is defined by a given symmetric positive-definite matrix, say C, i.e.

$$\|p\|_C = (p^T C p)^{\frac{1}{2}},$$

the solution of (4.11) is

$$p_k = -C^{-1}g_k. \tag{4.12}$$

If the two-norm is used, i.e. $\|p\| = (p^T p)^{\frac{1}{2}}$, the solution is just the negative gradient

$$p_k = -g_k. \tag{4.13}$$

Because it solves the problem (4.11) with respect to the two-norm, the negative gradient (4.13) is termed the *direction of steepest descent*. When p_k is always taken as (4.13), the model algorithm becomes the *steepest-descent method*.

Unless the gradient vanishes, the steepest-descent direction is clearly a descent direction, since the vectors p_k and g_k are trivially bounded away from orthogonality (note that the directional derivative is such that $g_k^T p_k = -g_k^T g_k < 0$). Consequently, any of the step-length criteria mentioned earlier may be combined with the steepest-descent algorithm to yield a method with guaranteed convergence.

Unfortunately, a proof of global convergence for an algorithm does *not* ensure that it is an efficient method. This principle is illustrated very clearly by considering the *rate* of convergence of the steepest-descent method. The usual first step in a rate-of-convergence analysis is to examine the behaviour of a method on a quadratic function. The special properties of a quadratic result in a simplified analysis; furthermore, some general properties of a method can usually be deduced from its performance on a quadratic, since every smooth function behaves like a quadratic in a sufficiently small region (see Section 3.2.3).

Let $F(x)$ be the quadratic function $c^T x + \frac{1}{2} x^T G x$, where c is a constant vector and G is a symmetric positive-definite matrix. If the steepest-descent method is applied to F, using an exact line search to determine the step length, its rate of convergence is *linear*. If $\lambda_{\max}$ and $\lambda_{\min}$ are the largest and smallest eigenvalues of G, then it can be shown that

$$F(x_{k+1}) - F(x^*) \approx \frac{(\lambda_{\max} - \lambda_{\min})^2}{(\lambda_{\max} + \lambda_{\min})^2} \left(F(x_k) - F(x^*) \right)$$

$$= \frac{(\kappa - 1)^2}{(\kappa + 1)^2} \left(F(x_k) - F(x^*) \right),$$

where κ denotes $\mathrm{cond}(G)$, the spectral condition number of G.

The striking feature of this result is that the asymptotic error constant, which gives the factor of reduction in the error at each step, can be *arbitrarily close to unity*. For example, if $\kappa = 100$ (so that G is not particularly ill-conditioned), the error constant is $(99/101)^2 \approx .96$, so that there is a very small gain in accuracy at each iteration. In practice, the steepest-descent method typically requires hundreds of iterations to make very little progress towards the solution. A similar poor result holds for the rate of convergence of the steepest-descent method on a general function.

Figure 4j depicts the solution path of an implementation of the steepest-descent method when applied to Rosenbrock's function (Example 4.2). The curved lines correspond to lines of equal function value; the linear segments correspond to the step taken within a given iteration. An accurate line search was performed at each iteration. The algorithm was started at the point $(-1.2, 1.0)^T$. Note that the algorithm would have failed in the vicinity of the point $(-0.3, 0.1)^T$ but for the fact that the linear search found, by chance, the *second* minimum along the search direction. Several hundred iterations were performed close to the new point without any perceptible change in the objective function. The algorithm was terminated after 1000 iterations.

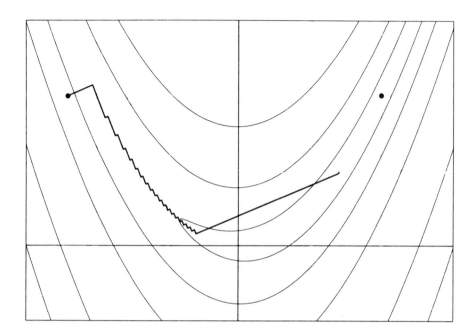

Figure 4j. Solution path of a steepest-descent algorithm on Rosenbrock's function. The linear segments correspond to the step taken within a given iteration. Note that the algorithm would have failed in the vicinity of the point $(-0.3, 0.1)^T$ but for the fact that the linear search found, by chance, the *second* minimum along the search direction. Several hundred iterations were performed close to the new point without any perceptible change in the objective function. The algorithm was terminated after 1000 iterations.

This example, which is only mildly ill-conditioned, illustrates that the existence of a convergence proof for the mathematical definition of an algorithm does not imply that an implementation of the algorithm will converge in an acceptable number of iterations.

We emphasize that a linear rate of convergence does not in itself condemn an algorithm; unfortunately, it is frequently true of linearly convergent methods that progress becomes arbitrarily slow. The remainder of this chapter will be concerned with methods designed to display a *superlinear* rate of convergence.

Notes and Selected Bibliography for Section 4.3

For a strict mathematical definition of the term "sufficient decrease" and many fundamental theorems concerning convergence, see Ortega and Rheinboldt (1970). For a detailed discussion of step-length algorithms based upon safeguarded polynomial interpolation see Gill and Murray (1974e). Algorithms based upon the condition (4.6) have been proposed by Goldstein and Price (1967) and Fletcher (1970a). Wolfe (1969) and Powell (1976a) use the condition

$$g(x_k + \alpha p_k)^T p_k \geq \gamma g_k^T p_k, \tag{4.14}$$

for some γ, $0 < \gamma < 1$, to ensure that α_k is not too small. Shanno and Phua (1976) give a line search based on the conditions (4.14) and (4.8) for $\eta > \mu$.

For certain algorithms (such as the conjugate-gradient algorithm of Section 4.8.3.3), it may be appropriate for η to be less than μ. In this case, difficulties may arise with step-length algorithms designed to satisfy *pairs* of conditions such as (4.7) (or (4.10)) and (4.8). For example, the set $\Gamma(\eta, \mu)$ may be empty. Although this circumstance is rare, it occurs in a non-trivial number of instances, particularly when $\|g_k\|$ is large during early iterations or when F is badly scaled (for example, a penalty or barrier function, which are discussed in Section 6.2.1). A complete discussion of the general case in which η may take *any* value in $[0, 1)$ is beyond the scope of this book. However, we note that it is always possible to compute a step length that gives a sufficient decrease in $F(x)$ and satisfies (4.7) or (4.10) for *some* value of η. For further details the reader is referred to Gill, Murray, Saunders and Wright (1979).

The use of the steepest-descent direction as a basis of a multivariate minimization algorithm dates back to Cauchy (1847), and has been the subject of extensive analysis. For details concerning the rate of convergence of the steepest-descent method, see, for example, Kantorovich and Akilov (1964) and Luenberger (1973).

A large class of methods for unconstrained optimization are based on a slightly different model algorithm. In particular, the step length α_k is nearly always taken as unity, so that the new iterate is defined by $x_{k+1} = x_k + p_k$. In order to ensure that the descent condition holds, it may thus be necessary to compute *several* trial vectors before finding a satisfactory p_k. Methods of this type are often termed *trust-region methods* (in contrast to *step-length-based methods*). We shall give a brief introduction to trust-region methods in the Notes for Section 4.4. A trust-region method for the nonlinear least-squares problem will also be discussed in Section 4.7.3.

4.4. SECOND DERIVATIVE METHODS

4.4.1. Newton's Method

The algorithms to be discussed in this section are the first of several classes of methods based on a *quadratic model* of F, in contrast to the scaled *linear* model function of the classical steepest-descent method. There are two major justifications for choosing a quadratic model: its simplicity and, more importantly, the success and efficiency in practice of methods based on it.

If first and second derivatives of F are available, a quadratic model of the objective function can be obtained by taking the first three terms of the Taylor-series expansion about the current point, i.e.

$$F(x_k + p) \approx F_k + g_k^T p + \frac{1}{2} p^T G_k p. \tag{4.15}$$

Within the context of the model algorithm, it is helpful to formulate the quadratic function (4.15) in terms of p (the step to the minimum) rather than the predicted minimum itself. The minimum of the right-hand side of (4.15) will be achieved if p_k is a minimum of the quadratic function

$$\Phi(p) = g_k^T p + \frac{1}{2} p^T G_k p. \tag{4.16}$$

Recall from Section 3.2.3 that a stationary point p_k of (4.16) satisfies the linear system

$$G_k p_k = -g_k. \tag{4.17}$$

A minimization algorithm in which p_k is defined by (4.17) is termed *Newton's method*, and the solution of (4.17) is called the *Newton direction*.

If G_k is positive definite, only one iteration is required to reach the minimum of the model function (4.15) from *any* starting point (note that the step length α_k is unity). Therefore, we expect good convergence from Newton's method when the quadratic model (4.15) is accurate. For a general nonlinear function F, Newton's method converges *quadratically* to x^* if x_0 is sufficiently close to x^*, the Hessian matrix is positive definite at x^*, and the step lengths $\{\alpha_k\}$ converge to unity.

The local convergence properties of Newton's method make it an exceptionally attractive algorithm for unconstrained minimization. A further benefit of the availability of second derivatives is that the sufficient conditions for optimality can be verified (see Section 3.2.2). In fact, Newton's method is often regarded as the standard against which other algorithms are measured, and much effort is devoted to the attempt to devise algorithms that can approach the performance of Newton's method. However, as in the univariate case, difficulties and even failure may occur because the quadratic model is a poor approximation to F outside a small neighbourhood of the current point.

Consider the case when G_k is positive definite, so that the quadratic model (4.15) has a unique minimum. In this case, the solution of (4.17) is guaranteed to be a descent direction, since $g_k^T p_k = -g_k^T G_k^{-1} g_k < 0$. Moreover, if the condition number of G_k is bounded by a constant that is independent of x, the solution of (4.17) is bounded away from orthogonality to the negative gradient. Hence, when G_k is positive definite with uniformly bounded condition number for all k, a globally convergent algorithm can be developed by taking the Newton direction as p_k, and choosing α_k to satisfy one of the sets of criteria discussed in Section 4.3.2.1. (In older texts, a Newton method that includes a step-length procedure is sometimes called a "damped" Newton method, but this terminology is becoming obsolete.) A step-length procedure must be included because a step of unity along the Newton direction will not necessarily reduce F, even though it is the step to the minimum of the model function. However, we emphasize that Newton's method will converge quadratically only if the step lengths converge sufficiently fast to the "natural" value of unity. It is interesting that the Newton direction can be regarded as a direction of "steepest descent" if the norm in (4.11) is defined as $\|p\| = (p^T G_k p)^{\frac{1}{2}}$ (cf. (4.12)).

If G_k is not positive definite, the quadratic model function need not have a minimum, nor even a stationary point. Recall from Section 3.2.3 that $\Phi(p)$ is unbounded below if G_k is indefinite, and that a unique stationary point will exist only if G_k is non-singular. When G_k is singular, there will be a stationary point only if g_k lies in the range of the columns of G_k.

There is no universally accepted definition of "Newton's method" when G_k is indefinite, because researchers do not agree about the interpretation of the local quadratic model with respect to choosing a search direction. For example, since the local quadratic model function is unbounded along certain directions, it could be argued that the "best" search direction would be some combination of these vectors. Numerous strategies have been developed to produce an efficient descent method for the indefinite case; a method that does not use the Newton direction under all circumstances will be termed a *modified Newton method*.

An unavoidable difficulty when G_k is indefinite is that there is a conflict between the quadratic model and the original nonlinear function, since the model function indicates that an *infinite* step should be taken from x_k. Thus, *there is no "natural" scaling in the indefinite case*. One can view the problem of scaling in terms of the search direction or the step length. However, the effect is the same in either case, and there is no obvious strategy for scaling the search direction or selecting the initial trial step length. Consequently, many function evaluations may be required in order to ensure that F sustains a sufficient decrease at x_{k+1}. *Any method for treating an indefinite Hessian must include some explicit or implicit procedure for scaling the step.*

4.4.2. Strategies for an Indefinite Hessian

One popular strategy in modified Newton methods is to construct a "related" positive-definite matrix $\bar{G}_k$ when G_k is indefinite. The search direction is then given by the solution of

$$\bar{G}_k p_k = -g_k. \tag{4.18}$$

Because $\bar{G}_k$ is positive definite, the solution of (4.18) is guaranteed to be a descent direction. Furthermore, when the original Hessian is positive definite, $\bar{G}_k$ is simply taken as G_k.

The formula (4.18) is not applicable in the case when x_k is a saddle point, where g_k vanishes and G_k is indefinite. In this eventuality, p_k is taken as a *direction of negative curvature*, i.e. a vector that satisfies

$$p_k^T G_k p_k < 0.$$

When G_k is indefinite, such a direction must exist, and F can be decreased by moving along it. A direction of negative curvature is also an advisable choice for p_k when $\|g_k\|$ is non-zero, but pathologically small.

If the condition number of $\bar{G}_k$ is uniformly bounded, the choice of search direction (4.18) can be combined with one of the step-length procedures given in Section 4.3.2.1 to yield a globally convergent modified Newton algorithm.

It is important to remember that the search direction in a modified Newton algorithm is obtained by solving the linear system (4.18). Furthermore, we have implicitly assumed that somehow it will be possible to determine whether G_k is positive definite (which is *not* obvious except in very special cases). All the modified Newton methods to be discussed are based on some form of *matrix factorization* that reveals whether G_k is positive definite, and can be adjusted to produce $\bar{G}_k$.

4.4.2.1. Methods based on the spectral decomposition.

When the eigensystem of G_k is available (see Section 2.2.5.4), it may be used to construct a related positive-definite matrix. Let

$$G_k = U \Lambda U^T = \sum_{i=1}^n \lambda_i u_i u_i^T, \tag{4.19}$$

where Λ is diagonal and U is orthonormal. The form (4.19) indicates that the transformation G_k involves two complementary subspaces. The matrix U_+ will denote the matrix whose columns are the eigenvectors $\{u_i\}$ from (4.19) that correspond to "sufficently positive" eigenvalues (with $\mathcal{U}_+$ the associated subspace); U_- will denote the matrix whose columns consist of the remaining eigenvectors (with an analogous meaning for $\mathcal{U}_-$). Note that any positive linear combination of columns of U_- corresponding to *negative* eigenvalues of G_k is a direction of negative curvature.

When constructing $\bar{G}_k$, one possible strategy is to retain the original structure of G_k *within* $\mathcal{U}_+$; this ensures that $\bar{G}_k = G_k$ when G_k is sufficiently positive definite. In order to achieve this aim, we define $\bar{G}_k$ as

$$\bar{G}_k = U \bar{\Lambda} U^T, \tag{4.20}$$

where $\bar{\Lambda}$ is a diagonal matrix with $\bar{\lambda}_i$ equal to λ_i if λ_i is sufficiently positive. The elements of $\bar{\Lambda}$ that correspond to columns of U_- can be chosen in various ways, depending on the properties that are desired for $\bar{G}_k$. Note that using a positive value of $\bar{\lambda}_i$ to replace a negative value of λ_i necessarily causes the effect of $\bar{G}_k$ to be "opposite" to that of G_k in the associated portion of $\mathcal{U}_-$.

To illustrate two of the possibilities, consider the following example with a diagonal Hessian matrix.

Example 4.6. Let the Hessian matrix be

$$G_k = \begin{pmatrix} 1 & 0 \\ 0 & -5 \end{pmatrix}, \quad \text{with gradient} \quad g_k = \begin{pmatrix} 2 \\ 5 \end{pmatrix}.$$

The Hessian matrix has (trivially) one negative and one positive eigenvalue, with columns of the identity as eigenvectors. The unmodified Newton direction, defined by the solution of (4.17), is given by $(-2, 1)^T$ which is a saddle point of the model function (4.15), but is *not* a descent direction.

Firstly, it might seem reasonable to choose $\bar{G}_k$ as the "closest" sufficiently positive-definite matrix to G_k. With this strategy, any λ_i less than some positive quantity δ that defines "positive" is replaced by $\bar{\lambda}_i = \delta$. For Example 4.6, this gives

$$\bar{G}_k = \begin{pmatrix} 1 & 0 \\ 0 & \delta \end{pmatrix}.$$

With this choice of $\bar{G}_k$ in (4.18), the direction of search is $(-2, -5/\delta)^T$. It is a general feature of this approach that the search direction is dominated by vectors in $\mathcal{U}_-$, and the norm of the search direction will usually be large. In effect, since the quadratic model is unbounded below, it can be "minimized" by an infinite step along *any* direction of negative curvature, and the positive-definite portion of the Hessian can be ignored. The size of the norm of p_k does not necessarily pose computational difficulties; for example, p_k can be re-scaled or an upper bound can be imposed on the step length in the linear search, as in (4.9).

A second approach to defining $\bar{G}_k$ is to *reverse* the portion of G_k in $\mathcal{U}_-$ (still leaving the influence of $\mathcal{U}_+$ unchanged). With this definition, the search direction includes the negative of the step to the maximum in $\mathcal{U}_-$. For Example 4.6, this strategy would give the modified matrix

$$\bar{G}_k = \begin{pmatrix} 1 & 0 \\ 0 & 5 \end{pmatrix},$$

and the norm of p_k from (4.18) is identical to that of the unmodified direction.

The matrix $\bar{G}_k$ defined by (4.20) is positive definite, and has a bounded condition number if the largest eigenvalue of G_k is bounded. For an indefinite G_k, the modified matrix satisfies $\|\bar{G}_k - G_k\| \le 2|\lambda_{\min}|$, where $\lambda_{\min}$ is the most negative eigenvalue of G_k. The search direction computed from (4.18) and (4.20) may be a direction of either positive or negative curvature.

Computation of the complete eigensystem of G_k usually requires between $2n^3$ and $4n^3$ arithmetic operations. Hence, methods based on this approach have been superseded by more efficient methods (such as the method discussed at the end of the next section) that approximate vectors in $\mathcal{U}_+$ and $\mathcal{U}_-$ without the expense of computing the spectral decomposition.

*4.4.2.2. Methods based on the Cholesky factorization. Any symmetric positive-definite matrix may be written as a Cholesky factorization LDL^T, where L is unit lower-triangular and D is a positive diagonal matrix (see Section 2.2.5.2). The j-th column of the matrix L is defined from columns 1 through $j - 1$ by the equations

$$d_j = g_{jj} - \sum_{s=1}^{j-1} d_s l_{js}^2, \tag{4.21a}$$

$$l_{ij} = \frac{1}{d_j}\left(g_{ij} - \sum_{s=1}^{j-1} d_s l_{js} l_{is} \right). \tag{4.21b}$$

Similar formulae exist for the row-wise computation.

By analogy with the modification of the eigensystem discussed in Section 4.4.2.1, it might seem reasonable to create a modified Newton method as follows: form the Cholesky factorization of G_k, and then define $\bar{G}_k$ as $L\bar{D}L^T$, where $\bar{d}_i = \max\{|d_i|, \delta\}$. However, this approach has two major defects. Firstly, the Cholesky factorization of a symmetric indefinite matrix may not exist (see the example in Section 2.2.5.2). Secondly, if a Cholesky factorization of an indefinite matrix does exist, its computation is in general a numerically unstable process, since the elements of the factors may be unbounded, even if G_k is well-conditioned. Furthermore, $\bar{G}_k$ can differ from G_k by an arbitrarily large amount when G_k is only "slightly" indefinite.

Example 4.7. Consider the matrix

$$G_k = \begin{pmatrix} 1 & 1 & 2 \\ 1 & 1+10^{-20} & 3 \\ 2 & 3 & 1 \end{pmatrix},$$

whose eigenvalues are approximately 5.1131, -2.2019 and .0888 (the quantity 10^{-20} is added to the diagonal in order to ensure that the Cholesky factors are bounded; it does not affect the smallest eigenvalue to this precision).

The exact Cholesky factors of Example 4.7 are

$$L = \begin{pmatrix} 1 & 0 & 0 \\ 1 & 1 & 0 \\ 2 & 10^{20} & 1 \end{pmatrix} \quad \text{and} \quad D = \begin{pmatrix} 1 & 0 & 0 \\ 0 & 10^{-20} & 0 \\ 0 & 0 & -(3+10^{20}) \end{pmatrix},$$

so that the approach just described results in $\bar{G}_k$ such that $\|G_k - \bar{G}_k\|_F$ is of order 10^{20}. By contrast, $\|G_k - \bar{G}_k\|_F \approx 4.4038$ using the second eigensystem approach described in Section 4.4.2.1.

An alternative, numerically stable method is to construct $\bar{G}_k$ from a *modified Cholesky factorization* of G_k. With this approach, Cholesky factors L and D are computed, subject to two requirements: all elements of D are strictly positive, and the elements of the factors satisfy a uniform bound, i.e. for $k = 1, \ldots, n$ and some positive value β it holds that

$$d_k > \delta \quad \text{and} \quad |r_{ik}| \leq \beta, \quad i > k, \tag{4.22}$$

where the auxiliary quantities r_{ik} are defined by $l_{ik}\sqrt{d_k}$, and have been introduced for convenience of exposition (the choice of β will be discussed later in this section). The matrices L and D are computed by implicitly increasing the diagonal elements of the original matrix during the factorization in order to satisfy (4.22).

We shall describe the j-th step of the factorization. Assume that the first $j-1$ columns of the modified Cholesky factorization have been computed, and that (4.22) holds for $k = 1, \ldots, j-1$. First, we compute

$$\gamma_j = |\xi_j - \sum_{s=1}^{j-1} d_s l_{js}^2|, \tag{4.23}$$

where ξ_j is taken as g_{jj}. The trial value $\bar{d}$ is given by

$$\bar{d} = \max\{\gamma_j, \delta\},$$

where δ is a small positive quantity. To test whether $\bar{d}$ is acceptable as the j-th element of D, we test whether the values of r_{ij} computed from (4.21b) (with d_j taken as $\bar{d}$) would satisfy (4.22). If so, d_j is set to $\bar{d}$, and the elements of the j-th column of L are obtained from r_{ij}. However, if any r_{ij} is greater than β, d_j is given by formula (4.23), with ξ_j replaced by $g_{jj} + e_{jj}$, where the positive scalar e_{jj} is chosen so that the maximum $|r_{ij}|$ is equal to β.

When this process is completed, the matrices L and D are the Cholesky factors of a positive-definite matrix $\bar{G}_k$ that satisfies

$$\bar{G}_k = LDL^T = G_k + E,$$

where E is a non-negative diagonal matrix whose j-th element is e_{jj}. Thus, the positive-definite matrix $\bar{G}_k$ can differ from the original Hessian only in its diagonal elements.

For a given G_k, the diagonal correction E clearly depends on β. It is desirable for β to be large enough so that G_k will not be modified unnecessarily. When G_k is positive definite, (4.21a) implies that, for $i = 1, \ldots, n$ and each j ($j \le i$), it holds that $l_{ij}^2 d_j \le g_{ii}$. Thus, β should satisfy $\beta^2 \ge \gamma$, where γ is the largest in magnitude of the diagonal elements of G_k, to ensure that E will be identically zero if G_k is sufficiently positive definite.

An upper bound on β is also imposed to preserve numerical stability and prevent excessively large elements in the factors; allowing an infinite value of β is equivalent to choosing d_j as the modulus of the j-th element of D in the unmodified LDL^T factors (when they exist). If $n > 1$, it can be shown that

$$\|E(\beta)\|_\infty \le \left(\frac{\xi}{\beta} + (n-1)\beta\right)^2 + 2\left(\gamma + (n-1)\beta^2\right) + \delta,$$

where ξ is the largest in modulus of the off-diagonal elements of G_k. This bound is minimized when $\beta^2 = \xi/\sqrt{n^2 - 1}$. Thus, β is chosen to satisfy

$$\beta^2 = \max\{\gamma, \xi/\sqrt{n^2 - 1}, \epsilon_M\},$$

where the machine precision ϵ_M is included in order to allow for the case where $\|G_k\|$ is small.

The modified Cholesky factorization is a numerically stable method that produces a positive-definite matrix differing from the original matrix only in its diagonal elements. The diagonal modification is *optimal* in the sense that an *a priori* bound upon the norm of E is minimized, subject to the requirement that sufficiently positive-definite matrices are left unaltered. In practice, the actual value of E is almost always substantially less than the *a priori* bound.

Moreover, the norm of the modification may be reduced further if symmetric interchanges are used. At the j-th step of the factorization, the obvious interchange strategy is to choose the j-th row and column as those that yield the largest value of γ_j (4.23). In this case, the modified Cholesky factorization satisfies

$$P^T G_k P + E = LDL^T,$$

where P is a permutation matrix. The modified Cholesky factorization with interchanges has the additional property that the modification is invariant to the order of the variables.

We shall summarize the modified Cholesky factorization with interchanges by presenting a step-by-step description of the algorithm. No additional storage is required beyond that required to store G_k, since the triangular factors overwrite the relevant parts of G_k as they are computed. During the computation of the j-th column of L, the algorithm utilizes the auxiliary quantities $c_{is} = l_{is}d_s$ for $s = 1, \ldots, j$, $i = j, \ldots, n$. These numbers may be stored in the array G_k until they are overwritten by the appropriate elements of L.

Algorithm MC (*The Modified Cholesky Factorization*).

MC1. [Compute the bound on the elements of the factors.] Set $\beta^2 = \max\{\gamma, \xi/\nu, \epsilon_M\}$, where $\nu = \max\{1, \sqrt{n^2-1}\}$, and γ and ξ are the maximum magnitudes of the diagonal and off-diagonal elements of G_k.

MC2. [Initialize.] Set the column index j to 1. Define $c_{ii} = g_{ii}$, $i = 1, \dots, n$.

MC3. [Find the maximum prospective diagonal and perform row and column interchanges.] Find the smallest index q such that $|c_{qq}| = \max_{j \le i \le n} |c_{ii}|$. Interchange all information corresponding to rows and columns q and j of G_k.

MC4. [Compute the j-th row of L. Find the maximum modulus of $l_{ij} d_j$.] Set $l_{js} = c_{js}/d_s$ for $s = 1, \dots, j-1$. Compute the quantities $c_{ij} = g_{ij} - \sum_{s=1}^{j-1} l_{js} c_{is}$ for $i = j+1, \dots, n$ and set $\theta_j = \max_{j+1 \le i \le n} |c_{ij}|$ (if $j = n$, define $\theta_j = 0$).

MC5. [Compute the j-th diagonal element of D.] Define $d_j = \max\{\delta, |c_{jj}|, \theta_j^2/\beta^2\}$ and the diagonal modification $E_j = d_j - c_{jj}$. If $j = n$, exit.

MC6. [Update the prospective diagonal elements and the column index.] Set $c_{ii} = c_{ii} - c_{ij}^2/d_j$, for $i = j+1, \dots, n$. Set j to $j+1$ and go to step MC3. ∎

The modified Cholesky factorization can be computed in approximately $\frac{1}{6}n^3$ arithmetic operations (about the same number as the unmodified factorization in the positive-definite case). We illustrate the modified Cholesky factorization with six-digit arithmetic on the matrix of Example 4.7, where all results are given to four significant figures. The value of β^2 is 1.061 and the factors are

$$L = \begin{pmatrix} 1 & 0 & 0 \\ .2652 & 1 & 0 \\ .5303 & .4295 & 1 \end{pmatrix}, \quad D = \begin{pmatrix} 3.771 & 0 & 0 \\ 0 & 5.750 & 0 \\ 0 & 0 & 1.121 \end{pmatrix} \quad \text{and} \quad E = \begin{pmatrix} 2.771 & 0 & 0 \\ 0 & 5.016 & 0 \\ 0 & 0 & 2.243 \end{pmatrix},$$

giving $\|G_k - \bar{G}_k\|_F = \|E\|_F \approx 6.154$.

A direction of negative curvature can also be computed from the modified Cholesky factorization. Let s be the index of the smallest of the $n - j$ quantities c_{jj} at step MC3 of the modified Cholesky factorization. When G_k is indefinite, the value of c_{ss} is negative, and the vector p defined by

$$L^T p = e_s$$

is a direction of negative curvature. With the matrix of Example 4.7, $s = 3$ and the corresponding vector p satisfies

$$\frac{p^T G_k p}{\|p\|_2^2} \approx -1.861.$$

Figure 4k illustrates the behaviour of a second-derivative method based upon the modified Cholesky factorization applied to Rosenbrock's function (Example 4.2). Note that, except for the first iteration, the method follows the base of the valley in an almost "optimal" number of steps, given that piecewise linear segments are used.

Notes and Selected Bibliography for Section 4.4

Many of the early modified Newton methods were not numerically stable (see Murray, 1972a, for a detailed discussion of these methods). The eigenvector-eigenvalue approach of Section 4.4.2.1 was suggested by Greenstadt (1967). The modified Cholesky factorization is due to Gill and Murray (1974a).

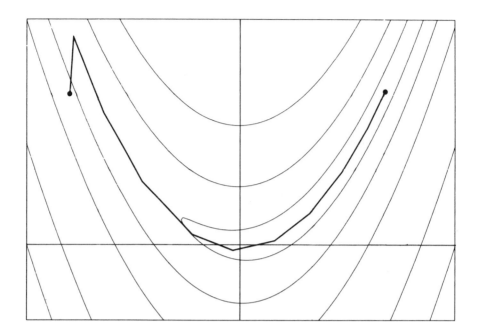

Figure 4k. Solution path of a modified Newton algorithm on Rosenbrock's function. Except for the first iteration, the method follows the base of the valley in an almost "optimal" number of steps, given that piecewise linear segments are used.

It is often thought that modified Newton methods perform well only close to the solution. This belief has caused many "hybrid" algorithms to appear in which a different method (usually steepest descent) is used until the iterates are close to the solution. It is our experience that there is little justification for a hybrid approach, since a carefully implemented modified Newton method will make good progress at points remote from the solution. As an illustration of this observation, note that the steepest-descent method failed on Example 4.2 (see Figure 4j) when the iterates were far from the solution, whereas the Newton method worked satisfactorily (see Figure 4k).

A modified Newton algorithm can be based on the symmetric *indefinite* factorization of Bunch and Parlett (1971). For any symmetric matrix G_k, we can write

$$P^T G_k P = LBL^T,$$

where P is a permutation matrix, L is a unit lower-triangular matrix and B is *block-diagonal*, with blocks of order one or two. Symmetric interchanges are performed to preserve symmetry and numerical stability.

A benefit of the symmetric indefinite factorization is that it allows one to determine the *inertia* of G_k (the inertia is a triple of integers, giving, respectively, the number of positive eigenvalues, the number of negative eigenvalues, and the number of zero eigenvalues). The inertia of B is the same as that of G_k (although the eigenvalues themselves are not the same). The 2×2 blocks in B are always constructed to have one positive eigenvalue and one negative eigenvalue, so that the number of positive eigenvalues of G_k is equal to the number of 1×1 positive blocks plus the number of 2×2 blocks.

More and Sorensen (1979) have suggested that the symmetric indefinite factorization be used to define $\bar{G}_k$. Given the spectral decomposition of B ($B = U\Lambda U^T$), a matrix $\bar{B}$ can be defined as in the algorithm of Section 4.4.2.1; thus $\bar{B} = U\bar{\Lambda}U^T$, where $\bar{\lambda}_i = \max\{|\lambda_i|, \delta\}$, and

$$P^T \bar{G}_k P = LU\bar{\Lambda}U^T L^T.$$

With this approach, a positive-definite G_k is unaltered. The symmetric indefinite factorization may be computed in about $\frac{1}{6}n^3$ arithmetic operations and $O(n^3)$ comparisons. A similar factorization that requires $O(n^2)$ comparisons has been suggested by Fletcher (1976) and Bunch and Kaufman (1977).

Any numerically stable algorithm for factorizing a symmetric indefinite matrix can be used as the basis for a modified Newton algorithm. For other possible factorizations, see Aasen (1971) and Kaniel and Dax (1979).

If G_k is indefinite, there are many ways of choosing the weight to give to directions of negative curvature. In the text we have described one of the simplest options: set p_k to be the solution of (4.18) everywhere that the norm of the gradient is large relative to the magnitude of the objective function; otherwise, use a direction of negative curvature. In this case a direction of negative curvature will be computed only near a saddle point. More generally, the direction of search may be defined as

$$p_k = \phi_1 p_1 + \phi_2 p_2,$$

where p_1 is the solution of (4.18), p_2 is a direction of negative curvature and ϕ_1 and ϕ_2 are non-negative scalars. We adopt the convention that ϕ_2 is zero and ϕ_1 is unity if G_k is positive definite (since no direction of negative curvature exists in this case). Fletcher and Freeman (1977) propose that a zero value of ϕ_1 should be used as often as possible; Graham (1976) suggests a weighted linear combination based upon the magnitude of the correction to the Hessian. More general schemes that define ϕ_1 and ϕ_2 as functions of a step length α_k have been suggested by McCormick (1977), More and Sorensen (1979) and Goldfarb (1980).

We mentioned briefly in the Notes for Section 4.3 that nonlinear optimization methods can be categorized into two broad classes: "step-length-based methods" and "trust-region methods". Although we consider mainly step-length-based methods in this book, it is appropriate at this point to discuss the basic features of techniques based on the model trust region. The idea of the model trust-region approach is to accept the minimum of the quadratic model only as long as the quadratic model adequately reflects the behaviour of F. Usually, the decision as to whether the model is acceptable is based on the norm of the computed search direction.

The most common mathematical formulation of this idea provides x_{k+1} as $x_k + p_k$, where p_k is the solution of the constrained subproblem:

$$\begin{aligned} &\underset{p \in \Re^n}{\text{minimize}} && g_k^T p + \frac{1}{2}p^T G_k p \\ &\text{subject to} && \|p\|_2 \leq \Delta, \end{aligned} \qquad (4.24)$$

for some Δ. It can be shown that, if λ is a scalar such that the matrix $G_k + \lambda I$ is positive semi-definite, the solution of the equations

$$(G_k + \lambda I)p = -g_k \qquad (4.25)$$

solves the subproblem (4.24) if either $\lambda = 0$ and $\|p\|_2 \leq \Delta$, or $\lambda \geq 0$ and $\|p\|_2 = \Delta$. Thus, if Δ is large enough, the solution of (4.24) is simply the Newton direction (i.e., the solution of (4.25) with $\lambda = 0$). Otherwise, the restriction on the norm will apply, and $\|p\|_2 = \Delta$. The search direction is typically found by solving (4.24) for trial values of Δ and evaluating F at the resulting trial points. A vector p such that $F(x_k + p)$ is sufficiently less than F_k must exist for small enough Δ since the second-order term of the model function may be made small compared to the first-order term. As $\Delta \to 0$, $\|p\|_2 \to 0$ and p becomes parallel to the steepest-descent direction.

The idea of defining a region of trust for the search direction was suggested for nonlinear least-squares problems by Levenberg (1944) and Marquardt (1963) (see Section 4.7.3). The application of the technique to general nonlinear problems was considered by Goldfeld, Quandt and Trotter (1966).

At the k-th iteration, some trust-region methods begin by attempting to solve (4.25) with $\lambda = 0$. If G_k is indefinite or $\|p\| > \Delta$ the algorithm proceeds to find an approximate solution of the nonlinear equation

$$\phi(\lambda) \equiv \|(G_k + \lambda I)^{-1} g_k\|_2 = \Delta. \qquad (4.26)$$

If G_k is replaced by its spectral decomposition, ϕ^2 can be written in the form

$$\phi(\lambda)^2 = \sum_{i=1}^{n} \left(\frac{u_i^T g_k}{\lambda + \lambda_i} \right)^2,$$

where $\{u_j\}$ and $\{\lambda_j\}$ are the eigenvectors and eigenvalues of G_k. Clearly, ϕ^2 is a rational function with poles occurring at a subset of the eigenvalues of $-G_k$.

Hebden (1973) has suggested a method for computing a root of (4.26) using safeguarded rational approximation (see Section 4.1.1.4). Hebden avoids the computation of the spectral decomposition by computing $\phi(\lambda)$ and $d\phi/d\lambda$ using the Cholesky factorization of $G_k + \lambda I$. If the matrix $G_k + \lambda I$ is indefinite, a scalar μ is computed such that $G_k + \lambda I + \mu I$ is not indefinite. Subject to certain safeguards, the scalar $\lambda + \mu$ is used as the next estimate of the root of (4.26). On average, this method requires about two Cholesky factorizations to find an approximate root of (4.26). If G_k is indefinite, at least two factorizations are required regardless of the accuracy of λ.

Once a satisfactory solution of (4.26) and the corresponding search direction p have been found, the point $x_k + p$ is accepted as x_{k+1} if $F(x_k + p)$ is sufficiently lower than F_k (compared to the change predicted by the quadratic model). If the reduction in F is unacceptable, p is rejected and Δ is made smaller by some rule — for example, Δ can be updated based on cubic interpolation using the function and gradient at the points x_k and $x_k + p$. If the reduction in F is acceptable, Δ may be made larger using a similar scheme. For details of various schemes for updating the size of the trust region, see Fletcher (1971a, 1980), Gay (1979a), Hebden (1973), Moré (1977) and Sorensen (1980a). For more details concerning the computation of the search direction, see Gay (1979b).

It is important to note that step-length-based methods and trust-region methods have many features in common. (i) If the function is well-behaved, both types of method are designed to become equivalent to Newton's method as the solution is approached. (ii) The search direction is implicitly defined by a scalar that is adjusted according to the degree of agreement between the predicted and actual change in F. In a step-length-based method, this scalar is the step length α_k; in a trust-region method, this scalar is the size of the trust region Δ. A step-length-based algorithm computes α_k as an approximation to a "target value" (the step to the minimum along p_k). A trust-region method does not have a specific target value for Δ, but adjusts Δ according to the "quality" of previous evaluations of F. (iii) If G_k is indefinite and $\|g_k\|$ is small or zero, both methods must compute a direction of negative curvature from a factorization of the Hessian matrix or a modified version of it. (iv) If G_k becomes indefinite (making it difficult to define an *a priori* estimate of $\|x_{k+1} - x_k\|$), both methods compute x_{k+1} based on information from the positive-definite part of G_k. With a step-length-based method this is done directly, by making the appropriate modification of the Hessian. With a trust-region method, $\|x_{k+1} - x_k\|$ is implicitly determined by the size of steps that were taken when G_k was positive definite.

A good implementation of a trust-region method should include many features of a step-length-based method — and *vice versa*. For example, a step-length-based algorithm should always be implemented so that $\|x_{k+1} - x_k\|_2$ is bounded by a positive scalar Δ (see Section 4.3.2.1). Similarly, a good trust-region method will use safeguarded polynomial interpolation to adjust the scalar Δ. The similarity of the algorithms is most clearly demonstrated in the following situation. Suppose that both algorithms are started at a point x_k at which G_k is positive definite. In addition, assume that the Newton direction p is such that $\|p\|_2 \leq \Delta$ and $F(x_k + p) > F_k$. The step-length-based algorithm will compute a point $\hat{\alpha}$ ($\hat{\alpha} < 1$) by safeguarded cubic interpolation, in an attempt to achieve a sufficient decrease in F. The trust-region algorithm will also compute $\hat{\alpha}$, but will use it to define a smaller value of Δ. At this stage the algorithms differ. The step-length algorithm computes the objective function at $x_k + \hat{\alpha}p_k$, whereas the trust-region method finds a new direction of search by solving the nonlinear equation (4.26) with the smaller value of Δ.

Notwithstanding the similarities between the methods, there are important differences in how the second-order information is utilized. A step-length-based algorithm usually attempts to leave G_k unchanged in the subspace spanned by the eigenvectors with positive eigenvalues. By contrast, a trust-region method alters the effect of the transformation G_k in all directions, since the eigenvalues of $G_k + \lambda I$ are the eigenvalues of G_k shifted by λ. Thus, even if G_k is positive-definite, the trust-region search direction may be defined by a "modified" Hessian.

The two-norm is chosen to restrict the magnitude of the search direction in (4.24) so that the constrained subproblem is relatively easy to solve. However, if we are prepared to invest more effort in the computation of the trial values of p, other norms can be selected. For example, Fletcher (1972a) has proposed the *method of hypercubes* in which simple bound constraints of the form $l_i \leq p_i \leq u_i$, are imposed to limit each component of p. In this case, p must be found by solving a bound-constraint quadratic program.

4.5. FIRST DERIVATIVE METHODS

4.5.1. Discrete Newton Methods

An algorithm that is essentially equivalent in practice to Newton's method can be defined by *approximating* the Hessian matrix by finite-differences of the gradient. Such an algorithm is appropriate in many circumstances. For example, it may be impossible or difficult to compute analytic second derivatives for the given function F.

A forward-difference approximation to the i-th column of G_k is given by the vector

$$y_i = \frac{1}{h}\big(g(x_k + he_i) - g(x_k)\big),$$

where the scalar h is the *finite-difference interval*, and the i-th unit vector e_i is the *finite-difference vector*. In this section, we assume for simplicity that a single interval is used to compute all the required finite-differences. In practice, a *set* of intervals $\{h_i\}$, $i = 1, \ldots, n$ should be specified, using the techniques described in Section 8.6.

The matrix Y whose i-th column is y_i will not, in general, be symmetric, and hence the matrix used to approximate the Hessian is

$$\hat{G}_k = \frac{1}{2}(Y + Y^T).$$

A method in which $\hat{G}_k$ replaces G_k in (4.17) is termed a *discrete Newton method*. Clearly, any modification technique designed for an indefinite G_k can equally well be applied to an indefinite $\hat{G}_k$.

The selection of a finite-difference interval raises some interesting issues, which highlight certain distinctions between the theory and practice of optimization. With exact arithmetic, a discrete Newton method will achieve quadratic convergence only if h goes to zero as $\|g\|$ does. However, an arbitrarily small value of h would be disastrous numerically because of the loss of significant digits in elements of $\hat{G}_k$ due to cancellation error (see Section 2.1.5). Thus, the value of h should be small enough to give satisfactory convergence, but large enough to ensure that the finite-difference approximation retains adequate accuracy. Fortunately, the choice of h does not appear to be too critical to the success of these methods in most practical problems (in contrast to the use of finite-difference approximations to the gradient itself, which we shall consider later in Section 4.6). See Section 8.6 for a discussion of how to select the finite-difference interval.

Discrete Newton methods typically converge "quadratically" until the limiting accuracy has been attained (at which point no additional precision in the solution could be achieved even if the exact Hessian were available). A detailed discussion of achievable accuracy is given in Section 8.2.2.

The iterates of a discrete Newton algorithm applied to Rosenbrock's function (Example 4.2) are shown in Figure 4l. This figure illustrates the striking similarity between a discrete Newton method and a Newton method with exact second derivatives (*cf.* Figure 4k), since there is no discernible difference between the two figures.

Discrete Newton methods retain the substantial advantages of Newton's method — rapid local convergence and the ability to detect and move away from a saddle point. They have the disadvantage that n gradient evaluations are required to approximate the Hessian when columns of the identity are used as finite-difference vectors. Hence, for $n > 10$ or so, discrete Newton methods tend to be less "efficient" than other first-derivative methods to be discussed later. However, in Section 4.8.1 we shall see that discrete Newton methods can be extremely efficient when the Hessian has a known sparsity pattern or structure.

4.5.2. Quasi-Newton Methods

4.5.2.1. Theory. We have seen that the key to the success of Newton-type methods (the methods of Section 4.4 and 4.5.1) is the curvature information provided by the Hessian matrix, which allows a local quadratic model of F to be developed. *Quasi-Newton methods* are based on the idea of *building up* curvature information as the iterations of a descent method proceed, using the observed behaviour of F and g. Note that this is in direct contrast to Newton-type methods, where *all* the curvature information is computed at a single point. The theory of quasi-Newton methods is based on the fact that *an approximation to the curvature of a nonlinear function can be computed without explicitly forming the Hessian matrix.*

Let s_k be the step taken from x_k, and consider expanding the gradient function about x_k in a Taylor series along s_k:

$$g(x_k + s_k) = g_k + G_k s_k + \cdots$$

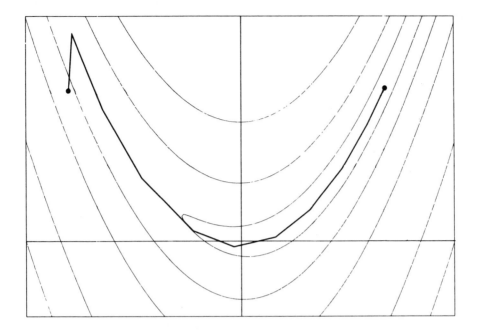

Figure 4l. Solution path of a discrete Newton algorithm on Rosenbrock's function. This figure illustrates the striking similarity between a discrete Newton method and a Newton method with exact second derivatives, since there is no discernible difference between this figure and Figure 4k.

The curvature of F along s_k is given by $s_k^T G_k s_k$, which can be approximated using only first-order information:

$$s_k^T G_k s_k \approx \big(g(x_k + s_k) - g_k\big)^T s_k. \tag{4.27}$$

This relationship would be *exact* for the quadratic model function of (4.15).

At the beginning of the k-th iteration of a quasi-Newton method, an *approximate Hessian matrix* B_k is available, which is intended to reflect the curvature information already accumulated. If B_k is taken as the Hessian matrix of a quadratic model function, the search direction p_k is the solution of a linear system analogous to (4.17):

$$B_k p_k = -g_k. \tag{4.28}$$

The initial Hessian approximation B_0 is usually taken as the identity matrix if no additional information is available. With this choice, the first iteration of a quasi-Newton method is equivalent to an iteration of the steepest-descent method.

After x_{k+1} has been computed, a new Hessian approximation B_{k+1} is obtained by *updating* B_k to take account of the newly-acquired curvature information. An *update formula* is a definition of B_{k+1} of the form

$$B_{k+1} = B_k + U_k, \tag{4.29}$$

where U_k is the *update matrix*. Let the vector s_k denote the change in x during the k-th iteration ($s_k \equiv x_{k+1} - x_k \equiv \alpha_k p_k$), and let y_k denote the change in gradient ($y_k \equiv g_{k+1} - g_k$). The standard condition required of the updated Hessian approximation is that it should approximate the curvature of F along s_k. Based on (4.27), B_{k+1} is thus required to satisfy the so-called *quasi-Newton condition*

$$B_{k+1}s_k = y_k. \tag{4.30}$$

During a single iteration, new information is obtained about the second-order behaviour of F along only one direction; thus, we would expect B_{k+1} to differ from B_k by a matrix of low rank. In fact, the quasi-Newton condition can be satisfied by adding a *rank-one* matrix to B_k. Assume that

$$B_{k+1} = B_k + uv^T, \tag{4.31}$$

for some vectors u and v. From the quasi-Newton condition (4.30):

$$B_{k+1}s_k = (B_k + uv^T)s_k = y_k, \quad \text{or} \quad u(v^Ts_k) = y_k - B_ks_k,$$

and therefore u must be in the direction $y_k - B_ks_k$. We assume that y_k is not equal to B_ks_k (otherwise, B_k would already satisfy the quasi-Newton condition). For any vector v such that v^Ts_k is non-zero, the vector u is given by $(1/v^Ts_k)(y_k - B_ks_k)$, and B_{k+1} is defined as

$$B_{k+1} = B_k + \frac{1}{v^Ts_k}(y_k - B_ks_k)v^T. \tag{4.32}$$

Given any vector w that is orthogonal to s_k, the rank-one matrix zw^T annihilates s_k. Therefore, the quasi-Newton condition (4.30) will continue to hold if further rank-one matrices of the form zw^T are added to B_{k+1} (although the elements of B_{k+1} will, of course, be altered by each additional matrix). Since the quasi-Newton condition does not uniquely determine the update matrix U_k, further conditions are usually imposed to make B_{k+1} have certain desirable properties.

Symmetry. Since the Hessian matrix is symmetric, it seems reasonable to require that each approximate Hessian matrix be symmetric also. Therefore, we seek updates that possess the property of *hereditary symmetry*, i.e., B_{k+1} is symmetric if B_k is symmetric. For a rank-one update, the requirement of symmetry inheritance uniquely determines the update. In order for the update (4.31) to maintain the symmetry of B_k, v must be a multiple of u. The rank-one update (4.32) then becomes

$$B_{k+1} = B_k + \frac{1}{(y_k - B_ks_k)^Ts_k}(y_k - B_ks_k)(y_k - B_ks_k)^T,$$

where $y_k - B_ks_k$ and $(y_k - B_ks_k)^Ts_k$ are non-zero. This update is termed the *symmetric rank-one update*.

Since there is only one symmetric rank-one update, we must allow rank-two update matrices in order to investigate other updates with hereditary symmetry. One technique for developing a symmetric update is the following. Given a symmetric matrix B_k, define $B^{(0)} = B_k$, and generate an updated $B^{(1)}$ using the general rank-one update (4.31), i.e.

$$B^{(1)} = B^{(0)} + uv^T, \quad v^Ts_k \neq 0.$$

The matrix $B^{(1)}$ satisfies the quasi-Newton condition but is not symmetric; hence, we symmetrize $B^{(1)}$ to obtain $B^{(2)}$:

$$B^{(2)} = \frac{1}{2}(B^{(1)} + B^{(1)T}).$$

However, since $B^{(2)}$ will not in general satisfy the quasi-Newton condition, the process is repeated. In this fashion, we generate a sequence of updated matrices: for $j = 0, 1, \ldots,$

$$B^{(2j+1)} = B^{(2j)} + \frac{1}{v^T s_k}(y_k - B^{(2j)} s_k)v^T,$$

$$B^{(2j+2)} = \frac{1}{2}(B^{(2j+1)} + B^{(2j+1)T}).$$

The sequence $\{B^{(j)}\}$ has a limit, given by

$$B_{k+1} = B_k + \frac{1}{v^T s_k}\left((y_k - B_k s_k)v^T + v(y_k - B_k s_k)^T\right)$$
$$- \frac{(y_k - B_k s_k)^T s_k}{(v^T s_k)^2} vv^T. \tag{4.33}$$

The update matrix in (4.33) is of rank two, and is well-defined for any v that is not orthogonal to s_k. The rank-two update analogous to the symmetric rank-one update can be derived by setting v equal to $y_k - B_k s_k$. The symmetric update (4.33) in which $v = s_k$ is termed the *Powell-Symmetric-Broyden (PSB) update*.

When v is taken as y_k, (4.33) becomes the well-known *Davidon-Fletcher-Powell (DFP) update*

$$B_{k+1} = B_k - \frac{1}{s_k^T B_k s_k} B_k s_k s_k^T B_k + \frac{1}{y_k^T s_k} y_k y_k^T + (s_k^T B_k s_k) w_k w_k^T, \tag{4.34}$$

where

$$w_k = \frac{1}{y_k^T s_k} y_k - \frac{1}{s_k^T B_k s_k} B_k s_k.$$

It can be verified by direct substitution that the vector w_k is orthogonal to s_k. Hence, any multiple of the rank-one matrix $w_k w_k^T$ may be added to B_{k+1} without affecting the satisfaction of the quasi-Newton condition (4.30). This observation leads to the *one-parameter family of updates*

$$B_{k+1}^\phi = B_k - \frac{1}{s_k^T B_k s_k} B_k s_k s_k^T B_k + \frac{1}{y_k^T s_k} y_k y_k^T + \phi_k(s_k^T B_k s_k) w_k w_k^T, \tag{4.35}$$

where the scalar ϕ_k depends on y_k and $B_k^\phi s_k$.

Considerable research has been performed in order to determine whether a particular choice of ϕ_k leads to a "best" update. However, much of this work is of purely theoretical interest, and our concern is only with the conclusions of this research. It is now generally believed that the most effective update from this class is the formula corresponding to the choice $\phi_k \equiv 0$. The resulting update formula, which is termed the *Broyden-Fletcher-Goldfarb-Shanno (BFGS) update*, is given by

$$B_{k+1} = B_k - \frac{1}{s_k^T B_k s_k} B_k s_k s_k^T B_k + \frac{1}{y_k^T s_k} y_k y_k^T. \tag{4.36}$$

If the direction of search is computed from the equations (4.28), an important simplification occurs in the one-parameter update formulae, since $B_k s_k = -\alpha_k g_k$. For example, the BFGS formula becomes

$$B_{k+1} = B_k + \frac{1}{g_k^T p_k} g_k g_k^T + \frac{1}{\alpha_k y_k^T p_k} y_k y_k^T. \tag{4.37}$$

If an exact linear search is made at each iteration, updates from the one-parameter family satisfy a very special relationship. Let $F(x)$ be any twice-continuously differentiable function, and assume that x_0 and B_0 are given. Let $\{x_k\}$, $\{B_k\}$, $\{p_k\}$ and $\{\alpha_k\}$ denote the sequences generated by the BFGS method, with $\{x_k^\phi\}$, $\{B_k^\phi\}$, $\{p_k^\phi\}$, and $\{\alpha_k^\phi\}$ the corresponding values for any member of the one-parameter family. If each of the sequences $\{B_k\}$ and $\{B_k^\phi\}$ is well-defined, and, for all k, α_k and α_k^ϕ are the minima of F which are nearest to the point $\alpha = 0$, then

$$x_k^\phi = x_k \quad \text{and} \quad B_k = B_k^\phi + \left(\frac{\phi_{k-1}}{g_{k-1}^T p_{k-1}^\phi}\right) g_k g_k^T. \tag{4.38}$$

Thus, under the stated conditions, *all the methods generate identical points*. This remarkable result indicates that the elements of a quasi-Newton approximation to the Hessian will not necessarily resemble those of the true Hessian.

The equivalence of the iterates generated by different update formulae with exact linear searches means that the number of iterations required to minimize the same function will be identical. However, the number of *function evaluations* tends to differ significantly among updates. This difference is attributable primarily to the fact that the trial step length used to initiate the calculation of α_k is generally closer to the univariate minimum for one update than another.

Positive definiteness. Since a stationary point x^* of F is a strong local minimum if the Hessian matrix at x^* is positive definite, it would seem desirable for the approximating matrices $\{B_k\}$ to be positive definite. Furthermore, if B_k is positive definite, the local quadratic model has a unique local minimum, and the search direction p_k computed from (4.28) is a descent direction. Thus, it is usual to require that the update formulae possess the property of *hereditary positive definiteness* — i.e., if B_k is positive definite, B_{k+1} is positive definite.

Hereditary positive-definiteness for an update can be proved in several ways; we shall outline a proof for the BFGS formula. If B_k is positive definite, there must exist a non-singular matrix R such that $B_k = R^T R$. The formula (4.36) may be written as

$$B_{k+1} = R^T W R, \tag{4.39}$$

where the matrix W is given by

$$W = I - \frac{1}{\bar{s}^T \bar{s}} \bar{s}\bar{s}^T + \frac{1}{\bar{y}^T \bar{s}} \bar{y}\bar{y}^T, \tag{4.40}$$

with $\bar{s} = Rs_k$ and $\bar{y} = (R^T)^{-1} y_k$. Equation (4.39) indicates that B_{k+1} will be positive-definite if W is positive-definite. Let $\Pi(A)$ denote the product of the eigenvalues of the matrix A. If we compute the product of the eigenvalues of the matrices on both sides of (4.39), we have

$$\Pi(B_{k+1}) = \Pi(R^T W R)$$
$$= \left(\Pi(R)\right)^2 \Pi(W)$$

(see Section 2.2.3.4). Since W is a rank-two modification of the identity, it has $n - 2$ unit eigenvalues if $n \geq 3$. Let λ_1 and λ_2 denote the two remaining eigenvalues. Examination of (4.40) indicates that the eigenvectors corresponding to λ_1 and λ_2 are linear combinations of $\bar{s}$ and $\bar{y}$.

By direct substitution, we can verify that

$$\lambda_1 + \lambda_2 = \frac{(\bar{y}^T \bar{s} + \bar{y}^T \bar{y})}{\bar{y}^T \bar{y}},$$

$$\lambda_1 \lambda_2 = \frac{\bar{y}^T \bar{s}}{\bar{s}^T \bar{s}}.$$

Since B_k is positive definite, it holds that $\bar{s}^T \bar{s} = s_k^T B_k s_k > 0$, and consequently both λ_1 and λ_2 will be positive if $\bar{y}^T \bar{s}$ is positive. Since $\bar{y}^T \bar{s} = y_k^T s_k$, the BFGS update has the property of hereditary positive-definiteness if and only if

$$y_k^T s_k > 0. \tag{4.41}$$

The condition (4.41) can always be satisfied by performing a "sufficiently accurate" linear search at each iteration. To see why, note that

$$y_k^T s_k = \alpha_k (g_{k+1}^T p_k - g_k^T p_k).$$

Since p_k is a descent direction, the step length α_k and the term $-g_k^T p_k$ are both positive; the (possibly negative) term $g_{k+1}^T p_k$ can be made as small in magnitude as necessary by increasing the accuracy of the linear search (since $g_{k+1}^T p_k$ vanishes when α_k is a univariate minimum along p_k).

Finite termination on quadratics. The quasi-Newton condition forces the Hessian approximation to produce a specified curvature along the search direction at a particular iteration, since $B_{k+1} s_k = y_k$. However, subsequent modifications may destroy this property, and hence it is sometimes required that the curvature information from certain previous iterations should be retained, i.e. that for some values of j $(j \le k)$

$$B_{k+1} s_j = y_j. \tag{4.42}$$

If (4.42) holds for n linearly independent vectors $\{s_j\}$, a quasi-Newton method will terminate in a finite number of iterations when F is a quadratic function with Hessian matrix G. Let S be the matrix whose i-th column is s_i. Then (4.42) becomes

$$B_{n+1} S = GS,$$

since $y_i = Gs_i$. Therefore, if S is non-singular, $B_{n+1} = G$.

If any update from the one-parameter family is applied to a quadratic function, and an exact linear search is performed at each iteration, then for $j = 0, \ldots, n-1$,

$$B_k s_j = y_j, \quad j < k;$$
$$s_j^T G s_i = 0, \quad i \ne j.$$

Hence, although matrices from the one-parameter family are not necessarily equal at any intermediate iteration, they all become equal to G at iteration $n+1$ under the conditions stated above.

4.5.2.2. Implementation. In order to obtain the search direction p_k in either a Newton-type or quasi-Newton method, it is necessary to solve a linear system. With the Newton-type methods already described, a new Hessian matrix is computed at every iteration, and (4.17) is solved by refactorizing G_k. In a quasi-Newton method, however, the matrix at a given iteration is a low-rank modification of the matrix from the previous iteration, and hence it might seem wasteful to compute a new factorization (see Section 2.2.5.7).

For this reason, the earliest quasi-Newton methods were formulated in terms of maintaining an approximation to the *inverse Hessian matrix*. Since a low-rank modification of a matrix generates a low-rank modification of its inverse (see Section 2.2.3.3), a theoretically equivalent update for the inverse Hessian can be produced from any of the quasi-Newton updates for the Hessian. If H_k is a quasi-Newton approximation to the inverse Hessian, the search direction p_k is defined by

$$p_k = -H_k g_k, \tag{4.43}$$

and the quasi-Newton condition is

$$H_{k+1} y_k = s_k. \tag{4.44}$$

The inverse Hessian approach might seem to offer an advantage in terms of the number of arithmetic operations required to perform an iteration, since solving (4.28) from scratch would require of order n^3 operations, compared with order n^2 to form the matrix-vector product $H_k g_k$. However, this seeming defect of maintaining an approximation to the Hessian can be eliminated by updating the *Cholesky factors* of the matrix, rather than an explicit representation of the elements of the matrix. If the Cholesky factors $L_k D_k L_k^T$ of B_k are available, the system (4.28) can be solved in order n^2 operations. Furthermore, the factors L_{k+1} and D_{k+1} of the updated matrix B_{k+1} can be determined for a low-rank quasi-Newton update in a number of operations comparable to the number required to compute H_{k+1} (see Section 2.2.5.7).

When the Cholesky factorization of the approximate Hessian is available, we can obtain a convenient estimate of the condition number of B_k. For example, it can be shown that, if $d_{\max}$ and $d_{\min}$ are the largest and smallest diagonal elements of D_k, then $\text{cond}(B_k) \geq d_{\max}/d_{\min}$. When the minimization is completed, an estimate of the condition number can be used to give an indication of whether or not the algorithm has converged successfully (see Section 8.3.3). On those occasions when the matrix B_k is so ill-conditioned that the value of p_k is likely to have no correct figures, the procedures for computing and updating the Cholesky factors can be modified so that the condition number of the Hessian approximation does not exceed a fixed upper bound; this property is useful in proving global convergence (see Section 4.5.2.3).

The use of the Cholesky factorization allows one to avoid a very serious problem that would otherwise arise in quasi-Newton methods: *the loss (through rounding errors) of positive definiteness in the Hessian (or inverse Hessian) approximation*. This aspect of quasi-Newton updates is often overlooked in theoretical discussions, but is critical to success, especially for the most difficult problems. For example, use of the BFGS update (4.37) with (4.41) satisfied should ensure in theory that all Hessian (or inverse Hessian) approximations remain positive definite. However, in practice it is not uncommon for rounding errors to cause the updated matrix to become singular or indefinite.

Example 4.8. Consider a simple two-dimensional example when the approximate Hessian is the identity

$$B_k = \begin{pmatrix} 1 & 0 \\ 0 & 1 \end{pmatrix}.$$

Assume that the calculations are being performed on a machine with five-digit decimal precision. When $s_k = (1, 10^{-3})^T$ and $y_k = (0, 1)^T$, it holds that $y_k^T s_k > 0$. However, the *computed* updated matrix has a zero element in the $(1, 1)$-th position, and is therefore not positive definite. With exact arithmetic, the $(1, 1)$-th element of the updated matrix would be a small positive number, which has been lost through rounding errors.

The loss of positive definiteness is undesirable not only because p_k may no longer be a descent direction, but also because the accumulated curvature information may be destroyed through addition of a spurious update matrix. Unfortunately, a loss of positive definiteness may not be obvious until several iterations have passed, since p_k may be a descent direction by chance even if B_k has become indefinite. The strategy of resetting B_k to the identity would produce a descent direction, but has the effect of discarding useful information. Hence, in implementing a quasi-Newton method, *it is important to retain the numerical positive definiteness of the Hessian (or inverse Hessian) approximation.*

When a quasi-Newton method is implemented using updates to the Cholesky factors, it is possible to avoid losing positive-definiteness. If the elements of D_k are positive at every iteration, the Hessian approximation is guaranteed to be positive definite. Furthermore, any loss of positive-definiteness must be revealed during the updating process, and hence can be detected in the iteration at which it occurs.

Figure 4m illustrates the behaviour of a quasi-Newton algorithm applied to Rosenbrock's function (Example 4.2). The algorithm was implemented with the BFGS update, and an accurate line search was performed at each iteration. Note that, like Newton's method, the algorithm makes good progress at points remote from the solution. The worst behaviour occurs near the origin, where the curvature is changing most rapidly.

*4.5.2.3. Convergence; least-change characterization. In this section, we briefly summarize some of the many interesting properties of quasi-Newton methods. The interested reader should consult the references mentioned in the Notes for further details.

If a twice-continuously differentiable function F has bounded second derivatives and the level set $L(F(x_0))$ is bounded, global convergence to a stationary point can be proved for a quasi-Newton method if every B_k is positive definite with a bounded condition number, and if one of the step-length criteria discussed in Section 4.3.2.1 is used to choose α_k. The restrictions on B_k ensure that the search direction remains sufficiently bounded away from orthogonality with the negative gradient. The matrices $\{B_k\}$ generated by an implementation that updates the Cholesky factorization (see Section 4.5.2.2) can be made to satisfy these requirements by including a procedure that explicitly bounds the condition number.

If the condition number of each approximate Hessian is not bounded explicitly, it has been possible to prove convergence only by imposing more restrictive conditions on the method used and the problem being solved. For example, the sequence generated by a member of the one-parameter family will be convergent to a stationary point if exact linear searches are performed and the eigenvalues of the Hessian matrix are bounded above and below by finite positive numbers. The BFGS formula is currently the only member of the one-parameter family for which convergence can be proved with the step-length criteria (4.7) and (4.8). However, the somewhat artificial conditions on the eigenvalues of the Hessian must still be imposed. (The reader should note that the restrictions on the class of objective functions are more likely to be an indication of the difficulty of proof than a symptom of the inadequacy of quasi-Newton methods. In practice, quasi-Newton methods are convergent for a larger class of functions than that defined by the restrictions of the convergence theorems.)

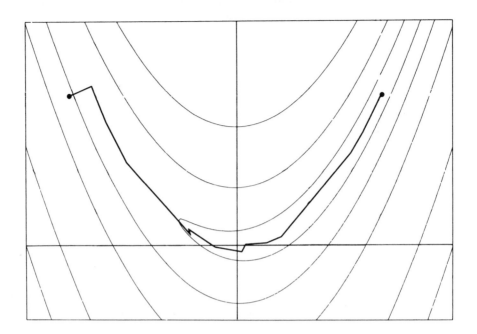

Figure 4m. Solution path of a BFGS quasi-Newton algorithm on Rosenbrock's function. Like Newton's method, the algorithm makes good progress at points remote from the solution. The worst behaviour occurs near the origin, where the curvature is changing most rapidly.

Much research has also been performed concerning the *rate* of convergence of quasi-Newton methods. Again, we give only a very brief synopsis of selected results, and suggest a study of the references for details.

Several proofs of convergence begin by assuming that the iterative procedure is defined by

$$x_{k+1} = x_k + p_k, \tag{4.45}$$

where p_k satisfies

$$B_k p_k = -g_k;$$

note that the step length α_k is taken to be unity. Certain other mild assumptions are also made about F, and $\{B_k\}$ is defined by one of several quasi-Newton updates (including the BFGS, PSB and DFP). Then, if $G(x^*)$ is positive definite, and x_0 and B_0 are "close enough" to x^* and $G(x^*)$, respectively, the sequence (4.45) is well-defined and converges linearly to x^*. A property of the matrices $\{B_k\}$ that is used to verify this local convergence is termed *bounded deterioration*; for example, for the updates mentioned, it can be shown that

$$\|B_{k+1} - G(x^*)\| \le \gamma_1 \|B_k - G(x^*)\| + (1 + \gamma_2) \max\{\|x_k - x^*\|, \|x_{k+1} - x^*\|\},$$

for some non-negative constants γ_1 and γ_2, where the choice of norm varies with the update formula.

In considering whether quasi-Newton methods converge superlinearly, we note that the following property holds for a superlinearly convergent sequence $\{x_k\}$

$$\lim_{k \to \infty} \frac{\|x_{k+1} - x_k\|}{\|x_k - x^*\|} = 1. \tag{4.46}$$

For example, consider the sequence converging quadratically to zero: 10^{-4}, 10^{-8}, 10^{-16}, etc.

If the iterates defined by (4.45) converge locally and linearly, convergence is *superlinear* if and only if the sequence $\{B_k\}$ satisfies

$$\lim_{k \to \infty} \frac{\|(B_k - G(x^*))(x_{k+1} - x_k)\|}{\|x_{k+1} - x_k\|} = 0. \tag{4.47}$$

One of the most interesting properties of these updates is that it is *not* necessary for B_k to converge to $G(x^*)$ in order for (4.47) to be satisfied, and therefore the iterates (4.45) can converge superlinearly even if the Hessian approximation does not converge to the true Hessian. The property (4.47) can be verified for the BFGS, PSB and DFP updates under suitable assumptions.

If step lengths other than unity are allowed, the sequence $\{\alpha_k\}$ must converge to unity at a sufficiently fast rate if superlinear convergence is to be verified. This is the reason for the frequent advice to initiate a step-length procedure for a quasi-Newton method with the "natural" step of unity.

An interesting property of certain quasi-Newton updates is that the update matrix U_k in (4.29) is the solution of an optimization problem: to find the minimum-norm perturbation U_k to B_k such that the matrix $B_k + U_k$ has certain properties (e.g., satisfies the quasi-Newton condition, is symmetric, is positive definite, etc.). For example, the PSB update can be derived by finding the update matrix U_k that solves

$$\begin{aligned} \text{minimize} \quad & \|U\|_F^2 = \sum_{i=1}^{n} \sum_{j=1}^{n} U_{ij}^2 \\ \text{subject to} \quad & U s_k = y_k - B_k s_k \\ & U = U^T, \end{aligned} \tag{4.48}$$

where $\|U\|_F$ denotes the Frobenius norm of the matrix U. Other updates can be defined by problems similar to (4.48), with different choices of norms.

Notes and Selected Bibliography for Section 4.5

The first quasi-Newton method was suggested by Davidon (1959), who called it a "variable metric" method. This name arises because of the interpretation of the direction defined by (4.28) as the step to the minimum of a quadratic model function. If the objective function is a quadratic with constant positive-definite Hessian G, the matrix G defines a "metric" or "norm", as noted in Section 4.3.2.2. If G is known, the quadratic function can be minimized in a single iteration. However, since a changing positive-definite approximation to G is used to define the search direction at every iteration, the norm also varies — hence the description "variable metric".

Davidon's method was publicized and improved by Fletcher and Powell (1963) (hence the name "DFP" formula), who showed that exact termination is achieved on a quadratic function with exact linear searches. The PSB formula is due to Powell (1970a). The BFGS formula was independently suggested by Broyden (1970), Fletcher (1970a), Goldfarb (1970) and Shanno (1970).

Since 1963, there has been an ever-expanding interest in quasi-Newton methods, and there is a vast literature on all aspects of the subject. See Avriel (1976), Brodlie (1977a), Dennis and Moré (1977), and Fletcher (1980) for further discussion and references.

Quasi-Newton methods are closely related to many other methods, such as conjugate-gradient methods (see Nazareth, 1979, and Section 4.8.3).

The use of the Cholesky factorization as an aid to the implementation of a numerically stable quasi-Newton method was suggested by Gill and Murray (1972) (see also Fletcher and Powell, 1974; Gill, Murray and Saunders, 1975; and Brodlie, 1977b). Further steps can be taken to avoid an unnecessary loss of precision when the norm of one of the rank-one terms is large compared to the magnitude of the total correction. For example, any rank-two correction may be written in the form

$$B_{k+1} = B_k + uu^T - vv^T,$$

where $u^T v = 0$ (see Gill and Murray, 1978c).

Much of the work on convergence of quasi-Newton methods is due to Broyden, Dennis, Moré, Powell and Stoer (see Broyden, 1967, 1970; Broyden, Dennis and Moré, 1973; Powell, 1971, 1975, 1976c; and Stoer, 1975, 1977). The characterization (4.46) of superlinear convergence is due to Dennis and Moré (1974). A summary of related results and a good bibliography are contained in Dennis and Moré (1977) and Brodlie (1977b). The remarkable result (4.38), that members of the one-parameter family of updates generate identical points is due to Dixon (1972a, b). The "least-change" characterization of quasi-Newton updates was first derived by Greenstadt (1970). For more details, see Dennis and Moré (1977) and Dennis and Schnabel (1979, 1980).

Much of the research in quasi-Newton methods has been concerned with determining the properties of updates that are known to be effective empirically in an attempt to isolate those properties that produce a good convergence rate. The idea is to then to choose a value of the parameter ϕ in (4.35) that incorporates this desirable property in a precise sense. For example, one of the most striking features of quasi-Newton methods is that they perform more efficiently (in terms of the number of function evaluations required) when an exact linear search is *not* performed. Several authors have attempted to explain this phenomenon. Fletcher (1970a) noted that every member of the one-parameter family of updates (4.35) can be written in the form

$$B_k^\phi = (1 - \phi)B_k^0 + \phi B_k^1,$$

where B_k^0 and B_k^1 are the approximate Hessian matrices produced by the BFGS and DFP updates. Fletcher defined the *convex class of formulae* as those updates for which $\phi \in [0, 1]$ and showed that members of this class have the property that, when F is quadratic, the eigenvalues of the matrix

$$R_k = G^{\frac{1}{2}} B_k^{-1} G^{\frac{1}{2}} \tag{4.49}$$

tend monotonically to unity as k increases *regardless of the step length taken.*

Other workers have sought to define extra conditions on the approximate Hessian so that an update is defined by a unique choice of ϕ. For example, Davidon (1975) and Oren and Spedicato (1976) have suggested updating formulae that minimize a bound on the condition number of the updated approximate Hessian. Similarly, Oren and Luenberger (1974) introduce a further parameter so that if exact linear searches are performed on a quadratic function, the condition numbers of the matrices (4.49) are monotonically decreasing (see also Oren, 1974a, b). Some recent work has been concerned with constructing formulae that give finite termination on a class of functions more general than quadratics; see Davidon (1979) and Sorensen (1980b). Davidon (1975) has suggested a class of methods that give exact termination on quadratics without exact linear searches. Spedicato (1975) has proposed a family of updates that are invariant with respect to a certain nonlinear scaling.

4.6. NON-DERIVATIVE METHODS FOR SMOOTH FUNCTIONS

In many practical problems, it may be impossible or difficult to compute even the gradient vector of $F(x)$. For example, the function value may be the result of a complex sequence of calculations, such as a simulation. When only function values are available, it is essential for the user to determine whether the function F is really smooth, even though its derivatives cannot be computed (in contrast to the case when the derivatives do not exist). As mentioned in Section 4.2.1, it is undesirable to use a method designed for non-smooth problems on a smooth function, in light of the considerable efficiencies that result when information is accumulated about the curvature of the function. It seems reasonable to suppose that useful information can also be developed based only on function values, with resulting improvements in algorithmic efficiency.

4.6.1. Finite-difference Approximations to First Derivatives

When minimizing a smooth function whose derivatives are not available, an obvious strategy is to use a first-derivative method, but to replace the exact gradient $g(x)$ with a *finite-difference approximation*. Unfortunately, this adaptation is non-trivial, and it is essential to consider some rather complicated decisions with respect to the calculation of the approximate gradient.

4.6.1.1. Errors in a forward-difference approximation. For simplicity of description, we shall consider the error in estimating the first derivative of the twice-continuously differentiable *univariate* function $f(x)$. The most common approximation is a *forward-difference formula* (as described in Section 2.3.5). In this case, $f'(x)$ is approximated by the quantity

$$\varphi_F(f, h) = \frac{f(x + h) - f(x)}{h}, \tag{4.50}$$

where "F" denotes "forward difference".

When using a finite-difference formula like (4.50), there are three sources of error in the approximation to $f'(x)$.

Truncation error. The *truncation error* consists of the neglected terms in the Taylor series, namely

$$\varphi_F(f, h) - f'(x) = \frac{h}{2} f''(\xi) \equiv T_F(h), \tag{4.51}$$

where ξ is a point in the interval $[x, x + h]$ (see Section 2.3.4).

Condition error. In practice, the computed function values to be used in calculating φ_F will be subject to error. Let $\hat{f}(x)$ and $\hat{f}(x + h)$ denote the computed values of $f(x)$ and $f(x + h)$; we shall assume that these values satisfy

$$\hat{f}(x) = f(x) + \sigma \quad \text{and} \quad \hat{f}(x + h) = f(x + h) + \sigma_h,$$

where σ and σ_h are the absolute errors in f at x and $x + h$ (see Section 2.1.6). If the inexact function values are used in (4.50), and no other errors are made, then the computed value of φ_F is given by

$$\varphi_F(\hat{f}, h) = \frac{\hat{f}(x + h) - \hat{f}(x)}{h}$$

and hence

$$\varphi_F(\hat{f}, h) = \frac{f(x + h) - f(x)}{h} + \frac{\sigma_h - \sigma}{h}$$

$$= \varphi_F(f, h) + C(\varphi_F, h).$$

The error $C(\varphi_F, h)$ in the value of $\varphi_F(\hat{f}, h)$ due to inaccurate values of f is termed the *condition error* (sometimes known as *cancellation error*; see Section 2.1.5). The condition error satisfies

$$C(\varphi_F, h) = \frac{2\psi_F \max\{|\sigma|, |\sigma_h|\}}{h}, \tag{4.52}$$

where $|\psi_F| \leq 1$.

When $|f|$ is not small, σ can be written in terms of a *relative* error in f, i.e.

$$\sigma = \epsilon f(x) \quad \text{and} \quad \sigma_h = \epsilon_h f(x + h),$$

where $|\epsilon| \leq \epsilon_R$ and $|\epsilon_h| \leq \epsilon_R$. The condition error can then be expressed as

$$C(\varphi_F, h) = \frac{2}{h} \theta_F M_F \epsilon_R, \tag{4.53}$$

where $|\theta_F| \leq 1$ and $M_F = \max\{|f(x)|, |f(x + h)|\}$.

Rounding error. Given $\hat{f}(x)$ and $\hat{f}(x + h)$, the calculation of φ_F involves rounding errors in performing subtraction and division. However, these errors are generally *negligible* with respect to the truncation and condition errors, and we shall therefore consider only truncation and condition error in the subsequent analysis.

4.6.1.2. Choice of the finite-difference interval. When approximating $f'(x)$ by $\varphi_F(\hat{f}, h)$, the error in the computed approximation can be viewed as the sum of the truncation error (4.51) and the condition error (4.52) (or (4.53)). We observe that the truncation error is a linear function of h and the condition error is a linear function of $(1/h)$, and hence that changes in h will tend to have opposite effects on these errors.

Example 4.9. As an illustration, consider the function

$$f(x) = (e^x - 1)^2 + \left(\frac{1}{\sqrt{1 + x^2}} - 1\right)^2, \tag{4.54}$$

which has been evaluated for various values of h at the point $x = 1$, using short precision on an IBM 370. The smallest value of h that will register a change in x during floating-point addition is the machine precision $\epsilon_M = 16^{-5} \approx .95 \times 10^{-6}$. The function was computed with h values increasing from ϵ_M in multiples of ten. Table 4a contains the results of the computation. The first

Table 4a

Condition and truncation errors in φ_F with $\epsilon_M = .953674 \times 10^{-6}$

h	$\hat{f}(x)$	$\hat{f}(x+h)$	$|T(h)|$	$|C(\varphi_F, h)|$	$\varphi_F(\hat{f}, h)$
ϵ_M	$.303828 \times 10^1$	$.303828 \times 10^1$	$.115697 \times 10^{-4}$	$.254867 \times 10^1$	$.700000 \times 10^1$
$10\epsilon_M$	$.303828 \times 10^1$	$.303837 \times 10^1$	$.115711 \times 10^{-3}$	$.487710 \times 10^{-1}$	$.950000 \times 10^1$
$10^2\epsilon_M$	$.303828 \times 10^1$	$.303919 \times 10^1$	$.115718 \times 10^{-2}$	$.298130 \times 10^{-1}$	$.952000 \times 10^1$
$10^3\epsilon_M$	$.303828 \times 10^1$	$.304739 \times 10^1$	$.115790 \times 10^{-1}$	$.223475 \times 10^{-2}$	$.955800 \times 10^1$
$10^4\epsilon_M$	$.303828 \times 10^1$	$.313045 \times 10^1$	$.116520 \times 10^0$	$.275015 \times 10^{-3}$	$.966490 \times 10^1$

three columns contain the values of h, the computed function value at $x = 1$, and the computed function value at $x + h$. The fourth column contains $|T(h)|$, the magnitude of the truncation error that would be incurred by using the exact $\varphi_F(h)$ (calculated in double precision) to approximate f'. The fifth column contains the magnitude of the condition error, which was calculated using the exact value of φ_F. The final column contains the computed values of $\varphi_F(\hat{f}, h)$. The exact value of $f'(x)$ (rounded to six figures) is $.954866 \times 10^1$.

Ideally, the finite-difference interval h would be chosen to yield the smallest error in φ_F, i.e., to minimize the sum of the truncation and condition errors. For Example 4.9, it can be seen from Table 4a that the "optimal" h at the given point lies between $100\epsilon_M$ and $1000\epsilon_M$. Unfortunately, it is impossible in general to compute the value of h that yields the smallest error in φ_F, since the quantities that appear in (4.51) and (4.53) are unknown (unless, as in the case of Table 4a, additional precision is available and the exact first derivative is known).

In practice, the finite-difference interval is chosen so as to minimize the following *computable* *bound* on the error

$$\frac{h}{2}|\Phi| + \frac{2}{h}\bar{C},$$

where Φ is an estimate of $f''(\xi)$ and $\bar{C}$ is a bound on the condition error (a method for computing Φ and $\bar{C}$ using function values only is given in Section 8.6). Assuming that Φ is non-zero, the "optimal" interval $\hat{h}$ is then given by

$$\hat{h} = \sqrt{\frac{4\bar{C}}{|\Phi|}}. \tag{4.55}$$

Using (4.55), we can compute an estimate of $\hat{h}$ for the function (4.54) of Example 4.9, using the exact value of $f''(x)$ for Φ, and the magnitude of the exact condition error for $\bar{C}$. At the point $x = 1$, $\hat{h} = 4.497 \times 10^{-4}$; at the point $x = 50$, $\hat{h} = 1.436 \times 10^{-3}$.

4.6.1.3. Estimation of a set of finite-difference intervals. The analysis given in Section 4.6.1.2 can be applied when the gradient $g(x)$ is to be approximated by a vector $\hat{g}(x)$, whose j-th component is defined by

$$\hat{g}_j(x_k) = \frac{1}{h_j}\big(F(x_k + h_j e_j) - F(x_k)\big).$$

In this case, a *set* of intervals $\{h_j\}$ must be obtained. Since the procedure described in Section 8.6 requires at least two evaluations of F in order to estimate $\hat{h}$ from (4.55), it is generally not considered efficient to estimate a new set of finite-difference intervals at every point in a minimization. Rather, a set of intervals is computed at the initial point x_0, using a procedure such as the one described in Section 8.6.

Let $\hat{h}_j$ denote the estimate of the optimal interval for the j-th variable, computed at the point x_0; note that $\hat{h}_j$ is an absolute interval. We then obtain a set of *relative* intervals $\{\delta_j\}$, defined by

$$\delta_j = \frac{\hat{h}_j}{1 + \sigma_j|(x_0)_j|},$$

where σ_j satisfies $0 \leq \sigma_j \leq 1$, and reflects the variation in F with x_j; in general, σ_j can be taken as unity (see Section 8.6 for a discussion of this point). The required absolute interval for the j-th variable at iteration k is then

$$h_j = \delta_j(1 + \sigma_j|(x_k)_j|). \tag{4.56}$$

It is important to note that there may not exist a fixed set of intervals that are appropriate throughout the range of points where gradients must be approximated during a minimization. Nonetheless, in most practical applications it is adequate to examine properties of the function in some detail at a point that typifies those for which the gradient will be required. In the unlikely event that a minimization algorithm fails because of a poor choice of finite-difference interval, it can be restarted in order to produce intervals more suitable for the neighbourhood in which failure occurred.

For some functions, a near-optimal value of h_j can be derived by inspection. This topic will be discussed in Section 8.6.

4.6.1.4. The choice of finite-difference formulae. The forward-difference formula requires only one additional evaluation of F for each component of the gradient, and will usually provide approximate gradients of acceptable accuracy unless $\|g(x)\|$ is small. Since $\|g_k\|$ approaches zero at the solution of an unconstrained problem, this means that *the forward-difference approximation will eventually be unreliable*, even for a well-scaled problem with carefully chosen finite-difference intervals. Unfortunately, in general the relative error in the forward-difference approximation will become unacceptably large *before the solution has achieved the maximum possible accuracy*. Hence, the forward-difference approximation may be inappropriate during the last few iterations.

It may also be necessary to abandon the use of the forward-difference formula even if x_k is far from optimal. For example, a forward-difference approximation should not be used any time the difference in function values is small relative to h_j, since large condition error will tend to invalidate the computed result. In addition, a more accurate gradient approximation should be used if the step length at any iteration is so small that the change in x is less than the perturbation associated with a finite-difference calculation.

When the forward-difference formula is not sufficiently accurate, a *central-difference approximation* $\bar{g}(x_k)$ can be used, whose j-th component is given by

$$\bar{g}_j(x_k) = \frac{1}{2h_j}\big(F(x_k + h_j e_j) - F(x_k - h_j e_j)\big). \tag{4.57}$$

The central-difference formula has the property that the associated truncation error is of order h_j^2 (see Section 2.3.5), but the condition error remains of order $(1/h_j)$.

When a switch to central differences is made because the forward-difference approximation is not sufficiently accurate, in general the finite-difference interval $\bar{h}_j$ for the central-difference formula should be *larger* than the interval used for the forward-difference formula. When F is well-scaled in the sense of Section 8.7, a good choice for $\bar{h}_j$ is

$$\bar{h}_j = (\hat{h}_j)^{\frac{2}{3}}.$$

When a central-difference approximation is used, we can define a set of relative intervals $\{\bar{\delta}_j\}$ as in the forward-difference case. The value of h_j to be used in (4.57) is then given by a formula analogous to (4.56), using $\bar{\delta}_j$ instead of δ_j.

A central-difference approximation requires *two* additional function evaluations in order to approximate each component of the gradient. Hence, it is not worthwhile to use central differences if the forward-difference approximation is adequate, and the switch to central differences should not be made until the errors in the forward-difference approximation have become unacceptable. If central differences must be used because of poor local properties of F when x_k is far from optimal, it is advisable to return to the forward-difference approximation if possible.

4.6.2. Non-Derivative Quasi-Newton Methods

The methods of Section 4.5 that can be adapted most successfully to the non-derivative case are the quasi-Newton methods of Section 4.5.2. When properly implemented, finite-difference quasi-Newton methods are extremely efficient, and display the same robustness and rapid convergence as their counterparts with exact gradients. However, it is *highly inadvisable to apply without modification a method based on analytic gradients* (i.e., to "pretend" that the approximate gradients are exact). Errors in the gradient can have a substantial effect on performance, and hence the logic of a quasi-Newton method must be modified as indicated in Section 4.6.1 when exact derivatives are not available. In particular, a finite-difference quasi-Newton method should be able to adjust the form of the derivative approximation and the finite-difference interval in response to the behaviour of F at x_k.

A finite-difference quasi-Newton algorithm will differ in other respects from a method for exact gradients because of the n (or $2n$) function evaluations required to obtain a gradient approximation. In particular, under the standard assumption that the cost of a function evaluation dominates the overhead of the method, a quasi-Newton algorithm based on finite-differences should be structured so as to reduce the number of times that the gradient must be approximated. Thus, the step length algorithm should not require the evaluation of the gradient at trial points during the line search. In addition, a more accurate line search should generally be performed at each iteration, since increased accuracy in the line search tends to reduce the overall number of iterations (and hence to decrease the number of gradient approximations).

The iterates of a finite-difference quasi-Newton algorithm applied to Rosenbrock's function (Example 4.2) are shown in Figure 4n. The BFGS update was used, with an accurate linear search at each iteration. Outside the neighbourhood of the solution, the performance of the method is almost identical to that of the regular BFGS method (*cf.* Figure 4m). Closer to the solution, additional iterations are performed in this case because of the inadequacy of the forward-difference formula employed to compute the gradient.

Notes and Selected Bibliography for Section 4.6

There have been many non-derivative methods for minimizing smooth functions that are not based on computing finite differences of the gradient vector; see, for example, Powell (1964), Greenstadt (1972), and Brent (1973a). However, it is now generally accepted that such methods are less efficient than finite-difference quasi-Newton methods.

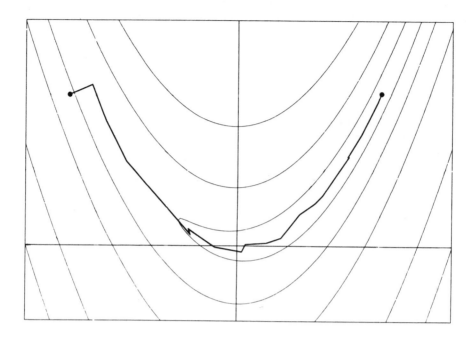

Figure 4n. Solution path of a finite-difference BFGS quasi-Newton algorithm when applied to Rosenbrock's function. Outside the neighbourhood of the solution, the performance of the method is almost identical to that of the regular BFGS method (*cf.* Figure 4m). Closer to the solution, additional iterations are performed because of the inadequacy of the forward-difference formula employed to compute the gradient.

Stewart (1967) has suggested a finite-difference quasi-Newton method in which an interval is computed from (4.55) at *every* iteration with $|\Phi|$ estimated by the diagonal elements of a DFP approximation to the Hessian matrix (see (4.34)). Stewart's procedure is based on the assumption that the diagonal elements of the approximate Hessian are good order-of-magnitude estimates of the exact second derivatives. Curtis and Reid (1974) present a method for computing a finite-difference interval for use with the central-difference formula. For more information on finite-difference calculations within minimization routines, see Gill and Murray (1972).

The general problem of finding approximate derivatives by finite differences is known as *numerical differentiation*. A modern discussion of methods for numerical differentiation (including a more complete exposition of the role of condition error in general finite-difference formulae) is given by Lyness (1977). See also Anderssen and Bloomfield (1974), Dahlquist and Björck (1974), Lyness and Moler (1967), Lyness and Sande (1971), Oliver and Ruffhead (1975) and Oliver (1980). Many automatic differentiation routines attempt to find the interval such that the total error in a finite-difference approximation is minimized. Among those algorithms proposed for the central-difference case, see Dumontet and Vignes (1977) and Stepleman and Winarsky (1979). In the type of application discussed here, it is not necessary to compute a derivative to the accuracy associated with a numerical differentiation technique. Moreover, numerical differentiation procedures usually require a significant number of function evaluations.

The use of approximate derivatives is extremely successful in quasi-Newton methods. The extension of other methods to the non-derivative case is less satisfactory. A Newton-type method can be developed when only function values are available by approximating both the Hessian and gradient by finite differences (see Mifflin, 1975). However, this approach is generally considered impractical, since $O(n^2)$ function evaluations are required at each iteration. Furthermore, extreme care must be exercised in choosing the finite-difference interval in order to obtain adequate accuracy in both derivative approximations. This approach might be feasible under some circumstances when the Hessian is sparse (see Section 4.8.1).

4.7. METHODS FOR SUMS OF SQUARES

4.7.1. Origin of Least-Squares Problems; the Reason for Special Methods

In a large number of practical problems, the function $F(x)$ is a sum of squares of nonlinear functions

$$F(x) = \frac{1}{2} \sum_{i=1}^{m} f_i(x)^2 = \frac{1}{2} \|f(x)\|_2^2. \tag{4.58}$$

The i-th component of the m-vector $f(x)$ is the function $f_i(x)$, and $\|f(x)\|$ is termed the *residual* at x. (The $\frac{1}{2}$ has been included in (4.58) in order to avoid the appearance of a factor of two in the derivatives.)

Problems of this type occur when fitting model functions to data, i.e. in *nonlinear parameter estimation*. If $\phi(x, t)$ represents the desired model function, with t an independent variable, then each individual function $f_i(x)$ is defined as $\phi(x, t_i) - y_i$, where the data points $\{y_i\}$ may be subject to experimental error. The independent variables $\{x_i\}$ can be interpreted as parameters of the problem that are to be manipulated in order to adjust the model to the data. If the model is to have any validity, we can expect that $\|f(x^*)\|$ will be "small", and that m, the number of data points, will be much greater than n. (If the latter condition is not true, then an arbitrary model will give a close fit to the data.)

A slightly different situation occurs if we wish to determine the free parameters of a system so that the output follows some specific continuous performance profile. Under these circumstances, there is no inherent experimental error, since the "data" are usually defined as a continuous function and the residuals are likely to be small at the optimum. This type of problem may be expressed as

$$\underset{x \in \Re^n}{\text{minimize}} \int_{t_0}^{t_1} \left(\phi(x, t) - \mathcal{F}(t) \right)^2 dt,$$

where $\mathcal{F}(t)$ is the "target" function. When the integral is discretized using a suitable quadrature formula, we obtain the least-squares problem:

$$\underset{x \in \Re^n}{\text{minimize}} \sum_{i=1}^{m} \left(\bar{\phi}(x, t_i) - \bar{\mathcal{F}}(t_i) \right)^2,$$

where $\bar{\phi}$ and $\bar{\mathcal{F}}$ incorporate the weights of the quadrature scheme.

Although the function (4.58) can be minimized by a general unconstrained method, in most circumstances the properties of (4.58) make it worthwhile to use methods designed specifically for the least-squares problem. In particular, the gradient and Hessian matrix of (4.58) have a special structure. Let the $m \times n$ Jacobian matrix of $f(x)$ be denoted by $J(x)$, and let the matrix $G_i(x)$

denote the Hessian matrix of $f_i(x)$. Then

$$g(x) = J(x)^T f(x); \qquad (4.59a)$$
$$G(x) = J(x)^T J(x) + Q(x), \qquad (4.59b)$$

where $Q(x) = \sum_{i=1}^{m} f_i(x) G_i(x)$. From (4.59b) we observe that the Hessian of a least-squares objective function consists of a special combination of first- and second-order information.

Least-squares methods are typically based on the premise that eventually the first-order term $J(x)^T J(x)$ of (4.59b) will *dominate* the second-order term $Q(x)$. This assumption is *not* justified when the residuals at the solution are very large — i.e., roughly speaking, when the residual $\|f(x^*)\|$ is comparable to the largest eigenvalue of $J(x^*)^T J(x^*)$. In such a case, one might as well use a general unconstrained method. For many problems, however, the residual at the solution *is* small enough to justify the use of a special method.

4.7.2. The Gauss-Newton Method

Let x_k denote the current estimate of the solution; a quantity subscripted by k will denote that quantity evaluated at x_k. From (4.59), the Newton equations (4.17) become

$$(J_k^T J_k + Q_k) p_k = -J_k^T f_k. \qquad (4.60)$$

Let p_N denote the solution of (4.60) (the Newton direction).

If $\|f_k\|$ tends to zero as x_k approaches the solution, the matrix Q_k also tends to zero. Thus, the Newton direction can be *approximated* by the solution of the equations

$$J_k^T J_k p_k = -J_k^T f_k. \qquad (4.61)$$

Note that the system (4.61) involves only the *first derivatives* of f, and *must be compatible*.

The solution of (4.61) is a solution of the *linear least-squares problem*

$$\underset{p \in \Re^n}{\text{minimize}} \; \frac{1}{2} \|J_k p + f_k\|_2^2, \qquad (4.62)$$

and is unique if J_k has full column rank. The vector that solves (4.62) is called the *Gauss-Newton direction*, and will be denoted by p_{GN}. The method in which this vector is used as a search direction is known as the *Gauss-Newton method*.

If J_k is of *full column rank*, the Gauss-Newton direction approaches the Newton direction as $\|Q(x_k)\|$ tends to zero, in the following sense: if $\|Q(x_k)\| = \epsilon$ for a sufficiently small positive scalar ϵ, then

$$\frac{\|p_N - p_{GN}\|}{\|p_N\|} = O(\epsilon).$$

Consequently, if $\|f(x^*)\|$ is zero and the columns of $J(x^*)$ are linearly independent, *the Gauss-Newton method can ultimately achieve a quadratic rate of convergence*, despite the fact that only first derivatives are used to compute p_{GN}.

Early implementations of the Gauss-Newton method typically formed the explicit matrix $J_k^T J_k$ and computed p_{GN} by solving the equations (4.61). The disadvantage of this approach is that the condition number of $J_k^T J_k$ is the square of that of J_k, and consequently unnecessary error may occur in determining the search direction. For example, on a machine with twelve-figure accuracy, if J_k has a condition number of 10^6, we would expect to lose only six figures of precision in computing the Gauss-Newton direction; however, if the equations (4.61) are used, there may be *no* correct figures in the computed solution.

Ill-conditioning is a common feature of nonlinear least-squares problems derived from parameter estimation problems because the underlying mathematical model is often ill-defined. Unnecessary exacerbation of the conditioning can be avoided by solving the linear least-squares problem (4.62) using the complete orthogonal factorization (Section 2.2.5.3) or the singular-value decomposition (Section 2.2.5.5).

If J_k does not have full column rank, the Gauss-Newton method must be implemented with great care to avoid unnecessary failures. Firstly, if J_k is rank-deficient, the solution of (4.62) is not unique. A satisfactory choice of p_k in this case is the solution of (4.62) of minimum Euclidean length. This vector can be computed using either the complete orthogonal factorization or the singular-value decomposition.

Any implementation that uses a minimum-norm solution to (4.62) in the rank-deficient case must include a strategy for estimating the rank of J_k. The procedures described in Section 2.2.5.3 for computing the complete orthogonal factorization of a rank-deficient matrix assumed that the rank of the matrix was known *a priori* to be r, and that the first r columns of the matrix were linearly independent. In practice, the rank of the Jacobian will be unknown, and it is therefore necessary to estimate the rank during the course of the computation.

Unfortunately, the definition of "rank" in the context of computation with floating-point arithmetic is problem-dependent. The question can never be resolved in a specific case without making an explicit judgement about the scaling, i.e. a determination as to which quantities can be considered as "negligible".

Example 4.10. To illustrate the complexity of the issue, consider the matrix

$$J = \begin{pmatrix} 1 & 1 \\ 0 & \epsilon \end{pmatrix},$$

where ϵ is not zero, but is small relative to unity. Mathematically, the two columns are linearly independent, and the matrix has rank two. In practice, however, the second vector may be a computed version of the first, so that numerically the two columns should be considered "equivalent", even though one is not an exact multiple of the other; in this event, the matrix has "rank" one. Thus, the decision as to whether these two vectors are linearly independent depends on whether or not the value of ϵ is "negligible" *in this problem*.

With exact arithmetic, linear dependence among a set of columns would reveal itself during the Householder reduction to upper-triangular form described in Section 2.2.5.3. If the $(k+1)$-th column were a linear combination of the previous k columns, components $k+1$ through m of the transformed dependent column (the "remaining column") would be exactly zero. With finite-precision computation, one might accordingly hope that the norm of a remaining column in the QR factorization would be "small" if that column were "nearly" linearly dependent on the previous columns. However, this hope is not realized, since it is equivalent to expecting that an ill-conditioned triangular matrix will have at least one small diagonal element. Triangular matrices exist that have no "small" diagonal elements, yet are arbitrarily badly conditioned. Consequently, a strategy based on the expectation of a "small" remaining column in the presence of near linear-dependence cannot be guaranteed.

The situation is slightly more straightforward with the singular-value decomposition, since near rank-deficiency must be revealed by small singular values. However, there may still be difficulties. For example, consider a matrix with singular values 1, 10^{-1}, $10^{-2}, \ldots,$ 10^{-10}, $10^{-11}, \ldots,$ 10^{-20}. Here it is necessary to make a fairly arbitrary decision as to which singular value is "negligible", and the estimated rank may vary considerably.

The determination of rank is critical in the Gauss-Newton method because its performance is highly sensitive to the estimate of rank. To illustrate this, consider the following example.

Example 4.11. Let J and f be defined by

$$J = \begin{pmatrix} 1 & 0 \\ 0 & \epsilon \end{pmatrix} \quad \text{and} \quad f = \begin{pmatrix} f_1 \\ f_2 \end{pmatrix},$$

where ϵ is small relative to unity and f_1 and f_2 are of order one. If J is judged to be of rank two, the search direction is given by

$$-\begin{pmatrix} f_1 \\ f_2/\epsilon \end{pmatrix}. \tag{4.63}$$

However, if we consider J to be of rank one, the search direction will be

$$-\begin{pmatrix} f_1 \\ 0 \end{pmatrix}. \tag{4.64}$$

The directions (4.63) and (4.64) are almost orthogonal, and (4.63) is almost orthogonal to the gradient vector. Hence, a change of unity in the estimated rank has generated two nearly orthogonal search directions.

It is unfortunate but true that no single strategy for rank estimation consistently yields the best performance from the Gauss-Newton method. When F is actually close to an ill-conditioned quadratic function, the best strategy is to allow the maximum possible estimate of the rank. On the other hand, there are several reasons for underestimating the rank. When J_k is nearly rank-deficient, a generous estimate of the rank tends to cause very large elements in the solution of (4.62). A lower estimate of the rank often produces a solution of more reasonable size, yet causes a negligible increase in the residual of (4.62). Another reason for tending to favour a conservative estimate of the rank is that the perturbation analysis of the least-squares problem shows that the relative change in the exact solution can include a factor $(\text{cond}(J_k))^2$ for an incompatible right-hand side. Thus, a smaller estimate of the rank may avoid the need to solve an ill-conditioned problem to obtain p_k.

4.7.3. The Levenberg-Marquardt Method

A popular alternative to the Gauss-Newton method is the *Levenberg-Marquardt method.* The Levenberg-Marquardt search direction is defined as the solution of the equations

$$(J_k^T J_k + \lambda_k I)p_k = -J_k^T f_k, \tag{4.65}$$

where λ_k is a non-negative scalar. A unit step is always taken along p_k, i.e., x_{k+1} is given by $x_k + p_k$. It can be shown that, for some scalar Δ related to λ_k, the vector p_k is the solution of the constrained subproblem

$$\underset{p \in \Re^n}{\text{minimize}} \quad \frac{1}{2}\|J_k p + f_k\|_2^2$$

$$\text{subject to} \quad \|p\|_2 \leq \Delta.$$

Hence, the Levenberg-Marquardt algorithm is of the trust-region type discussed in the Notes at the end of Section 4.4, and a "good" value of λ_k (or Δ) must be chosen in order to ensure descent. If λ_k is zero, p_k is the Gauss-Newton direction; as $\lambda_k \to \infty$, $\|p_k\| \to 0$ and p_k becomes parallel to the steepest-descent direction. This implies that $F(x_k + p_k) < F_k$ for sufficiently large λ_k.

Let $p_{\mathrm{LM}}(\lambda_k)$ denote the solution of (4.65) for a specified value of x_k, where λ_k is positive. If J_k is rank-deficient, in general

$$\frac{\|p_{\mathrm{N}} - p_{\mathrm{LM}}(\lambda_k)\|}{\|p_{\mathrm{N}}\|} = O(1),$$

regardless of the size of $\|Q(x)\|$ or λ_k.

*4.7.4. Quasi-Newton Approximations

Both the Gauss-Newton and Levenberg-Marquardt methods are based on the assumption that ultimately $J_k^T J_k$ is a good approximation to (4.59b), i.e. that Q_k can be neglected. This assumption is not justified for so-called *large-residual* problems, in which $\|f(x^*)\|$ is not "small". Some care must be used in defining this problem class. Assuming that $\|G_i(x)\|$ is of order unity for $i = 1, \ldots, m$, the matrix $Q(x)$ will always be significant in the Hessian of (4.58) if the residual $\|f(x^*)\|$ exceeds the small eigenvalues of $J(x^*)^T J(x^*)$. We therefore define a large-residual problem as one in which the optimal residual is large relative to the *small* eigenvalues of $J(x^*)^T J(x^*)$, but *not* with respect to its largest eigenvalue (see Section 4.7.1). If the optimal residual is too large, no advantage is gained by exploiting the least-squares nature of the objective function.

Assuming that only first derivatives of $f(x)$ are available, one possible strategy for large-residual problems is to include a quasi-Newton approximation M_k of the unknown second derivative term $Q(x)$. Application of quasi-Newton updates is more complicated than in ordinary unconstrained optimization because in the least-squares case a part of the Hessian at x_{k+1} is known exactly. The search direction with a quasi-Newton approximation to Q_k is given by

$$(J_k^T J_k + M_k)p_k = -J_k^T f_k.$$

The condition analogous to (4.30) imposed on the updated approximation M_{k+1} is

$$(J_{k+1}^T J_{k+1} + M_{k+1})s_k = y_k,$$

where $s_k = x_{k+1} - x_k$ and $y_k = J_{k+1}^T f_{k+1} - J_k^T f_k$. Note that M_{k+1} depends on J_{k+1} as well as the previous approximation M_k.

Any of the updating formulae discussed in Section 4.5.2 may be used to construct M_k. The following formula is based upon the BFGS update (4.36):

$$M_{k+1} = M_k - \frac{1}{s_k^T W_k s_k} W_k s_k s_k^T W_k + \frac{1}{y_k^T s_k} y_k y_k^T, \qquad (4.66)$$

where $W_k = J_{k+1}^T J_{k+1} + M_k$.

When M_{k+1} satisfies (4.66), the combined matrix $J_{k+1}^T J_{k+1} + M_{k+1}$ will be positive definite if $J_{k+1}^T J_{k+1} + M_k$ is positive definite. This property is useful asymptotically when $J_k^T J_k$ is approximately equal to $J_{k+1}^T J_{k+1}$. However, $J_k^T J_k + M_k$ may be indefinite at points far from the solution, and hence some care is needed to ensure that the search direction is a descent direction.

Superlinear convergence can be proved for certain quasi-Newton updates in the least-squares case. However, it is important to note that the properties of hereditary positive-definiteness and n-step termination described in Section 4.5.2 do not apply in the least-squares case because we are approximating only *part* of the Hessian. The difficulty of blending exact and approximated curvature information may explain why algorithms based on a quasi-Newton approximation of $Q(x)$ do not generally converge as rapidly in practice as their counterparts for general unconstrained minimization.

*4.7.5. The Corrected Gauss-Newton Method

If the Gauss-Newton method converges, it often does so with remarkable rapidity and can even be more efficient (in terms of function evaluations) than Newton's method itself. In this section we shall define an algorithm that can be viewed as a modification of the Gauss-Newton method which allows convergence for rank-deficient and large-residual problems.

Consider the singular-value decomposition of J_k:

$$J_k = U\begin{pmatrix} S \\ 0 \end{pmatrix}V^T,$$

where $S = \text{diag}(\sigma_1, \sigma_2, \ldots, \sigma_n)$ is the matrix of singular values, ordered such that $\sigma_i \geq \sigma_{i+1}$ $(1 \leq i \leq n-1)$; U is an $m \times m$ orthonormal matrix; and V is an $n \times n$ orthonormal matrix.

Substituting into the Newton equations (4.60) and cancelling the non-singular matrix V, we obtain

$$(S^2 + V^T Q_k V)V^T p_{\text{N}} = -S\bar{f}, \tag{4.67}$$

where $\bar{f}$ denotes the first n components of the m-vector $U^T f_k$. The Gauss-Newton equation (4.61) arises by *ignoring* the matrix $V^T Q_k V$ in (4.67). The equation for the Gauss-Newton direction can thus be written as

$$S^2 V^T p_{\text{GN}} = -S\bar{f}.$$

When S is non-singular, p_{GN} is given by

$$p_{\text{GN}} = -V S^{-1} \bar{f}.$$

Difficulties with the Gauss-Newton method occur when Q_k is not "negligible" — for example, when J_k is rank-deficient (i.e., S is singular), or in large-residual problems. The idea of the *corrected Gauss*-Newton method is to split the singular values of J_k into two groups by choosing an integer r $(0 \leq r \leq n)$ that represents, roughly speaking, the number of "dominant" singular values; r will be termed the *grade* of J_k, and is determined by the corrected Gauss-Newton algorithm. Various vectors and matrices are then partitioned based on the grade. The matrix V will be partitioned into V_1 (the first r columns), and V_2 (the last $n - r$ columns). Similarly, S_1 will denote the diagonal matrix whose diagonal elements are $\sigma_1, \ldots, \sigma_r$ (where $\sigma_r > 0$), and S_2 is given by $\text{diag}(\sigma_{r+1}, \ldots, \sigma_n)$; $\bar{f}_1$ and $\bar{f}_2$ denote the first r components and the last $n - r$ components of $\bar{f}$ respectively.

Every n-vector p can be written as a linear combination of the columns of V; in particular, we may write the Newton direction as

$$p_{\text{N}} = V_1 p_1 + V_2 p_2. \tag{4.68}$$

If we substitute (4.68) into the first r equations of (4.67) we obtain

$$(S_1^2 + V_1^T Q_k V_2)p_1 + V_1^T Q_k V_2 p_2 = -S_1 \bar{f}_1.$$

If all the terms involving Q_k are assumed to be $O(\epsilon)$ for some small ϵ and are therefore ignored, we may solve for $\bar{p}_1$, an $O(\epsilon)$ approximation to the vector p_1:

$$\bar{p}_1 = -S_1^{-1} \bar{f}_1.$$

The vector $V_1 \bar{p}_1$ is termed the *Gauss-Newton direction in the subspace spanned by V_1*.

The last $n - r$ Newton equations are given by

$$V_2^T Q_k V_1 p_1 + (S_2^2 + V_2^T Q_k V_2)p_2 = -S_2 \bar{f}_2,$$

If $\bar{p}_1$ is substituted for p_1 in these equations, an $O(\epsilon)$ approximation to p_2 is the solution of

$$(S_2^2 + V_2^T Q_k V_2)\bar{p}_2 = -S_2 \bar{f}_2 - V_2^T Q_k V_1 \bar{p}_1. \tag{4.69}$$

The corrected Gauss-Newton direction is then defined as

$$p_C = V_1 \bar{p}_1 + V_2 \bar{p}_2.$$

The Newton direction (4.68) can be considered as a corrected Gauss-Newton direction with $r = 0$ and consequently p_C "interpolates" between the Gauss-Newton direction and the Newton direction.

In the corrected Gauss-Newton method, the grade is updated at each iteration based on whether a satisfactory decrease in the objective function is attained. The idea is to maintain r at the value n as long as adequate progress is made with the Gauss-Newton direction. Obviously, the choice of the grade is important in the definition of the corrected Gauss-Newton direction. However, since the grade is updated based on the progress being made, it is less critical than the choice of rank in the Gauss-Newton method.

Three versions of a corrected Gauss-Newton method can be developed. Firstly, if exact second derivatives are available, Q_k can be used explicitly; secondly, a finite-difference approximation to $Q_k V_2$ can be obtained by differencing the gradient along the columns of V_2; and finally, a quasi-Newton approximation to Q_k may be used. In any corrected Gauss-Newton algorithm, some technique such as the modified Cholesky factorization (Section 4.4.2.2) must be used to solve the equations (4.69) (or (4.67) when the grade is zero) in order to ensure a descent direction.

4.7.6. Nonlinear Equations

A problem closely related to the nonlinear least-squares problem (and to general unconstrained optimization) is that of *solving a system of nonlinear equations*, i.e. finding x^* such that

$$f(x^*) = 0, \tag{4.70}$$

where $f(x)$ is an n-component function. The problems are related because the gradient of a smooth nonlinear function vanishes at a local minimum.

A Newton method can be defined for the nonlinear system (4.70), exactly as in the univariate case (Section 4.1.1.2). From the Taylor-series expansion of f about a point x_k, we can obtain a *linear approximation to f*, i.e.

$$f(x^*) \approx f(x_k) + J(x_k)(x^* - x_k). \tag{4.71}$$

The *Newton step* p_N is an approximation to $x^* - x_k$, and is defined by equating the right-hand side of (4.71) to zero. Thus, p_N satisfies a system of equations analogous to (4.1) and (4.17):

$$J(x_k)p_N = -f(x_k), \tag{4.72}$$

and the next iterate is given by $x_{k+1} = x_k + p_N$. Under standard assumptions about the non-singularity of $J(x^*)$ and the closeness of x_k to x^*, the sequence $\{x_k\}$ converges quadratically to x^*.

The same idea of using a linear approximation to a nonlinear function can be applied even when there are fewer equations than unknowns (so that $J(x_k)$ in (4.72) is not square). In this case, the equations (4.72) do not uniquely determine the Newton step.

Numerous methods have been developed specially for solving nonlinear equations. Many of these methods are based upon the Newton equations (4.72). However, methods for nonlinear least-squares problems are also applicable, since $J(x)^T f(x)$, the gradient of the sum of squares (4.58), vanishes when $f(x)$ is zero. The decision to use a nonlinear least-squares method rather than a method designed specifically for nonlinear equations should be made carefully. If $J(x)$ is singular, a least-squares method may converge to a point that is not a solution of (4.70), since the gradient (4.59a) may vanish even when $f(x)$ does not. Nonetheless, if the definition of the nonlinear equations contains small inaccuracies that prevent (4.70) from having a solution, yet give a small residual for (4.58), the least-squares approach will be preferable. Two major factors in the choice of method are the efficiency of the least-squares procedure when m is equal to n and the ability of the method to avoid squaring the condition number of the Jacobian.

Notes and Selected Bibliography for Section 4.7

See Dennis (1977) and Ramsin and Wedin (1977) for a general survey of methods for nonlinear least-squares problems.

The Gauss-Newton method has been considered by numerous authors, including Ben-Israel (1967), Fletcher (1968), and Wedin (1974). For some discussion of the defects of the Gauss-Newton method, see Powell (1972) and Gill and Murray (1976a). The use of the QR factorization to determine the solution of a linear least-squares problem was suggested by Businger and Golub (1965). For a discussion of the relationship between least-squares solutions and the pseudo-inverse, see Peters and Wilkinson (1970). Background information concerning the singular-value decomposition can be found in Golub and Reinsch (1971) and Lawson and Hanson (1974). We have not discussed how preconditioned conjugate-gradient methods (see Section 4.8.5) may be used to include the second-order part of the Hessian; this idea was suggested by Ruhe (1979) in the context of "accelerating" the Gauss-Newton method.

The Levenberg-Marquardt algorithm was proposed independently by Levenberg (1944) and Marquardt (1963). Fletcher (1971a) has proposed an algorithm that adjusts the value of λ_k in (4.65) according to the relationship between the actual and predicted change in the sum of squares. Alternatively, Δ may be adjusted and the implicitly defined value of λ_k can be computed using a variant of Hebden's method that is specially tailored to least-squares problems (see Hebden, 1973; Moré, 1977; and the Notes for Section 4.4).

Quasi-Newton methods for the nonlinear least-squares problem are given, for example, by Dennis (1973), Betts (1976), Dennis, Gay and Welsch (1977) and Gill and Murray (1978a). The corrected Gauss-Newton method is due to Gill and Murray (1978a).

It is sometimes worthwhile using a method that takes advantage of special structure in the functions $\{f_i\}$. For example, some of the variables may appear *linearly* in the $\{f_i\}$, or the Jacobian may be structured so that the problem may be transformed into several independent problems of smaller dimension. Methods for *separable least-squares problems* were first suggested by Golub and Pereyra (1973). Other references can be found in Kaufman and Pereyra (1978) and Ruhe and Wedin (1980).

Many authors have considered methods for nonlinear equations; see, for example, Broyden (1965), Powell (1970b), Boggs (1975), Brent (1973b), Gay and Schnabel (1978), and Moré (1977).

4.8. METHODS FOR LARGE-SCALE PROBLEMS

When n becomes very large, two related difficulties can occur with the general methods described in previous sections: the computation time required may become too long to justify solving the problem; and, more critical, there may not be enough storage locations, either in core or on auxiliary storage, to contain the matrix needed to compute the direction of search.

Fortunately, the nature of certain large-scale problems allows effective solution methods to be developed. The Hessian matrices of many large unconstrained problems have a very high proportion of zero elements, since it is often known *a priori* that certain variables have no nonlinear interaction (which implies that the corresponding elements of the Hessian are zero). The ratio of non-zero elements to the total number of elements in a matrix is termed its *density*. A matrix with an insignificant number of zero elements is called *dense*; a matrix with a high proportion of zero entries is called *sparse*. In practice, density tends to decrease as the size of the problem (the number of variables) increases, and the number of non-zero elements often increases only linearly with n.

Besides having a low density, the matrices that occur in large problems tend nearly always to have a *structure* or pattern. In a structured matrix the non-zero and zero elements are not scattered at random, but are known to occur in certain positions due to the nature of the problem and the relationships among the variables.

In this section, we consider how very large problems might be solved by taking advantage of known sparsity and structure in the Hessian matrix to reduce the requirements of computation and storage. For example, a multiplication involving a zero element need not be executed, and a large block of zeros in a structured matrix need not be stored.

4.8.1. Sparse Discrete Newton Methods

Discrete Newton methods have a fundamental advantage over other methods when the Hessian matrix is sparse with a known fixed sparsity pattern or structure.

In the dense case, the columns of the identity are used as *finite-difference vectors*, and one gradient evaluation is therefore required to form each column of the Hessian approximation (see Section 4.5.1). When the sparsity pattern of the Hessian is known and symmetry is assumed, it is possible to choose special finite-difference vectors that allow a finite-difference approximation to $G(x)$ to be computed with many fewer than n evaluations of the gradient.

For example, suppose that $G(x)$ is tri-diagonal:

$$G = \begin{pmatrix} \times\times & & & & & & & \\ \times\times\times & & & & & & \\ & \times\times\times & & & & & \\ & & \times\times\times & & & & \\ & & & \ddots & \ddots & \ddots & \\ & & & & \times & \times\times & \\ & & & & & \times\times\times & \\ & & & & & & \times\times \end{pmatrix},$$

and that we compute the vectors

$$y_i = \frac{1}{h\|z_i\|_2}\big(g(x_k + h z_i) - g(x_k)\big), \quad i = 1, 2,$$

where $z_1 = (1, 0, 1, 0, \ldots)^T$, $z_2 = (0, 1, 0, 1, \ldots)^T$, and h is an appropriate finite-difference interval. Let $y_{1,i}$ denote the i-th component of y_1, and similarly for y_2. The vectors y_1 and y_2 are approximations to the sums of odd and even columns of G_k, respectively. Therefore,

$$y_{1,1} \approx \frac{\partial^2 F}{\partial^2 x_1}; \quad y_{2,1} \approx \frac{\partial^2 F}{\partial x_1 \partial x_2}; \quad y_{1,2} \approx \frac{\partial^2 F}{\partial x_1 \partial x_2} + \frac{\partial^2 F}{\partial x_2 \partial x_3}; \quad \text{etc.}$$

Thus, for example,

$$y_{1,2} - y_{2,1} \approx \frac{\partial^2 F}{\partial x_2 \partial x_3}.$$

In this fashion, all the elements of G_k can be approximated with only two evaluations of the gradient, *regardless of the value of* n.

This technique is not restricted to band matrices. Consider, for example, a matrix with "arrowhead" structure

$$G = \begin{pmatrix} \times & & & & & & \times \\ & \times & & & & & \times \\ & & \times & & & & \times \\ & & & \times & & & \times \\ & & & & \ddots & & \vdots \\ & & & & & \times & \times \\ & & & & & & \times\times \\ \times\times\times\times & & \times & & \times\times \end{pmatrix}.$$

Because of symmetry, a finite-difference approximation $\hat{G}$ can be obtained with only two evaluations of the gradient, by using the finite-difference vectors $z_1 = (1, 1, \ldots, 1, 0)^T$, and $z_2 = (0, 0, \ldots, 0, 1)^T$. A finite-difference of the gradient along z_1 gives the first $n - 1$ diagonal elements of $\hat{G}$; the remaining non-zero elements are obtained by a finite-difference along z_2.

Discrete Newton methods for sparse problems usually consist of two stages: a preprocessing algorithm that permutes the rows and columns for a given sparsity pattern in order to achieve a structure that can be utilized to reduce the number of gradient evaluations; and the minimization proper. Since the saving of one gradient evaluation during the computation of $\hat{G}_k$ will be reflected in every iteration of the minimization, it is usually worthwhile to expend considerable effort in finding a suitable set of finite-difference vectors.

It is sometimes overlooked that the computation of a sparse Hessian approximation $\hat{G}_k$ is only the first of two steps required to obtain the search direction. The second is to solve a sparse linear system that involves $\hat{G}_k$ (see Section 4.5.1):

$$\hat{G}_k p_k = -g_k. \tag{4.73}$$

As in the dense case, some strategy must be used to ensure that p_k is a satisfactory descent direction. The critical new difficulty that arises in the sparse case is that $\hat{G}_k$ must be *transformed* in order to solve (4.73) with a matrix factorization. When elementary transformations are applied to a matrix, an element that was originally zero may become non-zero in the factors; this phenomenon is known as *fill-in*. Fill-in is significant in sparse problems because only the non-zero elements are stored, and hence additional storage is required for each new non-zero element.

Because of fill-in, the existence of a sparse Hessian approximation is not sufficient to ensure that (4.73) can be solved efficiently (if at all). For example, with the modified Cholesky factorization (Section 4.4.2.2), the columns of the factor L_k are linear combinations of the columns of $\hat{G}_k$, and hence L_k can be quite dense even when $\hat{G}_k$ is very sparse. A factorization of $\hat{G}_k$ can be used to solve (4.73) only if excessive fill-in does not occur.

If the Hessian matrix is structured and has a pattern of non-zero elements that produces a manageable amount of fill-in during the Cholesky factorization, a discrete modified Newton method is an extremely effective minimization technique. The convergence rate is generally quadratic (subject to the restrictions noted in Section 4.5.1 concerning the limiting accuracy), and convergence to a strong local minimum can usually be verified.

*4.8.2. Sparse Quasi-Newton Methods

Certain quasi-Newton updates can be derived in terms of finding the minimum-norm correction to the current Hessian approximation, subject to satisfying the quasi-Newton condition and possibly other conditions as well — e.g., the problem (4.48) characterizes the PSB update.

When the Hessian of the original problem is sparse, it is possible to add the further constraint to (4.48) that the updated matrix should retain a specified sparsity pattern. We define the set of indices $\mathcal{N}$ as $\{(i,j) \mid G_{ij} = 0\}$, so that $\mathcal{N}$ represents the specified sparsity pattern of the Hessian, and we assume that B_k has this sparsity pattern. The extended minimization problem to be solved for a sparse update matrix is then

$$
\begin{aligned}
\text{minimize} \quad & \|U\|_F^2 = \sum_{i=1}^{n} \sum_{j=1}^{n} U_{ij}^2 \\
\text{subject to} \quad & U s_k = y_k - B_k s_k \\
& U = U^T, \\
& U_{ij} = 0 \text{ for } (i,j) \in \mathcal{N}.
\end{aligned}
\qquad (4.74)
$$

Let $\sigma^{(j)}$ denote the vector s_k which has imposed upon it the sparsity pattern of the j-th column of B_k. Some lengthy and complicated analysis can be used to show that the solution of (4.74) is given by

$$
U_k = \sum_{j=1}^{n} \lambda_j \left(e_j \sigma^{(j)T} + \sigma^{(j)} e_j^T \right),
\qquad (4.75)
$$

where e_j is the j-th unit vector and λ is the vector of Lagrange multipliers associated with the subproblem (4.74). The vector λ is the solution of the linear system

$$
Q\lambda = y_k - B_k s_k,
\qquad (4.76)
$$

where

$$
Q = \sum_{j=1}^{n} \left(\sigma_j^{(j)} \sigma^{(j)} + \|\sigma^{(j)}\|_2^2 e_j \right) e_j^T.
$$

The matrix Q is symmetric and has the same sparsity pattern as B_k; Q is positive definite if and only if $\|\sigma^{(j)}\| > 0$ for all j. Note that the update matrix U_k (4.75) is of rank n, and that the linear system (4.76) must be solved from scratch in order to compute the update. Furthermore, it is not possible to impose the property of hereditary positive-definiteness on the sparse update.

In order to compute the sparse update, the linear system (4.76) must be solved for λ. In addition, since the update matrix is not of low rank (as it is in the dense case), (4.28) must be solved from scratch to compute the search direction. Therefore, a sparse quasi-Newton method requires the solution of two sparse linear systems, and storage for Q as well as B_k.

It is possible to obtain the sparse analogue of *any* quasi-Newton formula using a similar analysis. In this case a minimum-norm correction is required that annihilates the unwanted non-zero elements in the usual quasi-Newton update. The resulting sparse update is then of identical form to (4.75), but with the vector λ defined as the solution of the equations

$$Q\lambda = y_k - B_k s_k - U s_k,$$

where U is any quasi-Newton update with the sparsity pattern of B_k imposed upon it.

Sparse quasi-Newton methods are still in an early stage of development, and a significant amount of research remains to be done. At the moment, computational results indicate that, in terms of the number of function evaluations required, currently available sparse quasi-Newton methods are less effective than discrete Newton methods. Moreover, current sparse quasi-Newton methods lose three advantages of their dense counterparts: maintenance of positive definiteness, the ability to recur B_k in invertible form, and the lower overhead per iteration compared to a Newton-type method.

4.8.3. Conjugate-Gradient Methods

Conjugate-gradient-type methods form a class of algorithms that generate directions of search without storing a matrix, in contrast to all the methods discussed thus far. They are essential in circumstances when methods based on matrix factorization are not viable because the relevant matrix is too large or too dense.

4.8.3.1. Quadratic functions. Suppose that we wish to find the minimum of the quadratic function

$$\Phi(x) = c^T x + \frac{1}{2} x^T G x, \tag{4.77}$$

where G is symmetric and positive definite. Assume that x_k is an approximation to the minimum of Φ. Given $k + 1$ linearly independent vectors $p_0, p_1, \ldots, p_k$ that span the subspace $\mathcal{P}_k$, let P_k be the matrix with columns $p_0, p_1, \ldots, p_k$.

The minimum of Φ *over the manifold* $x_k + \mathcal{P}_k$ is defined as the solution of the minimization problem

$$\underset{w \in \Re^{k+1}}{\text{minimize}} \; F(x_k + P_k w).$$

If $x_k + P_k w$ is substituted into the expression (4.77), we find that the optimal w minimizes the quadratic function

$$w^T P_k^T g_k + \frac{1}{2} w^T P_k^T G P_k w, \tag{4.78}$$

where $g_k = \nabla\Phi(x_k) = c + Gx_k$. The function (4.78) is minimized at the point

$$w = -(P_k^T G P_k)^{-1} P_k^T g_k,$$

and consequently x_{k+1} (the point in the manifold where Φ is minimized) is given by

$$x_{k+1} = x_k - P_k(P_k^T G P_k)^{-1} P_k^T g_k. \tag{4.79}$$

The iterates in this procedure display several interesting properties. Firstly, g_{k+1}, the gradient of Φ at x_{k+1}, is orthogonal to the vectors $\{p_i\}$ (the columns of P_k). Note that

$$
\begin{aligned}
P_k^T g_{k+1} &= P_k^T (c + Gx_{k+1}) \\
&= P_k^T g_k - P_k^T G P_k (P_k^T G P_k)^{-1} P_k^T g_k \\
&= 0.
\end{aligned}
$$

Therefore,

$$g_{k+1}^T p_i = 0, \quad i = 0, \ldots, k.$$

Hence, if each previous iterate x_j, $j = 1, \ldots, k$ has been obtained by minimizing Φ over the linear manifold $x_{j-1} + P_{j-1}$, by induction it holds that for $j = 1, \ldots, k$

$$g_j^T p_i = 0, \quad j > i. \tag{4.80}$$

The expression (4.79) thus becomes

$$x_{k+1} = x_k + \gamma P_k(P_k^T G P_k)^{-1} e_k, \tag{4.81}$$

where e_k is the k-th column of the identity matrix and $\gamma = -g_k^T p_k$.

Great simplifications occur in (4.81) when the matrix $P_k^T G P_k$ is diagonal. Suppose that the $k+1$ vectors $\{p_j\}$, $j = 0, \ldots, k$, are *mutually conjugate with respect to the matrix* G, i.e. that for $i = 0, \ldots, k$ and $j = 0, \ldots, k$

$$p_i^T G p_j = 0, \quad i \neq j. \tag{4.82}$$

When (4.82) holds, (4.79) becomes

$$x_{k+1} = x_k + \alpha_k p_k, \tag{4.83}$$

where $\alpha_k = -g_k^T p_k / p_k^T G p_k$. The value α_k is the step to the minimum of Φ along p_k. Thus, (4.83) indicates that the $(k+1)$-th vector p_k can be considered as a *search direction* in the usual model algorithm for unconstrained optimization, when the iterates are obtained by minimizing Φ in the manifold defined by successive conjugate directions.

By definition of Φ,

$$g_{i+1} - g_i = G(x_{i+1} - x_i) = \alpha_i G p_i. \tag{4.84}$$

Let y_i denote the vector $g_{i+1} - g_i$. Then the *conjugacy condition* $p_i^T G p_j = 0$ is equivalent to the *orthogonality condition* $y_i^T p_j = 0$; we shall use these conditions interchangeably.

A set of mutually conjugate directions can be obtained by taking p_0 as the steepest-descent direction $-g_0$ and computing each subsequent direction as a linear combination of g_k and the previous k search directions,

$$p_k = -g_k + \sum_{j=0}^{k-1} \beta_{kj} p_j. \tag{4.85}$$

When the direction p_k is defined by (4.85), g_k is a linear combination of $p_0, p_1, \ldots, p_k$, and thus for $1 \le i \le k$, $g_i \in \mathcal{P}_k$. Since $p_0 = -g_0$, it also holds trivially that $g_0 \in \mathcal{P}_k$. Therefore, from (4.80)

$$g_k^T g_i = 0, \quad i < k. \tag{4.86}$$

Furthermore, p_k can be constructed to be conjugate to $p_0, \ldots, p_{k-1}$. Pre-multiplying (4.85) by $p_i^T G$ and using (4.82) and (4.84), we obtain for $i = 0, \ldots, k-1$:

$$p_i^T G p_k = -p_i^T G g_k + \sum_{j=0}^{k-1} \beta_{kj} p_i^T G p_j$$

$$= -\frac{1}{\alpha_k} (g_{i+1} - g_i)^T g_k + \beta_{ki} p_i^T G p_i. \tag{4.87}$$

The relationship (4.86) implies that the first term on the right-hand side of (4.87) vanishes for $i < k-1$; thus, to make p_k conjugate to p_i for $i < k-1$, we can simply choose β_{ki} to be zero. Since only the coefficient $\beta_{k,k-1}$ is non-zero, we shall drop the first subscript on β. To obtain the value of β_{k-1} that ensures that p_k is conjugate to p_{k-1}, we pre-multiply (4.85) by y_{k-1}^T and apply the orthogonality condition that $y_{k-1}^T p_k = 0$; this gives

$$0 = -y_{k-1}^T g_k + \beta_{k-1} y_{k-1}^T p_{k-1},$$

or

$$\beta_{k-1} = \frac{y_{k-1}^T g_k}{y_{k-1}^T p_{k-1}}. \tag{4.88}$$

Therefore, p_k can be written as

$$p_k = -g_k + \beta_{k-1} p_{k-1}, \tag{4.89}$$

where β_{k-1} is given by (4.88). Note that the orthogonality of the gradients and the definition of p_k implies the following alternative (equivalent) definitions of β_{k-1}:

$$\beta_{k-1} = \frac{y_{k-1}^T g_k}{\|g_{k-1}\|_2^2} \quad \text{or} \quad \beta_{k-1} = \frac{\|g_k\|_2^2}{\|g_{k-1}\|_2^2}.$$

4.8.3.2. The linear conjugate-gradient method. The conjugate-gradient method described in Section 4.8.3.1 can also be viewed as a method for solving a set of positive-definite symmetric linear equations; in fact, it was originally derived for this purpose. If the conjugate-gradient algorithm is applied to minimize the quadratic function $c^T x + \frac{1}{2} x^T G x$, where G is symmetric and positive definite, it computes the solution of the system

$$Gx = -c. \tag{4.90}$$

When used to solve the system (4.90), the algorithm is known as the *linear conjugate-gradient method*. The remarkable feature of the conjugate-gradient method is that it computes the solution of a linear system using only products of the matrix with a vector, and does *not* require the elements of the matrix explicitly. For example, if G were given by $R^T R$, it would not be necessary to form this product to apply the conjugate-gradient method.

When describing the linear conjugate-gradient method, it is customary to use the notation r_j (for "residual") for the gradient vector $c + Gx_j$. To initiate the iterations, we adopt the convention that $\beta_{-1} = 0$, $p_{-1} = 0$. Given x_0 and $r_0 = c + Gx_0$, each iteration includes the following steps for $k = 0, 1, \ldots$:

$$
\begin{aligned}
p_k &= -r_k + \beta_{k-1} p_{k-1}; \\
\alpha_k &= \frac{\|r_k\|_2^2}{p_k^T G p_k}; \\
x_{k+1} &= x_k + \alpha_k p_k; \\
r_{k+1} &= r_k + \alpha_k G p_k; \\
\beta_k &= \frac{\|r_{k+1}\|_2^2}{\|r_k\|_2^2}.
\end{aligned}
\tag{4.91}
$$

In theory, the linear conjugate-gradient algorithm will compute the exact solution of (4.90) within a fixed number of iterations. In particular, the method has the property that, if exact arithmetic is used, convergence will occur in m ($m \leq n$) iterations, where m is the number of distinct eigenvalues of G. Hence, the linear conjugate-gradient method might be considered as a *direct* method for solving (4.90). However, in practice rounding errors rapidly cause the computed directions to lose conjugacy, and the method behaves more like an *iterative* method. The method may converge very quickly if the eigenvalues of G are clustered into groups of approximately equal value; due to the adverse effects of rounding error, considerably more than n iterations may be required if G has a general eigenvalue structure.

4.8.3.3. General nonlinear functions. When the gradient is available, a generalization of the conjugate-gradient method can be applied to minimize a nonlinear function $F(x)$ without the need to store any matrices. The only significant modification to the method is that α_k must be computed by an iterative process rather than in closed form. The finite termination property of the method on quadratics suggests that the definition (4.89) should be abandoned after a cycle of n linear searches, and that p_k should then be set to the steepest-descent direction $-g_k$. This strategy is known as *restarting* or *resetting*. Restarting with the steepest-descent direction is based upon the questionable assumption that the reduction in $F(x)$ along the restart direction will be greater than that obtained if the usual formula were used. The conjugate-gradient algorithm that restarts with the steepest-descent direction every n iterations is sometimes known as the *traditional conjugate-gradient method*.

The iterates of an implementation of the traditional conjugate-gradient method applied to Rosenbrock's function (Example 4.2) are shown in Figure 4o. Although the method is not intended for problems with such a small value of n, the figure is useful in illustrating the cyclic nature of the method. Note that the method requires considerably fewer iterations than the method of steepest descent, yet more iterations than the BFGS quasi-Newton method (*cf.* Figures 4j and 4m).

The difficulty and cost of finding the exact minimum of $F(x)$ along p_k have resulted in many implementations of the traditional conjugate-gradient method that allow inexact linear searches.

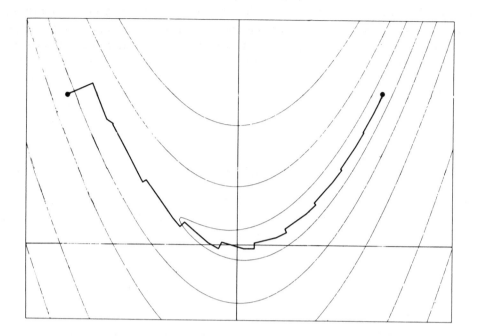

Figure 4o. Solution path of a conjugate-gradient algorithm on Rosenbrock's function. Although the method is not intended for problems with such a small value of n, the figure is useful in illustrating the cyclic nature of the traditional conjugate-gradient method.

However, care must then be exercised to ensure that p_k is a descent direction. If the formula (4.89) for p_k is pre-multiplied by g_k^T, we find that

$$g_k^T p_k = -g_k^T g_k + \beta_{k-1} g_k^T p_{k-1}.$$

If an exact linear search was made at the previous iteration, the direction p_k must be a descent direction, since $g_k^T p_{k-1}$ vanishes and hence $g_k^T p_k$ is just the negative quantity $-g_k^T g_k$. However, if an inexact linear search was made, the quantity $\beta_{k-1} g_k^T p_{k-1}$ may be positive and larger than $-g_k^T g_k$, in which case p_k will not be a descent direction. A typical remedy for such an eventuality is to restart the algorithm with p_k as the steepest-descent direction. Consequently, use of an arbitrary "crude" line search may cause the efficiency of the algorithm to be severely impaired by the use of a large number of steepest-descent iterations.

This problem is easily overcome if an additional check is made during the computation of the sequence $\{\alpha^j\}$ of trial steps generated by a safeguarded step-length algorithm. Let $\bar{g}_{k+1}$, $\bar{p}_{k+1}$ and $\bar{\beta}_k$ denote the quantities g_{k+1}, p_{k+1} and β_k respectively, computed at the point $x_k + \alpha^j p_k$. The step α^j will be considered as a candidate for the step-length α_k only if the condition

$$-\bar{g}_{k+1}^T \bar{p}_{k+1} \geq \sigma \|\bar{g}_{k+1}\|_2 \|\bar{p}_{k+1}\|_2 \tag{4.92}$$

is satisfied, where σ is a small positive scalar. If (4.92) is not satisfied at any trial point, an exact linear search should be performed to obtain α_k. The check (4.92) can be made efficiently because of the useful property of conjugate-gradient methods whereby the initial directional derivative for the next iteration can be computed relatively cheaply during the computation of the sequence $\{\alpha^j\}$. Note that $\bar{p}_{k+1}$ need not be computed explicitly, since $\bar{g}_{k+1}^T \bar{p}_{k+1}$ and $\|p_{k+1}\|$ can be obtained from $\bar{g}_{k+1}^T \bar{g}_{k+1}$, $\bar{\beta}_k$ and $\bar{g}_{k+1}^T p_k$. If the first value of α^j satisfies the condition (4.92), there is no extra cost involved in this test, since the associated quantities are used elsewhere in the algorithm.

*4.8.3.4. Conjugate-gradient methods with restarts. When periodic restarts with the steepest-descent direction are used within a conjugate-gradient algorithm, the reduction at the restart iteration is often poor compared with the reduction that would have occurred without restarting. Although it would seem useful if a cycle of n iterations could commence with the last direction of the previous cycle, this idea must be applied with some care.

If the initial direction p_0 in the linear conjugate-gradient method is taken as an arbitrary vector, the required conjugacy relations may not hold because g_0 is no longer a linear combination of the search directions. The following recurrence relation must be used to ensure that the successive directions are conjugate:

$$p_k = -g_k + \beta_{k-1} p_{k-1} + \gamma_k p_0, \tag{4.93}$$

where β_{k-1} is given by (4.88) and $\gamma_k = y_0^T g_k / y_0^T p_0$. This formula can be extended to nonlinear problems by computing a cycle of n directions: for $k = 0, \ldots, n-1$,

$$p_k = -g_k + \beta_{k-1} p_{k-1} + \gamma_k p_t, \tag{4.94}$$

where $\gamma_k = y_t^T g_k / y_t^T p_t$. The direction p_t is known as the *restart direction*, and is the last direction of the previous cycle along which a linear search was made.

In a nonlinear conjugate-gradient algorithm, p_k as defined by (4.94) may not be a descent direction, even if exact linear searches are made; this implies that the method may generate a poor direction of descent. Steps must therefore be taken to discard this direction, if necessary, in favour of the usual conjugate-gradient direction. A typical requirement for a "good" search direction is that p_k should be "sufficiently downhill"; for example, we may impose a condition analogous to (4.92), i.e. that

$$-g_k^T p_k \geq \rho \|p_k\|_2 \|g_k\|_2,$$

for some positive value ρ. If this requirement is not satisfied, a new cycle commences with p_{k-1} as the restart direction and with p_k recomputed from (4.89).

If an inexact linear search is used, the condition (4.92) should be checked at trial step lengths with the traditional conjugate-gradient direction defined by (4.89). This will ensure that, if the direction computed from the three-term formula (4.94) is discarded, the traditional conjugate-gradient direction will be a descent direction.

*4.8.3.5. Convergence. For a wide class of functions, the traditional conjugate-gradient method with exact linear searches and exact arithmetic is *n-step superlinearly convergent*, i.e.

$$\lim_{j \to \infty} \frac{\|x_{nj+n} - x^*\|}{\|x_{nj} - x^*\|} = 0.$$

The proof of n-step superlinear convergence is critically dependent upon the use of restarting. Algorithms that do not contain a restarting strategy almost always converge linearly. However, in practice rounding errors may destroy the superlinear convergence property. Our experience is that the conjugate-gradient method is nearly always linearly convergent, regardless of whether or not restarting takes place. The exceptions occur only in very special circumstances, such as when $F(x)$ is a well-conditioned quadratic function. In any event, the term "n-step superlinear convergence" has very little meaning when n is large, because of the large number of iterations required for the asymptotic convergence theory to hold; for example, it may be impractical to carry out even $2n$ iterations. In our view, it is unsatisfactory when a conjugate-gradient method requires more than approximately $5n$ iterations to achieve a meaningful level of accuracy.

When the Hessian matrix of F at the solution has eigenvalues that are clustered into sets containing eigenvalues of similar magnitude, the problem may be solved in significantly fewer than n iterations; this happens with certain types of penalty functions that occur in nonlinearly constrained optimization (see Section 6.2.1). Problems without this property tend to require anything from between n and $5n$ iterations, with a typical figure of $2n$ iterations. Although conjugate-gradient-type algorithms are far from ideal, they are currently the *only* reasonable method available for a general problem in which the number of variables is extremely large.

*4.8.4. Limited-Memory Quasi-Newton Methods

In this section we consider a class of methods that give a descent direction under restrictions upon the step length much milder than those for a nonlinear conjugate-gradient method. *Limited-memory quasi-Newton methods* are based upon the idea of computing p_k as $-Mg_k$, where M is a positive-definite matrix obtained by updating the identity matrix with a limited number of quasi-Newton corrections (*cf.* (4.43)). Although the direction of search is equivalent to the product of a matrix and a vector, the matrix is never stored explicitly; rather, only the vectors that define the updates are retained.

Different methods can be developed by varying the number of updating vectors stored and the choice of quasi-Newton updating formula. The matrix M for the "one-step" limited-memory BFGS update is given by the BFGS formula for the *inverse Hessian*, with the previous approximation taken as the identity matrix. The search direction is then given by

$$p_k = -g_k + \frac{1}{y_{k-1}^T s_{k-1}}\left(s_{k-1}^T g_k y_{k-1} + y_{k-1}^T g_k s_{k-1}\right)$$
$$- \frac{s_{k-1}^T g_k}{y_{k-1}^T s_{k-1}}\left(1 + \frac{y_{k-1}^T y_{k-1}}{y_{k-1}^T s_{k-1}}\right)s_{k-1}. \tag{4.95}$$

Inspection of (4.95) shows that if the one-step limited-memory BFGS formula is applied with an exact linear search, p_k is identical to the direction obtained from the conjugate-gradient method, since $s_{k-1}^T g_k$ vanishes. Hence, limited-memory quasi-Newton methods can generate mutually conjugate directions, but only if exact linear searches are made. To verify this, note that satisfaction of the quasi-Newton condition (4.44) for the inverse Hessian matrix means that $s_{k-1} = My_{k-1}$. Consequently, from the definition of p_k

$$y_{k-1}^T p_k = -y_{k-1}^T M g_k$$
$$= -s_{k-1}^T g_k.$$

The vectors y_{k-1} and p_k will thus be orthogonal only if $g_k^T s_{k-1}$ vanishes, i.e., if an exact linear search is made.

One important benefit from using limited-memory quasi-Newton formulae is that the algorithm generates only descent directions if the inner product $y_j^T s_j$ is positive for every pair y_j and s_j used in the updating formulae. To ensure that these inner products will always be positive, the step-length should be computed using a step-length algorithm that guarantees a reduction in the directional derivative (see Section 4.3.2.1).

*4.8.5. Preconditioned Conjugate-Gradient Methods

*4.8.5.1. **Quadratic functions.** Suppose that the linear conjugate-gradient method is to be used to solve the linear system $Gx = -c$, where G is symmetric and positive definite. In Section 4.8.3.2 it was stated that, with exact arithmetic, the number of iterations required to solve this system is equal to the number of distinct eigenvalues of G. Therefore, the rate of convergence should be significantly improved if the original system can be replaced by an equivalent system in which the matrix has many unit eigenvalues. The idea of *preconditioning* is to construct a transformation to have this effect on G.

Let W be a symmetric, positive-definite matrix. The solution of $Gx = -c$ can be found by solving the system

$$W^{-\frac{1}{2}} G W^{-\frac{1}{2}} y = -W^{-\frac{1}{2}} c, \qquad (4.96)$$

and forming $x = W^{-\frac{1}{2}} y$. Let R denote the matrix $W^{-\frac{1}{2}} G W^{-\frac{1}{2}}$; then R has the same eigenvalues as $W^{-1}G$, since $W^{-\frac{1}{2}} R W^{\frac{1}{2}} = W^{-1}G$ (see Section 2.2.3.4). The idea is choose W so that as many as possible of the eigenvalues of $W^{-1}G$ are close to unity. This is roughly equivalent to choosing W so that the condition number of $W^{-1}G$ is as small as possible; the matrix W is known as the *preconditioning matrix*.

In practice, the computation is arranged so that only W needs to be computed (not $W^{\frac{1}{2}}$). To see this, consider one iteration of the linear conjugate-gradient method (4.91) applied to the system (4.96), with the iterate denoted by u_k and the residual by v_k. Then, if we substitute $x_k = W^{-\frac{1}{2}} u_k$ and $r_k = W^{\frac{1}{2}} v_k$, the vectors u_k and v_k no longer appear. The final recurrence relations for the preconditioned conjugate gradient method are then the following. Define $r_0 = c + Gx_0$ (by convention, $\beta_{-1} = 0$ and $p_{-1} = 0$); for $k = 0, 1, \ldots,$

$$
\begin{aligned}
p_k &= -W^{-1} r_k + \beta_{k-1} p_{k-1}; \\
\alpha_k &= \frac{r_k^T W^{-1} r_k}{p_k^T G p_k}; \\
x_{k+1} &= x_k + \alpha_k p_k; \\
r_{k+1} &= r_k + \alpha_k G p_k; \\
\beta_k &= \frac{r_{k+1}^T W^{-1} r_{k+1}}{r_k^T W^{-1} r_k}.
\end{aligned}
\qquad (4.97)
$$

If the algorithm (4.97) is written in the format associated with minimizing the quadratic function $c^T x + \frac{1}{2} x^T G x$, we can derive a preconditioned form of the traditional conjugate-gradient method. At each iteration, p_k is obtained by solving the linear equations

$$Wz_k = -r_k$$

(recall that $r_k = g_k$) and setting

$$p_k = z_k + \beta_{k-1} p_{k-1},$$

where $\beta_{k-1} = -y_{k-1}^T z_k / y_{k-1}^T p_{k-1}$.

The preconditioning matrix W can be defined in several ways. For example, a suitable preconditioning matrix can be obtained by performing r steps of a limited-memory quasi-Newton method from the one-parameter family (4.35). If exact linear searches are made, the limited-memory matrix M satisfies the quasi-Newton condition for the r ($r \ll n$) pairs of vectors $\{s_j, y_j\}$, i.e.,

$$s_j = My_j.$$

Since for a quadratic function it holds that $Gs_j = y_j$, we observe that

$$s_j = MGs_j,$$

and the matrix MG has r unit eigenvalues with eigenvectors $\{s_j\}$. Therefore, M may be used as W^{-1}.

*4.8.5.2. Nonlinear functions. The idea of preconditioning may be extended directly to nonlinear problems. In this case the matrix W_k will vary from iteration to iteration; for example, a limited-memory approximate Hessian matrix may be used as W_k^{-1}.

A less sophisticated technique that has been applied successfully to general problems is a preconditioning based upon a diagonal scaling. Suppose that the direction of search is obtained from (4.28); then even if the off-diagonal elements of the approximate Hessian B_k are unknown, the *diagonal* elements can still be recurred, and may be used to precondition the conjugate-gradient method. Let γ_j and ψ_j denote the j-th elements of g_k and y_k respectively, and let $\Delta_k = \text{diag}(\bar{\delta}_1, \ldots, \bar{\delta}_n)$ and $\Delta_{k-1} = \text{diag}(\delta_1, \ldots, \delta_n)$ denote the approximate diagonal Hessians during the k-th and $(k-1)$-th iterations respectively. Then

$$\bar{\delta}_j = \delta_j + \frac{1}{g_{k-1}^T p_{k-1}}\gamma_j^2 + \frac{1}{\alpha_{k-1} y_{k-1}^T p_{k-1}}\psi_j^2$$

and

$$\Delta_k z_k = -g_k.$$

To ensure that the elements of Δ_k are positive, a diagonal is not updated if the result will be negative.

The use of the BFGS update to recur Δ_k has some theoretical justification since, in the quadratic case with exact linear searches, identical search directions are generated by the BFGS and conjugate-gradient algorithms. The motivation for using preconditioning on general nonlinear functions is to scale the search directions so that the initial step along p_k will be a better prediction of the minimum.

The limited-memory quasi-Newton methods can be generalized to accept diagonal preconditioning by starting the sequence of approximate Hessian matrices that generate W_k with a diagonal preconditioning matrix rather than the identity matrix. For example, the diagonally preconditioned one-step BFGS formula is given by

$$p_k = -\bar{g}_k + \frac{1}{y_{k-1}^T s_{k-1}}\left(s_{k-1}^T \bar{g}_k y_{k-1} + y_{k-1}^T \bar{g}_k s_{k-1}\right)$$
$$- \frac{s_{k-1}^T \bar{g}_k}{y_{k-1}^T s_{k-1}}\left(1 + \frac{y_{k-1}^T y_{k-1}}{y_{k-1}^T s_{k-1}}\right)s_{k-1},$$

where $\bar{g}_k = \Delta_k^{-1} g_k$.

*4.8.6. Solving the Newton Equations by Linear Conjugate Gradients

In some situations (see Section 4.8.1), it may be possible to compute a sparse Hessian G_k, but impossible to store its Cholesky factors. In other circumstances, it may be impractical to compute G_k itself, but reasonable to form its product with a vector. For example, when G_k is the product of sparse matrices, the explicit matrix G_k is likely to be dense even though its factors are sparse; however, the product of G_k with a vector may be computed efficiently.

Several variants of the linear conjugate-gradient method of Section 4.8.3.2 can be applied to "solve" the linear system that arises in Newton's method

$$G_k p_k = -g_k, \tag{4.98}$$

when it is practical to compute a *relatively small* number of matrix-vector products involving G_k.

For example, we may perform only a *limited number* of iterations of the linear conjugate-gradient method, and then take the final iterate of the truncated sequence as the search direction. This has been termed a *truncated Newton method* because the linear conjugate-gradient iterations are terminated ("truncated") before the full n steps that would theoretically be required to compute the exact solution. If a single linear iteration is used, p_k will be the steepest-descent direction $-g_k$; if a full n iterations are performed (with exact arithmetic), p_k will be the solution of (4.98). Thus, a truncated Newton algorithm computes a vector that interpolates between the steepest-descent direction and Newton direction. If G_k is positive definite and the initial iterate of the linear conjugate-gradient scheme is the steepest-descent direction $-g_k$, all succeeding linear iterates will be directions of descent with respect to F.

A truncated Newton method will be successful only if a good direction can be produced in a *small* number of linear conjugate-gradient iterations, and hence the use of preconditioning is essential. However, preconditioning can produce another benefit that does not depend on the eigenvalue structure of the preconditioning matrix W. In many optimization methods, the search direction p_k is computed implicitly or explicitly as

$$p_k = -M g_k,$$

where M is a positive-definite matrix; for example, limited-memory quasi-Newton methods define M as a low-rank modification to the identity matrix (see Section 4.8.4). If the matrix M is used to precondition G_k, the vector $-M g_k$ is the first member of the linear conjugate-gradient sequence, and is far more likely to give a good reduction in the function than the negative gradient. In this case, the truncated Newton search direction interpolates between the Newton direction and the direction that would be given by a limited-memory quasi-Newton method.

Notes and Selected Bibliography for Section 4.8

The generalization of quasi-Newton updates to unsymmetric sparse matrices was first proposed by Schubert (1970). An alternative derivation is given by Avila and Concus (1979). The use of the problem (4.74) to define a sparse update in the symmetric case was first suggested by Powell (1976a), and the problem was solved independently by Toint (1977) and Marwil (1978). Shanno (1980) derived the sparse updating formula for each member of the one-parameter family; Thapa (1979) considerably simplified Shanno's derivation and extended the theory to include a larger class of updates. For more on the theory and practice of sparse updating see Toint (1978, 1979), Dennis and Schnabel (1979), Thapa (1980) and Powell (1981).

The technique of differencing along special vectors was first used in the context of solving nonlinear equations by Curtis, Powell and Reid (1974). The application of this method to minimization has been considered by Gill and Murray (1973a). Various forms of the method which take specific advantage of symmetry are given by Powell and Toint (1979). For a further discussion of row and column orderings for finite-difference methods, see Coleman and Moré (1980).

If the Hessian is sparse but has no sparse Cholesky factorization, an approximate (partial) factorization may be computed by ignoring fill-in in a systematic way. This factorization may be used to define the direction of search (see Thapa, 1980), or to precondition the Newton equations for a linear conjugate-gradient algorithm.

The linear conjugate-gradient method was devised by Hestenes and Stiefel (1952) and extended to nonlinear functions by Fletcher and Reeves (1964). Nazareth (1979) has shown that the traditional conjugate-gradient method and the BFGS method are theoretically identical when applied to a quadratic function with exact linear searches. For modern surveys of conjugate-gradient methods, see Gill and Murray (1979a), Fletcher (1980) and Hestenes (1980a).

Dixon (1975) and Nazareth (1977) propose methods that do not require an exact linear search in order to generate a set of conjugate directions. An unfortunate feature of these methods is that the search direction cannot be guaranteed to be a descent direction for an arbitrary nonlinear function. Surprisingly, despite the importance of the exact linear search in the *theoretical* derivation of conjugate-gradient methods, a carefully implemented step-length algorithm can give comparable or better numerical results (see Gill and Murray, 1979a).

The use of (4.93) as a conjugate-gradient formula with arbitrary initial direction was suggested by Beale (1972) and popularized by Powell (1977a). Limited-memory quasi-Newton formulae were suggested by Perry (1977) and Shanno (1978). Buckley (1978), Nazareth (1979) and Nocedal (1980) exploit the relationship between the BFGS method and the traditional conjugate-gradient method in order to define methods which retain as many quasi-Newton updates as allowed by available storage.

One of the earliest references to preconditioning for linear equations is given by Axelsson (1974). See Concus, Golub and O'Leary (1976) for details of various preconditioning methods derived from a slightly different viewpoint. For a description of how a limited-memory quasi-Newton formula may be used to precondition the traditional conjugate-gradient method, see Nazareth and Nocedal (1978). Nonlinear diagonal preconditioning using quasi-Newton updating was suggested by Gill and Murray (1979a).

For simplicity, we have discussed the solution of the Newton equations by the linear conjugate-gradient method with the implicit assumption that the Hessian matrix is positive definite. O'Leary (1980a) and Gill, Murray and Nash (1981) define modified conjugate-gradient algorithms that are more appropriate when F does not have a uniformly positive-definite Hessian. The truncated Newton approach of using intermediate vectors from the linear conjugate-gradient algorithm as search directions is due to Dembo and Steihaug (1980), who also suggested a scheme for terminating the iterations that retains the superlinear convergence properties of Newton's method (see also Dembo, Eisenstat and Steihaug, 1980). The use of preconditioning to give a search direction that interpolates between a nonlinear conjugate-gradient direction and the Newton direction is due to Gill, Murray and Nash (1981).

When no derivatives are available, a conjugate-gradient-type method can be applied with finite-difference approximations to the gradient (see Section 4.6.1). However, for a general problem it is likely that a very large number of function evaluations will be needed for convergence.

CHAPTER FIVE

LINEAR CONSTRAINTS

To throw away the key and walk away,
Not abrupt exile, the neighbours asking why
But following a line with left and right,
An altered gradient at another rate.

—W. H. AUDEN, *in "The Journey"* (1928)

In this chapter, we consider methods for optimization problems in which the acceptable values of the variables are defined by a set of linear constraints. We shall see that the special properties of linear functions allow effective methods to be developed by combining techniques from unconstrained optimization and linear algebra. We shall be concerned with *smooth objective functions* only; linearly constrained problems with non-differentiable objective functions can be solved using the algorithms to be discussed in Section 6.2.2.2. It was shown in Section 3.3 that the necessary and sufficient optimality conditions for a linearly constrained smooth problem involve the existence of Lagrange multipliers, and these multipliers also play a crucial role in the development of linearly constrained methods.

As in Chapter 4, we shall emphasize methods with which the authors have had extensive experience, and methods that provide special insights or background. References to other methods and to further details are given in the Notes at the end of each section. The authors of methods and results will be mentioned only in the Notes.

5.1. METHODS FOR LINEAR EQUALITY CONSTRAINTS

The problem to be considered in this section is the minimization of a smooth function subject to a set of linear *equality* constraints:

$$
\begin{array}{lll}
\text{LEP} & \underset{x\in\Re^n}{\text{minimize}} & F(x) \\[2mm]
& \text{subject to} & \hat{A}x = \hat{b},
\end{array}
$$

where F is twice-continuously differentiable, the matrix $\hat{A}$ is $t \times n$, and the i-th row of $\hat{A}$ contains the coefficients corresponding to the i-th constraint. We shall assume that LEP has a bounded solution and that F is bounded below in the feasible region. The optimality conditions for LEP are given in Section 3.3.1. It will generally be assumed for simplicity that the rows of $\hat{A}$ are linearly independent. In practice this assumption is reasonable, since linear dependence among a set of linear equality constraints implies either that some of the constraints can be omitted without altering the solution of the problem, or that there is no feasible point. (In fact, it is essential for an algorithm for linearly constrained optimization to include an initial test for linear dependence among the equality constraints.)

A necessary condition for x^* to be a local minimum of LEP is

$$g(x^*) = \hat{A}^T \lambda^*, \tag{5.1}$$

where λ^* is a t-vector of Lagrange multipliers. Let Z denote a matrix whose columns form a basis for the set of vectors orthogonal to the rows of $\hat{A}$. A condition equivalent to (5.1) is

$$Z^T g(x^*) = 0. \tag{5.2}$$

A second-order necessary condition for x^* to be a local minimum of LEP is that the matrix $Z^T G(x^*) Z$ should be positive semi-definite. The vector $Z^T g(x)$ will be called the *projected gradient* of F at x, and the matrix $Z^T G(x) Z$ will be called the *projected Hessian matrix*.

5.1.1. The Formulation of Algorithms

5.1.1.1. The effect of linear equality constraints. As noted in Section 3.3.1, the imposition of t independent linear equality constraints on an n-dimensional problem reduces the dimensionality of the optimization to $n - t$. This reduction in dimensionality can be described formally in terms of two subspaces — the t-dimensional subspace defined by the rows of $\hat{A}$, and the complementary subspace of vectors orthogonal to the rows of $\hat{A}$. Let Y denote any matrix whose columns form a basis for the range space of $\hat{A}^T$. The representation of Y is not unique; for example, Y may be taken as $\hat{A}^T$. Since Y and Z define complementary subspaces, every n-vector x has a unique expansion as a linear combination of the columns of Y and Z

$$x = Y x_Y + Z x_Z,$$

for some t-vector x_Y (the *range-space* portion of x) and an $(n-t)$-vector x_Z (the *null-space* portion of x).

Suppose that the solution x^* of LEP is given by $Y x_Y^* + Z x_Z^*$. Because x^* is feasible,

$$\hat{A} x^* = \hat{A}(Y x_Y^* + Z x_Z^*) = \hat{b}.$$

Since $\hat{A} Z = 0$, it follows that

$$\hat{A} Y x_Y^* = \hat{b}. \tag{5.3}$$

By definition of Y, the matrix $\hat{A} Y$ is non-singular, and thus from (5.3) we see that the t-vector x_Y^* is uniquely determined. A similar argument shows that *any* feasible point must have the same range-space portion x_Y^*, and that any vector with range-space component x_Y^* will satisfy the constraints of LEP. Hence, the constraints entirely determine the range-space portion of the solution of LEP, and only the null-space portion of x^* remains unknown. This is precisely the expected reduction in dimensionality to $n - t$. In fact, we shall see that computing the solution of the constrained problem LEP can be viewed as an *unconstrained* problem in the $n - t$ variables x_Z.

Example 5.1. Consider the single constraint

$$x_1 + x_2 + x_3 = 3,$$

for which $\hat{A} = (1\ 1\ 1)$. The matrix Y may be taken as $\hat{A}^T$, and one possible Z is

$$Z = \begin{pmatrix} -\frac{\sqrt{3}}{3} & -\frac{\sqrt{3}}{3} \\ \frac{3+\sqrt{3}}{6} & -\frac{3-\sqrt{3}}{6} \\ -\frac{3-\sqrt{3}}{6} & \frac{3+\sqrt{3}}{6} \end{pmatrix}. \tag{5.4}$$

Substituting into (5.3), we obtain $3x_Y^* = 3$, or $x_Y^* = 1$. Therefore, regardless of $F(x)$, the solution must be of the form

$$x^* = \begin{pmatrix} 1 \\ 1 \\ 1 \end{pmatrix} + \begin{pmatrix} -\frac{\sqrt{3}}{3} & -\frac{\sqrt{3}}{3} \\ \frac{3+\sqrt{3}}{6} & -\frac{3-\sqrt{3}}{6} \\ -\frac{3-\sqrt{3}}{6} & \frac{3+\sqrt{3}}{6} \end{pmatrix} x_Z^*,$$

where x_Z^* is a two-dimensional vector.

5.1.1.2. A model algorithm. The most effective algorithms for LEP generate a sequence of feasible iterates. Such algorithms are practical and efficient because a simple characterization can be developed of all feasible points. As noted in Section 3.3.1, the step p from any feasible point to any other feasible point must be orthogonal to the rows of $\hat{A}$. Therefore, if x_k is feasible, p_k (the search direction at iteration k) should satisfy

$$\hat{A}p_k = 0. \tag{5.5}$$

Any vector p_k that satisfies (5.5) must be a linear combination of the columns of Z. Therefore, the following equivalence holds

$$\hat{A}p_k = 0 \Leftrightarrow p_k = Zp_z \tag{5.6}$$

for some $(n-t)$-vector p_z, and linear combinations of the columns of Z comprise *all* feasible directions for LEP.

Using (5.6), the model algorithm from Section 4.3.1 can be adapted to solve LEP by generating a sequence of feasible iterates. Given a feasible starting point x_0, set $k \leftarrow 0$ and repeat the following steps:

Algorithm LE (*Model algorithm for solving LEP*).

LE1. [Test for convergence.] If the conditions for convergence are satisfied at x_k, the algorithm terminates with x_k as the solution.

LE2. [Compute a feasible search direction.] Compute a non-zero $(n-t)$-vector p_z. The direction of search, p_k, is then given by

$$p_k = Zp_z. \tag{5.7}$$

LE3. [Compute a step length.] Compute a positive scalar α_k, the *step length*, for which it holds that $F(x_k + \alpha_k p_k) < F(x_k)$.

LE4. [Update the estimate of the minimum.] Set $x_{k+1} \leftarrow x_k + \alpha_k p_k$, $k \leftarrow k+1$, and go back to step LE1. ∎

If the initial x_0 is feasible, all subsequent iterates are feasible because of the form (5.7) of the search direction. Since feasibility is automatic, the principles described in Section 4.3.1 can be applied directly to the choice of the vector p_z in step LE2 of this algorithm. In particular, when $Z^T g_k$ is non-zero, p_k should be a *descent direction* at x_k, i.e.

$$g_k^T Z p_z < 0.$$

Since unlimited movement along p_k will not violate the constraints, the step length α_k in step LE3 must be chosen to satisfy an appropriate set of conditions that ensure a "sufficient decrease" in F, exactly as in the unconstrained case (see Section 4.3.2.1).

Our primary concern in developing algorithms for solving LEP is to compute a "good" descent direction by applying the techniques described in Chapter 4 to the choice of p_z. However, it is essential not to rely on properties from unconstrained problems that do not necessarily carry over to linearly constrained problems. For example, in an unconstrained problem it might be reasonable to assume that the Hessian of F is positive definite, since this will usually be the case in a neighbourhood of the solution. In a linearly constrained problem, however, the full Hessian need not be positive definite, even at the solution.

Because of the equivalence (5.6), we shall refer to p_k and p_z interchangeably. However, it is important to remember that the search direction is an n-dimensional vector that is constructed to lie in a particular $(n-t)$-dimensional subspace. The vector p_k depends on the representation of Z as well as the choice of p_z; two techniques for representing Z will be discussed in Section 5.1.3.

5.1.2. Computation of the Search Direction

5.1.2.1. Methods of steepest descent. The idea of an unconstrained "steepest-descent" direction (see Section 4.3.2.2) can be applied to the problem LEP by seeking the feasible direction of steepest descent. Expanding F in a Taylor series about x_k along a general feasible vector Zp_z, we have

$$F(x_k + Zp_z) = F_k + g_k^T Zp_z + \frac{1}{2}p_z^T Z^T G_k Zp_z + \cdots \tag{5.8}$$

where F_k, g_k and G_k are the function value, gradient vector, and Hessian matrix evaluated at x_k.

By ignoring all terms in (5.8) beyond the linear, we can define two versions of the "steepest-descent" problem (4.11) — namely, to find the vector p_z that minimizes the linear term of (5.8), subject to a restriction on its norm. If we impose a norm restriction on the full search direction p_k, the problem is

$$\underset{p_z \in \Re^{n-t}}{\text{minimize}} \ \frac{g_k^T Zp_z}{\|Zp_z\|_2},$$

and, using (4.12) with $C = Z^T Z$, the corresponding "steepest-descent" direction is

$$p_k = -Z(Z^T Z)^{-1} Z^T g_k. \tag{5.9}$$

This definition of p_k is independent of the form of Z. If the normalization applies to p_z, the problem is

$$\underset{p_z \in \Re^{n-t}}{\text{minimize}} \ \frac{g_k^T Zp_z}{\|p_z\|_2},$$

and the associated p_k is

$$p_k = -ZZ^T g_k. \tag{5.10}$$

Note that (5.9) and (5.10) are equivalent when the columns of Z are orthonormal.

Both (5.9) and (5.10) are descent directions, and each may be combined with an appropriate step-length procedure to produce an algorithm with guaranteed convergence, exactly as with the steepest-descent method in the unconstrained case. However, the poor linear rate of convergence (see Section 4.3.2.2) still applies, and thus either algorithm is unacceptably slow. To achieve an improved rate of convergence, algorithms must take account of the second-order behaviour of the objective function.

5.1.2.2. Second derivative methods. Modified Newton algorithms analogous to those described in Section 4.4 can be developed if analytic second derivatives of F are available. The *Newton direction* for LEP is the direction that solves the constrained problem

$$\underset{p \in \Re^n}{\text{minimize}} \quad F_k + g_k^T p + \frac{1}{2} p^T G_k p \tag{5.11}$$

$$\text{subject to} \quad \hat{A} p = 0.$$

Because of the equivalence (5.6), the vector p_z associated with the Newton direction can be computed by ignoring terms beyond the quadratic in (5.8), thereby producing a local quadratic model of the behaviour of F restricted to the subspace spanned by the columns of Z. Let G_z denote the projected Hessian matrix $Z^T G_k Z$, and g_z denote the projected gradient $Z^T g_k$. The solution of (5.11) is given by $Z p_z$, where p_z is the solution of

$$G_z p_z = -g_z. \tag{5.12}$$

Note that the equation (5.12) for p_z is analogous to equation (4.17) for the full search direction in the unconstrained case (which corresponds to $Z = I$).

Example 5.2. Consider the equality-constrained problem

$$\underset{x \in \Re^3}{\text{minimize}} \quad x_1^2 x_2^3 + 4x_1^2 x_3^2 + x_2^4 x_3^2 + 3x_1 x_2 + 4x_2 x_3 + 5x_1 x_3 + x_1 x_3$$

$$\text{subject to} \quad x_1 + x_2 + x_3 = 3.$$

At the feasible point $x_0 = (-1, 5, -1)^T$, the solution of (5.12) (with Z given by (5.4)) is

$$p_z = \begin{pmatrix} .72450 \\ -.32483 \end{pmatrix} \quad \text{and hence} \quad p_0 = \begin{pmatrix} .23075 \\ -.64004 \\ .40929 \end{pmatrix}$$

(all numbers rounded to five figures). Note that $x_0 + \gamma p_0$ is feasible for *any* value of γ.

If G_z is positive definite and p_z is the solution of (5.12), the Newton direction $Z p_z$ is a descent direction, since

$$g_k^T Z p_z = -g_z^T G_z^{-1} g_z < 0.$$

If the projected Hessian matrix is not positive definite, the local quadratic approximation is unbounded below, or does not have a unique minimum (see Section 3.2.3). In such a case, it is necessary to modify the definition of p_z, in order to develop a sensible interpretation of the local quadratic model restricted to the null space. Any of the techniques described in Section 4.4.2 for the unconstrained context can be used to produce a satisfactory modified Newton algorithm for

LEP when G_z is indefinite. In particular, the vector p_z can be taken as the solution of

$$\bar{G}_z p_z = -g_z,$$

where $\bar{G}_z$ is the positive-definite matrix resulting from the modified Cholesky factorization of G_z (see Section 4.4.2.2).

Under conditions similar to those in the unconstrained case (e.g., a sufficiently close starting point and a positive-definite projected Hessian), the Newton algorithm in which the search direction is defined by (5.7) and (5.12) will converge quadratically. As in the unconstrained case, quadratic convergence will be achieved only if the sequence of step lengths $\{\alpha_k\}$ converges to unity at a sufficiently fast rate.

5.1.2.3. Discrete Newton methods. When first, but not second, derivatives of F are available, a discrete Newton method (see Section 4.5.1) can be applied to LEP. The important difference from the unconstrained case is that only the $(n-t)$-dimensional *projected* Hessian matrix G_z is needed to compute the Newton direction. Therefore, it is *not* necessary to approximate the full Hessian by finite-differences. Since

$$g(x_k + h_i z_i) \approx g_k + h_i G_k z_i,$$

a direct finite-difference approximation to G_z can be obtained by using the $n-t$ columns of Z as finite-difference vectors. Define the vectors $\{w_i\}$, $i = 1, \ldots, n-t$, as

$$w_i = \frac{1}{h_i}(g(x_k + h_i z_i) - g_k),$$

for some appropriate finite-difference interval h_i. Then w_i is an $O(h_i)$ approximation to $G_k z_i$. Let w_i be the i-th column of the $n \times (n-t)$ matrix W. A symmetric approximation to G_z is given by

$$\hat{G}_z \equiv \frac{1}{2}(Z^T W + W^T Z).$$

Since only $n-t$ evaluations of the gradient are required to compute $\hat{G}_z$, the projected Hessian can be computed very cheaply when the number of constraints is large.

If $\hat{G}_z$ is indefinite, the definition of p_z must be modified as described in Section 5.1.2.2 for the exact projected Hessian. The convergence of a discrete Newton method is almost identical to that of an exact Newton method, except very close to the solution (see Section 4.5.1).

5.1.2.4. Quasi-Newton methods. Quasi-Newton methods (see Section 4.5.2) can also be applied to the problem LEP. Various proposals have been made for this extension. The earliest suggestions were based on maintaining an $n \times n$ *singular* "quasi-Newton" approximation to the inverse Hessian; the structure of the singularity was intended to produce a search direction of the form (5.7). However, it proved unsatisfactory in practice to use a single matrix to represent the null space of $\hat{A}$ as well as the changing curvature information. Due to rounding errors introduced by quasi-Newton updates, the satisfaction of (5.7) by the computed directions tended to deteriorate after only a few iterations.

A more successful approach is to retain a *fixed* representation of Z (see Section 5.1.3 for details of how Z may be computed), and to update a $(n-t)$-dimensional quasi-Newton approximation to the *projected Hessian*. Let B_z denote the current quasi-Newton approximation to the projected Hessian; the vector p_z is then given by the solution of a system of equations analogous to (4.28) in the unconstrained case:

$$B_z p_z = -g_z. \tag{5.13}$$

The standard quasi-Newton update formulae (see Section 4.5.2.1) can be adapted to reflect curvature within the subspace defined by Z by using only projected vectors and matrices in the computation of the updated matrix. For simplicity of notation, we shall denote the approximate projected Hessian at x_k by B_z, and the updated approximation by $\bar{B}_z$. Let s_z denote $Z^T s_k$, and y_z denote $Z^T y_k$. Then the BFGS formula (*cf.* (4.37)) for the updated projected matrix $\bar{B}_z$ is

$$\bar{B}_z = B_z + \frac{1}{g_z^T p_z} g_z g_z^T + \frac{1}{\alpha_k y_z^T p_z} y_z y_z^T. \tag{5.14}$$

Projected versions of other quasi-Newton updates may be developed in a similar fashion.

This approach to applying quasi-Newton methods to LEP has the attractive feature that the property of hereditary positive-definiteness, which we have seen in Section 4.5.2 is desirable in the unconstrained case, can be carried over to the linear-constraint case. The projected Hessian matrix must be at least positive semi-definite at the solution, whereas the full Hessian matrix may be indefinite at every iteration. Furthermore, the search directions always satisfy (5.7) with the same accuracy (depending on the form of Z), and the quasi-Newton updates thus do not affect feasibility.

Exactly as in the unconstrained case, numerical stability can be ensured by retaining the Cholesky factorization of the projected Hessian rather than an explicit representation of its inverse (see Section 4.5.2.2). The Cholesky factorization of the approximate projected Hessian can be updated after each rank-one modification. This procedure ensures that the matrix B_z is always positive definite, and hence that the computed search direction is always a descent direction.

A non-derivative quasi-Newton method (see Section 4.6.2) can also be applied to LEP. Since only the *projected* gradient is required to perform the update (5.14), the vector $Z^T g_k$ can be approximated directly by taking finite-differences of F along the $n - t$ columns of Z rather than with respect to each variable (the selection of the finite-difference interval will be discussed in Section 8.6). As with discrete Newton methods, efficiency improves as the number of constraints increases.

5.1.2.5. Conjugate-gradient-related methods. The sparsity-exploiting methods of Sections 4.8.1 and 4.8.2 are less likely to be applicable for problem LEP than for unconstrained optimization, since, in general, the projected Hessian will be a dense matrix even if $\hat{A}$ and G_k are sparse (although there may be exceptions when the constraints have a very special structure). By contrast, the linear conjugate-gradient methods discussed in Section 4.8.6 are well suited to solving (5.12) or (5.13) without forming the projected matrices. If it is possible to store a sparse Hessian G_k and a sparse representation of Z, the necessary matrix-vector products of the form $Z^T G_k Z v$ can be computed relatively cheaply by forming, in turn, $v_1 = Zv$, $v_2 = G_k v_1$ and $v_3 = Z^T v_2$ (a similar procedure can be used if a sparse Hessian approximation is available). If a sparse Hessian is not available, the vector v_2 may be found using a single finite difference of the gradient along v_1.

There are several possibilities for the application of preconditioning, depending on the available information and on the speed with which the iterates of the linear conjugate-gradient method converge. A preconditioned truncated Newton method will be particularly effective when the *projected* Hessian matrix is small enough to be stored explicitly. In this case, a quasi-Newton approximation B_z can be used as a preconditioning matrix.

5.1.3. Representation of the Null Space of the Constraints

The apparent differences among algorithms for solving LEP often arise simply from different ways of representing the matrix Z. Note that the term "representation of Z" does not necessarily imply that the matrix Z is formed. It is simply a convenient shorthand for the process by which algorithms for LEP ensure that the constraints are satisfied at each iterate (and determine an initial feasible point).

5.1.3.1. The LQ factorization. The first technique for computing Z is based on the LQ factorization of $\hat{A}$ (see Section 2.2.5.3). Let Q be an $n \times n$ orthonormal matrix such that:

$$\hat{A}Q = (\, L \quad 0 \,), \tag{5.15}$$

where L is a $t \times t$ non-singular lower-triangular matrix. From (5.15) it follows that the first t columns of Q can be taken as the columns of Y, and the last $n-t$ columns of Q can be taken as the columns of the matrix Z. Note that $Z^T Z = I_{n-t}$, since Z is a section of an orthonormal matrix. The matrix Z (5.4) corresponding to Example 5.1 was obtained from the LQ factorization.

The precise order of the columns of Z is not important, and consequently the appropriate columns of Q may be selected in any order. However, it is useful to consider the columns of Q in a specific order when solving inequality-constrained problems (see Section 5.2.4.1 and the Notes for Section 5.2).

When Y is defined as described above, it holds that $\hat{A}Y = L$, and hence the vector x_Y^* is the solution of

$$L x_Y^* = \hat{b}. \tag{5.16}$$

An advantage of computing x_Y^* from (5.16) is that the condition number of L is no larger than that of $\hat{A}$. The initial feasible point can be taken as $Y x_Y^*$.

A fundamental advantage of using the LQ factorization is that the choice of Z *does not cause a deterioration in conditioning when solving LEP*. For example, with a Newton-type method, it can be shown that $\mathrm{cond}(G_z)$ depends upon the condition numbers of Z and G_k:

$$\mathrm{cond}(G_z) \leq \mathrm{cond}(G_k)\big(\mathrm{cond}(Z)\big)^2.$$

When $\mathrm{cond}(Z)$ is large, G_z may be very ill-conditioned, even when the matrix G_k is well conditioned. The matrix Z determined from the LQ factorization is such that $\mathrm{cond}(Z) = 1$. In this case we have

$$\mathrm{cond}(G_z) \leq \mathrm{cond}(G_k),$$

which implies that *the conditioning of the original problem is not made worse by the numerical method.*

With the LQ factorization technique, the $n \times n$ matrix Q may be formed explicitly by multiplying together the orthogonal transformations used to triangularize $\hat{A}$. However, this may be inefficient in terms of computation and storage if there are very few constraints, since many algorithms for LEP require only matrix-vector products involving Z, Z^T, and Y. In this case, the orthogonal transformations used to produce the factorization may be stored in compact form, and the needed matrix-vector products can be computed simply by applying these transformations to the appropriate vectors. The determination as to whether it is more efficient to form Q explicitly or retain only the transformations depends on the relative values of n and t.

5.1.3.2. The variable-reduction technique. A second form for Z arises from partitioning the constraint matrix $\hat{A}$ as:

$$\hat{A} = (V \quad U), \tag{5.17}$$

where V is a $t \times t$ non-singular matrix. (We have assumed for simplicity that V corresponds to the *first* t columns of $\hat{A}$; in general, however, V can be any appropriate subset of the columns of $\hat{A}$.)

If x is partitioned accordingly as $x = (x_V \quad x_U)$, the constraints of LEP can be written as

$$(V \quad U)\begin{pmatrix} x_V \\ x_U \end{pmatrix} = \hat{b},$$

or

$$Vx_V + Ux_U = \hat{b}.$$

Since V is non-singular, we have

$$x_V = V^{-1}(\hat{b} - Ux_U). \tag{5.18}$$

From (5.18), the t variables x_V corresponding to the columns of V can be regarded as "dependent" on the remaining $n - t$ "independent" variables x_U. For any x_U, the constraints of LEP are automatically satisfied if x_V is defined by (5.18). This technique for treating linear equality constraints is called a *variable-reduction method*. An initial feasible point is given by $x_U = 0$, $x_V = V^{-1}\hat{b}$.

When $\hat{A}$ is given by (5.17), the following matrix Z is orthogonal to the rows of $\hat{A}$:

$$Z = \begin{pmatrix} -V^{-1}U \\ I \end{pmatrix}. \tag{5.19}$$

For Example 5.1, the Z corresponding to (5.19) is

$$Z = \begin{pmatrix} -1 & -1 \\ 1 & 0 \\ 0 & 1 \end{pmatrix}.$$

When (5.19) is used to represent Z, in general *the explicit matrix Z is not formed*. The quantities needed to compute the search direction are matrix-vector products that involve Z and Z^T, and these may be obtained by solving systems of equations that involve V and V^T. Thus, only a factorization of V is required. A form similar to (5.19) is of great importance in large-scale linearly constrained optimization (see Section 5.6).

5.1.4. Special Forms of the Objective Function

5.1.4.1. Linear objective function. When $F(x)$ is a linear function (say, $F(x) = c^T x$ for some constant vector c), the condition (5.1) will hold only if c is a linear combination of the rows of $\hat{A}$. This will be true for an arbitrary c only if $\hat{A}$ is non-singular, i.e., there are n constraints. In this case, the solution is entirely determined by solving the constraint equations. Hence, any feasible point is unique, and no search direction needs to be computed. This rather trivial result for the equality case will be significant when we consider linear inequality constraints.

5.1.4.2. Quadratic objective function. A linearly constrained problem with a quadratic objective function is called a *quadratic programming* (QP) problem, or simply a *quadratic program*. In this section, we consider the following simple quadratic programming problem

$$\operatorname*{minimize}_{x \in \Re^n} \quad c^T x + \frac{1}{2} x^T G x \tag{5.20}$$

$$\text{subject to} \quad \hat{A} x = \hat{b}.$$

for a constant symmetric matrix G and vector c.

When $Z^T G Z$ is positive definite, the solution of (5.20) is unique. In this case, given a feasible point $\bar{x}$ (with gradient $\bar{g} = G\bar{x} + c$), the step $\bar{p}$ from $\bar{x}$ to x^* is the solution of the n-dimensional problem

$$\operatorname*{minimize}_{p \in \Re^n} \quad \bar{g}^T p + \frac{1}{2} p^T G p \tag{5.21}$$

$$\text{subject to} \quad \hat{A} p = 0.$$

However, as shown in Section 5.1.1.2, the solution of (5.21) is obtained by computing the $(n-t)$-vector $\bar{p}_z$ that solves the *unconstrained* problem:

$$\operatorname*{minimize}_{p_z \in \Re^{n-t}} \quad \bar{g}^T Z^T p_z + \frac{1}{2} p_z^T Z^T G Z p_z, \tag{5.22}$$

and is defined by the linear system:

$$Z^T G Z \bar{p}_z = -Z^T \bar{g}.$$

Thus, $\bar{p}$ is given by

$$\bar{p} = -Z (Z^T G Z)^{-1} Z^T \bar{g}, \tag{5.23}$$

and the solution of (5.20) is simply $x^* = \bar{x} + \bar{p}$.

From the optimality conditions discussed in Section 3.3.1, it holds that

$$g(x^*) = \bar{g} + G\bar{p} = \hat{A}^T \lambda^*, \tag{5.24}$$

where the t-vector λ^* defines the Lagrange multipliers for (5.21). *The system (5.24) must be compatible* when $\bar{p}$ is defined by (5.23), or, equivalently, as the solution of (5.21).

If the projected Hessian matrix is not positive definite, the observations of Section 3.2.3 apply to (5.22). In particular, if $Z^T G Z$ is indefinite, there is no finite solution to (5.20).

5.1.5. Lagrange Multiplier Estimates

At the solution to LEP, it holds that

$$g(x^*) = \hat{A}^T \lambda^* \tag{5.25}$$

for some t-vector λ^* of Lagrange multipliers. By definition, Lagrange multipliers may be defined at any constrained stationary point, where the overdetermined system (5.25) is compatible. However, *there are no Lagrange multipliers at a non-stationary point*, since (5.25) is *not* compatible in general. Nonetheless, as we shall see in Section 5.2.3, it is essential to have some means of *estimating* Lagrange multipliers at points for which (5.25) does not hold. In this section, we shall briefly consider possible forms for a Lagrange multiplier estimate λ_k computed at the iterate x_k. It is important that any such estimate should be *consistent*, i.e. λ_k should have the property that

$$x_k \to x^* \quad \text{implies} \quad \lambda_k \to \lambda^*. \tag{5.26}$$

In addition, in order to be computationally practicable, a method for estimating Lagrange multipliers should use the same factorization involved in representing Z.

5.1.5.1. First-order multiplier estimates. When the LQ factorization is used, the vector λ_L (for "least-squares") that solves the problem

$$\underset{\lambda \in \Re^t}{\text{minimize}} \ \|\hat{A}^T \lambda - g_k\|_2 \tag{5.27}$$

may be used as an estimate of the Lagrange multipliers. The formal representation of λ_L is

$$\lambda_L = (\hat{A}\hat{A}^T)^{-1}\hat{A}g_k.$$

However, this form is unsatisfactory for *computation* of λ_L, because the condition number of $\hat{A}\hat{A}^T$ is the square of that of $\hat{A}$; instead, λ_L should be computed as described in Section 2.2.5.3. In this case, the computational error in λ_L in the neighbourhood of a stationary point will be bounded by a factor that includes only the condition number of $\hat{A}$.

With the variable-reduction method for representing Z, the multiplier estimate will utilize the available factorization of the matrix V. If we assume that V comprises the first t columns of $\hat{A}$, a multiplier estimate λ_V (for "variable reduction") whose computation requires only a factorization of V is the solution of the first t equations in (5.25):

$$V^T \lambda_V = g_V,$$

where g_V is the vector of first t components of g_k. Note that if $x_k = x^*$, both λ_L and λ_V will be the exact vector of Lagrange multipliers.

Both λ_L and λ_V are called *first-order* estimates for the following reason. For sufficiently small ϵ, if $\|x_k - x^*\|$ is of order ϵ, the error in either estimate is also of order ϵ. This implies that, if all the Lagrange multipliers are non-zero, either λ_L or λ_V will provide an accurate estimate of λ^* within a sufficiently small neighbourhood of x^*. The size of the neighbourhood obviously depends on the nonlinearity of F, which affects the size of $\|g_k - g(x^*)\|$. Furthermore, roughly speaking, with λ_L the size of the neighbourhood decreases as the condition number of $\hat{A}$ increases; if the rows of $\hat{A}$ are nearly linearly dependent, the neighbourhood will be extremely small. When λ_V is used, the size of the neighbourhood depends on the condition number of V, which can be arbitrarily large even if $\hat{A}$ is very well conditioned. At the point x_k, the norm of the residual $\hat{A}^T \lambda_L - g_k$ is bounded by $\|g_k - g(x^*)\|$. When λ_V is used, the first t components of the residual $\hat{A}^T \lambda_V - g_k$ are zero, and the size of the remaining components depends on the condition number of V. These observations apply to the *exact* values of λ_L and λ_V; we stress this point because, although these estimates should be identical at x^*, their computed values are obviously affected by rounding error. The computational error in λ_L depends on the condition number of $\hat{A}$, and that in λ_V depends on the condition number of V.

Some of these observations are illustrated in the following example.

Example 5.3. Consider a problem in which

$$\hat{A}^T = \begin{pmatrix} 1 & 1 \\ 1+\epsilon & 1 \\ 0 & -1 \end{pmatrix}, \quad g(x^*) = \begin{pmatrix} 2 \\ 2+\epsilon \\ -1 \end{pmatrix} \quad \text{and} \quad g_k = \begin{pmatrix} 2+\epsilon \\ 2 \\ -1+\epsilon \end{pmatrix}.$$

Note that $\text{cond}(\hat{A})$ is small, and that $\|g(x^*) - g_k\|$ is also small. The exact Lagrange multipliers are $\lambda^* = (1,1)^T$. However, consider the effect when the estimates λ_L and λ_V are evaluated at x_k, with V^T taken as the first two rows of $\hat{A}^T$. For $\epsilon = 10^{-5}$, $\lambda_L = (1.00001, .999990)^T$ and $\lambda_V = (-1.0, 3.00001)^T$ (both estimates rounded to 6 figures). Even for a small perturbation, the estimate λ_V is extremely inaccurate because the chosen V is ill-conditioned.

5.1.5.2. Second-order multiplier estimates. Under certain conditions, higher-order multiplier estimates can be obtained using second-order information at x_k. Let q denote the step from x_k to x^*. By definition, the exact Lagrange multipliers satisfy the equation

$$g(x^*) = g(x_k + q) = \hat{A}^T \lambda^*.$$

Expanding $g(x)$ in a Taylor series about x_k, we obtain

$$\hat{A}^T \lambda^* = g_k + G_k q + O(\|q\|^2). \tag{5.28}$$

Since q is unknown, (5.28) cannot be used directly to estimate λ^*. However, if a vector p is available that approximates q, it follows that

$$\hat{A}^T \lambda^* = g_k + G_k p + O(\|q\|^2 + \|p - q\|). \tag{5.29}$$

Based on (5.29), let η_L be the least-squares solution of the overdetermined equations

$$\hat{A}^T \eta_L = g_k + G_k p. \tag{5.30}$$

If $\|q\| = O(\epsilon)$ for sufficiently small ϵ, and $\|p - q\| = O(\epsilon^2)$, it follows from (5.29) and (5.30) that η_L is a *second-order multiplier estimate*, i.e. $\|\eta_L - \lambda^*\| = O(\epsilon^2)$.

We emphasize that the improved accuracy of a second-order estimate depends critically on the choice of p in (5.30). Obviously, the right-hand side of (5.30) is a prediction of $g(x_k + p)$ for any p, and will be a good estimate of $g(x^*)$ only if p is sufficiently close to $x^* - x_k$.

For Example 5.2, the optimal solution is $x^* = (-.14916, 3.20864, -.05948)^T$, with multiplier $\lambda^* = .46927$ (all numbers rounded to five figures). At the point $x_0 = (2.0, -1.0, 2.0)^T$, where $F(x_0) = 74.000$, the multiplier estimates are $\lambda_L = 51.000$ and $\eta_L = 33.093$; observe that both estimates have the correct sign, but neither is accurate. However, it is interesting that at the improved point $x_1 = (.18790, 2.66681, .14529)^T$ (where $F(x_1) = 5.2632$), the multiplier estimates are $\lambda_L = 15.910$ and $\eta_L = -323.14$; thus, neither estimate is accurate, and the second-order estimate η_L has the wrong sign (even though the projected Hessian is positive definite). At the close-to-optimal point $x_3 = (-.15000, 3.20998, -.05998)^T$, we can see the improved accuracy of the second-order estimate ($\lambda_L = .40869$ and $\eta_L = .46922$).

If p is taken as the solution of the subproblem (5.11) associated with a Newton-type algorithm, the equations (5.30) are compatible by definition. In this case, the least-squares solution of (5.30) is equivalent to the solution of any t independent equations. Thus, when (5.30) is compatible, second-order estimates may be computed using the variable-reduction form of Z by solving

$$V^T \eta_V = \beta,$$

where β denotes the vector of first t elements of $g_k + G_k p$.

Notes and Selected Bibliography for Section 5.1

The idea of representing the solution of an equality-constrained problem in terms of different subspaces appears implicitly in almost all discussions of methods for these problems. A full discussion of different formulations for Z is given in Gill and Murray (1974c) (see also Fletcher, 1972b). The projected steepest-descent method was proposed and extensively analyzed by Rosen

(1960). The variable-reduction technique is due to Wolfe (1962) (see also McCormick, 1970a, b). Wolfe (1967) proposed that Z could be defined as the last $n - t$ columns of the inverse of the matrix

$$T = \left(\begin{array}{c} \hat{A} \\ V \end{array} \right), \tag{5.31}$$

where V is any $(n - t) \times n$ matrix such that T is non-singular. The most obvious choices for the rows of V are a subset of the inactive constraints. The Z defined by the orthogonal factorization of $\hat{A}$ is equivalent to using the last $n - t$ columns of Q for V in (5.31) (see Gill and Murray, 1974c).

If the size of an equality-constrained problem permits the use of an orthogonal factorization, the LQ factorization is the most numerically stable method for defining Z. If there are *inequality constraints*, a different (but closely related) orthogonal factorization should be used (see the Notes for Section 5.2). However, because the LQ factorization is probably more familiar to readers, we shall continue to use it to describe methods in the inequality case.

Davidon's original (1959) paper introduced the idea of extending quasi-Newton updates to the case of linear equality constraints through an appropriately singular inverse Hessian approximation. This approach was considered in detail for the inequality case by Goldfarb (1969). The approach in which only the projected Hessian is updated is due to Gill and Murray (1973b, 1974d, 1977a). More information on first- and second-order Lagrange multiplier estimates is given by Gill and Murray (1979b).

5.2. ACTIVE SET METHODS FOR LINEAR INEQUALITY CONSTRAINTS

The problem to be considered in this section is that of minimizing F subject to a set of linear *inequality* constraints

LIP	$\displaystyle\operatorname*{minimize}_{x \in \Re^n} \quad F(x)$
	subject to $\quad Ax \geq b,$

where A is $m \times n$. For simplicity we shall not treat the case where the constraints are a mixture of equalities and inequalities; algorithms for such a problem can be developed in a straightforward manner from those to be described. The optimality conditions for LIP are discussed in Section 3.3.2. Recall that only the constraints *active* at the solution are significant in the optimality conditions. Assume that t constraints are active at x^*, and let $\hat{A}$ denote the matrix whose i-th row contains the coefficients of the i-th active constraint. The first-order necessary conditions for x^* to be optimal are

$$g(x^*) = \hat{A}^T \lambda^*; \tag{5.32a}$$

and

$$\lambda^* \geq 0. \tag{5.32b}$$

A crucial distinction from the equality-constraint case is the restriction (5.32b) on the sign of the Lagrange multipliers. As shown in Section 3.3.2, a strictly negative Lagrange multiplier would imply the existence of a feasible direction of descent, thereby contradicting the optimality of x^*.

The problem LIP is inherently more complicated than the equality-constrained problem because the set of constraints (if any) that hold with equality at the solution is generally unknown.

We shall consider only so-called *active set* algorithms, which arise from the following motivation. If the correct active set were known *a priori*, the solution of LIP would also be a solution of the equality-constrained problem

$$\begin{array}{cc} \underset{x\in\Re^n}{\text{minimize}} & F(x) \\[4pt] \text{subject to} & \hat{A}x = \hat{b}. \end{array} \tag{5.33}$$

We have discussed in Section 5.1 the variety of efficient methods for solving problems like (5.33), and have observed that the presence of linear equality constraints actually reduces the dimensionality in which optimization occurs. Consequently, we wish to apply techniques from the equality-constraint case to solve LIP. To do so, we select a "working set" of constraints to be treated as *equality* constraints. The working set will contain a subset of the original problem constraints, and the ideal candidate for the working set would obviously be the correct active set! However, since this is not available, active set methods are based on developing a working set that is a *prediction* of the correct active set. Since the prediction of the active set could be wrong, an active set method must also include procedures for testing whether the current prediction is correct, and altering it if not. An essential feature of the active set methods considered here is that all iterates are feasible.

The working set typically includes only constraints that are exactly satisfied at the current point; however, all constraints that are exactly satisfied are not necessarily contained in the working set. We observe that some authors refer to the "working set" and "active set" interchangeably. However, we believe that it is important to distinguish the set of constraints that are used to define the search direction. We retain the name "active set methods" for historical reasons.

Note that our definition of an active set method implies that the problem LIP will be solved in two phases. The first phase involves the determination of a feasible point that exactly satisfies a subset of the constraints $Ax \geq b$. The second phase involves the generation of an iterative sequence of *feasible* points that converge to the solution of LIP.

5.2.1. A Model Algorithm

Let k denote the iteration number during a typical active set method. At any iterate x_k, there is a set of quantities associated with the current working set. In particular, t_k will denote the number of constraints in the working set, $\mathcal{I}_k$ will denote the set of indices of these constraints, $\hat{A}_k$ will denote the set of coefficients of these constraints at x_k (with $\hat{b}_k$ the vector of corresponding components of the right-hand side), and Z_k will denote a basis for the subspace of vectors orthogonal to the rows of $\hat{A}_k$. When the meaning is clear, we shall sometimes refer to the *matrix* $\hat{A}_k$ as "the working set".

We shall deliberately not specify the exact criteria for the decisions involved in an active set method, because the details vary significantly between methods. Our intention is rather to provide an overview of the logical steps associated with such a method.

Assume that we are given a feasible starting point x_0; techniques for computing such a point are discussed in Section 5.7. Set $k \leftarrow 0$, determine t_0, $\mathcal{I}_0$, $\hat{A}_0$, $\hat{b}_0$, and Z_0 and execute the following steps.

Algorithm LI (*Model active set algorithm for solving LIP*).

LI1. [Test for convergence.] If the conditions for convergence are satisfied at x_k, the algorithm terminates with x_k as the solution.

LI2. [Choose which logic to perform.] Decide whether to continue minimizing in the current subspace or whether to delete a constraint from the working set. If a constraint is to be deleted, go to step LI6. If the same working set is retained, go on to step LI3.

LI3. [Compute a feasible search direction.] Compute a non-zero $(n-t_k)$-vector p_z. The direction of search, p_k, is then given by

$$p_k = Z_k p_z. \tag{5.34}$$

LI4. [Compute a step length.] Compute $\bar{\alpha}$, the maximum non-negative feasible step along p_k. Determine a positive step length α_k, for which it holds that $F(x_k + \alpha_k p_k) < F(x_k)$ and $\alpha_k \leq \bar{\alpha}$. If $\alpha_k < \bar{\alpha}$, go to step LI7; otherwise, go on to step LI5.

LI5. [Add a constraint to the working set.] If α_k is the step to the constraint with index r, add r to $\mathcal{I}_k$, and modify the associated quantities accordingly. Go to step LI7.

LI6. [Delete a constraint from the working set.] Choose a constraint (say, with index s) to be deleted. Delete s from $\mathcal{I}_k$, and update the associated quantities. Go back to step LI1.

LI7. [Update the estimate of the solution.] Set $x_{k+1} \leftarrow x_k + \alpha_k p_k$, $k \leftarrow k+1$, and go back to step LI1. ∎

5.2.2. Computation of the Search Direction and Step Length

With an active set strategy, the search direction is constructed to lie in a subspace defined by the working set. The definition (5.34) is simply a convenient technique for expressing the relationship $\hat{A}_k p_k = 0$. Note that unlimited movement along p_k will not alter the values of any constraints in the working set. The $(n - t_k)$-vector p_z may be computed in step LI3 based on any of the methods described in Section 5.1.2 for the equality-constrained problem.

Recall that in the equality-constraint methods discussed in Section 5.1, the step length α_k was not affected by the constraints, and was based only on achieving a "sufficient decrease" in F. By contrast, when the constraints are *inequalities*, in order to retain feasibility of the next iterate it is necessary to ensure that the step length does not violate any constraint that is not in the working set. Thus, the step along p_k to the nearest constraint (if any) becomes an *upper bound* on α_k.

For simplicity of notation, we shall temporarily drop the subscript k on the quantities associated with the current iterate; thus, the current point is simply x, the search direction is p, $\mathcal{I}$ contains the indices of the constraints in the working set, and so on. Let i be the index of a constraint that is not in the working set at x, so that $i \notin \mathcal{I}$. If $a_i^T p \geq 0$, any positive move along p will not violate the constraint. If $a_i^T p$ is non-negative for all such constraints, the constraints that are not in the working set impose no restriction on the step length. On the other hand, if $a_i^T p < 0$, there is a critical step γ_i where the constraint becomes "binding", i.e., such that $a_i^T(x + \gamma_i p) = b_i$. The value of γ_i is given by

$$\gamma_i = \left\{ \frac{b_i - a_i^T x}{a_i^T p} \mid i \notin \mathcal{I} \text{ and } a_i^T p < 0 \right\}.$$

Let $\bar{\alpha}$ be defined as

$$\bar{\alpha} = \begin{cases} \min\{\gamma_i\} & \text{if } a_i^T p < 0 \text{ for some } i \notin \mathcal{I}, \\ +\infty & \text{if } a_i^T p \geq 0 \text{ for all } i \notin \mathcal{I}. \end{cases}$$

The value of $\bar{\alpha}$ is the maximum non-negative feasible step that can be taken along p, and is taken as an upper bound on the final step length α. The step-length procedure used in an active set method must therefore attempt to locate a step α_F that produces a sufficient decrease in F, according to one of the sets of criteria discussed in Section 4.3.2.1, *subject to the restriction that*

$\alpha_F \leq \bar{\alpha}$. If such a value α_F can be found, α is taken as α_F. However, it may happen that no such point exists; in this case, α is taken as $\bar{\alpha}$, even though $\bar{\alpha}$ does not satisfy the criteria usually required of a step length for unconstrained optimization.

We emphasize that it is *not* appropriate to compute a step length without restriction, using a step-length algorithm for unconstrained optimization, and then set α to $\bar{\alpha}$ if the unconstrained step length exceeds $\bar{\alpha}$. Firstly, it can happen that no value of α_F exists; in contrast to the unconstrained case, it is not reasonable to assume that F is bounded below outside the feasible region. Secondly, even if α_F exists, the function evaluations at trial step lengths that exceed $\bar{\alpha}$ are effectively wasted if α_F is rejected; it is preferable to restrict function evaluations to acceptable step lengths only. The safeguarded procedures described in Section 4.3.2.1 can be adapted to include an upper bound on the step length.

When $\alpha < \bar{\alpha}$, the working set at the next iterate is unaltered. When $\alpha = \bar{\alpha}$, the working set must be modified to reflect the fact that a new constraint is satisfied exactly. When several constraints are satisfied exactly at the new point, so that there is a "tie" in the nearest constraint, only one of these is added at this step (it may be necessary to add some of the other constraints during subsequent iterations). The effect of adding a constraint depends on the representation of Z and on the method used to compute p_z (see Section 5.2.4).

The logic of Step LI5 implies that a constraint is added to the working set as soon as it is "encountered" (i.e., restricts the step length). Since the constraint is satisfied exactly at the new iterate, this strategy ensures that the constraint will not be violated by the next search direction. However, feasibility can also be retained by other strategies. For example, a constraint with index i that is not in the working set, but is exactly satisfied at x, could be added to the working set only if $a_i^T p_{k+1} < 0$.

5.2.3. Interpretation of Lagrange Multiplier Estimates

The discussion in this section will contain a brief overview of the main issues involved in deleting constraints from the working set, with emphasis on the need for care in interpreting Lagrange multiplier estimates.

If the working set $\hat{A}_k$ is the correct active set, the solution of the original problem LIP is also a solution of the equality-constrained problem

$$\begin{aligned} \underset{x \in \Re^n}{\text{minimize}} \quad & F(x) \\ \text{subject to} \quad & \hat{A}_k x = \hat{b}_k. \end{aligned} \tag{5.35}$$

If x_k is optimal for (5.35), the Lagrange multipliers corresponding to the constraints in the working set are the solution of the *compatible* system

$$g_k = \hat{A}_k^T \hat{\lambda}. \tag{5.36}$$

If any component of $\hat{\lambda}$ in (5.36) is negative, $F(x)$ can be decreased by moving in a direction that is strictly feasible with respect to that constraint alone (in effect, deleting that constraint from the working set). This suggests that x_k is *not optimal* for LIP, and hence that the Lagrange multipliers of (5.35) can be used to determine which constraint(s) (if any) should be deleted from $\hat{A}_k$.

There is an inherent tradeoff in any strategy based on using (5.35) to delete a constraint. In particular, it seems inefficient to solve (5.35) with a high degree of accuracy. If a constraint will be deleted, the working set is *not* the correct active set, and (5.35) is the *wrong problem*. The decision about deletion is therefore based on *estimates* of the Lagrange multipliers of (5.35) (see Section 5.1.5). However, the known techniques for computing estimates of the multipliers of (5.35) are not guaranteed to produce even the correct sign of the estimate unless x_k is "close enough" to the solution of (5.35) (see Example 5.3). In addition, there is a danger of serious algorithmic inefficiency when a constraint is deleted based on a highly inaccurate multiplier estimate. The well-known phenomenon of *zigzagging* occurs when a constraint is repeatedly dropped from the working set at one iteration, only to be added again at a subsequent iteration. Zigzagging can cause slow progress to the solution, or even convergence to a non-optimal point. Furthermore, the process of deleting a constraint from the working set is expensive in terms of housekeeping operations. The overall conclusion from these observations is that a constraint should not be deleted without some assurance that a negative multiplier estimate is reliable.

Since there is no known technique for assuring that the sign of a multiplier estimate is correct, complex strategies have been devised that attempt to measure both the reliability of a multiplier estimate and the improvement that would result if the constraint were deleted from the working set. Fortunately, although a multiplier estimate may not resemble the true multiplier when x_k is far from the solution, its interpretation can be made consistent with that of an exact multiplier, in that a negative multiplier estimate implies that the objective function can be reduced by deleting the corresponding constraint.

However, when a constraint is deleted from the working set based on a negative multiplier estimate, it is *not* true that *every* definition of the search direction with the new working set will be feasible with respect to the deleted constraint. For example, the Newton direction is guaranteed to be feasible only if a second-order multiplier estimate is negative.

Example 5.4. Consider the quadratic program

$$\underset{x \in \Re^n}{\text{minimize}} \quad x_1^2 + 2x_2^2$$
$$\text{subject to} \quad -x_1 - x_2 \geq 1.$$

The single constraint has the effect of moving the solution from the unconstrained minimum at the origin to the point $x^* = (-2/3, -1/3)^T$. At x^*, the optimal Lagrange multiplier is $4/3$. At the point $x = (-3, 2)^T$, which lies exactly on the constraint, the least-squares first-order multiplier has the value $\lambda_L = -1$. However, if the constraint were deleted from the working set at this point, the resulting Newton search direction would be infeasible, since it steps to the *unconstrained* minimum of the quadratic objective function. The second-order multiplier estimate has the correct sign since second-order multipliers are trivially exact for a quadratic program.

When second derivatives are available, confidence in multiplier estimates may be increased by deleting a constraint only when there is an acceptable measure of agreement between first- and second-order estimates. If the estimates are not close, neither can be considered reliable. Eventual agreement is guaranteed asymptotically, since the difference between the two estimates tends to zero as x_k tends to x^*.

Suppose that the *LQ* factorization is used. When the second-order estimate η_L is computed by solving the subproblem (5.11) and the compatible equations (5.30), the amount of additional work required to compute the first-order estimate λ_L is negligible. Define the scalar δ_i as the difference between the i-th components of λ_L and η_L, i.e.

$$\delta_i = (\eta_L - \lambda_L)_i.$$

A possible test for sufficient agreement between multipliers $(\lambda_L)_i$ and $(\eta_L)_i$ is that

$$2|\delta_i| < \min\{|(\eta_L)_i|, |(\lambda_L)_i|\}.$$

Note that elements $(\lambda_L)_i$ and $(\eta_L)_i$ that satisfy this test must agree in sign.

The knowledge that a constraint *can* be deleted does not automatically imply that it *should* be deleted. A good general principle is to delete a constraint from the working set only if it is likely that a lower value of the function will be achieved following deletion than if the constraint were retained. Further conditions may be imposed before a constraint is deleted. For example, if the multiplier is small compared to the norm of the projected gradient on the current subspace, it is likely that better progress will be obtained by retaining the current working set.

*5.2.4. Changes in the Working Set

When a constraint is added to the working set, a new row is added to $\hat{A}_k$. (For simplicity, it is usually assumed that the new row becomes the last row.) When a constraint is deleted from the working set, one of the rows of $\hat{A}_k$ is removed. In either case, it would clearly be inefficient to recompute Z_k from scratch; rather, it is possible to modify the representation of Z_k (and sometimes other matrices) to correspond with the new working set. In the remainder of this section, we consider how to update Z_k and other matrices following a single change in the working set. To simplify notation, we shall drop the subscript k associated with the current iteration. The matrix $\hat{A}$ will represent the original working set, and $\bar{A}$ will denote the "new" working set. The integer t will denote the number of constraints in the working set before the modification.

*5.2.4.1. Modification of Z.

When the LQ factorization of $\hat{A}$ is used, orthogonal transformations will ordinarily be applied to produce the updated factors. When a new constraint is included in the working set, the factors may be updated exactly as described in Section 2.2.5.7 for the QR factorization. Let the new row, say $\hat{a}^T$, be added as the $(t+1)$-th row of $\hat{A}$. Then we have

$$\bar{A} = \begin{pmatrix} \hat{A} \\ \hat{a}^T \end{pmatrix} = \begin{pmatrix} L & 0 \\ \hat{a}^TQ \end{pmatrix} Q^T = (\bar{L} \quad 0)\bar{Q}^T. \tag{5.37}$$

The new matrix $\bar{Q}$ in (5.37) can be represented as the product of Q and a Householder matrix constructed to annihilate components $t+2$ through n of the row $\hat{a}^TQ$, while leaving components 1 through t unaltered, i.e. $\bar{Q}$ can be written as

$$\bar{Q} = QH, \tag{5.38}$$

where H is a Householder matrix. Thus, the first t columns of $\bar{Q}$ are identical to those of Q, and columns $t+1$ through n are linear combinations of the corresponding columns of Q. A similar reduction can be achieved with H composed of a set of plane rotations (see Section 2.2.5.3).

When the s-th constraint is deleted from the working set, we have

$$\bar{A}Q = (M \quad 0),$$

where rows 1 through $s-1$ of M are in lower-triangular form, and the remaining rows have an extra super-diagonal element. To reduce M to the desired lower-triangular form of $\bar{L}$, plane rotations are applied on the right of Q. These rotations do not affect the last $n-t$ columns of Q, and thus $\bar{Z}$ is given by the old matrix Z augmented by a single column

$$\bar{Z} = (Z \quad \hat{z}), \tag{5.39}$$

where the new column $\hat{z}$ is a linear combination of the first t columns of Q.

In the variable-reduction method for representing Z, $\hat{A}$ is given by

$$\hat{A} = (\,V \quad U\,),$$

and an LU factorization of the non-singular matrix V is usually stored. Let $\bar{V}$ denote the new V (associated with $\bar{A}$). When a row is added to $\hat{A}$, $\bar{V}$ will be a $(t+1)$-dimensional square non-singular matrix, which is the same as V except for the last row and column. The $(t+1)$-th column of $\bar{V}$ must be selected from among the last $n-t$ columns of $\bar{A}$ so that $\bar{V}$ is non-singular. When a row is deleted from $\hat{A}$, a row and column must be deleted from V to obtain $\bar{V}$. In either case, elementary transformations such as those used in forming the original LU factorization are constructed in order to eliminate the appropriate elements to restore the triangular form of the updated factors $\bar{L}$ and $\bar{U}$.

*5.2.4.2. Modification of other matrices. In addition to altering Z when the working set changes, certain matrices must also be modified, depending on which method is being used to compute p_z. We shall consider the effect of these changes when the LQ factorization is used to represent Z; similar update procedures can be applied with the variable-reduction method.

With Newton-type methods, adding a constraint to the working set causes no difficulties. Recall that any constraint to be added is selected at the end of an iteration, and hence the new Z will be available to form (or approximate) the projected Hessian at the beginning of the next iteration.

When a constraint is deleted, $\bar{Z}$ is given by (5.39), and the new projected Hessian (at the same point!) is

$$\begin{pmatrix} G_z & Z^T G\hat{z} \\ \hat{z}^T G Z & \hat{z}^T G\hat{z} \end{pmatrix}. \tag{5.40}$$

The vector $G\hat{z}$ can be formed explicitly or approximated by a single finite-difference. Since the modified Cholesky factors of G_z are known, the factors of the new projected Hessian can be computed with one additional step of the row-wise modified Cholesky algorithm of Section 2.2.5.2.

With a quasi-Newton method, a matrix B_z is available that represents curvature information within the subspace defined by the old Z. When a constraint is added to the working set, the form (5.38) of $\bar{Q}$ and the existing Cholesky factors of B_z can be used to obtain Cholesky factors of the (smaller) projected Hessian approximation. These procedures are rather complicated, and will not be discussed here (see the references for further details).

When a constraint is deleted, minimization will henceforth occur within a larger subspace; however, no curvature information is available along the new direction $\hat{z}$. Therefore, the new approximate Hessian matrix may be given by

$$\begin{pmatrix} B_z & 0 \\ 0 & \gamma \end{pmatrix}.$$

The value of γ is usually taken as unity, but other choices may be appropriate if additional scaling information is available. Alternatively, it might be possible in some circumstances to fill in the new row and column of the projected Hessian approximation as in (5.40), using an approximation to $G\hat{z}$ obtained by a single finite-difference of the gradient along $\hat{z}$. In this case, the Cholesky factorization may be updated as in the Newton-type methods described above.

Notes and Selected Bibliography for Section 5.2

Some published algorithms for special problem types do not fall precisely within the category of "active set" strategy described here. For example, Conn (1976) describes a single-phase method for linear programming (see Section 5.3.1) in which each iterate satisfies a set of constraints exactly, but is not necessarily feasible. In this case, the method is one of descent with respect to a *non-differentiable penalty function* (see Section 6.2.2), rather than the linear objective function. A similar method using the *LU* factorization is given by Bartels (1980).

There may be other differences in the way in which the search direction is computed. Some active set methods choose the search direction based on solving a more complicated subproblem. One class of methods is based upon solving an *inequality-constrained* quadratic program at each iteration. Consider the problem

$$\underset{x \in \Re^n}{\text{minimize}} \quad F_k + g_k^T p + \frac{1}{2} p^T B_k p \tag{5.41}$$
$$\text{subject to} \quad \hat{A}p \geq 0,$$

where B_k is some approximation to the Hessian matrix of F (*cf.* (5.11)). If the solution p_k of the inequality QP (5.41) is used as the search direction, complications arise because it may be necessary to store or represent the full matrix B_k (a discussion of methods for solving (5.41) is given in Section 5.3.2). Although only the matrix $Z_i^T B_k Z_i$ is required at iteration i of the quadratic subproblem, it is not known *a priori* which sets of constraints will define Z_i as the iterations proceed. Hence, most inequality QP methods assume that the full matrix B_k is available. In contrast to the positive-definiteness of the projected Hessian in the equality-constraint case, there is no presumption that B_k should be positive definite. In particular, the Hessian of F need not be positive definite, even at x^*. If B_k is indefinite, (5.41) may not have a bounded solution; furthermore, the solution of (5.41) for an indefinite B_k is not necessarily a descent direction for F, since it may happen that $g_k^T p_k > 0$. Even if the solution of the inequality QP is a strong local minimum and $Z_i^T B_k Z_i$ is positive definite during *every* iteration of the subproblem, descent is not assured. (For a further discussion of inequality-constrained subproblems, see Murray and Wright, 1980.)

A number of approaches have been proposed to overcome the inherent difficulties of solving an inequality-constrained QP. For bound-constrained problems, Brayton and Cullum (1977, 1979) have suggested computing the search direction from an initial simplified inequality-constrained QP. If the result is unsatisfactory, a more complicated inequality-constrained QP is solved; finally, if neither QP subproblem has succeeded in producing a satisfactory search direction, an alternative method is used that does not involve a QP. Although the results reported by Brayton and Cullum indicate that the first QP solution is acceptable most of the time, the reason that both may fail is the presence of a possibly indefinite matrix B_k in the QP formulation, so that neither QP can be guaranteed to yield a descent direction.

For the linear-constraint case, Fletcher (1972a) has suggested computing the search direction from a QP subproblem that includes all the original inequality constraints as well as additional bounds on each component of the search direction; this method is similar to that of Griffith and Stewart (1961), in which a linear programming subproblem is solved. The purpose of the extra constraints in Fletcher's method is to restrict the solution of the QP to lie in a region where the current quadratic approximation of F is likely to be reasonably accurate. The bounds on p are adjusted at each iteration if necessary to reflect the adequacy of the quadratic model. Fletcher's algorithm effectively includes the "trust region" idea that is used in other areas of optimization (see the Notes for Section 4.4).

For a general discussion of methods for linearly constrained problems, see Gill and Murray (1974c, 1977a). The Newton-type method of McCormick (1970b) uses the variable-reduction form of Z_k. The earliest quasi-Newton method for linear inequality constraints was due to Goldfarb (1969), and was based on extending Davidon's (1959) idea of updating a singular "quasi-Newton" approximation to the inverse Hessian. Goldfarb's method is one of a class of methods to be discussed in Section 5.4. For more details of the method in which the factors of a projected Hessian are recurred, see Gill and Murray (1973b, 1974d, 1977a). An error analysis of the methods for updating the orthogonal factorization of the matrix of constraints in the working set has been given by Paige (1980).

In practical computation, it is more convenient to use a variant of the LQ factorization for the matrix of constraints in the working set. Consider the TQ *factorization*

$$\hat{A}_k Q_k = (\, 0 \quad T_k\,),$$

where T_k is a "reverse" triangular matrix such that

$$t_{ij} = 0 \quad \text{for} \quad i + j \leq n$$

(see Gill *et al.*, 1980). The triangular matrix T can be interpreted as a lower-triangular matrix with its columns in reverse order. The TQ factorization has the same favourable numerical properties as the LQ factorization. The most important feature of the TQ factorization is that the first $n - t_k$ columns of Q_k define Z_k, i.e.

$$Q_k = (\, Z_k \quad Y_k \,).$$

The fundamental advantage of this factorization is that, if a row or column of $\hat{A}_k$ is added or deleted, the new column of Z_k or Y_k is in its correct position within Q_k. For example, if a constraint is deleted from the working set, the modification to the projected Hessian described in Section 5.2.4.1 requires that the new column of Z_k be added in the last position. This occurs naturally when updating the TQ factorization.

The number of iterations of a method can be significantly affected by the strategy used to select the working set. If a constraint is deleted from the working set, we are implicitly assuming that, in one step, a greater reduction in F will be achieved with the constraint omitted than would occur if the constraint were retained. For some methods (in particular, those to be discussed in Section 5.4), the relevant changes in F can be predicted to second order for each constraint with a negative multiplier. If these quantities are computed, we can ensure that a constraint is not deleted until a sufficiently good change in F is predicted to occur. Moreover, when a decision is made to delete a constraint, the constraint that yields the maximum predicted change in F can be deleted. If second derivatives are not available, *substitute tests* can be employed that approximate the relevant information. One of the first methods to use a substitute test in order to predict the change in F was suggested by Rosen (1961). For a complete description of these techniques and some cautionary remarks concerning their use, see Gill and Murray (1974d).

See Gill and Murray (1979b) for a more detailed discussion of techniques for testing the accuracy of Lagrange multiplier estimates, and of the conditions that must apply in order to guarantee a feasible descent direction for Newton-type and quasi-Newton methods. For details of various anti-zigzagging strategies in nonlinear programming see Wolfe (1966), McCormick (1969) and Zoutendijk (1970).

5.3. SPECIAL PROBLEM CATEGORIES

5.3.1. Linear Programming

The *linear programming* (LP) problem is the special case of LIP in which the objective function is linear, i.e.

$$
\boxed{\text{LP} \qquad \begin{aligned} \underset{x\in\Re^n}{\text{minimize}} \quad & c^T x \\ \text{subject to} \quad & Ax \geq b. \end{aligned}} \tag{5.42}
$$

The active set procedure discussed in Section 5.2.1 has many special properties when applied to the LP problem, and numerous efficiencies are possible because of the linearity of the objective function. As indicated in Section 5.1.4.1, an equality-constrained problem with a linear objective function and linearly independent constraints must in general contain n constraints in order for the solution to be unique. Since the LP problem will be solved as a sequence of equality-constrained subproblems, the number of constraints in the working set at each iteration will generally be equal to the number of variables.

We shall temporarily drop the subscript k associated with the current iteration. Let x denote the current iterate, with $\hat{A}$ the non-singular matrix of constraints in the working set. In standard LP terminology, x lies at a *vertex* of the feasible region, i.e. a point where n linearly independent constraints are satisfied exactly. Since x is optimal for the equality-constrained subproblem defined by the working set, the next step is to check the sign of the corresponding Lagrange multipliers $\hat{\lambda}$, which satisfy the non-singular system of linear equations

$$
\hat{A}^T \hat{\lambda} = c. \tag{5.43}
$$

If all elements of $\hat{\lambda}$ are strictly positive, x is the optimal solution of LP. However, if any component of $\hat{\lambda}$ is negative (say, $\hat{\lambda}_s < 0$), the objective function can be decreased by moving "off" the s-th constraint (i.e., deleting it from $\hat{A}$), but remaining "on" all the others in the working set. Thus, we seek a direction p such that

$$
\hat{a}_s^T p = \gamma, \qquad \gamma > 0, \tag{5.44}
$$

and

$$
\hat{a}_j^T p = 0, \qquad j \neq s. \tag{5.45}
$$

Since only the direction of p is significant, we may assume that $\gamma = 1$ in (5.44). Conditions (5.44) and (5.45) imply that p is the unique solution of the linear system

$$
\hat{A}p = e_s, \tag{5.46}
$$

where e_s is the s-th column of the identity matrix; therefore, p is the s-th column of $\hat{A}^{-1}$. Hence, when moving from vertex to vertex, the search direction is unique.

When $\hat{\lambda}_s$ is strictly negative, the linear objective function will decrease without bound along the direction p. Therefore, the maximum feasible step $\bar{\alpha}$ (to the nearest constraint not in the working set) will be taken, and the newly encountered constraint will be added to the working set. Many economies of computation are possible for LP problems. For example, the first-order prediction of the change in the objective function following deletion of a constraint is exact. Since $\hat{A}$ is non-singular, an active set method can be implemented using orthogonal transformations, but without storing Q. Alternatively, the LU factorization of $\hat{A}$ can be computed and updated, since each change in the working set results in a rank-one modification.

When applied to a linear programming problem, the general active set method described in Section 5.2.1 is the well-known *simplex method*, which was originally developed from quite a different point of view. The simplex method is one of the most famous of all numerical methods, and has been widely studied. One particularly interesting aspect of the simplex method is that there is a theoretical (finite) upper bound on the number of iterations required to reach the optimal solution, but it becomes astronomically large as the problem size increases. In practice, however, the simplex method is remarkably efficient, and the number of iterations tends to be a *linear* function of the problem size.

It is possible to define non-simplex steps if the current iterate is not a vertex (i.e., fewer than n constraints are in the working set) by computing a feasible descent direction as in a general active set method. For example, the search direction may be taken as the "steepest-descent" vector $p = -ZZ^Tc$ (cf. (5.10)). Note that when x is not a vertex, it should be possible to decrease the objective function without deleting a constraint from the working set.

The most common notation for linear programming problems is somewhat different than that presented here, and will be discussed in Section 5.6.1.

5.3.2. Quadratic Programming

The general form of a quadratic programming problem with only inequality constraints is

$$
\boxed{
\begin{array}{ll}
\text{QP} & \displaystyle \operatorname*{minimize}_{x \in \Re^n} \quad c^T x + \frac{1}{2} x^T G x \\[1em]
& \text{subject to} \quad Ax \geq b.
\end{array}
}
$$

for a constant matrix G and vector c.

Almost all major algorithms for quadratic programming are active set methods. This point is emphasized because even closely related algorithms may be described in widely different terms. For example, some QP methods are based on a "tableau" involving the constraints and the matrix G, others involve "pivots", and so on. In describing QP algorithms, we shall retain the notation introduced in Section 5.2.1. Thus, $\hat{A}_k$ will denote the working set at the iterate x_k, and Z_k will denote a matrix whose columns form a basis for the set of vectors orthogonal to the rows of $\hat{A}_k$.

5.3.2.1. Positive-definite quadratic programming. A positive-definite quadratic program is one in which the projected Hessian matrix $Z_k^T G Z_k$ is known *a priori* to be positive-definite at every iteration. In general, this knowledge will be available only if G itself is positive definite. A positive-definite projected Hessian matrix implies that the search direction is always well-defined, and that the quadratic function has a unique minimum in the subspace defined by Z_k.

The "search direction" at x_k is obtained by solving the Newton equations

$$
Z_k^T G Z_k p_z = -Z_k^T g_k, \tag{5.47}
$$

and setting $p_k = Z_k p_z$. Because of the quadratic nature of the objective function, there are only two choices for the step length. A step of unity along p_k is the exact step to the minimum of the function restricted to the null space of $\hat{A}_k$. If a step of unity can be taken, the next iterate will be a constrained stationary point with respect to the equality constraints defined by $\hat{A}_k$, and exact Lagrange multipliers can be computed to determine whether a constraint should be deleted. Otherwise, the step along p_k to the nearest constraint is less than unity, and a new constraint will be included in the working set at the next iterate.

As with linear programming, advantage can be taken of the special features of the quadratic objective function, so that the needed quantities are computed with a minimum of effort. Since G is constant, special techniques have been developed for updating factorizations of the projected Hessian matrix. For example, when a constraint is deleted from the working set, the projected Hessian matrix is augmented by a single row and column (see (5.40)). This change causes the upper-triangular matrix R in the Cholesky factorization to be augmented by a single column.

Other computations also simplify when the general active set method of Section 5.2.1 is applied to a quadratic program. In particular, consider calculation of the search direction when x_k is a non-optimal constrained stationary point, and the LQ factorization is used to represent Z_k. By definition

$$Z_k^T g_k = 0. \tag{5.48}$$

In this situation, the Lagrange multipliers are evaluated, and a constraint with a negative multiplier will be deleted from the working set. The updated matrix $\bar{Z}_k$ will be given by (5.39), and, from (5.48), g_k will be orthogonal to all columns of $\bar{Z}_k$ except the last one. Hence

$$\bar{Z}_k^T g_k = (\hat{z}^T g_k) e_{n-t+1}. \tag{5.49}$$

The search direction is obtained by solving the linear system (5.47). The coefficient matrix is in the form of a product of the Cholesky factors $R^T R$ of the updated projected Hessian matrix, and the right-hand side is the vector (5.49) (the projected gradient with the updated working set), so that (5.47) becomes:

$$R^T R p_z = -\gamma e_{n-t+1}, \tag{5.50}$$

where $\gamma = (\hat{z}^T g_k)$. The first step in solving (5.50) involves the lower-triangular matrix R^T. Because of the special form of the right-hand side of (5.50), the result is a multiple of e_{n-t+1}, and hence the vector p_z is the solution of

$$R p_z = \beta e_{n-t+1}, \tag{5.51}$$

for some scalar β.

Using the active set method described above, it can be shown that p_z *is always the solution of an upper-triangular system of the form* (5.51). This simplification in computing the search direction is typical of the efficiencies that are possible when an active set method is applied to a quadratic program.

5.3.2.2. Indefinite quadratic programming.

Complications arise in *indefinite quadratic programming problems*, in which the matrix $Z_k^T G Z_k$ is indefinite for some Z_k. In this case, it is not true that any constrained stationary point is a local minimum in the current null space. Furthermore, the direction defined by (5.47) is not necessarily a direction of descent for the quadratic function.

We wish to be able to compute a feasible direction of descent for the quadratic objective function even when the projected Hessian is indefinite. In doing so, it is desirable to retain the maximum amount of information in the present (indefinite) projected Hessian, in order to preserve the efficiencies associated with quadratic programming; therefore, standard techniques that *alter* an indefinite matrix (see Section 4.4.2) are not suitable. However, there is a danger of substantial numerical instability if care is not exercised in updating a factorization of an indefinite matrix.

If the projected Hessian is indefinite at a constrained stationary point defined by the working set $\hat{A}_k$, there must exist a vector p_z (a direction of negative curvature) such that

$$p_z^T Z_k^T G Z_k p_z < 0;$$

hence, if the problem contained no further constraints, the quadratic objective function would be unbounded below in the current subspace. However, assuming that the QP has a *finite* solution, a move along such a direction must encounter a constraint that is not already in the working set. Eventually, enough constraints must be added to the working set so that the projected Hessian becomes positive definite.

The algorithm for positive-definite QP described in Section 5.3.2.1 can be adapted to the indefinite case if we ensure that the initial iterate is either a vertex or a constrained stationary point where the projected Hessian is positive definite. Under these circumstances, the only way in which the projected Hessian can become indefinite is when a single constraint is *deleted* from the working set; the effect of this change on the Cholesky factors of the new projected Hessian is that R is augmented by a single column and the new last diagonal element of R is undefined (being the square root of a negative number). However, the definition (5.51) of the search direction is such that the last diagonal element of R affects only the *scaling* of p_z and not its direction. Consequently, without any loss of generality, we may use the modulus of the operand when computing the square root for the last diagonal element of R. In this special situation, the search direction defined by (5.47) is a direction of negative curvature, and hence can be used exactly as in the general algorithm.

Once a direction of negative curvature is computed, a constraint must be added during the next iteration and it can be shown that the addition of a new constraint to the working set cannot increase the number of negative eigenvalues in the projected Hessian. Suppose that such a constraint "exchange" takes place, and that the last diagonal element of R was previously altered as described above in order to avoid taking the square root of a negative number. It can be shown that the arbitrary value resulting from the constraint deletion does not propagate to any other elements, i.e. the factors of the "exchanged" projected Hessian will still have a single arbitrary last diagonal element. This result implies that when a constraint exchange results in a positive-definite projected Hessian, the last diagonal element of the triangular factor can be recomputed if it was previously set to an arbitrary value.

There might appear to be a danger of numerical instability in allowing an indefinite projected Hessian. Certainly the occurrence of an undefined quantity during the calculation of R implies that the usual bound on growth in magnitude in the elements of R from the positive-definite case (see Section 2.2.5.2) does not apply. However, after a sequence of constraint exchanges, it is possible only for the *last row* of R to be "contaminated" by growth. As in the case of a triangular factor with an arbitrary last diagonal element, the offending row may be recomputed as soon as a positive-definite projected Hessian is obtained.

This result justifies tolerating a very limited form of indefiniteness, as described above. However, the overall viewpoint of the algorithm is that the projected matrix should be kept "as positive definite as possible". Therefore, once the projected Hessian is indefinite, *no further constraints are deleted* until enough constraints have been added so that the projected Hessian has become positive definite.

There is no loss in generality for the user in the requirement that the initial point should be a vertex or a constrained minimum. In practice, the algorithm can be started at *any* feasible point — whether this point be found by the algorithm or specified by the user. If no feasible point is known, the procedure to be described in Section 5.7 can be applied to determine a feasible point

x. If the projected Hessian at x is not positive definite or x is not a vertex, artificial constraints are added by the algorithm to the working set. These constraints involve artificial variables $\{y_i\}$, and are of the form $y_i \geq x_i$ or $y_i \leq x_i$ (so that they are satisfied exactly at the starting point). Enough of these bounds can be added to the set of constraints so that the initial point satisfies the required conditions. These special constraints are "marked" and the direction of the inequalities is chosen so that they are deleted from the working set as soon as possible. Furthermore, because of the special form of the artificial constraints, no additional storage is required. As an example of how this scheme would work, consider the case where the feasible point procedure provides an "artificial" vertex x such that t ($t < n$) constraints of the original problem satisfy $\hat{A}x = \hat{b}$. Suppose that x is such that $Z^T G Z$ is positive definite, where $\hat{A}Z = 0$. As the $n - t$ artificial bounds are deleted, the QP method will essentially build the column-wise Cholesky factorization of the matrix $Z^T G Z$.

With this approach, an indefinite quadratic program can be solved using the same basic algorithm as for the positive-definite case. Moreover, if the algorithm is applied to a positive-definite quadratic program, the method is identical to that for the positive-definite case.

*5.3.3. Linear Least-Squares with Linear Constraints

A problem that is very closely related to quadratic programming arises from the need to minimize the two-norm of the residual of a set of linear equations subject to a set of linear constraints. Consider the *constrained linear least-squares problem*

$$
\begin{array}{ll}
\text{LLS} & \displaystyle\operatorname*{minimize}_{x \in \Re^n} \quad \frac{1}{2}\|Hx - d\|_2^2 \\[1em]
& \text{subject to} \quad Ax \geq b,
\end{array}
$$

where H is an $s \times n$ matrix. In many problems, $s \geq n$; in fact, it is usually true that $s \gg n$.

The problem LLS is simply a quadratic program of the form QP given in Section 5.3.2, with $G = H^T H$ and $c = H^T d$. However, methods can be developed to solve LLS that avoid explicit use of a matrix whose condition number is the square of that of H.

Given a feasible point x_k that exactly satisfies a set of linearly independent constraints (with associated matrix $\hat{A}_k$), the minimum residual subject to the equalities defined by $\hat{A}_k$ is achieved by taking the step p_k that solves (5.47), i.e.

$$
Z_k^T H^T H Z_k p_z = -Z_k^T H^T d_k,
$$

where $d_k = Hx_k - d$. These are just the normal equations for the unconstrained linear least-squares problem

$$
\operatorname*{minimize}_{p_z \in \Re^{n-t}} \frac{1}{2}\|HZ_k p_z - d_k\|_2^2.
$$

To simplify the discussion, we shall assume that $\hat{A}_k$ contains enough constraints to ensure that the matrix HZ_k is of full rank (this is equivalent to requiring that the least-squares subproblem has a unique solution).

Let P_k be an orthonormal matrix of dimension $s \times s$. Then

$$
\|HZ_k p_z - d_k\|_2^2 = \|P_k(HZ_k p_z - d_k)\|_2^2. \tag{5.52}
$$

Suppose that P_k is chosen to be a matrix that transforms HZ_k to upper-triangular form (see Section 2.2.5.3), i.e.

$$P_k H Z_k = \begin{pmatrix} R_k \\ 0 \end{pmatrix},$$

where R_k is a non-singular $(n - t_k) \times (n - t_k)$ upper-triangular matrix. Then, from (5.52),

$$\|P_k(HZ_k p_z - d_k)\|_2^2 = \left\| \begin{pmatrix} R_k \\ 0 \end{pmatrix} p_z - P_k d_k \right\|_2^2.$$

From the last expression, we see that the residual vector of the transformed problem will be minimized when the components of $R_k p_z$ are equal to the first $n - t_k$ components of $P_k d_k$. The minimum residual is therefore achieved for the vector p_z that is the unique solution of the linear system $R_k p_z = \bar{d}_k$, where $\bar{d}_k$ denotes the vector of first $n - t_k$ components of $P_k d_k$ (see Section 2.2.5.3).

An important feature of this method is that it is unnecessary to store the matrix P_k, since any orthogonal matrices applied on the left of HZ_k can be discarded after being applied to d_k. This scheme allows significant savings in storage when $s \gg n$.

If an orthogonal factorization of $\hat{A}_k$ is used to define the matrix Z_k, the vector $\bar{d}_k$ and the matrix R_k can be updated as constraints are added to and deleted from the working set. If a constraint is added to $\hat{A}_k$, Z_{k+1} is obtained by post-multiplying Z_k by a sequence of plane rotations (see Section 5.2.4.1). The rotations have the effect of creating non-zero elements below the diagonal of R_k, but the new elements may then be eliminated by applying another sequence of rotations from the left (these rotations implicitly define a new matrix P_{k+1}). If a constraint is deleted from the working set, the first $n - t_k$ columns of Z_{k+1} are identical to those of Z_k, but a new last column z_{n-t_k+1} is generated. The effect of this change on the matrix R_k is also to add a new last column, which is computed from the vector Hz_{n-t_k+1}. To avoid the necessity of storing the matrix H throughout the computation, it is more convenient to update the factorization

$$P_k H \tilde{Q}_k = U_k,$$

where $\tilde{Q}_k = (\, Z_k \quad Y_k \,)$ (a permutation of the orthogonal matrix associated with the LQ factorization of $\hat{A}_k$) and U_k is an $n \times n$ upper-triangular matrix. In this case, the matrix R_k is just the first $n - t_k$ rows and columns of U_k.

In practice, it is unreasonably restrictive to assume that the matrix HZ_k is of full rank for all Z_k. However, the algorithm given above is easily generalized to use the complete orthogonal factorization of HZ_k

$$P_k H Z_k V_k = \begin{pmatrix} R_k & 0 \\ 0 & 0 \end{pmatrix},$$

where R_k is an $r \times r$ upper-triangular matrix with r the rank of HZ_k (see Section 2.2.5.3).

Notes and Selected Bibliography for Section 5.3

The simplex method for linear programming was originated in 1947 by Dantzig (see Dantzig, 1963). Since that time, the algorithm has been refined to such an extent that problems containing several thousand constraints are solved routinely by commercial packages. The subject of large-scale linear programming will be discussed more fully in Section 5.6.1.

When the objective function is linear it is possible to be more sophisticated in the choice of which constraint to delete from the working set. Let $p^{(j)}$ denote the direction of search obtained by deleting a constraint with negative multiplier λ_j. First-order Lagrange multiplier estimates provide a first-order estimate of the change in F resulting from a unit step along $p^{(j)}$, and this prediction is *exact* for linear programming problems. If a unit step were taken along $p^{(j)}$, the change to the objective function would be

$$c^T p^{(j)} = \lambda_j.$$

Thus, the strategy of deleting the constraint with the most negative multiplier is based on trying to achieve the largest first-order change in F for a unit step. An alternative strategy is to delete the constraint that predicts the best first-order decrease in F for a *unit change in x*. This strategy can be applied relatively efficiently for linear programming (see Greenberg and Kalan, 1975; Goldfarb and Reid, 1977). In this case, the change in the objective per unit step is given by

$$c^T p^{(j)} / \|p^{(j)}\|_2 = \lambda_j / \|p^{(j)}\|_2,$$

and the quantities $\|p^{(j)}\|_2$ may be *recurred* from iteration to iteration. The resulting linear programming algorithm is known as the *steepest-edge* simplex method. Harris (1973) describes how to approximate the norms of $p^{(j)}$.

Some of the earliest methods for quadratic programming were viewed as modifications of linear programming techniques. The most successful of these methods were those of Beale (1959, 1967a) and, for the positive-definite case, Dantzig and Wolfe (see Wolfe, 1959; and Dantzig, 1963). The first "active set" QP methods of the type suggested here were suggested by Murray (1971a) and Fletcher (1971b) (Fletcher's method is discussed in the Notes for Section 5.4).

The QP method utilizing the Cholesky factorization of $Z_k^T G Z_k$ was suggested by Gill and Murray (1978b). In this method, the complete matrix Q_k associated with the LQ factorization of the matrix of constraints in the working set is stored. Methods can be derived that avoid storing the complete matrix Q_k, but this will always result in some loss of numerical stability. Murray (1971a) has proposed a method for indefinite QP in which only the matrix Y_k from the orthogonal factorization is stored and the Z_k is chosen so that $Z_k^T G Z_k$ is a diagonal matrix. Bunch and Kaufman (1980) give a similar algorithm in which $Z_k^T G Z_k$ is recurred as a block-diagonal matrix. (Murray's method and the Bunch-Kaufman method are identical when G is positive definite.) Powell (1980) has given a method for the positive-definite case in which Q_k is not stored at all, and the matrix Z_k is constructed from L_k and $\hat{A}_k$. Conn and Sinclair (1975) have proposed a single-phase quadratic programming algorithm that does not necessarily generate a feasible point at each iteration. For a comparison between various methods for quadratic programming, see Djang (1980).

The method for linear least-squares problems with linear constraints discussed in Section 5.3.3 is essentially due to Stoer (1971) (see also, Schittkowski and Stoer, 1979) with modifications suggested by Saunders (1980). Other methods for constrained linear least-squares problems are presented by Lawson and Hanson (1974). Methods for the solution of linearly constrained linear-ℓ_1 problems that use the orthogonal factorization techniques discussed here have been suggested by Bartels and Conn (1980).

*5.4. PROBLEMS WITH FEW GENERAL LINEAR CONSTRAINTS

The methods described in the preceding sections (which we shall call *null-space methods*) tend to improve in efficiency as the number of constraints in the working set *increases*. However, methods of this type may not be the best choice when the number of linear constraints is *small* compared to the number of variables, since the dimension of the null space will be much larger than the range space. An alternative approach is to design a *range-space method* based on a subproblem whose dimensionality is reduced as the number of constraints in the working set *decreases*. We shall discuss range-space methods for two forms of the objective function: a positive-definite quadratic function and a general nonlinear function.

*5.4.1. Positive-Definite Quadratic Programming

Consider the case of a quadratic programming problem in which the Hessian matrix is positive definite. At the feasible point x_k, suppose that there are t_k ($t_k \ll n$) linearly independent constraints that are satisfied exactly, with corresponding matrix $\hat{A}_k$. The direction p_k such that $x_k + p_k$ remains feasible with respect to $\hat{A}_k$ and minimizes the quadratic function in the appropriate subspace is the solution of the QP subproblem

$$\begin{array}{c} \underset{x\in\Re^n}{\text{minimize}} \quad g_k^T p + \frac{1}{2}p^T G p \\ \text{subject to} \quad \hat{A}_k p = 0, \end{array} \tag{5.53}$$

where $g_k = G x_k + c$ (cf. (5.21)). Let p_k and λ_k denote the solution of (5.53) and its vector of Lagrange multipliers. The optimality conditions for (5.53) imply that p_k and λ_k satisfy the $n + t_k$ linear equations

$$\begin{pmatrix} G & -\hat{A}_k^T \\ \hat{A}_k & 0 \end{pmatrix}\begin{pmatrix} p_k \\ \lambda_k \end{pmatrix} = \begin{pmatrix} -g_k \\ 0 \end{pmatrix}. \tag{5.54}$$

If the solution of (5.53) exists and is unique, the coefficient matrix

$$\begin{pmatrix} G & -\hat{A}_k^T \\ \hat{A}_k & 0 \end{pmatrix} \tag{5.55}$$

of (5.54) is non-singular. Moreover, if G is positive definite and $\hat{A}_k$ has full row rank, the *inverse* of (5.55) is of the form

$$\begin{pmatrix} G^{-1} - G^{-1}\hat{A}_k^T(\hat{A}_k G^{-1}\hat{A}_k^T)^{-1}\hat{A}_k G^{-1} & G^{-1}\hat{A}_k^T(\hat{A}_k G^{-1}\hat{A}_k^T)^{-1} \\ -(\hat{A}_k G^{-1}\hat{A}_k^T)^{-1}\hat{A}_k G^{-1} & (\hat{A}_k G^{-1}\hat{A}_k^T)^{-1} \end{pmatrix}$$

For a positive-definite quadratic program there are thus two alternative (equivalent) forms for the direction of search and the multipliers:

Null-space
$$p_k = -Z_k(Z_k^T G Z_k)^{-1}Z_k^T g_k$$
$$\lambda_k = (\hat{A}_k \hat{A}_k^T)^{-1}\hat{A}_k(G p_k + g_k)$$

Range-space
$$\lambda_k = (\hat{A}_k G^{-1}\hat{A}_k^T)^{-1}\hat{A}_k G^{-1}g_k.$$
$$p_k = G^{-1}(\hat{A}_k^T \lambda_k - g_k),$$

The formal expression for the range-space version of p_k given above is not appropriate for practical computation. Let L_k be the $t_k \times t_k$ lower-triangular matrix associated with the LQ factorization of $\hat{A}_k$ (5.15), and let Y_k be the matrix whose columns are the first t_k columns of the orthogonal factor Q_k. Substituting the expression $\hat{A}_k = L_k Y_k^T$ into the range-space equations, we obtain

$$p_k = G^{-1}(Y_k\theta - g_k), \tag{5.56}$$

where

$$\theta = (Y_k^T G^{-1} Y_k)^{-1} Y_k^T G^{-1} g_k. \tag{5.57}$$

Since Y_k is a section of an orthogonal matrix, the condition number of $Y_k^T G^{-1} Y_k$ is no worse than that of G. The computation of p_k from (5.56) and (5.57) may be implemented by storing the Cholesky factors of G (a computation that needs to be performed only once), and updating L_k, Y_k and the Cholesky factors of $Y_k^T G^{-1} Y_k$ as constraints are added to and deleted from the working set. If we assume that Y_k and the factors of G can be stored in approximately the same space as the unsymmetric matrix Q_k, the tradeoff in storage between the two formulations depends on the dimensions of the matrices $Z_k^T G Z_k$ and $Y_k^T G^{-1} Y_k$. If t_k is expected to be close to n, the null-space method requires less storage. If t_k is typically small, the range-space method should be used if storage is limited. If $t_k \approx \frac{1}{2}n$, the choice depends on other considerations. The null-space method is more numerically reliable, because of possible difficulties with cancellation error in computing p_k from (5.56). However, the range-space approach may be preferred if the matrix G has some specific structure of zero elements that can be used to reduce the storage of its Cholesky factors.

In many problems, the maximum size of the matrices $Z_k^T G Z_k$ and $Y_k^T G^{-1} Y_k$ can be specified in advance. For example, if there are 5 general constraints in a 50-variable problem, the matrix $Y_k^T G^{-1} Y_k$ will never be larger than 5×5. By contrast, suppose that G is a matrix of the form

$$G = \begin{pmatrix} G_{11} & 0 \\ 0 & 0 \end{pmatrix},$$

where G_{11} is an $r \times r$ non-singular matrix. If the QP (5.53) is to have a unique solution, the matrix $Z_k^T G Z_k$ must be positive definite; hence, there must be at least $n - r$ constraints in the working set, and the number of columns of Z_k can not exceed r.

*5.4.2. Second Derivative Methods

In this section, we shall discuss two methods that may be used for problems with a general nonlinear objective function and few linear constraints when second derivatives are available.

*5.4.2.1. A method based on positive-definite quadratic programming.
A Newton-type null-space method would typically be based on using the solution of the QP (5.53) (with $G = G_k$) as the search direction. However, some care is necessary in order to define the range-space formulation if G_k is indefinite.

The solution of the QP (5.53) is unaltered if G_k is replaced by the matrix

$$G_\nu \equiv G_k + \nu \hat{A}_k^T \hat{A}_k, \tag{5.58}$$

since $Z_k^T G_\nu Z_k = Z_k^T G_k Z_k$. Moreover, if $Z_k^T G_k Z_k$ is positive definite, there exists a finite scalar $\bar{\nu}$ such that G_ν is positive definite for all $\nu > \bar{\nu}$. This result implies that a range-space algorithm may be applied in the indefinite case by choosing a suitable value of ν and computing the modified Cholesky factorization (see Section 4.4.2.2) of G_ν. The search direction can then be computed from (5.56) and (5.57), using G_ν instead of G_k.

If F is well scaled (see Section 8.7) and the constraints are scaled so that $\|\hat{A}_k\|_\infty$ is of order unity, in general it will be acceptable to use the value

$$\nu = \frac{1}{\delta}(\|G_k\|_\infty + \epsilon_M), \qquad (5.59)$$

where $\delta \in [\epsilon_M^{1/2}, \epsilon_M^{1/3}]$, with ϵ_M the relative machine precision.

*5.4.2.2. A method based on an approximation of the projected Hessian. An alternative technique that is efficient when the dimension of Z_k is large relative to n is based on the following limiting relationship. As ν increases in (5.58), the inverse of G_ν approaches the following limit:

$$\lim_{\nu \to \infty} G_\nu^{-1} = Z_k(Z_k^T G_k Z_k)^{-1} Z_k^T, \qquad (5.60)$$

and

$$\|G_\nu^{-1} - Z_k(Z_k^T G_k Z_k)^{-1} Z_k^T\| = O(\frac{1}{\nu}).$$

This property can be used to define a feasible search direction of the form (5.34), by selecting a finite value of ν in (5.58). The vector p_ν is then defined as the solution of the equations

$$G_\nu p_\nu = -g_k. \qquad (5.61)$$

Because the search direction must be of the form (5.34), p_k is taken as

$$p_k = (I - Y_k Y_k^T)p_\nu \ (\equiv Z_k Z_k^T p_\nu).$$

If $Z_k^T G_k Z_k$ is positive definite and ν is sufficiently large, the direction p_k will be a descent direction. If $Z_k^T G_k Z_k$ is not positive definite, no finite value of ν exists such that G_ν is positive definite. In this situation, the use of the modified Cholesky factorization of G_ν to solve (5.61) will ensure that p_k is a descent direction.

This technique approximates the Newton direction $-Z_k(Z_k^T G_k Z_k)^{-1} Z_k^T g_k$ by the vector $-Z_k Z_k^T G_\nu^{-1} g_k$, which is in error by a term of order $(1/\nu)$. With the choice (5.59) of ν on a well-scaled problem, the accuracy achieved with this approximation is similar to that associated with a direct finite-difference approximation to the projected Hessian matrix $Z_k^T G_k Z_k$ (see Section 5.1.2.3). The condition number of G_ν increases with ν; however, the error in the computed p_ν tends to lie almost entirely in the subspace in which the error in the approximation (5.60) occurs.

Notes and Selected Bibliography for Section 5.4

The earliest applications of quasi-Newton methods to linearly constrained problems were all based upon the range-space solution of an equality-constrained quadratic subproblem. Suppose that the inverse of a matrix of the form (5.55) is partitioned so that

$$\begin{pmatrix} G_k & -\hat{A}_k^T \\ \hat{A}_k & 0 \end{pmatrix}^{-1} = \begin{pmatrix} H^* & C^{*T} \\ -C^* & K^* \end{pmatrix}, \qquad (5.62)$$

where H^*, C^* and K^* are matrices of dimension $n \times n$, $t_k \times n$ and $t_k \times t_k$ respectively. The quasi-Newton method of Goldfarb (1969) is a range-space method in which the matrix H^* is stored and updated explicitly in order to define p_k, and the first-order Lagrange multiplier estimates λ_L are computed using the explicit inverse of $\hat{A}_k \hat{A}_k^T$. Murtagh and Sargent (1969) have suggested a range-space method in which quasi-Newton approximations to the matrices G_k^{-1} and $(\hat{A}_k G_k^{-1} \hat{A}_k^T)^{-1}$ are recurred explicitly. For additional comments on these methods, see Gill and Murray (1974d).

Range-space methods for general linearly constrained minimization and indefinite quadratic programming are the subject of current research. Range-space methods for indefinite QP are less straightforward than their null-space counterparts. The problem is that the factorizations of G and $\hat{A}_k G^{-1} \hat{A}_k^T$ do not provide information concerning whether or not $Z_k^T G Z_k$ is positive definite. As a result, it is more difficult to distinguish between a constrained stationary point and a constrained minimum. However, if x_k is a constrained minimum, and a constraint is deleted from the working set, the point $x_k + p_k$ will be a constrained minimum unless p_k is a direction of negative curvature within the subspace defined by the reduced working set. Thus, if x_0 is a vertex of the feasible region, it is possible to detect whether or not subsequent iterates are constrained minima by examining the curvature of F along each search direction. If negative curvature is detected, a constraint must be added to the working set before any further constraints are deleted (see Section 5.3.2.2). This exchange scheme is the basis of Fletcher's (1971b) range-space QP algorithm. In Fletcher's method, the matrices H^* and C^* of (5.62) are updated explicitly. The range-space method for positive-definite QP that utilizes the orthogonal factorization of $\hat{A}_k$ is similar to a range-space quasi-Newton method described by Gill and Murray (1974d).

When solving positive-definite quadratic programs, range-space methods may be defined in which an extended matrix of the form (5.62) is factorized or inverted explicitly (see, for example, Bartels, Golub and Saunders, 1970). Many methods of this type are intended for large quadratic programs, and are based on simplex-type "pivoting" operations. Tomlin (1976) has implemented a version of Lemke's method (Lemke, 1965) in which an LU factorization of a matrix similar to (5.62) is updated.

See Gill (1975) for further discussion of range-space methods for general nonlinear functions.

For more information on the selection of the working set for range-space methods, see Lenard (1979).

5.5. SPECIAL FORMS OF THE CONSTRAINTS

When the constraints of a linearly constrained problem are known to have a special form, it is possible to develop methods that exploit the particular nature of the constraints. We shall consider two cases that are very common in practice: problems in which all the constraints are simple bounds on the variables, and problems in which the constraints are a mixture of bounds and general linear constraints.

5.5.1. Minimization Subject to Simple Bounds

An important special case of LIP occurs when the only constraints are simple bounds on the variables, so that the problem is

$$\begin{array}{ll} \underset{x \in \Re^n}{\text{minimize}} & F(x) \\ \text{subject to} & l \le x \le u. \end{array} \tag{5.63}$$

Note that there are $2n$ constraints in (5.63) if each bound is considered as a general inequality constraint. However, obvious advantage can be taken of the fact that a variable may be equal to at most one of the bounds (when l_i is not equal to u_i), and that any subset of the constraints defines a full-rank matrix. A bound constraint is "active" when a variable is equal to one of its bounds. The working set can be defined simply by partitioning a typical iterate x_k into its "free" components x_{FR} and its "fixed" components x_{FX}. Thus, using the notation defined in Section 5.2.1, t_k gives the number of variables currently fixed on one of their bounds; the rows of $\hat{A}_k$ are simply a selection of signed rows of the identity (corresponding to the fixed variables); and the columns of Z_k can be taken as the rows of the identity corresponding to the free variables.

We now describe a typical iteration, where for simplicity, we omit the subscript that indicates the iteration number. Let t denote the number of variables currently to be held fixed on one of their bounds. The search direction will be zero in components corresponding to the fixed variables, and only the $n-t$ "free" components need to be specified. The subscript "FR" will denote a vector or matrix of dimension $n-t$, whose elements are associated with the appropriately numbered free variables; the vector g_{FR} is equivalent to $Z^T g$, and so on.

All the algorithms discussed in Section 5.1.2 have a particularly simple form when applied to a bound-constrained problem. For example, the Newton equations (5.12) become

$$G_{\text{FR}} p_{\text{FR}} = -g_{\text{FR}}. \tag{5.64}$$

To apply a discrete Newton method with simple bound constraints, only the elements of G_{FR} need to be approximated by finite-differences. The definition of a bound-constraint quasi-Newton method is similarly straightforward. Let B_{FR} denote an $(n-t)$-dimensional approximation of the Hessian matrix with respect to the free variables. By analogy with (5.13), the vector p_{FR} is defined by the equations

$$B_{\text{FR}} p_{\text{FR}} = -g_{\text{FR}}.$$

Assuming that the set of fixed variables remains the same, the updated Hessian approximation using the BFGS update is given by

$$\bar{B}_{\text{FR}} = B_{\text{FR}} + \frac{1}{g_{\text{FR}}^T p_{\text{FR}}} g_{\text{FR}} g_{\text{FR}}^T + \frac{1}{\alpha y_{\text{FR}}^T p_{\text{FR}}} y_{\text{FR}} y_{\text{FR}}^T.$$

Changes in the working set with bound constraints simply involve fixing a variable on a bound when a constraint is added, and freeing a variable from its bound if a constraint is deleted. The special form of $\hat{A}$ for bound constraints makes it particularly easy to compute Lagrange multiplier estimates. Suppose that the j-th fixed variable is x_i. Then, for either of the first-order estimates discussed in Section 5.1.5.1, the multiplier estimate for a bound constraint is simply the appropriately signed component of the gradient corresponding to the given fixed variable. If $x_i = l_i$, then $\lambda_j = g_i$; if $x_i = u_i$, then $\lambda_j = -g_i$. If second derivatives are available, second-order estimates are defined as

$$\eta_L = \lambda + \bar{G} p_{\text{FR}},$$

where p_{FR} is the Newton direction resulting from (5.64), and $\bar{G}$ is the sub-matrix of G with rows corresponding to the fixed variables, and columns corresponding to the free variables. Note that no equations need to be solved in order to compute either type of multiplier estimate.

The efficiencies that are possible in computing the multiplier estimates may encourage a strategy in which more than one constraint is deleted from the working set at the same point. Thus, the structure of bound constraints can be exploited to the extent that a bound-constraint algorithm may produce a different sequence of iterates from that generated by a general active set method applied without regard to the special nature of the constraints.

An important feature of bound-constraint methods is that it is relatively straightforward to ensure that the objective function is computed at feasible points only when using finite-difference formulae. This is of crucial importance if $F(x)$ is not defined outside the feasible region. The modifications that must be made in order to generate feasible points concern only the finite-difference formulae for approximating the first and second derivatives of F. Suppose that F

is being minimized with a quasi-Newton method based on finite-difference approximations to the first derivatives. If the j-th variable x_j is fixed on a bound, an approximation to g_j, the corresponding component of the gradient, is required in order to compute the Lagrange multiplier estimate. Suppose that x_j is equal to its upper bound u_j. An appropriate finite-difference formula is

$$\bar{g}_j = \frac{1}{h_j}\big(F(x) - F(x - h_j e_j)\big).$$

This formula is used on the assumption that $u_j - l_j \geq h_j$; otherwise, the point $x - h_j e_j$ will violate the lower bound. If x_j is a fixed variable (i.e., $l_j = u_j$), the multiplier is not required.

Care must also be exercised in computing differences for free variables when a variable is close to a bound. In this case, the step should be taken in the opposite direction from the closest bound. For example, a central-difference formula that will not violate the lower bound of a free variable is:

$$\hat{g}_j = \frac{1}{2h_j}\big(4F(x + h_j e_j) - 3F(x) - F(x + 2h_j)\big),$$

(assuming that $u_j - l_j \geq 2h_j$).

At each iteration, the variables of interest are simply a subset of the original variables; hence, methods for bound-constrained minimization are more closely related to algorithms for unconstrained problems than to problems with general linear constraints. Even when the problem of interest is *unconstrained*, it is often helpful to introduce unnecessary but reasonable bounds on the variables, and then to solve the problem by means of a bound-constraint algorithm. The benefits of this procedure stem from two observations about most practical problems. Firstly, it is rare indeed for there to be no restrictions upon the expected size of each variable. The presence of bound constraints sometimes serves as a check on the problem formulation, since the unexpected occurrence of a variable on its upper or lower bound may reveal errors in specification of the objective function. Secondly, even if *no* bounds are active at the solution, their presence can prevent the function from being evaluated at unreasonable or nonsensical points during the iterations. Thus, it is advisable to solve even unconstrained problems using a bound-constraint method. If the method has been properly implemented, the user should suffer no loss in efficiency or ease of use.

*5.5.2. Problems with Mixed General Linear Constraints and Bounds

In the methods discussed thus far, we have not distinguished between general linear constraints and simple bounds (except for the case when all the constraints are simple bounds). In this section we briefly consider active set methods for problems in which the constraints are a mixture of bounds and general linear constraints.

At a typical iteration, the working set will include a mixture of general constraints and bounds. Suppose that $\hat{A}_k$ contains r bounds and $t - r$ general constraints; let the suffices "FR" and "FX" denote the vectors or matrices of components corresponding, respectively, to the variables that are free and the variables that are fixed on their bounds. If we assume for simplicity that the last r variables are fixed, the matrix of constraints in the working set can be written as

$$\hat{A}_k = \begin{pmatrix} \hat{A}_{\text{FR}} & \hat{A}_{\text{FX}} \\ 0 & I_r \end{pmatrix}, \tag{5.65}$$

where $\hat{A}_{\text{FR}}$ is a $(t - r) \times (n - r)$ matrix.

When $\hat{A}_k$ is of the form (5.65), it is possible to achieve economies in the calculation of the search direction and the Lagrange multipliers, since fixing r variables on their bounds simply has the effect of removing r columns from the general constraints.

Consider first the computation of the search direction. A suitable $n \times (n - t)$ matrix Z_k whose columns span the null space of the matrix $\hat{A}_k$ of (5.65) is given by

$$Z_k = \begin{pmatrix} Z_{\mathrm{FR}} \\ 0 \end{pmatrix},$$

where Z_{FR} is an $(n - r) \times (n - t)$ matrix whose columns form a basis for the null space of $\hat{A}_{\mathrm{FR}}$. The search direction appropriate for a method using the exact Hessian is defined by $(Z_{\mathrm{FR}} p_{\mathrm{FR}} \quad 0)$, where p_{FR} satisfies the equations

$$Z_{\mathrm{FR}}^T G_{\mathrm{FR}} Z_{\mathrm{FR}} p_{\mathrm{FR}} = -Z_{\mathrm{FR}}^T g_{\mathrm{FR}}$$

(*cf.* (5.12)). Note that a reduction in dimensionality has occurred that is equal to the number of variables fixed at their bounds.

Similar economies can be achieved during the computation of the Lagrange multiplier estimates. Let the vector of multipliers be partitioned into a $(t - r)$-vector $\hat{\lambda}_G$ (the multipliers corresponding to the general linear constraints) and an r-vector $\hat{\lambda}_B$ (corresponding to the fixed variables); the equations that define the multipliers are then

$$\hat{A}^T \hat{\lambda} = \begin{pmatrix} \hat{A}_{\mathrm{FR}}^T & 0 \\ \hat{A}_{\mathrm{FX}}^T & I_r \end{pmatrix} \begin{pmatrix} \hat{\lambda}_G \\ \hat{\lambda}_B \end{pmatrix} = \begin{pmatrix} \hat{A}_{\mathrm{FR}}^T \hat{\lambda}_G \\ \hat{A}_{\mathrm{FX}}^T \hat{\lambda}_G + \hat{\lambda}_B \end{pmatrix} = \begin{pmatrix} g_{\mathrm{FR}} \\ g_{\mathrm{FX}} \end{pmatrix}. \qquad (5.66)$$

Multiplier estimates corresponding to the general constraints can be obtained as the least-squares solution of the first $t - r$ equations of (5.66)

$$\hat{A}_{\mathrm{FR}}^T \hat{\lambda}_G \approx g_{\mathrm{FR}}.$$

The multipliers associated with the bound constraints may then be computed from the remaining equations of (5.66), i.e.

$$\hat{\lambda}_B = g_{\mathrm{FX}} - \hat{A}_{\mathrm{FX}}^T \hat{\lambda}_G.$$

The effect on $\hat{A}_k$ of changes in the working set depends on whether a general or a bound constraint is involved. If a *bound* constraint enters or leaves the working set, the *column* dimension of $\hat{A}_{\mathrm{FR}}$ and $\hat{A}_{\mathrm{FX}}$ alters; if a *general* constraint enters or leaves the working set, the *row* dimension of $\hat{A}_{\mathrm{FR}}$ and $\hat{A}_{\mathrm{FX}}$ alters.

Although it is possible to update a factorization of $\hat{A}_{\mathrm{FR}}$ as both rows and columns are added and deleted, few computer codes take advantage of the special structure of bound constraints in the factorization — mainly because of the increased complexity of the computer code. This practice may be justified if the additional work per iteration is insignificant compared to the cost of evaluating the objective function (which is usually the case for small problems). However, when solving linear or quadratic programs, the efficiency of a method tends to depend on the amount of housekeeping associated with performing each iteration. In particular, for a QP it is usually worthwhile to update the factors of $\hat{A}_{\mathrm{FR}}$ if there are likely to be a significant number of fixed variables at any stage of the solution process.

Notes and Selected Bibliography for Section 5.5

The algorithm presented in Section 5.5.1 for minimizing a nonlinear function subject to upper and lower bounds on the variables is due to Gill and Murray (1976b). An alternative technique is to eliminate the bounds using a transformation of the variables (see Section 7.4.1.1). However, *the resulting unconstrained problem should not be solved using a general unconstrained method* (see Sisser, 1981). Moreover, the reduction in dimensionality associated with active set methods does not occur with a transformation method.

Considerable efficiencies can be achieved if a quadratic function is minimized subject to only bound constraints (see Fletcher and Jackson, 1974; Gill and Murray, 1976b, 1977a). In this case, a "good" initial set of fixed variables can be chosen without significant computational effort. Moreover, if G is positive definite and if no constraint is added during the subsequent iteration, it is possible to compute the precise change in F resulting from a constraint deletion.

5.6. LARGE-SCALE LINEARLY CONSTRAINED OPTIMIZATION

5.6.1. Large-Scale Linear Programming

The *standard form* for a linear programming problem is

Standard form LP	$\begin{aligned} &\underset{x \in \Re^n}{\text{minimize}} \quad c^T x \\ &\text{subject to} \quad Ax = b \\ &\qquad\qquad\quad\ l \leq x \leq u. \end{aligned}$

The origin of the standard form is partly historical. However, we shall see that the standard form has certain significant advantages in terms of computational efficiency and storage for large-scale problems. Note that in the standard form all the *general* constraints are *equalities*, and that the only inequalities are upper and lower bounds on the variables.

Obviously, there is a complete equivalence between the form (5.42) and the standard form, since any problem stated in one form can be replaced by a problem in the other. Certain techniques are typically used to carry out the problem transformation. In particular, to obtain the standard form a general inequality constraint in (5.42) is replaced by an equality constraint that contains an additional bounded ("slack") variable. For example, the constraint $a^T x \geq b$ can be replaced by $a^T x - y = b$, where we impose the simple bound $y \geq 0$. Although it might appear that this transformation has the harmful effect of adding a large number of extra variables to the problem, in fact only a small amount of additional work and storage is associated with the slack variables (since they do not enter the objective function and are constrained only by simple bounds).

Suppose that there are m general equality constraints in the standard form. Under the usual assumption in LP that the matrix of constraints in the working set is non-singular, there must be $n - m$ variables fixed on their bounds at every iteration. (A feasible point where $n - m$ of the variables are on one of their bounds is called a *basic feasible solution* of an LP in standard form.)

We shall temporarily drop the subscript k associated with the current iteration. Without loss of generality, we assume that at the point x the *last* $n - m$ variables are fixed on one of their bounds. In this case the current point satisfies the equality constraints

$$\hat{A}x = \begin{pmatrix} B & N \\ 0 & I_{n-m} \end{pmatrix} \begin{pmatrix} x_B \\ x_N \end{pmatrix} = \begin{pmatrix} b \\ b_N \end{pmatrix}. \tag{5.67}$$

The matrix B in (5.67) is a selection of the columns of A that forms an $m \times m$ non-singular matrix known as the *column basis*. The columns of A that are in B are known as *basic columns*, and the variables in x defined by this partition of A are termed *basic variables* (denoted by x_B in (5.67)). Similarly, columns in N and their associated variables x_N are termed *non-basic*. The components of b_N in (5.67) are drawn from either l or u, depending on which bound is binding for each non-basic variable.

As described in Section 5.3.1, in an LP problem each iterate is optimal for the equality-constrained subproblem defined by $\hat{A}$. To determine whether x is optimal for the original LP problem, it is necessary to evaluate the Lagrange multipliers. Let the vector of multipliers be partitioned into an m-vector π (the multipliers corresponding to the equality constraints) and an $(n - m)$-vector σ (corresponding to the fixed variables); the equations (5.43) that define the multipliers are then

$$\hat{A}^T \hat{\lambda} = \begin{pmatrix} B^T & 0 \\ N^T & I_{n-m} \end{pmatrix} \begin{pmatrix} \pi \\ \sigma \end{pmatrix} = \begin{pmatrix} B^T \pi \\ N^T \pi + \sigma \end{pmatrix} = \begin{pmatrix} c_B \\ c_N \end{pmatrix}, \tag{5.68}$$

where c_B denotes the vector of elements of c corresponding to the basic variables, and c_N denotes the vector of elements of c corresponding to the non-basic variables.

The vector π is obtained by solving the non-singular linear system

$$B^T \pi = c_B. \tag{5.69}$$

Note that the components of π are not restricted in sign, since they are associated with equality constraints. The multipliers associated with the fixed variables may then be computed from the remaining equations of (5.68), i.e.

$$\sigma = c_N - N^T \pi.$$

In LP terminology, the Lagrange multipliers σ_j are known as the *reduced costs*, and the numerical process for computing them is known as *pricing*.

The vector σ indicates whether x is optimal (note that the elements of σ should be positive for variables fixed on their lower bounds, and negative for variables fixed on their upper bounds). If all the components of σ have the correct sign, the current point is optimal. Otherwise, one of the non-basic variables will be released from its bound, so that a bound constraint is deleted from the working set. Let the search direction p be partitioned as $(p_B \quad p_N)$, and suppose that the variable x_s is to be released from a lower bound. Then the component of p_N that corresponds to x_s will be unity, and all other components of p_N will be zero. The desired relationships (5.44) and (5.45) will hold with the given form of $\hat{A}$ when p satisfies

$$\begin{pmatrix} B & N \\ 0 & I_{n-m} \end{pmatrix} \begin{pmatrix} p_B \\ p_N \end{pmatrix} = e_s$$

(*cf.* (5.46)), so that p_B is given by the solution of

$$B p_B = -a_s, \tag{5.70}$$

where a_s is the s-th *column* of A.

Movement along p to decrease the objective function will cause a basic variable to encounter one of its bounds (i.e., a new bound constraint will be added to the working set). The effect of this change is that different columns of A will compose B and N at the next iteration. Each complete iteration of the simplex method has the effect of swapping a basic and a nonbasic variable. This is why LP terminology refers to a change in the working set as a "change of basis". Note that this is in contrast to the situation when the constraints are a mixture of general linear and simple bound constraints (see Section 5.5.2), in which both row and column changes in the basis are made. When the simplex method is implemented using the standard form, the matrix defined by the working set always contains a *fixed* number of rows corresponding to the general constraints, and thus B is of *fixed size*.

Most modern implementations of the simplex method for large-scale linear programming use the *LU* factorization for the solution of equations (5.69) and (5.70). At the first iteration, the factorization has the form

$$L^{-1}B = M_{m-1}\cdots M_2 M_1 B = U. \tag{5.71}$$

where each M_j is a lower-triangular elementary matrix and U is an upper-triangular matrix (for simplicity, we have assumed that no row or column interchanges are necessary to ensure numerical stability, *cf.* (2.40)). The solution of equations of the form $Bx = b$ is achieved by computing

$$y = M_{m-1}\cdots M_2 M_1 b,$$

and then solving

$$Ux = y,$$

by back substitution. (Note that forward or backward substitution on a triangular matrix can be performed just as easily whether the matrix is stored by rows or by columns.) Computation of y involves a forward pass through the transformations M_j, and is therefore called *FTRANL*. Computation of x requires a backward pass through either the rows or the columns of U, and is called *BTRANU*. (Similarly, solution of systems of the form $B^T z = c$ involves the operations *FTRANU* and *BTRANL*.)

The efficient solution of linear programs with a very large number of variables is possible only if the matrix A is sufficiently sparse (i.e., A has a sufficiently small number of non-zero entries). In this case the number of non-zero entries in the *LU* factorization of B will be such that the factors will fit into the available memory. The number of non-zeros in the factors depends heavily upon the order of the rows and columns. This implies that it is worth expending some effort to obtain an initial row and column ordering that does not lead to an excessive number of newly-generated elements. This process is sometimes known as *pre-assigned pivot selection*. An ideal example of this would occur if a sparse basis matrix could be permuted to be lower triangular — in which case no factorization would be necessary to solve (5.69) and (5.70). In general, pre-assignment procedures take advantage of the fact that basis matrices are usually *almost* triangular.

When columns are exchanged between B and N, the *LU* factorization must be updated. This results in the addition of new elementary matrices $\{M_j\}$ to the list for L and the formation of some new columns $L^{-1}a_j$. Since new non-zero elements may be generated in U during this updating, the data structure for U must be such that newly created nonzeros are included at the end of the existing data, yet can be readily accessed at the appropriate time during the solution of (5.69) or (5.70) (for more details of such schemes, the reader is referred to the references cited in the Notes).

As the iterations proceed, the amount of storage required for the factors increases. As columns enter or leave the basis, they do so in a fixed order prescribed by the path of the minimization. It is usually the case, therefore, that a factorization computed after a sequence of updates will contain far more non-zeros than a factorization computed from scratch with the aid of a sensible pre-assigned pivot scheme. After a certain number of column exchanges, it is eventually necessary to recompute the factorization. This process, which is generally known as *reinversion*, is usually enforced at fixed intervals unless it is triggered by a lack of storage for updates.

5.6.2. General Large-Scale Linearly Constrained Optimization

The problem of interest in this section is

$$
\begin{aligned}
\underset{x \in \Re^n}{\text{minimize}} \quad & F(x) \\
\text{subject to} \quad & Ax = b \\
& l \le x \le u.
\end{aligned}
\tag{5.72}
$$

We assume that the number of variables in (5.72) is "large" and that the matrix A is sparse. Note that the constraints of (5.72) are in exactly the same form as the standard form LP; the algorithm to be described relies heavily on the technology of large-scale linear programming.

As a rule, there are *fewer* algorithmic options for large problems, since many computational procedures that are standard for small problems become unreasonably expensive in terms of arithmetic and/or storage. However, in another sense the options for large problems are less straightforward because of the critical effect on efficiency of special problem structure and the details of implementation.

When solving large problems, it may be necessary to alter or compromise what seems to be an "ideal" or "natural" strategy. In fact, an approach that would not be considered for small problems may turn out to be the best choice for some large problems. Similarly, certain standard assumptions about the relative costs of portions of an algorithm become invalid in the large-scale case. For example, the measure of efficiency of an algorithm for dense unconstrained optimization is often taken as the number of evaluations of user-supplied functions (e.g., the objective function, the gradient) that are required to reach a specified level of accuracy. Although this measure is recognized to be overly simplistic, it is nonetheless a reasonable measure of effectiveness for most problems. This is because the number of arithmetic operations per iteration tends to be of order n^3 at most, and the amount of work required for storage manipulation is negligible. However, even for unconstrained problems of *moderate* size, the work associated with linear algebraic procedures and data structure operations tends to become significant with respect to the function evaluations.

At a typical iteration of an active set method applied to (5.72), the matrix $\hat{A}$ will contain all the rows of A and an additional set of rows of the identity that correspond to variables on one of their bounds. A critical difference from the LP case is that there is no a *priori* number of fixed variables. Any generalization of the division of the variables into basic and nonbasic must allow for variation in the number of nonbasic variables (fixed variables). Let r denote the number of fixed variables at the current iteration. Then the matrix A is (conceptually) partitioned as follows:

$$
A = (\, B \quad S \quad N \,).
\tag{5.73}
$$

The $m \times m$ "basis" matrix B is square and non-singular, and its columns correspond to the *basic* variables. The r columns of N correspond to the *nonbasic* variables (those fixed on their bounds). The $n - m - r$ columns of the matrix S correspond to the remaining variables, which will be termed *superbasic*. The number of superbasic variables indicates the number of degrees of freedom remaining in the minimization (since there are $m + r$ constraints in the working set). Note the contrast with the LP case of 5.6.1, in which S is null and the number of columns in N is always $n - m$.

At a given iteration, the constraints in the working set (rearranged for expository purposes) are given by

$$
\hat{A}x = \begin{pmatrix} B & S & N \\ 0 & 0 & I \end{pmatrix} \begin{pmatrix} x_B \\ x_S \\ x_N \end{pmatrix} = \begin{pmatrix} b \\ b_N \end{pmatrix},
\tag{5.74}
$$

where the components of b_N are taken from either l or u, depending on whether the lower or upper bound is binding.

When an active set method is applied to a dense problem, we have seen how to define a feasible descent direction Zp_Z. For large-scale problems, a similar procedure can be followed through appropriate definition of the components in each partition of the search direction.

A matrix Z that is orthogonal to the rows of $\hat{A}$ is given by

$$Z = \begin{pmatrix} -B^{-1}S \\ I \\ 0 \end{pmatrix}. \tag{5.75}$$

The matrices B^{-1} and Z are not computed explicitly. An advantage of (5.75) is that Z or Z^T may be applied to vectors using only a factorization of the square matrix B.

The form (5.75) of Z means that the partitioning of the variables into basic, nonbasic, and superbasic sets carries directly over into the calculation of the search direction p. If p is partitioned as $(p_B \quad p_S \quad p_N)$, the form (5.75) of Z implies that $p_N = 0$ and

$$Bp_B = -Sp_S. \tag{5.76}$$

Equation (5.76) shows that p_B can be computed in terms of p_S. Thus, the only variables to be adjusted are the superbasic variables, which act as the "driving force" in the minimization. The determination of the vector p_S is exactly analogous to that of p_Z in the methods discussed earlier, in that it completely specifies the search direction. Once p_S has been computed, p_B is obtained from (5.76).

*5.6.2.1. Computation of the change in the superbasic variables. The motivation that underlies the definition of a "good" choice of the vector p_S is similar to that discussed in Section 5.1.2 for defining p_Z. The most effective methods define p_S in terms of a *quadratic approximation* to the objective function, subject to the constraint (5.76). Thus, we wish to specify p_S so that the search direction "solves" an equality-constrained quadratic problem of the form

$$\begin{aligned} \underset{p \in \Re^n}{\text{minimize}} \quad & g^T p + \frac{1}{2} p^T H p \\ \text{subject to} \quad & \hat{A}p = 0, \end{aligned} \tag{5.77}$$

where H is an approximation to the Hessian matrix of $F(x)$. The vector p_S that corresponds to the solution of (5.77) can be computed using only the *projected matrix* $Z^T H Z$, by solving the linear system

$$Z^T H Z p_S = -Z^T g, \tag{5.78}$$

where Z is given by (5.75). Note that $Z^T g = -S^T B^{-T} g_B + g_S$.

If the equations (5.78) are to be solved using factorization, the dimension of the projected Hessian must be small enough to allow the factorization to be stored. Since the projected Hessian will generally be dense even for a sparse problem, this means that the number of constraints in the working set must be sufficiently large at every iteration. Fortunately, in many large-scale problems, an a *priori* lower bound exists on the number of constraints that will be satisfied exactly. For example, suppose that only a small number of the variables appear nonlinearly in the objective function (this is typical when a few nonlinearities are introduced into a problem that was originally a linear program). In this case, $F(x)$ is of the form

$$F(x) = f(\tilde{x}) + c^T x,$$

where $\tilde{x} = (x_1, x_2, \ldots, x_q)^T$, and $f(\tilde{x})$ denotes any differentiable nonlinear function. In this case, if the second-order sufficiency conditions hold, the dimension of the projected Hessian matrix at the solution is bounded by q.

Even when the dimension of the projected Hessian is not too large, Newton-type methods based on factorization are generally not practicable, because of the substantial amount of computation required to obtain a new projected Hessian at every iteration (recall that every operation with Z is equivalent to solving a linear system that involves B). By contrast, quasi-Newton methods can be adapted very effectively for large problems. As discussed in Section 5.2.4.2, the algorithm can update a Cholesky factorization of a quasi-Newton approximation to the *projected Hessian*.

There are large-scale problems for which the number of constraints active at the solution is not nearly equal to the number of variables. For example, many of the constraints may be present only to prevent nonsensical solutions, and the constraints should not be active at a "reasonable" answer. If a factorization of the $(n - t)$-dimensional matrix in (5.78) cannot be stored, it may be possible to solve the system (5.78) instead using a conjugate-gradient-type algorithm such as those described in Section 5.1.2.5. In particular, if second derivatives are available and the Hessian is sparse, a Newton-like method can be defined by taking H in (5.77) as the current Hessian G. It is assumed that a reasonable number of matrix-vector products of the form $Z^T G Z v$ can be computed relatively cheaply by forming, in turn, $v_1 = Z v$, $v_2 = G v_1$ and $v_3 = Z^T v_2$. A similar procedure can be used if a sparse approximation to the Hessian is available. In order to be practicable for large problems, any iterative method for solving (5.78) should converge rapidly (because of the expense associated with operations that involve Z). Furthermore, the method must be able to produce a satisfactory descent direction even if $Z^T H Z$ is indefinite.

*5.6.2.2. Changes in the working set.

As long as $\|Z^T g\|$ is "large", only the basic and superbasic variables are optimized. If one of these variables encounters a bound as the iterations proceed, it is "moved" into the set of nonbasic variables, and the working set is altered accordingly.

When $\|Z^T g\|$ is "small" (*cf.* (5.2)), it is considered that the current iterate is "nearly" optimal on the current working set. In this situation, we determine whether the objective function can be further reduced by releasing any nonbasic variable from its bound. This possibility is checked by computing Lagrange multiplier estimates from the system

$$\begin{pmatrix} B^T & 0 \\ S^T & 0 \\ N^T & I \end{pmatrix} \begin{pmatrix} \pi \\ \sigma \end{pmatrix} \approx \begin{pmatrix} g_B \\ g_S \\ g_N \end{pmatrix}. \tag{5.79}$$

(*cf.* (5.1)). We define the vectors π and σ as

$$\pi = B^{-T} g_B; \tag{5.80}$$

$$\sigma = g_N - N^T \pi. \tag{5.81}$$

The system (5.79) is compatible when $Z^T g = 0$, since in this case

$$g_S = S^T B^{-T} g_B = S^T \pi.$$

The vector σ thus provides a set of Lagrange multipliers for the bound constraints. If a nonbasic variable can be released from its bound, the iterations continue with an expanded superbasic set.

We now summarize the ways in which the procedures of an active set method for a sparse problem differ from those used in the dense case, although the motivation is identical. Firstly, the null space of $\hat{A}$ is defined in terms of *a partition of the variables*, rather than a matrix Z with orthonormal columns. The expression (5.75) for Z indicates that an ill-conditioned basis matrix B can affect the condition of all calculations in the algorithm, and may drastically alter the scaling of the variables. When the columns of Z are orthonormal, $\|Z^T g\|_2 \leq \|g\|_2$; otherwise, $Z^T g$ is "unscaled". Since an orthonormal Z is not practical (in terms of storage or computation) for most large-scale problems, additional numerical difficulties are likely in the computation of p and the projected Hessian approximation.

Secondly, the multiplier estimates computed from (5.80) and (5.81) are exact only when $Z^T g = 0$, and the neighbourhood in which their sign is correct depends on the *condition of B*. Hence, when $\|Z^T g\|$ is merely "small", it may be inefficient to release a variable based on the vector σ from (5.81). Although a feasible descent direction can be computed, the deleted constraint may very soon be re-encountered. This difficulty is less severe in the dense case, where the multiplier estimates computed with an orthonormal Z will in general have the correct sign in a much larger neighbourhood of a constrained stationary point because the size of the neighbourhood depends only on the condition of $\hat{A}$. The increased unreliability of Lagrange multiplier estimates is unavoidable when Z is given by (5.75), and must be taken into account in all large-scale optimization.

Finally, the cost of computing and updating the factorization of B is substantial, in terms of both arithmetic operations and storage manipulation. For many large-scale problems, the work associated with performing the steps of the algorithm dominates the cost of evaluating the nonlinear objective function.

Notes and Selected Bibliography for Section 5.6

The simplex method for linear programming has been implemented in commercial packages that routinely solve problems with thousands of variables. For an interesting history of the early development of commercial LP packages, see Orchard-Hays (1978a). For further details of the design and implementation of large-scale LP software, see Orchard-Hays (1968, 1978b, c), Beale (1975), Benichou *et al.* (1977), Greenberg (1978a, b) and Murtagh (1981).

Methods for linear programming differ mainly in the way in which the equations (5.69) and (5.70) are solved. The earliest implementations compute the inverse of the basis matrix B (see, for example, Orchard-Hays, 1968). Other implementations include the LU factorization (Bartels and Golub, 1969; Forrest and Tomlin, 1972); the LQ factorization with Q not stored (Gill and Murray, 1973c); the LQ factorization with Q stored as the product of a diagonal and orthogonal matrix (Gill, Murray and Saunders, 1975).

The advances in linear programming have largely been due to increased knowledge concerning the solution of large sparse systems of linear equations. One of the earliest schemes for selecting the row and column ordering of the basis B is due to Markowitz (1957). In this scheme, the rows and columns are searched during the LU factorization so that the potential fill-in within the rows and columns yet to be factorized is locally minimized. This scheme has been implemented by Reid (1975, 1976). Hellerman and Rarick (1971, 1972) have suggested a scheme in which the ordering of the rows and columns takes place *before* the factorization commences. Row and column permutations are applied to the basis matrix in order to give a matrix that is lower-triangular except for a number of diagonal blocks called *bumps* (see also Tarjan, 1972, and Duff and Reid, 1978). Each bump is then processed until it is lower triangular save for columns of non-zero elements called *spikes*. It can be shown that, in the subsequent LU factorization, fill-in occurs only within the spike columns.

Currently, the most efficient methods for large-scale linear programming are based on using the LU factorization. The two most important techniques for *updating* the LU factorization are due to Forrest and Tomlin (1972) (see also Tomlin, 1975a) and Bartels and Golub (see Bartels, 1971). Updating with the Forrest-Tomlin method results in the addition and deletion of a column from U. Thus, the scheme is suitable for implementations in which the matrix U is stored by columns. The updating procedure requires non-zeros to be deleted from U, but does not generate any fill-in within existing columns. Hence the outstanding advantage of the method is that it can be implemented efficiently using auxiliary or virtual storage for both L and U.

In contrast, the Bartels-Golub scheme uses row interchanges to ensure numerical stability during the updating. Since new non-zero elements are created in some of the rows of U, the scheme for storing the non-zeros must be designed so that elements can be inserted in the data structure without reordering the existing elements. This feature of the method implies that U must be held in core. An efficient Fortran implementation has been developed by Reid (1975, 1976), in which U is stored compactly by rows. The nonzeros in any one row are stored contiguously but neighbouring rows may reside in different areas of core. The implementation performs well in both real and virtual memory. As an alternative, Saunders (1976) has shown that, if the Hellerman-Rarick pre-assigned pivot strategy is used before the initial factorization, it is possible to predict the area of U in which fill-in will occur during subsequent updating. This partition of U is then stored explicitly in-core, thereby enabling the interchanges associated with the Bartels-Golub to be performed efficiently.

Certain important large-scale linear programming problems are characterized by special structure in the constraint matrix; for example, the matrices associated with models of dynamic systems tend to have a "staircase" structure. A discussion of these problems is beyond the scope of this book. The interested reader should consult the bibliography in Dantzig *et al.* (1981).

Recently, considerable interest has been generated by the publication (Khachiyan, 1979) of the result that a certain *ellipsoid algorithm* can find a feasible point for certain sets of linear inequalities in polynomial time. The algorithm, developed principally by Shor (see Shor, 1970, 1977; Shor and Gershovich, 1979) and Nemirovsky and Yudin (1979), defines an initial ellipsoid that encloses a finite volume of the feasible region, and proceeds by defining a sequence of shrinking ellipsoids, each of which contains this feasible region. It can be shown by contradiction that the center of one of the ellipsoids must eventually be a feasible point (if not, the volume of the ellipsoids would ultimately be smaller than that of the feasible region they contain). Thus, a feasible-point algorithm can be adapted to solve the linear programming problem, and conversely. The simplex method is potentially an exponential-time algorithm, and simple examples exist that elicit the algorithm's worst-case performance. It is therefore natural for the question to have been raised: for the solution of linear programs, could the ellipsoid algorithm prove to be superior to the simplex method? Despite the great *theoretical* importance of the Russian algorithm, the prospects for developing an efficient ellipsoid algorithm for large-scale linear programs are not good (see Gill *et al.* (1981a) for details of numerical experiments on some large-scale linear programs). For those readers who are interested in the vast volume of literature that has appeared since the publication of Khachiyan's paper, see, for example, Aspvall and Stone (1980), Gács and Lovász (1981), Goffin (1980), Lawler (1980) and Wolfe (1980a, b).

One of the few practical algorithms for large-scale optimization is the method of *approximation programming* (see Griffith and Stewart, 1961). This method is now often called *successive linear programming* and has been implemented in various forms, typically in conjunction with an existing mathematical programming system; see Baker and Ventker (1980), Beale (1974, 1978) and Batchelor and Beale (1976).

The term "superbasic variable" was coined by Murtagh and Saunders (1978), who originated the algorithm for large-scale linearly constrained algorithm described in Section 5.6.2. The Murtagh-Saunders method updates the Cholesky factorization of a BFGS quasi-Newton approximation to the projected Hessian, and switches to the traditional conjugate-gradient method when the factors cannot be stored in core. A portable Fortran code, MINOS, implementing the Murtagh-Saunders method is available from the Systems Optimization Laboratory, Stanford University (see Murtagh and Saunders, 1977).

The method of Marsten and Shanno (1979) (see also Marsten, 1978) is identical to the Murtagh-Saunders method except that a limited memory quasi-Newton method is used instead of the regular BFGS method. Also, the method is not necessarily restarted after a constraint is encountered. Buckley (1975) uses the $n \times n$ matrix T of (5.31) to define a quasi-Newton method for large problems. Buckley's method is most efficient when the number of general inequality constraints in the problem exceeds the number of variables.

*5.7. FINDING AN INITIAL FEASIBLE POINT

All the algorithms for problems with linear inequality constraints discussed in this chapter have been "feasible-point" methods — i.e., if the initial point x_0 is feasible, all subsequent iterates x_k are also feasible. In order to allow such methods to be applied when an initial feasible point is not known, we now consider techniques for determining a feasible point with respect to a set of linear inequality constraints.

The problem of finding a feasible point has been resolved in the case of linear programming by a technique known as *phase 1 simplex*. If there are m inequality constraints of the form

$$Ax \geq b,$$

we define an *artificial objective function*

$$\bar{F}(x) = - \sum_{j \in J} (a_j^T x - b_j),$$

where J is the set of indices of the constraints violated at x. The function $\bar{F}(x)$ is a *linear function*, and is equal to the *sum of infeasibilities* at x. Note that $\bar{F}$ is zero at any feasible point, and positive at an infeasible point. Therefore, a feasible point can be found by minimizing the function $\bar{F}(x)$, subject to the constraints $a_j^T x - b_j \geq 0$ for all $j \notin J$ (a linear programming problem).

The phase 1 linear program is typically solved by a slightly modified version of the simplex method. We mention two common modifications, which concern the step to be taken along the direction of search. Firstly, the largest step possible can be taken, subject to the restriction that the number of violated constraints should not be increased. Let x and p denote the current point and search direction; the set K contains the indices of constraints that are *strictly satisfied* at x. The step α to be taken along p is then defined as

$$\alpha = \min \left\{ \frac{b_i - a_i^T x}{a_i^T p} \mid i \in K \text{ and } a_i^T p < 0 \right\}$$

(see Section 5.2.2). The nearest constraint (say, with index r) can then be added to the working set as usual. The second strategy used in phase 1 is based on increasing the step along p until the artificial objective function ceases to be reduced. For example, if α is chosen as described above, but the directional derivative

$$-\sum_{j \in J} a_j^T p - a_r^T p$$

is negative at $x + \alpha p$, the objective function is still decreasing, and a larger step will be taken.

In either case, several violated constraints may become satisfied during a single iteration. Since zigzagging is not a problem when the objective function is linear, a constraint with a negative Lagrange multiplier can be deleted immediately from the working set. It can be proved that this procedure will converge under certain conditions on the constraints. If a vertex of the phase 1 subproblem is found such that all the Lagrange multipliers are positive, but some constraints are still violated, there is no feasible point with respect to the specified set of constraints.

When the original problem is a linear program, m must exceed n. Therefore, an initial vertex is constructed by selecting any set of n linearly independent constraints to be the initial working set (we shall consider this process in greater detail in Section 5.8). Hence, the simplex method can be applied directly to minimize the sum of infeasibilities. When the sum of infeasibilities has been reduced to zero, a feasible vertex will be available to begin solving the original linear program.

When m is less then n (which is common when the original F is a nonlinear function), it is not possible to apply the simplex method directly, since there may be no initial vertex to initiate the phase 1 simplex (even though there are feasible points). Under these circumstances, artificial vertices can be created by adding simple bounds on all the variables. However, this can have two adverse effects: the result of phase 1 can be an "unreasonable" initial point (since some of the artificial bounds must be satisfied exactly); or added bounds may exclude part of the feasible region.

It is possible to avoid this problem by using the non-simplex strategy of Section 5.3.1 to solve the phase 1 linear program. If $m \ll n$, it may not be practical to store the matrix Z (see Section 5.4). In this case, the projected steepest-descent direction $p = -ZZ^T\bar{c}$ (where $\bar{c}$ is the gradient of the artificial objective function) may be computed as $p = -(I - YY^T)\bar{c}$, with Y taken as a matrix that spans the range space of the rows of the matrix of constraints in the working set.

Notes and Selected Bibliography for Section 5.7

In commercial codes, the phase 1 simplex method has now completely replaced an earlier technique in which a modified LP with a known basic feasible solution is solved directly. The modified LP is obtained by adding a set of artificial variables and giving each new variable a large coefficient in the objective function. All the artificial variables are basic at the initial point; as the iterations proceed, their large coefficients cause them to be replaced by the natural variables. This method is known as the *big-M method* (see Dantzig, 1963). It is not reliable in practice.

For details of the implementation of the phase 1 simplex method in commercial codes, see Orchard-Hays (1968) and Greenberg (1978b).

*5.8. IMPLEMENTATION OF ACTIVE SET METHODS

*5.8.1. Finding the Initial Working Set

Although a linearly constrained problem can be regarded as being solved in two phases, where phase one is the feasibility algorithm and phase two the solution proper, an active set method can be regarded as a *single algorithm* in which the objective function is initially linear, but changes to the usual nonlinear function when a feasible point is found. Since a feasibility procedure is included, *any point* may be used as x_0. An important practical implication of a combined algorithm is that the *working set remains the same* when the objective function changes.

The efficiency of an algorithm for linearly constrained minimization is critically dependent upon the number of constraints in the working set. With null-space methods, the row dimension of $\hat{A}$ should be as large as possible; for range-space methods, the row dimension of $\hat{A}_k$ should be as small as possible. Since constraints are added and deleted one at a time, the size of the working set at the initial point x_0 greatly influences the efficiency of the subsequent computation.

In order to initiate the first phase of the optimization, an initial working set (the *crash set*) must be determined by a procedure that is called a *crash start*. The only mandatory feature of the initial working set is that all the linearly independent equality constraints should be included. The selection of any other constraints is determined primarily by the nature of the second phase of the optimization, i.e. whether a null- or range-space method is employed. If the second phase uses a null-space method, the initial working set should include as many constraints as possible. If a range-space method is employed, the initial working set should include the linearly independent equality constraints only.

We shall denote by $\hat{A}_C$ the matrix whose t_C linearly independent rows correspond to the constraints from the crash set. If $t_C < n$, there are infinitely many points such that

$$\hat{A}_c x = \hat{b}_c. \tag{5.82}$$

The first phase of the optimization can be started at the point $\hat{x}_0$ that is in some sense the solution of (5.82) "closest" to the user-supplied starting point x_0. For example, if $\hat{x}_0$ is defined as $x_0 + p$, then p can be taken as the vector of smallest two-norm such that $\hat{A}_c(x_0 + p) = \hat{b}_c$. If the *LQ* factorization of $\hat{A}_C$ is available, the minimum-norm solution of (5.82) can be computed directly, and we obtain

$$p = Y_c v, \quad \text{where} \quad L_c v = b_c - \hat{A}_c x_0,$$

where Y_C and L_C are appropriate portions of the *LQ* factorization.

There are many ways in which the crash set may be chosen for a null-space method. For example, $\hat{A}_C$ may include all the linearly independent equality constraints and a set of linearly independent inequality constraints $\{a_j\}$ such that

$$|a_j^T x_0 - b_j| \le \delta,$$

for some small scalar δ.

A more complicated form of crash is appropriate for large-scale linear programming, where the problem will usually be in standard form (see Section 5.6.1). In this case, a crash *column basis* B_C must be found. Crash procedures for large-scale optimization are usually more concerned with finding a column basis that is cheap to factorize than finding a basis that defines a basic feasible solution close to some user-supplied value. For example, a crash basis could be constructed entirely of slack variables, or columns could be selected to give B_C a triangular or almost triangular structure. Crash procedures in many production codes have several intermediate phases in which reduced costs computed from one easily invertible crash basis are used to select columns for another.

*5.8.2. Linearly Dependent Constraints

Linear dependence among the linear constraints is commonplace in practical problems, but it is not a serious impediment to practical computation. If the set of constraints that are satisfied exactly at an iterate x_k is linearly dependent, the boundary of the feasible region is well defined at such a point, but some of the constraints are redundant. Linear dependence among the linear constraints that are exactly satisfied at a point is commonly known as *degeneracy*, and such a point is called a *degenerate point*.

An important feature of the model algorithm for LIP given in Section 5.2.1 is that an attempt is always made to move from a current feasible point without necessarily including in the working set all the constraints that are satisfied exactly at x_k. A constraint is added to the working set only if a nonzero step would violate the constraint. In this way, constraints are added to the working set one at a time — even if many constraints are satisfied exactly at a given iterate. A fundamental advantage of this strategy is that, if $t_k < n$, the matrix of constraints in the working set need never be rank-deficient, and it is perfectly reasonable to assume that the normals of the constraints in the working set are linearly independent. To see this, assume that the vector a_{t+1} is the normal of a constraint that is exactly satisfied at x_k, but is not included in the working set, and that a_{t+1} is a linear combination of the rows of $\hat{A}_k$. The search direction p will satisfy $\hat{A}_k p = 0$, and it follows that $a_{t+1}^T p = 0$ also; thus, the vector p will not intersect the dependent constraint.

However, if a constraint is *deleted* from the working set at a degenerate point, p does not satisfy $\hat{A}_k p = 0$. Consequently, any positive step along p may violate one of the dependent constraints that was not in the working set; such a constraint must then be added to the working set. If x_k is not a vertex, a move can be made without deleting any more constraints from the working set. If x_k is a vertex, a constraint *must* be dropped from the working set and there is the chance that the sequence of working sets obtained by deleting and adding constraints may repeat itself after finitely many steps — a phenomenon known as *cycling*. Note that degeneracy is not a difficulty in itself; but when it is present, cycling is a possibility.

The resolution of degeneracy at a vertex, i.e. the computation of a search direction such that the objective function undergoes a strict decrease, is guaranteed if enough combinations of constraints are deleted from the working set. However, the resolution of degeneracy at a vertex is essentially a combinatorial problem whose solution may require a significant amount of computation. Fortunately, the occurrence of cycling is rare; when it does occur, simple heuristic strategies almost always succeed in breaking the deadlock.

5.8.3. Zero Lagrange Multipliers

If any of the exact Lagrange multipliers of a linearly constrained problem are zero or very close to zero, it becomes difficult to ascertain whether or not it is beneficial to delete a constraint from the working set. Note that the mere occurrence of near-zero Lagrange multiplier *estimates* does not necessarily cause problems. For example, if x_k is not close to a constrained stationary point, we can retain the same working set and hope that the magnitude of the offending multiplier estimate will increase. Similarly, if any Lagrange multiplier estimates are sufficiently negative (e.g., if $\lambda_k = (1, 0, -1)^T$), the constraint corresponding to the most negative estimate can safely be deleted.

However, serious problems occur when the smallest multiplier estimate is zero or near zero. To guarantee that a point x_k is optimal when zero Lagrange multipliers are present, it is necessary

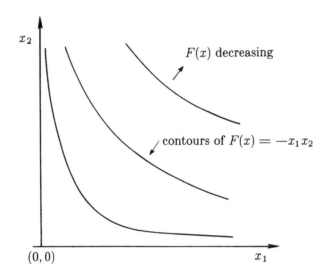

Figure 5a. The possible effect of several zero Lagrange multipliers. Two lower bound constraints are satisfied exactly at the origin, both with zero multipliers. Although neither constraint is active at the solution, deleting either one alone will not allow the objective function to be reduced by remaining on the other.

to examine higher-order derivatives that will not, in general, be available. Unlike a negative multiplier, which indicates that the corresponding constraint should not be active, a zero multiplier indicates that we are not able to determine from first-order information alone whether a lower function value exists on a reduced set of constraints. The difficulty in making this determination is in deciding which subset (if any) of constraints that are currently satisfied exactly, but have zero multipliers, should be deleted. Hence, any procedure for making this decision when there are many zero multipliers would be combinatorial in nature, and has the potential of requiring a significant amount of computation.

The following example illustrates that the difficulty can *not* be resolved simply by identifying *one* constraint that should not be active.

Example 5.5. Figure 5a illustrates the contours of the two-variable problem

$$\underset{x \in \Re^2}{\text{minimize}} \quad -x_1 x_2$$

$$\text{subject to} \quad 0 \le x_i \le 10, \quad i = 1, 2.$$

The two lower bound constraints are satisfied exactly at the origin, both with zero multipliers. Although neither lower bound constraint is active at the solution, deleting either one alone will not allow the objective function to be reduced by remaining on the other.

The errors made during the computation of the multipliers further complicate the issue. In practice, rounding errors mean that the computed Lagrange multipliers are rarely exactly zero. It is impossible to distinguish computationally whether the multipliers of the original problem are exactly zero, or are merely "small". Thus, if a computed multiplier is small but negative, we cannot be certain that the corresponding constraint should be deleted, since the estimate may be negative only because of rounding error.

Notes and Selected Bibliography for Section 5.8

The crash procedure that utilizes the orthogonal factorization is implemented in the Fortran programs for linearly constrained optimization available from the Systems Optimization Laboratory, Stanford University (see Gill *et al.*, 1980). These subroutines are designed for problems with mixed general linear and bound constraints, and update an orthogonal factorization of the matrix $\hat{A}_{FR}$ (see Section 5.5.2). (Note that such algorithms are trivially suitable for minimizing functions subject to only upper and lower bounds.)

Details of crash procedures for large-scale linear programming codes are given by Orchard-Hays (1968). Note that, since we would expect the number of general constraints to be small during the application of a range-space method, the crash basis for a phase 1 algorithm designed to find a feasible *vertex* must contain a large number of artificial constraints (see Section 5.3.2.2 for a definition of an artificial constraint).

Methods for handling degeneracy in linear programming include Charnes' perturbation technique (Charnes, 1952), the lexographic method of Dantzig, Orden and Wolfe (1955), and the *ad hoc* method of Wolfe (1963a). See also Harris (1973) and Benichou *et al.* (1977). Bland (1977) has suggested a finite version of the simplex method based upon a double least-index pivot selection rule. Perold (1981a, b) shows that the presence of degeneracy may allow savings to be made in the basis factorization.

Methods for dealing with near-zero Lagrange multipliers in general linearly constrained minimization are given by Gill and Murray (1977b). In linear programming terminology, zero Lagrange multipliers are sometimes called *degenerate dual variables*. In the linear programming case, zero multipliers are less difficult to handle, since a zero multiplier indicates the existence of a weak minimum. The worst that can happen is that constraints with small multipliers are deleted unnecessarily.

CHAPTER SIX

NONLINEAR CONSTRAINTS

Freedom and constraint are two aspects of the same necessity.

—ANTOINE DE SAINT-EXUPÉRY, in *La Citadelle* (1948)

In this chapter, we consider the minimization of a nonlinear function subject to a set of nonlinear constraints. Problems of this type are considerably more complicated than those with only linear constraints, and the methods reflect this increase in complexity. There tends to be a great diversity in methods for nonlinear constraints; even methods that are very similar in principle may differ significantly in detail.

As in Chapters 4 and 5, we shall emphasize methods with which the authors have had extensive experience, and methods that provide special insights or background. References to other methods and to further details are given in the Notes at the end of each section. The authors of methods and results will be mentioned only in the Notes.

In order to simplify the discussion, we shall treat two separate cases: problems in which all the constraints are *equalities*, and problems in which all the constraints are *inequalities*. A similar distinction was made in the discussion of optimality conditions for nonlinear constraints, and in the treatment of methods for linear constraints. It should be clear how to adapt the methods to be described to the case when nonlinear constraints are a mixture of equalities and inequalities.

For the most part, we shall not make a distinction between linear and nonlinear constraints in this chapter. In practice, it is almost always worthwhile to treat linear constraints separately, by the methods described in Chapter 5.

The two problem forms to be considered are the *equality-constrained problem*:

NEP	$\displaystyle\operatorname*{minimize}_{x\in\Re^n}\quad F(x)$
	subject to $\quad \hat{c}_i(x) = 0, \quad i = 1, \ldots, t,$

and the *inequality-constrained problem*:

NIP	$\displaystyle\operatorname*{minimize}_{x\in\Re^n}\quad F(x)$
	subject to $\quad c_i(x) \geq 0, \quad i = 1, \ldots, m.$

The optimality conditions for problems NEP and NIP are derived in Section 3.4. We shall refer to the set of functions $\{\hat{c}_i\}$ (or $\{c_i\}$) as the *constraint functions*. The objective function and the constraint functions taken together comprise the *problem functions*. The problem functions will be assumed to be at least twice-continuously differentiable, unless otherwise stated.

The reader will recall that the gradient of the constraint function $\hat{c}_i(x)$ is denoted by $\hat{a}_i(x)$, and its Hessian by $\hat{G}_i(x)$, with a similar convention for the function $c_i(x)$. The matrix $\hat{A}(x)$ denotes the matrix whose i-th row is $\hat{a}_i(x)^T$, and similarly for $A(x)$.

6.1. THE FORMULATION OF ALGORITHMS

6.1.1. Definition of a Merit Function

In the discussion of optimality conditions for problems NEP and NIP, it was observed that at a feasible point where a nonlinear constraint holds as an equality, in general the value of the constraint is altered by movement along *any* direction (i.e., in the terminology of Section 3.3.2, *there is no binding perturbation with respect to a nonlinear constraint*). This result introduces complications into the treatment of nonlinear constraints, and has a major implication for the design of algorithms for nonlinearly constrained problems.

With the methods discussed in Chapter 5 for the linear-constraint case, an initial feasible point is computed, and all iterates thereafter are feasible. This is possible because the search direction can be constructed so that the constraints in the current working set are automatically satisfied at all trial points computed during the iteration. In order to retain feasibility with respect to constraints that are not in the working set, the step is restricted by an explicit upper bound, so that no further constraints are violated by any trial step. Thus, only the objective function needs to be considered in an active set method for linear constraints when determining whether an "improvement" has occurred.

By contrast, when even one constraint function is nonlinear, it is *not* straightforward (and may even be impossible) to generate a sequence of iterates that exactly satisfy a specified subset of the constraints. If feasibility is not maintained, then in order to decide whether x_{k+1} is a "better" point than x_k, it is necessary to define a *merit function* that somehow balances the (usually) conflicting aims of reducing the objective function and satisfying the constraints. Algorithms for NEP and NIP that do not maintain feasibility necessarily include a definition — explicit or implicit — of a merit function. We shall see that the algorithms to be described in this chapter vary significantly in the criteria used to measure improvement at each iteration.

6.1.2. The Nature of Subproblems

6.1.2.1. Adaptive and deterministic subproblems. Optimization methods generate a sequence of iterates based on subproblems that are related in some way to the original problem. In all the methods described for unconstrained and linearly constrained optimization, an *iteration* is defined by two subproblems: the calculation of a search direction and of a step length. An important distinction between these subproblems involves their relationship with the generally unknown user-supplied functions. Calculation of the search direction may be viewed as a *deterministic* subproblem, in the sense that *the calculations performed do not depend on the function values*. Typically, the number of evaluations of user-supplied functions associated with a deterministic subproblem is known *a priori*; for example, with a discrete Newton method for a dense unconstrained problem, n differences of gradients are necessary to obtain a finite-difference approximation to the Hessian. The determination of a step length, on the other hand, is an *adaptive* subproblem, since a completely different sequence of calculations might be executed, depending on the function values encountered. In general, the number of evaluations of user-supplied functions required to solve an adaptive subproblem is not known in advance; for example, satisfaction of the "sufficient decrease" condition (see Section 4.3.2.1) depends on the function (and/or gradient) values at the trial points.

This distinction is significant because the subproblems associated with methods for nonlinearly constrained problems cover a wide range of complexity. In particular, some methods display the same structure as unconstrained and linearly constrained methods — namely, an iteration is composed of the deterministic calculation of a search direction followed by the adaptive calculation of a step length. However, other methods generate the next iterate by solving a complete general unconstrained or linearly constrained subproblem. In the latter case, each iteration involves *two levels* of adaptive subproblem, since the outermost subproblem is solved through a sequence of iterations that include an adaptive subproblem. The effect of these differences among methods is that the amount of work associated with performing an "iteration" varies enormously. One would expect an iteration that involves two levels of adaptive subproblem to be considerably more expensive in terms of evaluations of the problem functions than an iteration that includes only one adaptive subproblem.

In our discussion of methods, we shall be concerned primarily with the motivation for the outermost subproblem. When the outermost subproblem is a general unconstrained or linearly constrained problem, there is obviously a choice of solution method, depending on the information available about the problem functions and derivatives, the size of the problem, and so on. However, we shall not discuss the application of a particular solution method in these cases unless it is crucial to the definition of the algorithm.

6.1.2.2. Valid and defective subproblems. A second distinction between subproblems concerns the issue of *solvability*. This property is relevant because the subproblems associated with certain methods may fail to have a satisfactory solution, *even when the original problem itself is perfectly well posed*. We define a *valid subproblem* as one whose solution exists and is well defined. For example, solving a non-singular linear system is a valid subproblem. When the subproblem has no solution, or the solution is unbounded, we shall call the subproblem *defective*. An example of a defective subproblem would be the unconstrained minimization of a quadratic function with an indefinite Hessian, or the solution of an incompatible system of equations.

We emphasize this property of subproblems because it is characteristic of many methods for nonlinearly constrained problems that the initial formulation of a subproblem may be defective. In this case, it is clearly desirable to formulate an alternative subproblem, since otherwise the algorithm will fail. Unfortunately, for some types of subproblem, *the determination of validity is an unsolvable problem*; for example, there is no known way to determine whether a general unconstrained function is unbounded below. Hence, the option should be available of abandoning the effort to solve a subproblem if it appears that the subproblem is defective.

It is also possible that a defective subproblem is an accurate reflection of the fact that the original problem has no valid solution. A discussion of these issues is beyond the scope of this introductory chapter, but anyone who wishes to solve a nonlinearly constrained problem should be aware of the complicated situations that can occur.

6.2. PENALTY AND BARRIER FUNCTION METHODS

One approach to solving a nonlinearly constrained problem is to construct a function whose *unconstrained* minimum is either x^* itself, or is related to x^* in a known way. The original problem can then be solved by formulating a sequence of *unconstrained subproblems* (or possibly a single unconstrained subproblem). We shall refer to the generic objective function of such an unconstrained subproblem as Φ_U.

6.2.1. Differentiable Penalty and Barrier Function Methods

Differentiable penalty and barrier function methods are not generally considered to be as effective as the methods to be described in later sections. However, the motivation for these methods can be applied in a variety of contexts, and a knowledge of their properties is essential in order to understand related methods.

6.2.1.1. The quadratic penalty function. In general, an unconstrained minimum of F will occur at an infeasible point, or F may be unbounded below. Therefore, the solutions of any proposed sequence of unconstrained subproblems will converge to x^* only if Φ_U includes a term that ensures feasibility in the limit. One possible choice for such a term is a differentiable function that assigns a positive "penalty" for increased constraint violation.

To illustrate the ideas of a penalty function method, we consider the *quadratic penalty function*

$$P_Q(x, \rho) \equiv F(x) + \frac{\rho}{2}\hat{c}(x)^T\hat{c}(x), \tag{6.1}$$

where $\hat{c}(x)$ contains the constraints that are violated at x. (By convention, an equality constraint is always considered to be violated.) The non-negative value ρ is termed the *penalty parameter*, and $\hat{c}(x)^T\hat{c}(x)$ is called the *penalty term*.

The penalty term in (6.1) is continuously differentiable, but has discontinuous second derivatives at any point where an inequality constraint becomes exactly satisfied; hence, x^* will be a point of discontinuity in the second derivatives of P_Q for an inequality-constrained problem in which any constraints are active at the solution. However, this apparent defect of P_Q can be overcome by considering the inequality constraint $c_j(x) \geq 0$ to be violated if $c_j(x) < \epsilon$ for some small positive ϵ. With this definition of "violated constraints" in (6.1), no discontinuities in the second derivatives of P_Q will occur in a sufficiently small neighbourhood of x^*. (Although it is possible to define penalty terms with continuous second derivatives at the solution, they have no advantage in theory or practice.)

Under mild conditions it can be shown that there exists $x^*(\rho)$, an unconstrained minimum of (6.1), for which

$$\lim_{\rho \to \infty} x^*(\rho) = x^*. \tag{6.2}$$

It is significant that (6.2) is true even when the constraint qualification (see Section 3.4.1) does not hold.

We illustrate the effects of the penalty transformation (6.1) with a simple example.

Example 6.1. Consider the univariate problem

$$\begin{array}{ll} \underset{x \in \Re^1}{\text{minimize}} & x^2 \\ \text{subject to} & x - 1 = 0. \end{array}$$

For this problem, we have

$$P_Q(x, \rho) = x^2 + \frac{\rho}{2}(x - 1)^2$$

and

$$x^*(\rho) = \frac{\rho}{\rho + 2}.$$

The dashed line in Figure 6a represents the graph of $F(x)$; the solid line is the graph of $P_Q(x, \rho)$ for $\rho = 10$. Clearly,

$$\lim_{\rho \to \infty} x^*(\rho) = x^* = 1.$$

It might be thought that result (6.2) implies that x^* can be computed to any desired accuracy by simply choosing a very large value of the penalty parameter, and then carrying out a single unconstrained minimization of P_Q. However, this strategy is *inadvisable*. Let t denote the number of constraints active at x^*. If $0 < t < n$, the condition number of the Hessian matrix of $P_Q(x, \rho)$ evaluated at $x^*(\rho)$ increases as ρ becomes larger, with singularity occurring in the limit. In two dimensions, the ill-conditioning displays itself in nearly parallel contours of the penalty function.

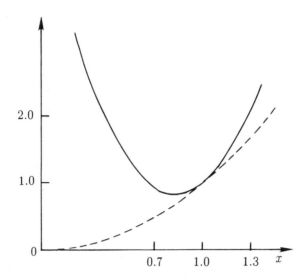

Figure 6a. The dashed line represents the graph of the objective function of Example 6.1, $F(x) = x^2$; the solid line represents the graph of the quadratic penalty function $P_Q(x, 10) = x^2 + 5(x - 1)^2$ corresponding to Example 6.1.

Example 6.2. Consider the following example, which will be used as an illustration in several later discussions.

$$\underset{x \in \mathfrak{R}^2}{\text{minimize}} \quad x_1 x_2^2$$

$$\text{subject to} \quad 2 - x_1^2 - x_2^2 \geq 0.$$

The single constraint is active at the the solution $x^* = (-.81650, -1.1547)^T$, with Lagrange multiplier $\lambda^* = .81650$ (all numbers rounded to five significant figures).

The first diagram in Figure 6b displays the contours of the objective function, the position of the contour line corresponding to a zero value of the constraint, and the location of the minimum. We observe that the original problem is well conditioned. Figure 6c shows the contours of the quadratic penalty function associated with Example 6.2 for two values of ρ ($\rho = 1$ and $\rho = 100$). The unconstrained subproblem is well-conditioned when $\rho = 1$; the ill-conditioning associated with a large penalty parameter is already apparent when $\rho = 100$.

The conditioning of the Hessian matrices is significant because the minima of successive penalty functions become increasingly poorly determined in the null space of the active constraint gradients. If the initial value of ρ is "too large", even a robust unconstrained algorithm will typically experience great difficulty in the attempt to compute $x^*(\rho)$ (see Section 8.3). Therefore, in order to solve NEP or NIP by a penalty function method, a *sequence of unconstrained problems* is solved, with increasing values of the penalty parameter. Usually, each successive $x^*(\rho)$ is used as the starting point for minimization with the next value of the penalty parameter, until acceptable convergence criteria for the solution of the original problem are satisfied.

We briefly describe a typical penalty function algorithm. Given an initial penalty parameter ρ_0, a positive integer K, and an initial point x_0, set $k \leftarrow 0$ and execute the following steps.

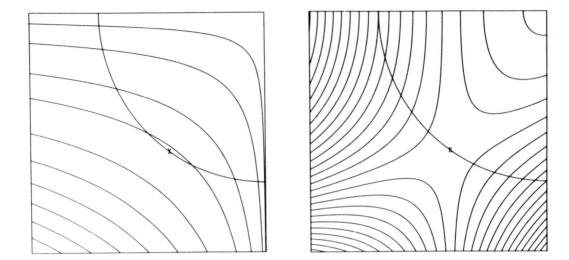

Figure 6b. The first diagram depicts the contours of the function $F(x) = x_1 x_2^2$ corresponding to Example 6.2. The contour line corresponding to a zero value of the constraint is superimposed. The single nonlinear constraint $2 - x_1^2 - x_2^2 \geq 0$ is active at the solution point $(-.81650, -1.1547)$ (marked with an "X"), with Lagrange multiplier $\lambda^* = .81650$. The second diagram depicts the contours of the Lagrangian function for the same problem, with the constraint treated as an equality.

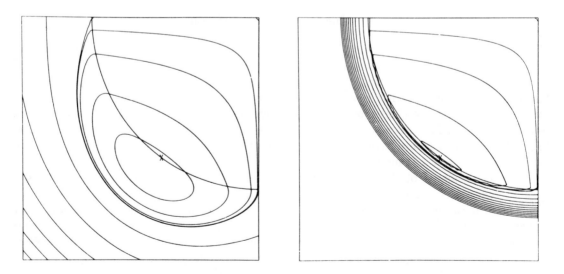

Figure 6c. The two diagrams represent the contours of the quadratic penalty function $P_Q(x, \rho)$ corresponding to Example 6.2 for two values of ρ ($\rho = 1$ and $\rho = 100$).

Algorithm DP (*Model algorithm with a differentiable penalty function*).

DP1. [Check termination conditions.] If x_k satisfies the optimality conditions, the algorithm terminates successfully. If $k > K$, the algorithm terminates with a failure.

DP2. [Minimize the penalty function.] With x_k as the starting point, execute an algorithm to solve the unconstrained subproblem

$$\underset{x \in \Re^n}{\text{minimize }} P_Q(x, \rho_k), \tag{6.3}$$

subject to safeguards to cope with unboundedness. Let x_{k+1} denote the best estimate of the solution of (6.3).

DP3. [Increase the penalty parameter.] Set ρ_{k+1} to a larger value than ρ_k, set $k \leftarrow k + 1$, and return to Step DP1. ∎

We have already emphasized the ill-conditioned nature of the subproblems associated with a penalty function method when $0 < t < n$ and ρ is large. We now consider two other properties of methods based on a differentiable penalty function.

Local minimum property. It is often overlooked that the effect of a penalty-type transformation is, *at best*, to create a *local minimum* of the penalty function which, for sufficiently large ρ, is "near" x^*. However, the subproblem (6.3) may be defective (see Section 6.1.2.2), and there are simple examples where the penalty function is unbounded below *for any value of the penalty parameter*. When P_Q is unbounded below, an unconstrained algorithm may experience difficulty in converging to the desired local minimum; for example, the iterates in the unconstrained subproblem may move outside the region of the local minimum introduced by the transformation. This property has important implications if a standard unconstrained method is to be used to solve the unconstrained subproblem. All the methods described in Chapter 4 were based on the assumption that the objective function is (at least) bounded below. However, since it is not at all unusual for a penalty function to be unbounded below, any unconstrained method to be used in minimizing a penalty function must include the ability to detect, and, if possible, to recover from unboundedness (hence, the qualification expressed in Step DP2 of the model algorithm). An unsafeguarded method might not terminate, or would expend a very large number of function evaluations before determining that the algorithm had failed. This implies that a standard algorithm should *not* be used as a "black box" in a penalty function method.

Lagrange multiplier estimates. Because of the special structure of a penalty function as a composite of the problem functions, much information about the original problem can be deduced from the properties of the sequence $\{x^*(\rho_k)\}$. In particular, estimates can be obtained of the Lagrange multipliers (if they exist). Since $x^*(\rho)$ is an unconstrained minimum of the differentiable penalty function, it holds that:

$$g + \rho \hat{A}^T \hat{c} = 0,$$

where g, $\hat{A}$, and $\hat{c}$ are evaluated at $x^*(\rho)$. The quantities

$$\lambda_i(\rho) = -\rho \hat{c}_i(x^*(\rho)) \tag{6.4}$$

are thus analogous to the Lagrange multipliers at x^*, since they are the coefficients in the expansion of g as a linear combination of the rows of $\hat{A}$. Under mild conditions, it can be shown for the penalty function P_Q that

$$\lim_{\rho \to \infty} \lambda_i(\rho) = \lambda_i^*, \tag{6.5}$$

and $\|\lambda^* - \lambda(\rho)\| = O(1/\rho)$. In the case of Example 6.1, where $\lambda_1^* = 2$, the value defined by (6.4) is given by $\lambda_1(\rho) = 2\rho/(\rho + 2)$.

When a penalty function method is used to solve an inequality-constrained problem, the relationships (6.4) and (6.5) imply that, if $\lambda_i^* > 0$, the minima of successive penalty functions occur at points that are *strictly infeasible* with respect to the constraints active at x^*. Thus, it is commonly stated that for a penalty function method, the "working set" is the "violated set". This property can be useful in identifying the active set. It also rules out the use of a penalty function method when feasibility is important; a method for this case is discussed in Section 6.2.1.2.

6.2.1.2. The logarithmic barrier function.

Since penalty function methods for NIP generate infeasible iterates, they are not appropriate for problems in which feasibility must be maintained. In this section, we introduce a class of *feasible-point* methods, in which only feasible iterates are generated (other feasible-point methods will be described in Sections 6.3 and 6.5.3.4). Feasible-point methods are essential in many practical applications where some (or all) of the problem functions may be undefined or ill-defined outside the feasible region. Furthermore, feasible-point methods are desirable in situations where a solution of only limited accuracy is required, since the optimization process will yield a feasible answer even if terminated prematurely; note that use of a penalty function method in this case would result in an infeasible solution.

The idea of a barrier function method is analogous to that of a penalty function method: to create a sequence of modified differentiable functions whose unconstrained minima will converge in the limit to x^*. Since the unconstrained minimum of $F(x)$ (if it exists) will, in general, occur at an infeasible point, the modified objective function for a feasible-point method *must include a term that prevents iterates from becoming infeasible* (in contrast to a penalty function, which merely adds a penalty for feasibility). To accomplish this, a "barrier" is created by adding to $F(x)$ a weighted sum of continuous functions with a positive singularity at the boundary of the feasible region. An unconstrained minimum of such a modified function must lie strictly inside the feasible region. As the weight assigned to the "barrier" term is decreased toward zero, the sequence of unconstrained minima should generate a strictly feasible approach to the constrained minimum. Barrier function transformations are used for inequality-constrained problems in which the feasible region has an interior, and generate iterates that remain *strictly feasible* with respect to the constraints. *Therefore, the barrier function transformation cannot be applied to an equality constraint.*

Only a single barrier function will be discussed here, to illustrate the features of these methods. For problem NIP, the *logarithmic barrier function* is:

$$B(x, r) = F(x) - r \sum_{i=1}^{m} \ln\big(c_i(x)\big). \tag{6.6}$$

The positive value r is termed the *barrier parameter*.

Barrier function methods are similar to penalty function methods in several ways. In particular, it can be shown that under mild conditions there exists $x^*(r)$, an unconstrained minimum of (6.6), such that

$$\lim_{r \to 0} x^*(r) = x^*.$$

Example 6.3. We illustrate the effect of a barrier function transformation on Example 6.1 with the constraint changed to an inequality.

$$\underset{x\in\Re^1}{\text{minimize}} \quad x^2$$

$$\text{subject to} \quad x - 1 \geq 0.$$

For this problem,

$$B(x, r) = x^2 - r \ln (x - 1)$$

and

$$x^*(r) = \frac{1}{2} + \frac{1}{2}\sqrt{1 + 2r}.$$

Many properties of barrier function methods are analogous to those described for differentiable penalty functions.

Inevitable ill-conditioning. Let t denote the number of constraints active at the solution. When $0 < t < n$, the Hessian matrix of the barrier function becomes increasingly badly conditioned at $x^*(r)$ as r approaches zero, with singularity occurring in the limit. Figure 6d displays the contours of the logarithmic barrier function corresponding to Example 6.2, for two values of r ($r = 0.2$ and $r = 0.001$). No contours have been drawn in the infeasible region, since the intent is for the barrier function to be evaluated only in the strict interior of the feasible region. Notice that the contour lines are close to being parallel for the smaller value of the barrier parameter. Because of the difficulties associated with computing an unconstrained minimum of (6.6) for small r, a *sequence* of subproblems must be solved when using a barrier function method.

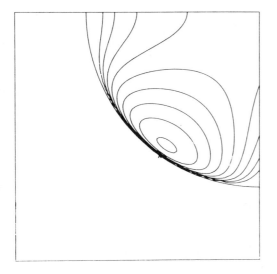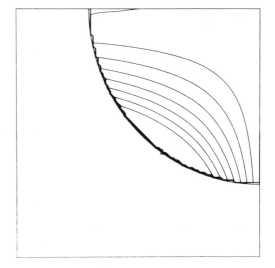

Figure 6d. The two diagrams represent the contours of the logarithmic barrier function $B(x, r)$ corresponding to the two-dimensional Example 6.2 for two values of r ($r = 0.2$ and $r = 0.001$).

Local minimum property. Since the barrier function transformation introduces only a local minimum of the barrier function, there is some danger of an unbounded subproblem. However, this is much less likely to happen than with a penalty function transformation, since the feasible region in many problems is closed. Nonetheless, any unconstrained method to be applied to a barrier function must be able to cope with unboundedness.

Inefficiency in the linear search. An additional difficulty is that the standard line search procedures for unconstrained optimization tend to be inefficient when applied to functions with singularities, for several reasons. Firstly, standard line searches assume that the function can be evaluated at any trial point; however, since the barrier function is undefined outside the feasible region, it is necessary to assign an essentially arbitrary value to the function if any trial point is infeasible. It is not possible to place an *a priori* upper bound on the step to the nearest nonlinear inequality constraint along a given direction (in contrast to the case of *linear* inequality constraints, where the maximum feasible step can be computed at the beginning of an iteration). Furthermore, if the estimate of the next trial step length is based on a polynomial model of the univariate behaviour of the function (see Section 4.1.2.3), the approximation of the barrier function is likely to be poor (particularly for small values of the barrier parameter, when the minimum becomes very close to the singularity). Fortunately, special line search procedures can be designed to take advantage of the known form of the singularity in a barrier function (see the references cited in the Notes).

Lagrange multiplier estimates. The properties of the sequence $\{x^*(r)\}$ as r approaches zero provide information about the Lagrange multipliers of the original problem. At $x^*(r)$, it holds that

$$g = \sum_{i=1}^{m} a_i \frac{r}{c_i}, \tag{6.7}$$

where g, c_i and a_i are evaluated at $x^*(r)$. Hence, the gradient of F at $x^*(r)$ is a non-negative linear combination of the gradients of *all* the constraints. This implies that the coefficients in the linear combination (6.7) are analogous to the Lagrange multipliers at x^*. If the i-th constraint is inactive, the coefficient approaches zero as r approaches zero; if $c_i(x)$ is active, then under mild conditions

$$\lim_{r \to 0} \frac{r}{c_i\big(x^*(r)\big)} = \lambda_i^*.$$

Note the contrast with a penalty function method for inequalities, in which eventually only the violated constraints appear in the optimality conditions for $x^*(\rho)$.

6.2.2. Non-Differentiable Penalty Function Methods

In Section 6.2.1.1 it was observed that, even for smooth, well-posed problems, methods based on a differentiable penalty function suffer from inevitable ill-conditioning and the need to solve a sequence of subproblems. Furthermore, there is no reason to insist on a differentiable penalty function if the problem functions themselves are not smooth. Therefore, an alternative approach is to create a *non-differentiable, but well-conditioned*, penalty function of which x^* is a local unconstrained minimum. We shall consider the definition and properties of such a penalty function in Section 6.2.2.1, and its application to non-smooth problems in Section 6.2.2.2.

6.2.2.1. The absolute value penalty function. The most popular non-differentiable penalty function is the *absolute value penalty function*:

$$P_A(x, \rho) = F(x) + \rho \sum_{i \in I} |\hat{c}_i(x)| \equiv F(x) + \rho \|\hat{c}(x)\|_1, \qquad (6.8)$$

where the vector $\hat{c}(x)$ represents the constraints violated at x (defined by the set of indices I). (Recall that $\|y\|_1 = \sum |y_i|$.)

Note that $P_A(x, \rho)$ has discontinuous derivatives at points where a constraint function vanishes; hence, unless there are no active constraints, x^* will necessarily be a point of discontinuity in the derivative of P_A. The crucial distinction between P_A and P_Q is that, under mild conditions, ρ need not become arbitrarily large in order for x^* to be an unconstrained minimum of P_A. Rather, there is a threshold value $\bar{\rho}$ such that x^* *is an unconstrained minimum of* P_A *for any* $\rho > \bar{\rho}$. For this reason, penalty functions like P_A are sometimes termed *exact penalty functions* (in contrast to a differentiable penalty function, for which $x^*(\rho)$ is not equal to x^* for any finite ρ); other exact penalty functions are mentioned in the Notes for Section 6.4.

To see the effect of such a transformation, let us re-consider Example 6.1:

$$\begin{array}{ll} \text{minimize} & x^2 \\ x \in \Re^1 & \\ \text{subject to} & x - 1 = 0, \end{array}$$

for which

$$P_A(x, \rho) = x^2 + \rho|x - 1|. \qquad (6.9)$$

For any $\rho > 2$, x^* is the unconstrained minimum of (6.9). The dashed line in Figure 6e represents the graph of $F(x)$. The solid line represents the graph of P_A with $\rho = 4$, and it is obvious that x^* is a local minimum.

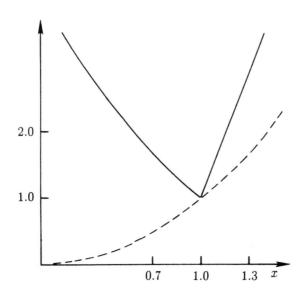

Figure 6e. The dashed line represents the graph of the objective function of Example 6.1, $F(x) = x^2$. The solid line represents the graph of the non-differentiable penalty function $P_A(x, 4) = x^2 + 4|x - 1|$ corresponding to Example 6.1.

We shall outline a typical algorithm for solving a constrained optimization problem with smooth problem functions using a non-differentiable penalty function (an algorithm for the case when the problem functions are not smooth is given in Section 6.2.2.2).

Given an initial point x_0, a positive integer K, and a penalty parameter ρ, set $k \leftarrow 0$ and execute the following steps.

Algorithm EP (*Model algorithm with an exact penalty function*).

EP1. [Check termination conditions.] If x_k satisfies the optimality conditions, the algorithm terminates with x_k as the solution. If $k > K$, the algorithm terminates with a failure.

EP2. [Increase the penalty parameter if necessary.] If $k > 0$, increase ρ.

EP3. [Minimize the penalty function.] With x_k as the starting point, execute an algorithm to solve the unconstrained subproblem

$$\underset{x \in \Re^n}{\text{minimize}} \ P_A(x, \rho), \tag{6.10}$$

subject to safeguards to ensure termination in case of unboundedness. Let x_{k+1} denote the best estimate of the solution of (6.10).

EP4. [Update the iteration count.] Set $k \leftarrow k + 1$ and return to Step EP1. ∎

Note that whenever ρ is "sufficiently large", the algorithm will terminate successfully in Step EP1 because x_k satisfies the optimality conditions for the original problem.

Methods based on non-differentiable penalty functions avoid *inevitable* ill-conditioning, since ρ does not need to be made arbitrarily large to achieve convergence to x^*. In fact, if an adequate value of ρ is used, x^* can be computed with a single unconstrained minimization. Unfortunately, $\bar{\rho}$ depends on quantities evaluated at x^* (which is of course unknown), and therefore the value of ρ to be used in (6.10) must be estimated, and adjusted if necessary (as in Step EP2). Difficulties can arise if an unsuitable value of the penalty parameter is chosen. If ρ is too small, the penalty function may be unbounded below, or the region in which iterates will converge to x^* may be very small. On the other hand, the unconstrained subproblem will be ill-conditioned if ρ is too large.

The effect of varying ρ can be observed in Figure 6f, which represents the contours of P_A corresponding to Example 6.2 for three values of ρ ($\rho = 1.2$, $\rho = 5$ and $\rho = 10$). Since the absolute value penalty term is weaker than the quadratic penalty term, the difficulties associated with creating only local minima are more severe with P_A than with P_Q. Note that, although the first unconstrained function shown in Figure 6f is the best-conditioned at the solution, an unconstrained method might fail to converge to x^* unless the initial point were close to x^*. (In fact, for *any* value of ρ, x^* is not the global minimum of P_A for Example 6.2.) The deterioration in conditioning as the penalty parameter increases can be observed by comparing the first and third functions shown in Figure 6f.

Since P_A is non-differentiable even at the solution, standard unconstrained methods designed for smooth problems cannot be applied directly. However, special algorithms have been designed to minimize P_A for smooth problems; references to such methods are given in the Notes. These methods take advantage of the many special features of P_A when the problem functions are smooth. In particular, information about the original nonlinearly constrained problem — for example, Lagrange multiplier estimates — can be used to improve efficiency of the algorithm for minimizing the non-differentiable penalty function, and for obtaining a good value of the penalty parameter.

 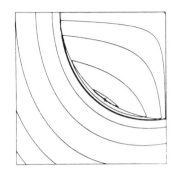 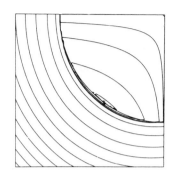

Figure 6f.   The three diagrams represent the contours of the absolute value penalty function $P_A(x, \rho)$ corresponding to Example 6.2 for three values of the penalty parameter ($\rho = 1.2$, $\rho = 5$, and $\rho = 10$).

6.2.2.2. A method for general non-differentiable problems. If any of the problem functions in NEP or NIP are discontinuous or have discontinuous derivatives, the absolute value penalty function is no longer a composite of smooth functions. Thus, a method designed for non-smooth functions (such as the polytope method described in Section 4.2.2) must be applied to minimize P_A; however, for certain problems in which the discontinuities have a known structure, it is possible to develop special methods (see Section 6.8).

The choice of the penalty parameter in P_A should balance the objectives of keeping the subproblem well-conditioned near the solution, yet encouraging a large region of convergence (in general, the larger the penalty parameter, the larger the region in which x^* is a local minimum). The following outline of an algorithm is therefore suggested for problems with non-structured discontinuities. Although the algorithm is similar to Algorithm EP (in Section 6.2.2.1), a significant difference is that the tests associated with steps of the algorithm — for example, finding a "satisfactory minimum" — are highly problem-dependent. In addition, x_k is not accepted as the solution until an attempt has been made to find a better point (in contrast to the case when the problem functions are smooth).

Given an initial point x_0, an initial penalty parameter ρ, a positive value $\rho_{\max}$, and a positive constant γ ($\gamma > 1$), set $k \leftarrow 0$ and execute the following steps.

Algorithm ND (*Model algorithm for constrained non-differentiable problems*).

ND1. [Solve an unconstrained subproblem.]   With x_k as the starting point, use the polytope method of Section 4.2.2 to try to compute an approximation to the minimum of $P_A(x, \rho)$. An upper bound should be set on the number of evaluations of the problem functions during this step, to ensure that the method will terminate. Let x_{k+1} denote the best point found by the polytope method.

ND2. [Increase the penalty parameter if necessary.]   If $k > 0$, go to Step ND3. If $\rho > \rho_{\max}$, the algorithm terminates with a failure. If Step ND1 failed to locate a satisfactory minimum of P_A, or if x_{k+1} is not feasible, set $\rho \leftarrow \gamma\rho$ and go back to Step ND1. Otherwise, go to Step ND4.

ND3. [Determine whether the penalty parameter can be decreased.] If x_{k+1} is not a "significantly better" solution than x_k, the algorithm terminates with the "better" of x_k and x_{k+1} as the solution.

ND4. [Decrease the penalty parameter.] Set $\rho \leftarrow \rho/\gamma$, $k \leftarrow k+1$, and go back to Step ND1. ∎

The polytope method is appropriate for minimizing P_A because its effectiveness is improved when the solution is a point of discontinuity in the derivative (which is always the case with P_A when any constraints are active). However, it is unlikely that the gradient of P_A is discontinuous in all directions, and thus there is some danger that the polytope method will collapse into a subspace. The reduction of the penalty parameter in Step ND4 is included in order to try to improve the accuracy of the minimum in all directions.

Notes and Selected Bibliography for Section 6.2

Penalty and barrier functions have a long history, and were probably used to solve engineering problems well before they gained mathematical respectability. They occur in many forms, and are often called by special names in different applications. For example, in crystallographic calculations a quadratic penalty term is added to the objective function as a "restraint", to ensure that the solution of an optimization problem is "not too far" from the constraints. The absolute value penalty function has been used for many years in structural and mechanical design problems.

The major work concerning differentiable penalty and barrier functions is the book *Nonlinear Programming: Sequential Unconstrained Minimization Techniques*, by Fiacco and McCormick (1968), which summarizes and extends a series of papers by Fiacco and McCormick in *Management Science*. This book serves as a primary source of results related to penalty and barrier function methods. In addition, it contains a detailed historical survey of these methods, to which the interested reader is referred.

Courant (1943) is usually credited with the first published reference in the mathematical literature to the quadratic penalty function. Frisch (1955) first described the logarithmic barrier function. Carroll (1959, 1961) proposed the *inverse* barrier function corresponding to NIP:

$$B_I(x, r) \equiv F(x) + r \sum_{i=1}^{m} \frac{1}{c_i(x)}.$$

Rosenbrock (1960) described how to define a modified objective function similar to a barrier function.

A general treatment of penalty and barrier functions is given by Zangwill (1965, 1967a), and a more recent overview is given by Ryan (1974).

The great interest in penalty and barrier function methods during the 1960's is attributable in large part to the development of powerful methods for unconstrained minimization (see Chapter 4 for a discussion of unconstrained optimization methods). The SUMT code of Fiacco and McCormick may well be the most widely used general code for nonlinearly constrained optimization.

The inevitable ill-conditioning of the Hessian matrices of differentiable penalty and barrier function methods was first reported by Murray (1967). A detailed discussion of the phenomenon is given by Murray (1969a, 1971b) and Lootsma (1969, 1970, 1972).

Special line search procedures for penalty and barrier functions have been given by, amongst others, Fletcher and McCann (1969), Murray (1969a), Lasdon, Fox, and Ratner (1973), and Murray and Wright (1976).

Penalty and barrier methods are sometimes used as the first phase of a "hybrid", or "two-phase", method, in which the second phase includes a method that exhibits good local convergence properties. For example, Ryan (1971) and Osborne and Ryan (1972) have suggested combining the quadratic penalty function with an augmented Lagrangian method (see Section 6.4). Rosen (1978) and Van der Hoek (1979) have proposed using a penalty function method to obtain a good initial point for a projected Lagrangian method (see Section 6.5).

The idea of devising a *non-differentiable* penalty function such that x^* is an unconstrained minimum for a *finite* value of the penalty parameter appears implicitly in Ablow and Brigham (1955), and was suggested for the convex problem by Pietrzykowski (1962) and Zangwill (1965, 1967a). A treatment of the properties of the absolute value penalty function for general problems is given by Pietrzykowski (1969). Other references concerned with properties of exact penalty functions of this type are Evans, Gould and Tolle (1973), Howe (1973), Bertsekas (1975a), Charalambous (1978), Han and Mangasarian (1979) and Coleman and Conn (1980a).

Methods based on minimization of non-differentiable penalty functions were initially regarded as impractical because standard unconstrained algorithms (which assume differentiability) could not be applied to solve the subproblem. However, much work has been devoted to special methods for minimizing the non-differentiable penalty function (and other functions in which the derivative discontinuities have a known structure). Various methods of this type are discussed in Zangwill (1967b), Conn (1973), Conn and Pietrzykowski (1977), Lemaréchal (1975), Mifflin (1977), Maratos (1978), Coleman (1979), and Mayne and Maratos (1979). The way in which many of these methods define the search direction is very closely related to that of a QP-based projected Lagrangian method (to be discussed in Section 6.5.3).

The absolute value penalty function has also been proposed for use as a *merit function* within QP-based projected Lagrangian methods (see Section 6.5.3.3), and some projected Lagrangian methods are based explicitly on minimizing the absolute penalty function. Further discussion of these methods is given in the Notes for Section 6.5.

Many methods for general non-smooth constrained problems have been developed for particular applications. These methods are typically based on function comparison techniques (see Section 4.2.1), and are often called *direct search methods*. Most such procedures are heuristic in nature, and few (if any) assurances can be given about their performance. Some direct search methods involve the unconstrained minimization of a penalty or barrier function (as does Algorithm ND of Section 6.2.2.2). The interested reader should consult the survey article of Swann (1974).

6.3. REDUCED-GRADIENT AND GRADIENT-PROJECTION METHODS

6.3.1. Motivation for Reduced-Gradient-Type Methods

Gradient-projection and reduced-gradient methods (which we shall refer to as *reduced-gradient-type methods*) are based on extending methods for linear constraints to the nonlinear case. Confusion often arises when methods of this type are discussed, since their essential similarity is obscured by presentation of algorithmic details rather than the underlying principles. The description here is designed to give insight into the overall motivation rather than to specify a particular algorithm.

Reduced-gradient-type methods for *nonlinear* constraints are motivated by the same idea as active set methods for linear constraints — to stay "on" a subset of the nonlinear constraints while reducing the objective function (in effect, the requirement of satisfying the constraints should reduce the dimensionality of the minimization). Achieving this objective requires the ability to adjust the variables so that the active constraints continue to be satisfied exactly at each trial point. When the constraints are linear, we have seen that feasibility *can* be maintained in this fashion by appropriate construction of the search direction (see Section 5.1); unfortunately, extension to the nonlinear case is not straightforward because there is no known procedure for remaining "on" a nonlinear constraint. In particular, some sort of iterative "correction" process is required to follow a curving constraint boundary.

To illustrate these ideas, assume that a two-variable problem contains the single equality constraint

$$5 + x_2^3 \exp\left(x_1^2 \sin(x_2 - x_1)\right) = 0. \tag{6.11}$$

The constraint (6.11) can be interpreted as defining a functional relationship between x_1 and x_2. Thus, one of the variables can be used to solve for the other, so that *only one degree of freedom remains in the minimization*. Given a value of x_1, the value of x_2 can be adjusted by a univariate zero-finding procedure (Section 4.1.1) in order to satisfy (6.11).

An ideal version of a reduced-gradient-type method would have the property that all constraints active at the solution are satisfied exactly at every iteration. The differences among reduced-gradient-type algorithms arise from the variety of techniques used to achieve the aims of staying feasible and reducing F.

Historically, the terms "gradient projection" and "reduced gradient" were used in connection with particular transformations. However, we prefer to use these terms in a more general sense, to refer to any transformation that "projects" F into the "reduced" subspace of points that satisfy the constraints.

6.3.2. Definition of a Reduced-Gradient-Type Method

We shall describe the general pattern of a typical iteration in a reduced-gradient method applied to a problem in which we assume for simplicity of presentation that the correct active constraints are known (some techniques for making this determination will be considered in Section 6.3.3). However, a model algorithm will not be given because of the many variations in strategy associated with each step. Since an iteration of a reduced-gradient-type method differs in pattern from the iterations described previously, we shall introduce a modified notation to emphasize the special structure of the iteration.

Let $\hat{c}(x)$ denote the set of t constraint functions active at the solution. Assume that x_k is the current iterate, and that $\hat{c}(x_k) = 0$. The next iterate will be a point x_{k+1} such that $\hat{c}(x_{k+1}) = 0$ and $F(x_{k+1})$ has sustained a "sufficient decrease" (see Section 4.3.2.1). The desired result of the iteration is the step from x_k to x_{k+1}, which we shall denote by s_k.

6.3.2.1. Definition of the null-space component. The first step of a typical iteration is to enforce satisfaction of the constraints by ensuring that

$$\hat{c}(x_k + s_k) = 0. \tag{6.12}$$

If only *linear constraints* were included in $\hat{c}$, the requirement (6.12) would remove t degrees of freedom from s_k (see Section 5.1.1). In order to extend this idea to nonlinear constraints, we make a *linear approximation* to $\hat{c}$, based on its Taylor-series expansion about x_k, i.e.

$$\hat{c}(x_k + s_k) \approx \hat{c}(x_k) + \hat{A}(x_k)s_k. \tag{6.13}$$

From (6.12) and (6.13), we see that a vector p_k that approximates s_k satisfies a set of *linear equality constraints*

$$\hat{A}_k p_k = 0, \tag{6.14}$$

where $\hat{A}_k$ denotes $\hat{A}(x_k)$. This restriction is analogous to the property (5.5) of the search direction in the linear-constraint case, with the difference that the matrix $\hat{A}_k$ in (6.14) varies from iteration to iteration because of constraint nonlinearities.

As observed in our discussion of linear constraints (see (5.6)), (6.14) implies that p_k must be of the form

$$p_k = Z_k p_z, \tag{6.15}$$

where Z_k is a matrix whose $n - t$ columns form a basis for the set of vectors orthogonal to the rows of $\hat{A}_k$, and p_z is an $(n - t)$-vector. Thus, since p_k lies entirely in the null space of $\hat{A}_k$, we observe the desired reduction to $n - t$ degrees of freedom (represented by the vector p_z).

The vector p_z is typically determined by minimizing an approximation to the objective function, expressed only in terms of p_z. For example, in early algorithms p_z was taken as the "steepest-descent" vector

$$p_z = -Z_k^T g_k. \tag{6.16}$$

(Note that the vector $Z_k^T g_k$ is the gradient of F projected into the subspace of vectors orthogonal to the gradients of the active constraints). An improved rate of convergence can be achieved by using second-order information to define p_z; in particular, p_z can be defined as the solution of

$$Z_k^T W_k Z_k p_z = -Z_k^T g_k, \tag{6.17}$$

where W_k is an approximation to the Hessian of the Lagrangian function.

A widely known method of this type is called the *generalized reduced-gradient* (GRG) method. The GRG method is based on a particular form of Z_k, namely the variable-reduction form given in Section 5.1.3.2. We assume that $\hat{A}_k$ has full rank, and that the first t columns are linearly independent; hence, $\hat{A}_k$ can be written as

$$\hat{A}_k \equiv (\, V \quad U \,),$$

where V is a $t \times t$ non-singular matrix. The associated variable-reduction form of Z_k is

$$Z_k = \begin{pmatrix} -V^{-1}U \\ I \end{pmatrix}. \tag{6.18}$$

Because of this form of Z_k, an important simplification occurs in the specification of p_k. Let the variables be partitioned according to the columns of $\hat{A}_k$, so that p_k is given by $(\, p_V \quad p_U \,)$ (and similarly for g_k). The constraints (6.12) imply that the t "dependent" variables p_V can be expressed in terms of the $n - t$ "independent" variables p_U. In this way, the reduction in dimensionality to $n - t$ is expressed *in terms of the original variables of the problem*, and p_z is simply p_U. The name of the generalized reduced-gradient method arises because the term *reduced gradient* is applied to the vector $-U^T V^{-T} g_V + g_U$, which is simply $Z_k^T g_k$ when Z_k is defined by (6.18).

6.3.2.2. Restoration of feasibility. If the constraints were purely linear, the step from x_k to x_{k+1} could be taken as γp_k for some scalar γ, where p_k is defined by (6.15). However, with *nonlinear* constraints, in general it is *not* true that $x_k + \gamma p_k$ will be feasible, and hence p_k may not be used as a "search direction" in the usual sense. The approach taken in a reduced-gradient-type method is typically to define s_k to be of the form

$$s_k = \gamma p_k + Y_k p_Y, \qquad (6.19)$$

where the columns of Y_k form a basis for the range space of $\hat{A}_k^T$.

The first condition to be satisfied by s_k is that $x_k + s_k$ must be feasible. Thus, it is necessary to find a value of γ in (6.19) and a corresponding p_Y, such that $x_k + s_k$ is feasible. We emphasize that p_Y depends on γ, and thus *the adjustment of γ is not a standard "line search"*.

The determination of γ to ensure feasibility usually proceeds by an iterative process in which, for each trial value of γ, we attempt to solve the system of t nonlinear equations

$$\hat{c}(x_k + s_k) \equiv \hat{c}(x_k + \gamma p_k + Y_k p_Y) = 0, \qquad (6.20)$$

for the t unknown components of p_Y. The calculation of p_Y for a given value of γ is an *adaptive subproblem*, since the constraint functions (and possibly their gradients) will be evaluated an unknown number of times while solving (6.20). Furthermore, (6.20) may be a defective subproblem, since it may be impossible to find a suitable p_Y — either because no such vector exists, or because the procedure for solving nonlinear equations encounters difficulties. Under mild conditions, it can be shown that a solution of (6.20) will exist if γ is sufficiently small.

For the GRG method, the matrix Y_k is given simply by

$$Y_k = \begin{pmatrix} I \\ 0 \end{pmatrix} \begin{matrix} \}t \\ \}n-t \end{matrix}.$$

Thus, in this case the calculation of s_k to satisfy (6.20) involves adjustment of the change p_V in the dependent variables x_V. (This is exactly the procedure discussed previously for satisfying (6.11).)

6.3.2.3. Reduction of the objective function. Merely finding γ and p_Y such that (6.20) holds does not necessarily constitute a successful iteration in a reduced-gradient-type method, since a further adjustment of γ may be required in order to satisfy a "sufficient decrease" condition such as

$$F(x_k + s_k) \le F(x_k) - \delta_k, \qquad (6.21)$$

for some positive scalar δ_k (see Section 4.3.2.1). Under mild conditions on p_z, it can be shown that (6.21) can be satisfied for a sufficiently small value of γ. When a value of γ is found for which (6.20) and (6.21) are satisfied, x_{k+1} is taken as $x_k + s_k$.

To summarize, the definition (6.19) of s_k involves the "linear" adjustment of p_z, and the "nonlinear" adjustment of p_Y. The determination of γ is in itself an adaptive subproblem, and includes the further adaptive subproblem of finding p_Y to satisfy (6.20).

Figure 6g represents the first two iterates when a reduced-gradient-type method is applied to Example 6.2, starting at the point labelled "0" and treating the constraint as an equality. The numbered points are the sequence of trial points at which the problem functions were evaluated (the points corresponding to "1" and "2" are outside the range of the diagram). The value of p_z was defined by (6.17). The points labelled "6" and "8" are the two feasible iterates with decreasing values of F. Note that for this problem, the first value of γ for which p_Y exists fortuitously satisfies (6.21) as well as (6.20).

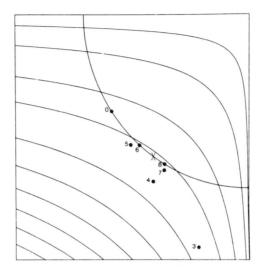

Figure 6g. An illustration of the performance of a reduced-gradient-type method when applied to Example 6.2. Each trial point is marked by a "•" and a number denoting the order in which it was computed. The second and third trial points are off the scale of the diagram. Note that the iterates (the points labelled "6" and "8") tend to approach the solution tangentially to the constraint.

6.3.2.4. Properties of reduced-gradient-type methods. Gradient-projection and reduced-gradient methods have been implemented in several software packages, and can be very successful on problems where the constraints are nearly linear. When the starting point is far from optimal and the constraints are highly nonlinear, requiring a strict standard of feasibility such as (6.20) may force the algorithm to take painfully small steps during the entire approach to the solution, since the step s_k must be small enough to allow continued feasibility while moving tangent to the constraints at x_k (as well as a reduction in the objective function). This phenomenon can be observed even in the simple example depicted in Figure 6g. Furthermore, from (6.14) and Figure 6g we see that the iterates tend to follow a "tangential" path to the solution, which implies that the underlying linear approximation of the constraints may be inadequate.

In order to avoid the inefficiencies that result from insisting on exactly satisfying a subset of the constraints, the tendency has been to allow increasingly greater freedom to violate the constraints before restoring feasibility. However, tolerance of a substantial level of infeasibility means that the methods have moved so far from the original motivation that they should be placed into a different category. In some instances, methods of this type can be interpreted as projected Lagrangian methods, which will be discussed in Section 6.5.

6.3.3. Determination of the Working Set

When a reduced-gradient-type method is applied to nonlinear *inequality* constraints, the general approach is to develop a form of "active set strategy", as in the case of linear inequality constraints. Such a strategy can be implemented in several different ways. However, regardless of the details of implementation, the essence of a reduced-gradient-type method applied to inequalities is *to satisfy exactly, at every iteration, the constraints that are believed to be active at the solution.* Note the contrast with barrier function methods (Section 6.2.1.2), which maintain feasibility by staying "off" all constraints, so that the active constraints are satisfied exactly only in the limit.

An active set strategy can be implemented by defining a "working set" of constraints at each iteration; in the context of a reduced-gradient-type method, the working set will be composed of constraints that are "exactly satisfied". The constraints in the working set are then considered as equalities for the purpose of computing p_z and correcting back to feasibility. The working set is modified as the iterations proceed.

The computation of s_k is more complicated for the inequality-constraint case. In particular, when determining γ in (6.20) it must be considered whether any inactive constraint is violated for a given trial value of γ. When the chosen value of γ causes a previously satisfied constraint to be violated, the value of γ must be iteratively adjusted so that the new constraint is exactly satisfied (and is then added to the working set at the next iteration).

In addition, it is necessary to be able to delete a constraint that has incorrectly been included in the working set. This determination is typically based on Lagrange multiplier estimates, which will be discussed in detail in Section 6.6. The decision as to when to delete a constraint, and the particular form of multiplier estimate, vary with the version of the reduced-gradient-type method.

Some reduced-gradient-type algorithms implement an active set strategy indirectly by converting all nonlinear inequality constraints to equalities through the addition of extra ("slack") variables that are subject to simple bounds. The effect of this approach is that all nonlinear constraints are always in the working set; whether the original inequality constraint is exactly satisfied depends on whether the associated slack variable lies on its bound.

Notes and Selected Bibliography for Section 6.3

The idea of a "reduced" or "projected" gradient first appeared with respect to *linear* constraints; a discussion of papers on this topic is given in the Notes for Sections 5.1 and 5.2. Rosen (1961) first suggested a "gradient projection" method for nonlinear constraints based on his similar method for linear constraints (Rosen, 1960). His method implicitly used an orthogonal Z_k to define p_z from (6.16).

A generalized reduced-gradient method was first proposed by Abadie and Carpentier (1965, 1969). The original GRG method explicitly uses the variable-reduction form (6.18) of Z_k. The GRG code of Abadie and his co-workers at Électricité de France has been widely used in solving practical problems (see, for example, Abadie and Guigou, 1970). In a computational study of Colville (1968), GRG methods were found to be among the most reliable and efficient at that time. Numerous versions of the GRG method have been developed; as observed in the discussion in Section 6.3.1, there is substantial room for flexibility in each step of a reduced-gradient-type procedure.

Subsequent work on reduced-gradient-type methods has been performed, amongst others, by Sargent and Murtagh (1973), Abadie (1978), and Lasdon and Waren (1978).

Reduced-gradient-type methods are known to encounter difficulties when it is necessary to follow the boundaries of highly nonlinear constraints (see Section 6.3.2). Various suggestions have been made to avoid this situation — in particular, it has been suggested that the iterates should not be required to satisfy the constraints with high accuracy. However, with this formulation, reduced-gradient-type methods can be regarded as a subset of projected Lagrangian methods (to be discussed in Section 6.5). Sargent (1974) presents a general overview of reduced-gradient-type methods, including some discussion of the relationship between these methods and projected Lagrangian methods.

6.4. AUGMENTED LAGRANGIAN METHODS

The class of *augmented Lagrangian methods* can be derived from several different viewpoints. One motivation consistently associated with these methods is to construct an unconstrained sub-problem with an objective function Φ_U such that: (i) ill-conditioning is not inevitable (in contrast to the methods described in Section 6.2.1); (ii) the function to be minimized is continuously differentiable (in contrast to the methods of Section 6.2.2).

6.4.1. Formulation of an Augmented Lagrangian Function

The methods to be described in the next two sections are derived from the optimality conditions discussed in Section 3.4, and the role of Lagrange multipliers is crucial. Our intent is to give an overview of the motivation for the definition of Φ_U. Hence, a detailed discussion of techniques for computing Lagrange multiplier estimates for nonlinearly constrained problems will be postponed until Section 6.6.

Augmented Lagrangian methods can be applied to both equality and inequality constraints. There are various possible strategies for treating inequality constraints with these methods. A common technique for treating inequality constraints is to use a *pre-assigned active set strategy*, where a determination is made at each iteration concerning which constraints are included in $\hat{c}$ (see Section 6.5.5). For simplicity, we shall initially assume that the correct set of active constraints is known; in the remainder of this section, $\hat{c}(x)$ will denote the set of constraint functions active at x^*. A discussion of other techniques used in applying augmented Lagrangian methods to inequality constraints will be given in the Notes.

We shall assume that the sufficient conditions for optimality described in Section 3.4 hold at x^*, so that

$$g(x^*) = \hat{A}(x^*)^T \lambda^*, \tag{6.22}$$

where $\hat{A}$ is the matrix whose t rows are the gradients of the constraints active at x^*. The vector λ^* is unique when $\hat{A}(x^*)$ has full rank.

Let $L(x, \lambda)$ denote the Lagrangian function

$$L(x, \lambda) \equiv F(x) - \lambda^T \hat{c}(x). \tag{6.23}$$

(Note that the definition (6.23) depends on the specification of the vector $\hat{c}$.) The relationship (6.22) can be interpreted as a statement that x^* is a *stationary point* (with respect to x) of the Lagrangian function when $\lambda = \lambda^*$, which might suggest that the Lagrangian function could be used as Φ_U. Unfortunately, x^* is not necessarily a *minimum* of the Lagrangian function (see the second diagram in Figure 6b). Thus, even if λ^* were known, the Lagrangian function would not be a suitable choice for the objective function of the subproblem.

Let $W(x, \lambda)$ denote the Hessian with respect to x of the Lagrangian function, i.e.

$$W(x, \lambda) \equiv G(x) - \sum_{i=1}^{t} \lambda_i \hat{G}_i(x).$$

Although $W(x^*, \lambda^*)$ may not be positive definite (nor even non-singular), we can expect the *projected Hessian of the Lagrangian function* — $Z(x^*)^T W(x^*, \lambda^*) Z(x^*)$ — to be positive definite, where $Z(x)$ denotes a matrix whose columns form a basis for the set of vectors orthogonal to the rows of $\hat{A}(x)$. Therefore, x^* is a *minimum of the Lagrangian function within the subspace of vectors orthogonal to the active constraint gradients.*

This property suggests that it is possible to construct a suitable Φ_U by *augmenting* the Lagrangian function through the addition of a term that retains the stationary properties of x^*, but alters the Hessian in the subspace of vectors defined by $\hat{A}(x^*)$. The most popular *augmented Lagrangian function* (and the only one to be considered here) is given by

$$L_A(x, \lambda, \rho) \equiv F(x) - \lambda^T \hat{c}(x) + \frac{\rho}{2} \hat{c}(x)^T \hat{c}(x), \qquad (6.24)$$

where ρ is a positive penalty parameter.

Both the quadratic penalty term of (6.24) and its gradient vanish at x^* (under the assumption that $\hat{c}$ includes the constraints active at x^*). Thus, if $\lambda = \lambda^*$, x^* is a stationary point (with respect to x) of (6.24). The Hessian matrix of the penalty term is $\sum_i \hat{c}_i(x) \hat{G}_i(x) + \hat{A}(x)^T \hat{A}(x)$. Since $\hat{c}(x^*)$ vanishes, the Hessian of the penalty term at x^* is simply $\hat{A}(x^*)^T \hat{A}(x^*)$, which is a positive semi-definite matrix with strictly positive eigenvalues corresponding to eigenvectors in the range of $\hat{A}(x^*)^T$. Thus, adding a positive multiple of the penalty term has the effect of increasing the (possibly negative) eigenvalues of W corresponding to eigenvectors in the range space of $\hat{A}(x^*)^T$, but leaving the other eigenvalues unchanged. Using this property, it can be shown under mild conditions that there exists a *finite* $\bar{\rho}$ such that x^* is an *unconstrained minimum* of $L_A(x, \lambda^*, \rho)$ for all $\rho > \bar{\rho}$. Furthermore, if $Z(x)$ is a matrix orthogonal to the rows of $\hat{A}(x)$, it holds that

$$Z(x^*)^T \nabla^2_{xx} L_A(x^*, \lambda^*, \rho) Z(x^*) = Z(x^*)^T W(x^*, \lambda^*) Z(x^*),$$

i.e., the projected Hessian of the Lagrangian function at x^* is unaltered by the penalty term.

We illustrate the effect of augmenting the Lagrangian function on a simple example.

Example 6.4. Consider the univariate problem

$$\begin{array}{ll} \underset{x \in \Re^1}{\text{minimize}} & x^3 \\ \text{subject to} & x + 1 = 0. \end{array}$$

The unique solution is $x^* = -1$, and $\lambda^* = 3$. Thus,

$$L(x, \lambda^*) = x^3 - 3(x + 1),$$

and x^* is not a local minimum of $L(x, \lambda^*)$. However, for all $\rho > \bar{\rho}$ ($\bar{\rho} = 6$), x^* is a local minimum of the augmented function:

$$L_A(x, \lambda^*, \rho) = x^3 - 3(x + 1) + \frac{\rho}{2}(x + 1)^2.$$

The solid line in Figure 6h is the graph of the objective function, the dotted line is the graph of the Lagrangian function, and the dashed line is the graph of the augmented Lagrangian function for $\rho = 9$. Note that this augmented Lagrangian function is unbounded below for *any* value of the penalty parameter.

6.4.2. An Augmented Lagrangian Algorithm

The discussion of Section 6.4.1 suggests that, under ideal circumstances, x^* could be computed by a single unconstrained minimization of the differentiable function (6.24). Unfortunately, this ideal

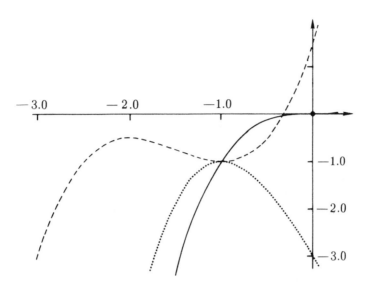

Figure 6h. The solid line is the graph of the objective function of Example 6.4, $F(x) = x^3$. The dotted line denotes the graph of the Lagrangian function, and the dashed line is the graph of the augmented Lagrangian $L_A(x, \lambda^*, 9)$.

can almost never be realized, since in general λ^* will not be available until the solution has been found. Hence, an augmented Lagrangian method must include a procedure for estimating the Lagrange multipliers. The crucial role of the multiplier estimate explains the alternative name for such an algorithm — the *method of multipliers.*

Several different algorithms can be developed from the properties of augmented Lagrangian functions. The fundamental characteristic associated with an augmented Lagrangian method is the n-dimensional *unconstrained minimization* of a differentiable function that involves Lagrange multiplier estimates and a penalty term, where the primary purpose of the penalty term is to make x^* an unconstrained minimum rather than a stationary point.

6.4.2.1. A model algorithm. In this section, we briefly describe one basic structure for an iteration in an augmented Lagrangian method. Before beginning the algorithm, the following are assumed to be available: an initial selection of the constraints to be included in $\hat{c}$; an initial estimate of the Lagrange multipliers, λ_0; a penalty parameter ρ; a positive integer K, which serves as an upper bound on the number of unconstrained minimizations to be performed; and an initial point x_0. Then set $k \leftarrow 0$ and perform the following steps.

Algorithm AL (*Model augmented Lagrangian algorithm*).

AL1. [Check termination criteria.] If x_k satisfies the optimality conditions, the algorithm terminates with x_k as the solution. If $k > K$, the algorithm terminates with a failure.

AL2. [Minimize the augmented Lagrangian function.] With x_k as the starting point, execute a procedure to solve the subproblem

$$\underset{x \in \Re^n}{\text{minimize}} \; L_A(x, \lambda_k, \rho), \tag{6.25}$$

including safeguards to cope with unboundedness. Let x_{k+1} denote the best approximation to the solution of (6.25).

AL3. [Update the multiplier estimate.] If appropriate, modify the specification of $\hat{c}$. Compute λ_{k+1}, an updated estimate of the Lagrange multipliers.

AL4. [Increase the penalty parameter if necessary.] Increase ρ if the constraint violations at x_{k+1} have not decreased sufficiently from those at x_k.

AL5. [Update the iteration count.] Set $k \leftarrow k + 1$, and go back to Step AL1.   ∎

6.4.2.2. Properties of the augmented Lagrangian function. The following properties are important in any method based on the augmented Lagrangian function L_A.

Local minimum property. As with the penalty functions described in Section 6.2, at best x^* is guaranteed to be only a *local* minimum of L_A. Furthermore, L_A may be unbounded below for any value of ρ, so that the subproblem (6.25) may be defective. Hence, a standard unconstrained algorithm should not be applied without safeguards (as indicated in Step AL2 of the model algorithm). Although increasing the penalty parameter often overcomes the difficulty, no strategy can ensure convergence to x^*.

Choice of the penalty parameter. The choice of a suitable penalty parameter can be complicated even if the augmented Lagrangian function is bounded below. Obviously, ρ must be large enough so that x^* is a local minimum of L_A. However, there are difficulties with either a too-large or too-small value of ρ. The phenomenon of an ill-conditioned subproblem occurs if ρ becomes too large, as in the penalty case. However, if the current ρ is too small, there is not only the danger noted above of unboundedness of the augmented function, but also the possibility of ill-conditioning in the subproblem. Since increases in ρ are designed to make an indefinite matrix become positive definite, each negative eigenvalue of the Hessian of the Lagrangian function at the solution will correspond to a *zero eigenvalue* of the Hessian of L_A for some value of ρ. Hence, the Hessian of L_A will be ill-conditioned for certain ranges of ρ.

Figure 6i shows the contours of $L_A(x, \lambda^*, \rho)$ corresponding to Example 6.2 for three values of ρ, where the single constraint has been assumed to be active. The second function ($\rho = 0.2$) is quite well conditioned. The first ($\rho = 0.075$) and third ($\rho = 100$) functions show the effects of too-small and too-large values of ρ.

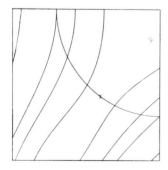 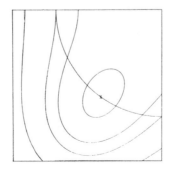 

Figure 6i.   The three diagrams represent the contours of the augmented Lagrangian function $L_A(x, \lambda^*, \rho)$ corresponding to Example 6.2 (with the constraint treated as an equality) for three values of ρ ($\rho = 0.075$, $\rho = 0.2$ and $\rho = 100$).

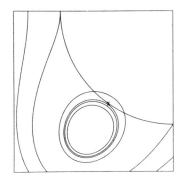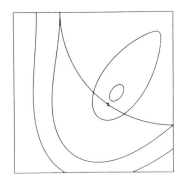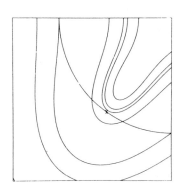

Figure 6j. The three diagrams depict the contours of the augmented Lagrangian function $L_A(x, \lambda, \rho)$ corresponding to Example 6.2 (with the constraint treated as an equality) for three values of the Lagrange multiplier estimate ($\lambda = 0.5$, $\lambda = 0.9$ and $\lambda = 1.0$). The value of ρ is fixed at 0.2.

The critical role of Lagrange multiplier estimates. It is essential to recall that x^* is not necessarily a stationary point of (6.24) except when λ is equal to λ^*, and thus augmented Lagrangian methods will converge to x^* only if the associated multiplier estimates converge to λ^*.

Figure 6j shows the contours of the augmented Lagrangian function (6.24) corresponding to Example 6.2 (with the constraint treated as an equality) for three values of the Lagrange multiplier estimate ($\lambda = 0.5$, $\lambda = 0.9$ and $\lambda = 1.0$). The penalty parameter has been set to 0.2, the value that gives the best-conditioned subproblem when $\lambda = \lambda^*$. This example illustrates the dramatic effect on the augmented Lagrangian function of even a small change in the estimate of the Lagrange multiplier. The third figure, which corresponds to $\lambda = 1.0$, illustrates the need for caution in applying an unsafeguarded method for unconstrained optimization, since L_A has no local minima (the subproblem has become defective). *Note that the multiplier estimate need not be particularly inaccurate to produce a defective subproblem.*

It can be shown that the rate of convergence to x^* of the iterates $\{x_k\}$ from the model algorithm of Section 6.4.2.1 will be no better than the rate of convergence of the multiplier estimates $\{\lambda_k\}$ to λ^*. This result is quite significant, since it implies that even a quadratically convergent technique applied to solve (6.25) will not converge quadratically to x^* unless a second-order multiplier estimate (see Section 6.6.2) is used.

In early versions of augmented Lagrangian methods, the following technique was suggested for updating the multiplier estimate. At the solution (say, $\bar{x}$) of (6.25), it holds that

$$g(\bar{x}) - \hat{A}(\bar{x})^T \lambda_k + \rho \hat{A}(\bar{x})^T \hat{c}(\bar{x}) = 0. \qquad (6.26)$$

Therefore, a new multiplier estimate may be defined as

$$\lambda_{k+1} = \lambda_k - \rho \hat{c}(\bar{x}), \qquad (6.27)$$

since from (6.26) we see that the vector on the right-hand side of (6.27) contains the coefficients in the expansion of $g(\bar{x})$ as a linear combination of the rows of $\hat{A}(\bar{x})$. (It will be shown in Section 6.6.1 that (6.27) can, in some circumstances, define a first-order multiplier estimate.)

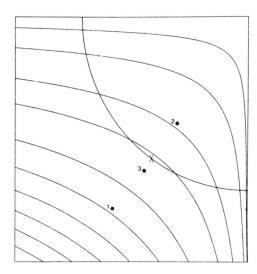

Figure 6k. The first three iterates of an augmented Lagrangian method when applied
to Example 6.2, with the constraint treated as an equality. Each iterate is marked
with a "•" and a number denoting the order in which it was computed. Each iterate
is the result of an unconstrained minimization.

Figure 6k shows the first three iterates that result when Algorithm AL is applied to Example
6.2 (with the constraint treated as an equality), for $\lambda_0 = 0$ and $\rho = 0.6$; the updated multiplier
estimate in Step AL3 is given by (6.27). The use of (6.27) yields a linearly convergent sequence
$\{\lambda_k\}$, and hence Algorithm AL will converge only linearly with this update. For Example 6.2,
the sequence $\{\|x_k - x^*\|_2\}$, $k = 3, \ldots, 8$, is 1.69×10^{-1}, 9.17×10^{-2}, 4.61×10^{-2}, 2.41×10^{-2},
1.23×10^{-2}, and 6.39×10^{-3}; the error is being reduced linearly, by a factor of approximately
two at each iteration. When Algorithm AL is applied to Example 6.2 with the same choices of
λ_0 and ρ given above, but with a second-order multiplier update (see Section 6.6.2), the sequence
$\{\|x_k - x^*\|_2\}$, $k = 1, \ldots, 6$, is 5.61×10^{-1}, 1.88×10^{-1}, 2.75×10^{-2}, 6.59×10^{-4}, 3.87×10^{-7},
and 3.19×10^{-13}; the second-order convergence is obvious.

*6.4.3. Variations in Strategy

The basic algorithm described in Section 6.4.2.1 represents only one application of the properties
of an augmented Lagrangian function. There are obviously many variations in strategy that
might be included in an algorithm that relies on the relationship (6.22).

     A criticism of the model algorithm of Section 6.4.2.1 is that the effort required to obtain
an accurate solution of the general unconstrained subproblem (6.25) may be unjustified if the
penalty parameter is wildly out of range, or the Lagrange multiplier estimate is significantly in
error. Furthermore, the model algorithm requires a complete unconstrained minimization with
respect to x before performing any update to the Lagrange multiplier estimate. Since limited
accuracy of the multiplier estimates restrains the ultimate rate of convergence of $\{x_k\}$ to x^*, it
might seem desirable to obtain improved multiplier estimates at more frequent intervals.

Based on these observations, augmented Lagrangian methods can be devised in which there is a range of flexibility in the accuracy with which (6.25) is solved. In the most extreme version of this strategy, x and λ are updated at similar intervals: a *single iteration* is performed of a method for solving (6.25); x is altered accordingly; a new estimate of λ is obtained; and the augmented subproblem is reformulated. With this strategy, the algorithm is, in effect, posing a *different subproblem at each iteration*; and since unconstrained methods are typically based on a quadratic model of the objective function, such an augmented Lagrangian method could be interpreted not in terms of the general function L_A, but rather in terms of a *quadratic approximation to L_A*. In this case, the method effectively becomes a QP-based projected Lagrangian method (to be discussed in Section 6.5.3).

Notes and Selected Bibliography for Section 6.4

It is not clear when the idea of augmenting a Lagrangian function first originated. Certainly, augmented Lagrangians date back to the mid-1940's (see Hestenes, 1946, 1947). A history of the development of these methods is given by Hestenes (1979). The use of augmented Lagrangian functions as a computational tool is of more recent origin. The approach discussed here generally follows that of Hestenes (1969). An alternative derivation of augmented Lagrangian methods was given independently by Powell (1969). Powell's description involved a quadratic penalty term with a "shift" and "weight" corresponding to each constraint. In Powell's algorithm (for problem NEP), the function whose unconstrained minimum is sought is

$$F(x) + \frac{1}{2} \sum_{i=1}^{t} \sigma_i \big(\hat{c}_i(x) - \theta_i \big)^2,$$

where σ_i is a positive weight and θ_i is a shift for the i-th constraint. When all the weights are equal, this function differs by a constant from the augmented Lagrangian function (6.24). Both Hestenes and Powell suggested the multiplier update (6.27). Another early paper on augmented Lagrangian methods is that of Haarhoff and Buys (1970). Surveys of the properties of augmented Lagrangian methods are given in, for example, Bertsekas (1975b), Mangasarian (1975), Bertsekas (1976a, b) and Fletcher (1977). A more general treatment of the concept of augmentability is given in Hestenes (1980b). Augmented Lagrangian functions other than (6.24) can be derived; see, for example, Boggs and Tolle (1980).

Augmented Lagrangian methods can be applied to inequality constraints in various ways. If $\hat{c}(x)$ is defined as the set of constraints violated at x_k, the augmented Lagrangian function has discontinuous derivatives at the solution (if any constraints are active). Buys (1972) and Rockafellar (1973a, b, 1974) suggested the following generalization of the augmented Lagrangian function for an inequality-constrained problem:

$$L_A(x, \lambda, \rho) = F(x) + \sum_{i=1}^{m} \begin{cases} -\lambda_i c_i(x) + \dfrac{\rho}{2} c_i(x)^2, & \text{if } c_i(x) \le \dfrac{\lambda_i}{\rho}; \\[2mm] -\dfrac{\rho}{2} \lambda_i^2, & \text{if } c_i(x) > \dfrac{\lambda_i}{\rho}. \end{cases} \tag{6.28}$$

The vector λ in this definition is an "extended" multiplier vector, with a component corresponding to *every* constraint. The function (6.28) and its gradient are continuous at any point where a constraint changes from "active" (i.e., $c_i \le \lambda_i/\rho$) to "inactive" ($c_i > \lambda_i/\rho$). There are discontinuities in the second derivatives at such points, but the hope is that they will occur far from the solution, and will not impede the performance of the unconstrained algorithm used to minimize the augmented Lagrangian function. The analogue of the Hestenes/Powell update (6.27) for the augmented Lagrangian function (6.28) is

$$\bar{\lambda}_i = \lambda_i - \min(\rho c_i, \lambda_i),$$

where $\bar{\lambda}_i$ denotes the updated value of the i-th multiplier. Using this update, the new multiplier for any "active" inequality constraint will remain non-negative, and the updated multiplier for any "inactive" inequality constraint will be zero. Buys (1972) suggests other techniques for more frequent updates of the multipliers for inequality-constrained problems. Inequality constraints may also be treated by using a pre-assigned active set strategy to select the constraints that should be included in $\hat{c}$ (see Section 6.5.5).

Since the convergence of augmented Lagrangian methods depends critically on sufficiently accurate multiplier estimates, much attention has been devoted to developing methods in which the multiplier estimates have the same rate of convergence as the iterates. When a Newton-type method is applied to solve the unconstrained subproblem (6.25), a *second-order* multiplier estimate should be used in order to achieve the expected quadratic convergence (see Section 6.6.2). If a superlinearly convergent quasi-Newton method is used to solve the unconstrained subproblem (6.25), it is desirable to update the multipliers with a superlinearly convergent update. The rate of convergence of augmented Lagrangian methods with various multiplier updates is considered by, for example, Buys (1972), Bertsekas (1975b), Kort (1975), Bertsekas (1976b), Byrd (1976, 1978), Kort and Bertsekas (1976), Tapia (1977, 1978), Glad (1979), and Glad and Polak (1979).

Many researchers have been concerned with modifying augmented Lagrangian methods so that multiplier estimates are updated at more frequent intervals, rather than at the solution of the unconstrained subproblem (6.25). Methods of this type are discussed in the references given in the previous paragraph.

When only a *single iteration* of an unconstrained method for solving the subproblem (6.25) is performed, followed by a multiplier update, the procedure becomes similar to a QP-based projected Lagrangian method (see Section 6.5.3). In fact, Tapia (1978) has shown that under certain conditions, the iterates generated by a quasi-Newton method applied to minimize (6.25), combined with a suitable multiplier update, are identical to those obtained by a certain QP-based projected Lagrangian method. However, the conceptual difference between an unconstrained subproblem and a linearly constrained subproblem remains significant in distinguishing the two approaches.

Many variations of augmented Lagrangian methods have been developed. In particular, a combination of an augmented Lagrangian technique and a reduced-gradient-type method has been described in the papers by Miele, Cragg, Iyer and Levy (1971), and Miele, Cragg, and Levy (1971).

The selection and interpretation of the penalty parameter ρ have been the study of much interest and research. Of particular interest is the region of convergence of the iterates as a function of ρ (see Bertsekas, 1979).

A related approach to constrained problems has been to construct an *exact penalty function* based on the augmented Lagrangian function. It was observed in Section 6.4.2.2 that x^* is an unconstrained minimum of (6.24) only if $\lambda = \lambda^*$. This suggests the idea of having the multiplier estimates appear in the augmented Lagrangian function as a *continuous function of x*, i.e. λ becomes $\lambda(x)$. If $\lambda(x)$ converges to λ^* as x converges to x^*, then for sufficiently large ρ, x^* will be an unconstrained minimum of

$$F(x) - \lambda(x)^T \hat{c}(x) + \frac{\rho}{2} \hat{c}(x)^T \hat{c}(x). \tag{6.29}$$

Thus, like the absolute value penalty function, the function (6.29) is an exact penalty function (see Section 6.2.2.1). This idea was introduced by Fletcher (1970b), and is discussed in Fletcher and Lill (1970), Lill (1972) and Fletcher (1973). One difficulty with the approach is that the form of the function $\lambda(x)$ involves first derivatives of the problem functions (see Section 6.6.1), and hence the gradient of (6.29) is very complicated; however, a simplified form of $\lambda(x)$ can be developed by exploiting special properties of the solution. Further discussion of these methods can be found in the references given above.

6.5. PROJECTED LAGRANGIAN METHODS

6.5.1. Motivation for a Projected Lagrangian Method

6.5.1.1. Formulation of a linearly constrained subproblem. When the sufficient conditions given in Section 3.4 hold, x^* is a minimum of the Lagrangian function within the subspace of vectors orthogonal to the active constraint gradients. This property suggests that x^* can be defined as the solution of a *linearly constrained* subproblem, whose objective function (denoted by Φ_{LC}) is related to the Lagrangian function, and whose linear constraints are chosen so that minimization occurs only within the desired subspace. By restricting the space of minimization, there should be no need to augment the Lagrangian function simply to convert x^* from a stationary point to a minimum. The class of *projected Lagrangian methods* includes algorithms that contain a sequence of linearly constrained subproblems based on the Lagrangian function. Because Φ_{LC} is meant to be based on the Lagrangian function, its definition will include estimates of the Lagrange multipliers (exactly as with Φ_U for an augmented Lagrangian method).

Since a linearly constrained subproblem is itself a constrained optimization problem, it seems reasonable that the Lagrange multipliers of the subproblem should provide estimates of the multipliers of the original problem. Let λ_k denote the multiplier estimate used to define the subproblem at x_k. A desirable property of the subproblem is that if $x_k = x^*$ and $\lambda_k = \lambda^*$, x^* should solve the subproblem, and λ^* should be its Lagrange multiplier.

6.5.1.2. Definition of the subproblem. In order to simplify the initial presentation, we shall assume that the correct active constraints can somehow be determined, and that the vector $\hat{c}$ contains only the constraints active at the solution. Techniques for determining the active set will be considered in Section 6.5.5.

Assume that q is the step to x^* from the non-optimal point x_k, and hence that

$$\hat{c}(x^*) = \hat{c}(x_k + q) = 0. \tag{6.30}$$

In order to compute an approximation to x^* (say, $\bar{x}$), consider the Taylor-series expansion of $\hat{c}$ about x_k (written in a slightly non-standard form):

$$\hat{c}(x^*) = \hat{c}_k + \hat{A}_k(x^* - x_k) + O(\|x^* - x_k\|^2) = 0, \tag{6.31}$$

where $\hat{c}_k$ and $\hat{A}_k$ denote $\hat{c}(x_k)$ and $\hat{A}(x_k)$. Using the linear approximation to $\hat{c}$ from (6.31), we obtain the following set of linear constraints to be satisfied by $\bar{x}$:

$$\hat{A}_k(\bar{x} - x_k) = -\hat{c}_k. \tag{6.32}$$

The constraints (6.32) specify that $\bar{x}$ should lie in the hyperplane corresponding to a zero of the linearized approximation to the nonlinear constraints at x_k. Note that (6.32) is analogous to the definition of a Newton step to the solution of the nonlinear equations (6.30) (*cf.* (4.72) in Section 4.7.6).

Our initial discussion will concern subproblems of the form

$$\begin{array}{ll} \underset{x \in \Re^n}{\text{minimize}} & \Phi_{LC} \\ \text{subject to} & \hat{A}_k x = -\hat{c}_k + \hat{A}_k x_k. \end{array} \tag{6.33}$$

The subproblem (6.33) may be defective because of incompatible constraints. However, this can happen only when $\hat{A}_k$ is rank-deficient, or when there are more than n constraints in $\hat{c}$. We observe also that the solution of (6.33) depends on the current iterate, since the constraints include quantities evaluated at x_k.

Assuming that (6.33) is a valid subproblem, we shall denote its solution and Lagrange multiplier by x_k^* and λ_k^*, respectively. The first-order optimality conditions for (6.33) are then

$$\hat{A}_k x_k^* = -\hat{c}_k + \hat{A}_k x_k; \tag{6.34a}$$

and

$$\nabla \Phi_{LC}(x_k^*) = \hat{A}_k^T \lambda_k^*. \tag{6.34b}$$

In addition, a second-order necessary condition is that $Z_k^T \nabla^2 \Phi_{LC}(x_k^*) Z_k$ must be positive semi-definite, where Z_k denotes (as usual) a matrix whose columns span the subspace of vectors orthogonal to the rows of $\hat{A}_k$.

The subproblem (6.33) depends on the specification of Φ_{LC}. In the next two sections, we consider two classes of projected Lagrangian methods, corresponding to two forms for Φ_{LC}.

6.5.2. A General Linearly Constrained Subproblem

6.5.2.1. Formulation of the objective function. It might seem at first that Φ_{LC} could simply be taken as the current approximation to the Lagrangian function, i.e.

$$F(x) - \lambda_k^T \hat{c}(x). \tag{6.35}$$

This choice of Φ_{LC} would satisfy the condition that, if $x_k = x^*$ and $\lambda_k = \lambda^*$, x^* would be a solution of (6.33). However, since the gradient of (6.35) *vanishes* at x^* if $\lambda_k = \lambda^*$, the Lagrange multiplier λ_k^* would be zero (rather than λ^*), and hence (6.35) is not acceptable as Φ_{LC}.

A closely related definition of Φ_{LC} such that the subproblem (6.33) does satisfy the stated conditions is

$$F(x) - \lambda_k^T \hat{c}(x) + \lambda_k^T \hat{A}_k x. \tag{6.36}$$

From the optimality condition (6.34b), we observe that x_k^* remains a solution of (6.33) if the gradient of Φ_{LC} at x_k^* includes any term of the form $\hat{A}_k^T v$ for some vector v (so long as the projected Hessian matrix of Φ_{LC} at x^* is not indefinite). However, the Lagrange multipliers of the subproblem are clearly altered by such a modification. The gradient of the term $\lambda_k^T \hat{A}_k x$ in (6.36) is simply $\hat{A}_k^T \lambda_k$, and its Hessian is zero. Thus, when Φ_{LC} is taken as (6.36), $x_k = x^*$ and $\lambda_k = \lambda^*$, it follows that $x_k^* = x^*$ and $\lambda_k^* = \lambda^*$, as required.

6.5.2.2. A simplified model algorithm. In order to discuss a specific method, we state a simplified projected Lagrangian algorithm. However, we shall see that it is not suitable as a general-purpose method, and shall discuss certain variations in strategy that produce a similar algorithm with improved robustness.

Given an initial point x_0 and an initial Lagrange multiplier vector λ_0, set $k \leftarrow 0$ and repeat the following steps.

Algorithm SP (*Simplified projected Lagrangian algorithm*).

SP1. [Check termination criteria.] If x_k satisfies the optimality conditions, the algorithm terminates with x_k as the solution.

SP2. [Solve the linearly constrained subproblem.] With x_k as the starting point, execute a procedure to solve the subproblem

$$\underset{x \in \Re^n}{\text{minimize}} \quad F(x) - \lambda_k^T \hat{c}(x) + \lambda_k^T \hat{A}_k x$$
$$\text{subject to} \quad \hat{A}_k x = -\hat{c}_k + \hat{A}_k x_k, \tag{6.37}$$

subject to safeguards to cope with unboundedness.

SP3. [Update the multiplier estimate.] Set x_{k+1} to the best approximation to the solution of (6.37), and λ_{k+1} to the Lagrange multiplier vector associated with the subproblem (6.37). (Note that this step assumes that (6.37) is a valid subproblem.) Set $k \leftarrow k + 1$, and go back to Step SP1. ∎

Solving the subproblem. The subproblem (6.37) involves the minimization of a general nonlinear function subject to linear equality constraints, and can be solved using an appropriate technique, such as those described in Section 5.1. The choice of solution method will depend on the information available about the problem functions (i.e., the level and cost of derivative information), and on the problem size.

A method based on (6.37) might be considered less straightforward than an augmented Lagrangian method because a linearly constrained subproblem is more difficult to solve than an unconstrained subproblem. However, this argument is not applicable if the original problem includes any linear constraints that are to be treated by the methods of Chapter 5. Even if an augmented Lagrangian technique is used to handle the nonlinear constraints, a linearly constrained subproblem must be solved in any case. Thus, since most practical problems include a mixture of linear and nonlinear constraints, it can be argued that the complexity of the linearly constrained subproblem (6.37) associated with the nonlinear constraints is no greater than if a method were applied in which the nonlinear constraints were included in an unconstrained subproblem.

Local convergence. It can be shown that, when the sufficient conditions for optimality given in Section 3.4 are satisfied at x^*, and (x_0, λ_0) is sufficiently close to (x^*, λ^*), the sequence $\{(x_k, \lambda_k)\}$ generated by Algorithm SP converges *quadratically* to (x^*, λ^*). Furthermore, if $\|x_k - x^*\|$ and $\|\lambda_k - \lambda^*\|$ are of order ϵ for sufficiently small ϵ, the errors in x_{k+1} and λ_{k+1} are of order ϵ^2; thus, the use of a first-order multiplier estimate to define the objective function of (6.37) does not restrict the rate of convergence to be linear (in contrast to the augmented Lagrangian method discussed in Section 6.4.2.2).

Figure 6l shows the first three iterates when Algorithm SP is applied to Example 6.2, with x_0 taken as $(-1.5, -1.6)^T$, and $\lambda_0 = 0$. Note that each displayed iterate is the solution of the subproblem (6.37), and hence is the result of several "inner" iterations. For this problem, the sequence $\{\|x_k - x^*\|_2\}$, $k = 1, \ldots, 4$, is 2.71×10^{-1}, 3.20×10^{-2}, 3.57×10^{-4}, and 4.50×10^{-8}, thereby demonstrating the quadratic convergence in x.

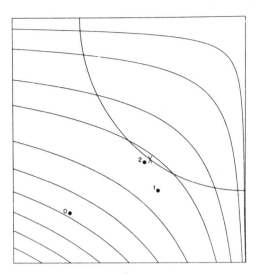

Figure 6l.   The first three iterates of a projected Lagrangian method when applied to Example 6.2. Each iterate is the solution of a linearly constrained subproblem.

*6.5.2.3. Improvements to the model algorithm. Despite the excellent local convergence properties of the model algorithm of Section 6.5.2.2, it suffers from a lack of robustness. In particular, the use of the subproblem (6.37) is questionable unless the values of x_0 and λ_0 are sufficiently close to the optimal values. The derivation of (6.37) indicates that its formulation is critically dependent on the optimality conditions (6.22) and (6.30), which hold only at x^* and λ^*. If x_k and λ_k are not accurate, the subproblem (6.37) may be defective, so that x_k^* does not exist. Even when (6.37) is valid, its solution may be farther from x^* than x_k, and hence the simplified method SP may fail to converge.

In order to develop an algorithm that will converge to x^* under more general conditions on x_0 and λ_0, various strategies have been suggested. The most popular approach is to attempt to find a suitably close starting point by some other means, and then to initiate the algorithm described above.

An obvious technique for generating a "close" point is to apply *a quadratic penalty function method* (see Section 6.2.1.1), with *relatively small* values of the penalty parameter. With this approach, the hope is that, for a reasonable value of ρ, $x^*(\rho)$ will be "close" to x^*; $x^*(\rho)$ may then be taken as the starting point for the subproblem (6.37), and good initial estimates of the Lagrange multipliers can be obtained — for example, using (6.4) or one of the procedures to be discussed in Section 6.6. A second approach is to solve a linearly constrained subproblem in which Φ_{LC} is an *augmented Lagrangian function* (see Section 6.4.1), and set the penalty parameter to zero when the iterates are believed to be "close" to x^*. (An algorithm for large-scale problems based on the latter approach will be described in Section 6.7.1.)

With either approach, it is important to make a "good" choice of the penalty parameter used in defining the penalty function or augmented Lagrangian function. If the penalty parameter must be very large in order to find a point that is "sufficiently close" to x^*, the usual problem of ill-conditioning may occur. In addition, the decision as to when to "switch" to Algorithm SP (i.e., to use (6.36) as Φ_{LC}) is non-trivial. Safeguards should also be included for recovering from an incorrect judgement that a sufficiently close point has been found.

6.5.3. A Quadratic Programming Subproblem

6.5.3.1. Motivation. It might be argued that a defect of the approach described in Section 6.5.2 is that the effort required to solve (6.37) is not justified by the resulting improvement in x_{k+1}. If x_k is a poor estimate of the optimal value, the constraints of (6.37) may not serve as a good prediction of the step to a zero of the constraint functions. If λ_k is not sensible, the objective function in (6.37) may bear no relation to the Lagrangian function at the solution.

An alternative approach is to use the properties of the optimum *to pose a simpler subproblem.* Toward this end, it is desirable for the linearly constrained subproblem to be deterministic rather than adaptive (see Section 6.1.2), so that in general less work is required to solve the subproblem. However, a possible drawback of posing a simplified subproblem may be that certain desirable properties hold at the solution of the more complex subproblem, but do not necessarily hold if only a simplified subproblem is solved. Therefore, it is necessary to exercise caution when undertaking such a modification.

We consider a class of methods in which Φ_{LC} is specialized to be a *quadratic function,* so that the subproblem of interest is a *quadratic program.* When the subproblem is a quadratic program, it is customary for the solution of the subproblem to be taken as the *step* from x_k to x_{k+1} (rather than as x_{k+1} itself). In order to simplify our overview of the motivation for the QP subproblem, we shall temporarily assume that *the correct active set has been determined* (a more detailed discussion of the treatment of inequality constraints will be given in Section 6.5.5). The subproblems to be described can be applied whenever an accurate prediction can be made of the active set. It will be apparent that these algorithms have many similarities to the methods for linear constraints described in Chapter 5.

Let t denote the number of active constraints, and let $\hat{c}(x)$ be defined as the vector of active constraint functions. At iteration k, a typical QP subproblem has the following form

$$\underset{p \in \Re^n}{\text{minimize}} \quad d_k^T p + \frac{1}{2} p^T H_k p \tag{6.38a}$$

$$\text{subject to} \quad C_k p = b_k. \tag{6.38b}$$

We shall refer to methods for nonlinear constraints that involve an explicit QP subproblem like (6.38) as *QP-based methods;* such methods are also called *sequential quadratic programming methods.*

Assuming that (6.38) is a valid subproblem, its solution will be denoted by p_k. Since (6.38) is a constrained optimization problem, a Lagrange multiplier vector (which we shall denote by η_k) exists and satisfies

$$H_k p_k + d_k = C_k^T \eta_k. \tag{6.39}$$

If x_k is close to x^*, a QP subproblem can be developed from exactly the same motivation given in Section 6.5.1.2. In particular, making a *linear approximation* to $\hat{c}$ from its Taylor-series expansion about x_k results in the set of constraints

$$\hat{A}_k p = -\hat{c}_k, \tag{6.40}$$

where $\hat{c}_k$ is the vector of active constraint values at x_k, and the rows of $\hat{A}_k$ contain their gradients.

Similarly, the quadratic function (6.38a) can be interpreted as a *quadratic approximation to the Lagrangian function*. The matrix H_k is almost always viewed as an approximation of the Hessian of the Lagrangian function. One might therefore assume that the vector d_k in (6.38a) should be taken as the gradient of the Lagrangian function, $g_k - \hat{A}_k^T \lambda_k$, where λ_k is the current estimate of λ^*. However, since the solution of (6.38) is unaltered if the vector d_k includes any linear combination of the rows of $\hat{A}_k$, d_k is usually taken simply as g_k, the gradient of F at x_k. This modification means that *the multipliers η_k (6.39) of the QP subproblem can be taken as estimates of the multipliers of the original problem*, based on reasoning similar to that in Section 6.5.2.1.

6.5.3.2. A simplified model algorithm. In order to have a specific algorithm to discuss, we describe a simplified model algorithm based on a quadratic programming subproblem. However, as with the simplified model algorithm given in Section 6.5.2.2, we shall see that this algorithm must be modified in order to be successful in solving general problems.

Given an initial point x_0 and an initial Lagrange multiplier vector λ_0, set $k \leftarrow 0$ and repeat the following steps.

Algorithm SQ (*Simplified QP-based projected Lagrangian algorithm*).

SQ1. [Check termination criteria.] If x_k satisfies the optimality conditions, the algorithm terminates with x_k as the solution.

SQ2. [Solve the quadratic programming subproblem.] Let p_k denote the solution of the quadratic program

$$
\begin{aligned}
\underset{p \in \Re^n}{\text{minimize}} \quad & g_k^T p + \frac{1}{2} p^T H_k p \\
\text{subject to} \quad & \hat{A}_k p = -\hat{c}_k.
\end{aligned}
\tag{6.41}
$$

(The definition of H_k is assumed to depend on x_k and on λ_k.)

SQ3. [Update the estimate of the solution] Set $x_{k+1} \leftarrow x_k + p_k$, compute λ_{k+1}, set $k \leftarrow k+1$, and go back to Step SQ1. ∎

Solving the subproblem. The solution p_k of (6.41) can be computed directly, and may be conveniently expressed in terms of a matrix Y_k whose columns span the range space of $\hat{A}_k^T$, and a matrix Z_k whose columns span the set of vectors orthogonal to the rows of $\hat{A}_k$. (Techniques for computing such matrices are discussed in Section 5.1.3.) Let p_k be written as

$$
p_k = Y_k p_Y + Z_k p_Z.
\tag{6.42}
$$

The vector p_Y is determined by the constraints of (6.41), since we have

$$
\hat{A}_k p_k = \hat{A}_k Y_k p_Y = -\hat{c}_k.
\tag{6.43}
$$

If (6.41) has a solution, the equations (6.43) must be compatible. Let r be the rank of $\hat{A}_k$; then the matrix $\hat{A}_k Y_k$ must contain an $r \times r$ non-singular submatrix, and hence p_Y is uniquely determined by any r independent equations of (6.43). The vector p_Z is determined by minimization of the quadratic objective function of (6.41), and solves the linear system

$$
Z_k^T H_k Z_k p_Z = -Z_k^T (g_k + H_k Y_k p_Y).
\tag{6.44}
$$

The Lagrange multiplier η_k of (6.41) satisfies the (necessarily compatible) equations

$$H_k p_k + g_k = \hat{A}_k^T \eta_k. \tag{6.45}$$

The vector p_k defined by (6.42), (6.43), and (6.44) is a valid solution of (6.41) only if $Z_k^T H_k Z_k$ is positive definite. If $Z_k^T H_k Z_k$ is indefinite, p_k is *not* a minimum of (6.41) (in fact, the solution is unbounded). An advantage of computing p_k from (6.43) and (6.44) is that the determination of whether the solution is well defined can be made during the calculations.

There are many mathematically equivalent expressions for the solution of (6.41). In particular, the optimality conditions for (6.41) can be expressed as a system of *linear equations* in p_k and η_k, namely

$$\begin{pmatrix} H_k & -\hat{A}_k^T \\ \hat{A}_k & 0 \end{pmatrix} \begin{pmatrix} p_k \\ \eta_k \end{pmatrix} = \begin{pmatrix} -g_k \\ -\hat{c}_k \end{pmatrix}. \tag{6.46}$$

Note that equations (6.46) are similar to the equations (5.54) associated with a range-space method for linearly constrained optimization (see Section 5.4).

Local convergence properties. A local convergence result can be proved for algorithm SQ if H_k is taken as W_k, the current approximation to the Hessian of the Lagrangian function based on exact second derivatives, i.e.

$$H_k = W_k \equiv G(x_k) - \sum_{i=1}^{t} (\lambda_k)_i \hat{G}_i(x_k). \tag{6.47}$$

In particular, the sequence $\{(x_k, \lambda_k)\}$ generated by Algorithm SQ with H_k taken as W_k converges *quadratically* to (x^*, λ^*) if (i) (x_0, λ_0) is sufficiently close to (x^*, λ^*); (ii) the sufficient conditions for optimality given in Section 3.4 hold at x^*; and (iii) λ_{k+1} is taken as η_k (6.45). Exactly as for the model algorithm SP of Section 6.5.2.2, the use of a first-order multiplier estimate to define W_k does not restrict the convergence rate to be linear; in particular, if $\|x_k - x^*\|$ and $\|\lambda_k - \lambda^*\|$ are of order ϵ for sufficiently small ϵ, the errors in x_{k+1} and λ_{k+1} are of order ϵ^2.

Figure 6m displays the first three iterates when Algorithm SQ is applied to Example 6.2, starting at $x_0 = (-1.5, -1.6)^T$, with $\lambda_0 = 0$ and H_k defined by (6.47). Note that the sequence is very similar to that in Figure 6l, but that each iterate in Figure 6m requires only the solution of the linear systems (6.43) and (6.44); however, (6.44) involves the Hessian matrices of the problem functions. The sequence $\{\|x_k - x^*\|_2\}$, $k = 1, \ldots, 4$, is 2.03×10^{-1}, 1.41×10^{-2}, 8.18×10^{-5}, and 4.95×10^{-10} (the quadratic convergence is apparent).

The quadratic convergence under these conditions can be explained by interpreting the subproblem (6.41) as the definition of a *Newton step* in x and λ with respect to the system of nonlinear equations defined by the optimality conditions at x^*, namely

$$g(x^*) - \hat{A}(x^*)^T \lambda^* = 0 \tag{6.48a}$$

and

$$\hat{c}(x^*) = 0. \tag{6.48b}$$

If we differentiate the functions in (6.48) at (x_k, λ_k) with respect to the unknowns (x^*, λ^*), the Newton equations (*cf.* (4.72)) are precisely the linear system (6.46) with H_k equal to W_k.

Unfortunately, exactly as with the simplified Algorithm SP of Section 6.5.2.2, Algorithm SQ will, in general, fail to converge except under the restrictive conditions mentioned above. This

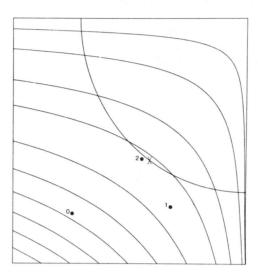

Figure 6m. The first three iterates of a QP-based projected Lagrangian method when applied to Example 6.2. Each iterate is the solution of an equality-constrained quadratic program involving the Hessian of the Lagrangian function.

is not surprising in view of its interpretation as a Newton method with respect to the nonlinear equations (6.48), since it was observed in Sections 4.1 and 4.4 that an unsafeguarded Newton method is not a robust algorithm, even in the univariate case. Newton's method is "ideal" in terms of its local convergence properties, but not in terms of guaranteed convergence.

In the next two sections, we shall consider modifications to Algorithm SQ that improve its robustness.

6.5.3.3. Use of a merit function. We have emphasized from the beginning of our discussion of QP-based methods that their derivation is based on conditions that hold only in a small neighbourhood of the optimum, and hence that the significance of the subproblem (6.41) is questionable when x_k is not close to x^* (or λ_k is not close to λ^*). In order to ensure that x_{k+1} is, in some well-defined sense, a "better" point than x_k, the solution of the QP subproblem can be used as a *search direction*. The next iterate is then defined as

$$x_{k+1} = x_k + \alpha_k p_k, \tag{6.49}$$

where p_k is the result of the QP subproblem, and α_k is a step length chosen to yield a "sufficient decrease" (see Section 4.3.2.1) in some suitably chosen *merit function* that measures progress toward x^*. We shall denote the merit function by Φ_M.

Several different choices of Φ_M have been suggested, including a quadratic penalty function (Section 6.2.1.1), an exact penalty function (Section 6.2.2), and an augmented Lagrangian function (Section 6.4.1). The merit function must satisfy several criteria. Firstly, Φ_M should provide a sensible measure of the progress toward the solution. Secondly, we shall see in the next section that different formulations of the QP subproblem may be desirable in order to improve the robustness of a QP-based method. Whatever the chosen formulation of the subproblem, it must be ensured that a decrease in the merit function can be guaranteed at each iteration. Thirdly, it is important in terms of efficiency that the calculation of Φ_M should not be too expensive. Finally, it is desirable that the requirement of decreasing the merit function should not restrict the rate of convergence of the QP-based method. For example, if H_k is taken as W_k, quadratic convergence will be achieved only if the merit function allows the step lengths $\{\alpha_k\}$ to converge sufficiently fast to unity.

*6.5.3.4. Other formulations of the subproblem. We have observed that the formulation (6.41) is derived from conditions that hold at x^*, and hence that there is no reason to expect the subproblem to be meaningful when x_k is not close to x^*. Therefore, we shall briefly consider formulations of the subproblem that may differ from (6.41) when x_k is far from optimal.

A subproblem based on the quadratic penalty function. A QP subproblem can be formulated from the properties of the trajectory of points $x^*(\rho)$ generated by a quadratic penalty function method (see Section 6.2.1.1). Assume that x_k lies on the penalty trajectory. It can be shown that, under mild conditions, the step p to the point $x^*(\rho)$ further along the trajectory (nearer to x^*) is approximately the solution of the quadratic program

$$
\begin{aligned}
&\underset{p \in \Re^n}{\text{minimize}} \quad g_k^T p + \frac{1}{2} p^T H_k p \\
&\text{subject to} \quad \hat{A}_k p = -\hat{c}_k - \left(\frac{1}{\rho}\right)\lambda_k,
\end{aligned}
\tag{6.50}
$$

where H_k is an approximation to the Hessian of the Lagrangian function, and λ_k is the current Lagrange multiplier estimate. This suggests an algorithm in which p_k is taken as the solution of (6.50), and the quadratic penalty function is used as Φ_M. Note that as ρ becomes arbitrarily large, (6.50) and (6.41) will have the same solution. Furthermore, *no ill-conditioning in the subproblem is introduced by a large penalty parameter*, in contrast to the methods of Section 6.2.1.1. With a method based on the subproblem (6.50), the penalty parameter is chosen to reflect the estimated distance to the solution, and becomes arbitrarily large as the iterates approach the solution.

A subproblem based on the logarithmic barrier function. A QP-based method can be defined for inequality-constrained problems in which strict feasibility must be maintained. The QP subproblem in this case is based on the trajectory of points $x^*(r)$ generated by a logarithmic barrier function method (see Section 6.2.1.2). Assume that x_k lies on the barrier trajectory. It can be shown that, under mild conditions, the step p to the point $x^*(r)$ further along the trajectory (nearer to x^*) is approximately the solution of the QP subproblem

$$
\begin{aligned}
&\underset{p \in \Re^n}{\text{minimize}} \quad g_k^T p + \frac{1}{2} p^T H_k p \\
&\text{subject to} \quad \hat{A}_k p = -\hat{c}_k + \delta_k,
\end{aligned}
\tag{6.51}
$$

where H_k is an approximation of the Hessian of the Lagrangian function, and the i-th component of δ_k is given by $r/(\lambda_k)_i$, with λ_k the current multiplier estimate. Note that (6.51) becomes equivalent to (6.41) as r approaches zero. Thus, a feasible-point algorithm can be defined in which the search direction p_k is given by the solution of (6.51), and the logarithmic barrier function is used as Φ_M in determining α_k in (6.49). In this way, all iterates remain feasible if x_0 is feasible. The barrier parameter in the subproblem is chosen to reflect the distance from x_k to x^*. Note that *no ill-conditioning in the subproblem is introduced as r approaches zero*.

Interpretation of alternative subproblems. In order to retain the desirable local convergence rate of a QP-based method, the linear constraints of any QP subproblem should converge sufficiently rapidly to the form (6.40) as x_k approaches x^*. However, because of the unpredictability of nonlinear constraints, it is desirable for the specification of the subproblem to be flexible enough to adapt to different conditions that may hold as the iterations proceed. The alternative subproblems (6.50) and (6.51) reflect such flexibility through *shifted linear constraints*. We now indicate how the shifted constraints can be interpreted geometrically as the imposition of more general properties on the linearized constraints.

For either the quadratic penalty or logarithmic barrier function, the trajectory of approach to the solution is *non-tangential*. If x_k is a local minimum of the penalty or barrier function, and a step of unity is taken along the vector p_k resulting from (6.50) or (6.51), the shifted constraints have the effect that the next iterate *stops short* of the step to a zero of the linearized constraint functions; hence, the iterates typically do not exactly satisfy any subset of the constraints. This property tends to improve the accuracy of a linear approximation to the constraints.

The shifted constraints have a different interpretation at other points. Let $\hat{c}_i$ denote a constraint function that is believed to be active at the solution, based on some sensible prediction of the active set (see Section 6.5.5). If $\hat{c}_i$ is very small, but x_k is demonstrably far from optimal, it is generally *undesirable* to attempt to reduce the magnitude of $\hat{c}_i$, since this would lead to an inefficiency of the "boundary-following" type described in the Notes for Section 6.3. In contrast to (6.40), the corresponding linear constraint of (6.50) or (6.51) tends to move the next iterate away from the boundary of the feasible region.

*6.5.4. Strategies for a Defective Subproblem

With any of the techniques mentioned for formulating a linearly constrained subproblem, the possibility exists that the subproblem will be defective. In such an instance, it is essential to *formulate a different subproblem* whose solution is well-defined. We shall give only a brief overview of possible strategies; the reader should consult the references cited in the Notes for further details.

*6.5.4.1. Incompatible linear constraints.

When the constraints of the original problem are *linear*, the methods described in Chapter 5 effectively solve a sequence of linearly constrained subproblems; in this case, the constraints of the subproblem can be incompatible only if the original problem has no solution. With nonlinear constraints, however, *the constraints of the subproblem may be incompatible even when the original problem has a well posed solution*. In particular, the constraints

$$\hat{A}_k p = d_k$$

can be incompatible whenever the rows of $\hat{A}_k$ are linearly dependent (i.e., when $\hat{A}_k$ is rank-deficient or contains more than n rows). *We emphasize that this situation is not uncommon in practice.*

The likelihood of incompatibility can be reduced by including some flexibility in specifying the right-hand side of the constraints of the subproblem. If the original problem has a well-conditioned solution, this type of modification should not impair the ultimate rate of convergence; as observed in Section 6.5.3.4, the specific form of the constraints is crucial only when x_k is very close to x^*. In addition, the constraints to be included in $\hat{c}_k$ may be chosen subject to an acceptable measure of linear independence with respect to the gradients of the constraints already selected. Alternatively, the search direction may not be required to satisfy the linear constraints exactly.

For example, p_k may be taken as a solution of

$$\underset{p \in \Re^n}{\text{minimize}} \, \|\hat{A}_k p - \hat{d}_k\|_2^2.$$

In this case, p_k will satisfy a perturbed set of linear constraints of the form

$$\hat{A}_k p_k = \hat{d}_k + \hat{e}_k,$$

for some "minimal" $\|\hat{e}_k\|$.

Because p_k is undefined when the constraints of the subproblem are incompatible, it follows that there may be difficulties in defining p_k when the matrix $\hat{A}_k$ is ill-conditioned, or "nearly" rank-deficient. Even though the intersection of a set of nonlinear constraints may be well-conditioned, the current linear approximation to the constraints may be ill-conditioned. Ill-conditioning in $\hat{A}_k$ has two main effects. Firstly, the range-space portion p_Y of the search-direction becomes unreliable (see (6.43)). Secondly, the Lagrange multiplier estimates are likely to be poor, since $\hat{A}_k$ is used in their computation (see Section 6.6).

*6.5.4.2. Poor approximation of the Lagrangian function.
Another form of defective subproblem occurs when the subproblem has an unbounded solution. For a QP-based method, this will happen when the projected matrix $Z_k^T H_k Z_k$ is indefinite.

With the methods described in Section 6.5.2, there is no finite procedure for determining whether the subproblem (6.37) (or a related subproblem) has a bounded solution. Hence, in any method applied to solve the subproblem, it is essential to include safeguards to cope with unboundedness. In particular, the method should be required to terminate after a specified number of evaluations of the problem functions.

When the matrix $Z_k^T H_k Z_k$ is indefinite in a QP-based method, the situation is analogous to the appearance of an indefinite Hessian in Newton-type methods for unconstrained and linearly constrained optimization. In those cases, the subproblem can be altered in various ways — for example, a related positive-definite matrix may be used to define the search direction (see Section 4.4.2). However, a general difficulty with *any* technique for treating an indefinite Hessian is that there is no natural scaling of the search direction. For unconstrained and linearly constrained subproblems, the step to the next iterate is suitably scaled by selecting a step length that ensures a sufficient decrease in the objective function. However, with a projected Lagrangian method for nonlinear constraints, the search direction includes not only the null-space portion p_Z determined by the objective function of the subproblem, but also the range-space portion p_Y determined by the linearized constraints. Hence, a poorly scaled p_Z may dominate a perfectly reasonable p_Y (and vice versa). The difficulty of adjusting the relative scaling of the two portions of the search direction is inherent in any procedure that modifies the subproblem.

*6.5.5. Determination of the Active Set

We now turn to the topic of the determination of the active set when a projected Lagrangian method is applied to a problem with *inequality constraints*. We assume in this section that *all* the constraints are inequalities of the form $c(x) \geq 0$, and that $A(x)$ denotes the matrix whose i-th row is the gradient of $c_i(x)$. We shall consider only subproblems that are posed in terms of p_k (such as (6.41)) rather than x_k (such as (6.37)); the same observations apply in either case.

We shall discuss two contrasting strategies that have been adopted to handle inequality constraints; obviously, many intermediate strategies can be developed. Firstly, the linearly constrained subproblem can be posed with only *equality constraints*, which represent a subset of the constraints of the original problem. All the subproblems described thus far in Section 6.5 have been of this type. A second strategy is to pose the linearly constrained subproblem with only *inequality constraints*, which represent *all* the constraints of the original problem.

*6.5.5.1. An equality-constrained subproblem. In order to pose an equality-constrained subproblem, some prior determination must be made as to which constraints are to be included in the working set that defines the search direction (in effect, to be treated as equalities). The ideal choice for the working set would clearly be the correct set of active constraints; hence, the procedure for deciding on the selection of equality constraints in the subproblem will be called a *pre-assigned active set strategy*.

With the active set methods for *linear* constraints described in Chapter 5, a typical subproblem treats as equalities a set of constraints that are satisfied exactly at the current iterate. By contrast, with nonlinear constraints it is usual for the active constraints to be satisfied exactly only in the limit (except with reduced-gradient-type methods). Thus, the criteria used to select the active set are more complicated in the nonlinear-constraint case. A typical pre-assigned active set strategy examines the constraint functions at x_k with respect to properties that are known to hold for the active constraints in a neighbourhood of the the solution. The active set selection may also be based on the known behaviour of the merit function Φ_M; for example, when the quadratic penalty function (6.1) is used as Φ_M, the active constraints will tend to be *violated* (see Section 6.2.1.1). In some circumstances, the Lagrange multipliers of the previous subproblem may be used to predict which constraints are active. Another strategy is to select a trial pre-assigned working set; the Lagrange multipliers of the corresponding subproblem may then be examined, and a different working set may be constructed. This process could be interpreted as analogous to "deleting" a constraint in the linear-constraint case.

A benefit of posing the subproblem with only linear equality constraints is that, in general, the subproblem will be *easier to solve* than one with inequality constraints. However, it is essential that the algorithm should not fail if the prediction of the active set is wrong. In particular, the merit function (see Section 6.5.3.3) should always reflect *all* the constraints, even though the search direction is defined only by the current working set.

*6.5.5.2. An inequality-constrained subproblem. Any of the formulations mentioned thus far may be used to define a subproblem with inequality constraints of the form

$$A_k p \geq d_k, \tag{6.52}$$

where A_k denotes $A(x_k)$. In this case, the *active set of the subproblem can serve as a prediction of the active set of the original problem*.

It can be shown that in a neighbourhood of x^*, under certain conditions, the active set of a projected Lagrangian method in which the subproblem constraints are defined by (6.52) (with d_k taken as $-c_k$) will make a correct prediction of the active set of the original problem. This result is clearly essential in order to place any reliance on the active set of the subproblem as an estimate of the correct active set. However, this result is not sufficient to favour this strategy over one in which some other technique is used to predict the active set, since any sensible pre-assignment strategy should give a correct prediction when x_k is very close to x^*. It remains to be determined whether the active set prediction from an inequality-constrained subproblem is more accurate than that resulting from a reasonable pre-assigned active set strategy (see the comments of Section 6.6.3).

The general difficulties associated with a defective subproblem noted in Section 6.5.4 also exist with any formulation based on (6.52), but may be more complicated to resolve. In particular, an iterative procedure is necessary in order to determine whether a feasible point exists for (6.52); if the constraints (6.52) are incompatible, it is unclear how best to modify the subproblem.

Notes and Selected Bibliography for Section 6.5

The idea of linearizing nonlinear constraints occurs in many algorithms for nonlinearly constrained optimization, including the reduced-gradient-type methods described in Section 6.3. The method of approximation programming (MAP) of Griffith and Stewart (1961) involves a linear programming subproblem based on linear approximations to the objective and constraint functions; methods of this type are also called *sequential linear programming* (SLP) methods. One difficulty with this approach is that the solution to an LP subproblem will lie at a vertex of the (linearized) constraint set, and this is unlikely to be true of the solution to the original problem. Hence, some strategy must be devised to make the constraints of the subproblem reflect the nature of the true solution (e.g., by adding bounds on the variables).

Cutting plane methods (for convex programming problems) involve a sequence of subproblems that contain linearized versions of nonlinear constraints (see, for example, Kelley, 1960; Topkis and Veinott, 1967). However, these methods tend to be inefficient on general problems, and also suffer from numerical difficulties due to increasing ill-conditioning in the set of linearized constraints. Section 6.8.2.1 contains a brief discussion of convex programming problems.

Rosen and Kreuser (1972) and Robinson (1972) proposed methods based on solving a general linearly constrained subproblem of the form (6.33). Robinson (1972, 1974) showed that such a procedure is quadratically convergent under certain conditions. Rosen (1978) suggested a two-phase algorithm in which a quadratic penalty function method is used to find a point sufficiently close to x^* before switching to a method based on (6.37). Best *et al.* (1981) have suggested a version of this approach in which provision is made to return to the penalty phase if the proper rate of convergence is not achieved. Murtagh and Saunders (1980) have proposed a method in which the objective function of the subproblem is related to an augmented Lagrangian function; see Section 6.7 and the Notes for Section 6.7 for a more complete discussion of this method.

Van der Hoek (1979) and Best *et al.* (1981) have suggested methods in which the linearly constrained subproblems contain only equality constraints, which are selected from the original constraints using a pre-assigned active set strategy. It can be shown that this procedure retains the desired properties of quadratic convergence if the initial iterate is sufficiently close to x^*, and if the correct active set is used to define the subproblem.

To the best of our knowledge, the first suggestion of using a QP subproblem to solve a nonlinearly constrained problem was made for the special case of convex programming (see Section 6.8.2.1) by Wilson (1963), in his unpublished Ph.D. thesis. Wilson's method was subsequently described and interpreted by Beale (1967b).

In Wilson's method, an inequality-constrained QP is solved at each iteration, with linear constraints

$$A_k p \geq -c_k. \tag{6.53}$$

The quadratic function is an approximation to the Lagrangian function in which the exact Hessians of F and $\{c_i\}$ are used, and the QP multipliers from the *previous iteration* serve as Lagrange multiplier estimates. The new iterate is given by $x_k + p_k$, where p_k is the solution of the QP, so that no line search is performed.

Murray (1969a, b) proposed an inequality-constrained QP subproblem similar to (6.50), which was derived from the limiting behaviour of the solution trajectory of the quadratic penalty function. He suggested several possibilities for H_k, including a quasi-Newton approximation, and also several alternative multiplier estimates. The idea of "partially" solving the subproblem was also proposed — namely, to perform only a limited number of iterations (possibly even one) to solve the inequality-constrained QP; in this way, a linearized inequality constraint with a negative multiplier would be "deleted" from the working set. Murray suggested that the solution of the subproblem should be used as a *search direction*, and a *line search* be performed at each iteration with respect to the quadratic penalty function (which therefore serves as a "merit function").

Biggs (1972, 1974, 1975) presented a variation of Murray's method in which the QP subproblem was of the form (6.50), and contained equality constraints only. Biggs also proposed some special multiplier estimates.

Garcia Palomares and Mangasarian (1976) suggested a QP-based method derived from the application of quasi-Newton techniques to solve the system of nonlinear equations (6.48). Han (1976, 1977a, b) revived the idea of obtaining the search direction by solving an inequality-constrained QP with constraints (6.53), as in Wilson's method. He suggested quasi-Newton updates to an approximation to the Hessian of the Lagrangian function, but assumed that the full Hessian of the Lagrangian function was everywhere positive definite. In Han's method, the Lagrange multipliers of the QP sub-problem from the previous iteration are used as estimates of the multipliers of the original problem. Han's algorithm is shown to have superlinear convergence under certain conditions. Han also suggested the use of the non-differentiable "exact" penalty function P_A (6.8) as a merit function within a line search.

Powell (1977b, 1978) proposed an inequality-constrained QP procedure similar to Han's in which a positive-definite quasi-Newton approximation to the Hessian of the Lagrangian function is retained, even when the full matrix is indefinite. He also showed that this method will converge superlinearly under certain assumptions.

The best way to apply quasi-Newton techniques within a QP-based projected Lagrangian method is unclear. Applying quasi-Newton updates in the nonlinear-constraint case is more complicated than in the linear-constraint case (Section 5.1.2.4) for at least two reasons. Firstly, $\hat{A}_k$ changes completely at every iteration, *even if the working set remains constant*. Secondly, the function which we are attempting to approximate by a quadratic undergoes a *discrete change in its derivatives* when the working set is altered. Nonetheless, various strategies have been suggested for using quasi-Newton updates of the Hessian of the Lagrangian function (or of the projected Hessian of the Lagrangian function), and these methods appear to perform well in practice; see, for example, Powell (1977b), Murray and Wright (1978), and Schittkowski (1980).

Certain difficulties that may arise in QP-based methods when the absolute value penalty function is used as a merit function are described by Maratos (1978) and Chamberlain (1979). For a discussion of the merits of different formulations of the QP subproblem in projected Lagrangian methods, see Murray and Wright (1980). Chamberlain *et al.* (1980) discuss some issues in the use of P_A as a merit function.

The feasible-point QP-based method in which the subproblem is defined by (6.51) is described by Wright (1976) and Murray and Wright (1978). An implementation of the QP-based methods derived from (6.50) and (6.51) for problems with both linear and nonlinear constraints is described in Gill *et al.* (1980).

As observed in the Notes for Section 6.2, many methods for minimizing a non-differentiable penalty function are of the same form as QP-based methods, and the definition of the search direction is equivalent to (6.42). In particular, the methods of Coleman (1979) and Coleman and Conn (1980b, c) are of this type.

The idea of splitting the search direction into two orthogonal components as in (6.42) appears in many algorithms for nonlinearly constrained optimization — for example, in the methods of Bard and Greenstadt (1969), Luenberger (1974), Mayne and Polak (1976), and Heath (1978). It is also implicit in the reduced-gradient-type methods discussed in Section 6.3.

The approach of solving a nonlinearly constrained problem by applying Newton's method to a sequence of systems of nonlinear equations (*cf.* (6.48)) is discussed by Tapia (1974a, b).

For an overview of projected Lagrangian methods, see Fletcher (1974, 1977) and Murray (1976).

6.6. LAGRANGE MULTIPLIER ESTIMATES

The success and efficiency of the methods described in Sections 6.3, 6.4 and 6.5 depend on accurate estimates of the Lagrange multipliers. Recall that in the linear-constraint case, Lagrange multiplier estimates are typically used only for inequality-constrained problems, in order to decide whether to delete a constraint from the working set; the multiplier estimates do not enter the subproblem objective function or affect the rate of convergence. However, within Lagrangian-based methods for nonlinear constraints (such as those of Sections 6.3, 6.4, and 6.5), *Lagrange multiplier estimates are essential in the definition of the subproblem objective function, even when all the constraints are equalities.* Furthermore, the rate of convergence of a Lagrangian-based method depends critically on the use of sufficiently accurate multiplier estimates.

The Lagrange multiplier λ^* contains the coefficients in the expansion of $g(x^*)$ as a linear combination of the gradients of the active constraints, i.e.

$$g(x^*) = \hat{A}(x^*)^T \lambda^*, \tag{6.54}$$

where $\hat{A}$ is defined by the gradients of the constraints $\hat{c}$ that satisfy

$$\hat{c}(x^*) = 0. \tag{6.55}$$

Thus, for any nonlinearly constrained problem, x^* is the solution and λ^* is the Lagrange multiplier of a related *equality-constrained* problem that includes only the constraints active at x^*:

$$\begin{array}{cc} \underset{x \in \Re^n}{\text{minimize}} & F(x) \\ \text{subject to} & \hat{c}(x) = 0. \end{array} \tag{6.56}$$

(Obviously, (6.56) is equivalent to the original problem when only equality constraints are present.) Conditions (6.54) and (6.55) are satisfied only at a constrained stationary point of (6.56), and *there are no Lagrange multipliers* at any other points. At a typical iterate x_k, we shall therefore be concerned with the calculation of λ_k, an *estimate* of the Lagrange multipliers of (6.56).

With the active set methods for linear constraints described in Chapter 5, Lagrange multiplier estimates are calculated only at feasible points where a subset of the constraints are satisfied exactly. However, when solving a nonlinearly constrained problem, x_k will generally not be feasible; furthermore, no subset of the constraints will typically be satisfied exactly.

6.6.1. First-Order Multiplier Estimates

In Sections 6.6.1 and 6.6.2, we shall assume that $\hat{c}_k$ contains the constraints that are believed to be active at x_*^*, and that $\hat{A}_k$ contains their gradients; $\hat{c}_k$ thus plays the role of a *working set* in the definition of Lagrange multiplier estimates.

Condition (6.54) suggests that the vector λ_L (for "least-squares") that solves the least-squares problem

$$\underset{\lambda}{\text{minimize}} \; \|\hat{A}_k^T \lambda - g_k\|_2^2 \tag{6.57}$$

would provide an estimate λ_k of the multipliers of (6.56). (The formulation (6.57) is analogous to (5.27), except that $\hat{A}_k$ varies with x_k.) The vector λ_L is consistent in the sense of (5.26), and is always defined; however, it is unique only when $\hat{A}_k$ has full rank. (Stable techniques for solving (6.57) are described in Section 2.2.5.3.)

The residual of (6.57) is zero when the overdetermined equations

$$\hat{A}_k^T \lambda \approx g_k \tag{6.58}$$

are compatible. However, unless $\hat{c}(x_k) = 0$, λ_L is only an *estimate* of the multipliers of (6.56), even when the residual of (6.57) is zero. We emphasize this point because *(6.58) is compatible at any point where the gradient of F is a linear combination of the columns of $\hat{A}_k^T$.* Points with this property arise automatically when executing certain algorithms. In particular, (6.58) is compatible when x_k is a local minimum of the augmented Lagrangian function (6.24); this explains why the very simple form of multiplier estimate defined by (6.27) is equivalent to λ_L when evaluated at special values of x_k. (A first-order multiplier estimate that allows for the fact that $\hat{c}_k$ is non-zero is $\lambda_L - (\hat{A}_k \hat{A}_k^T)^{-1} \hat{c}_k$.)

A variable-reduction multiplier estimate λ_V may be defined as in the linear-constraint case (see Section 5.1.5), as the solution of

$$V^T \lambda_V = g_v, \tag{6.59}$$

where V is a $t \times t$ non-singular submatrix of $\hat{A}_k$, and g_v contains the corresponding components of g_k. When (6.58) is compatible, λ_V is equivalent to λ_L.

If $\|x_k - x^*\|$ is of order ϵ for sufficiently small ϵ, and $\hat{A}(x^*)$ has full rank, then $\|\lambda_L - \lambda^*\|$ and $\|\lambda_V - \lambda^*\|$ are of order ϵ. Hence, λ_L and λ_V are called *first-order estimates of λ^*.* Roughly speaking, the accuracy of λ_L depends on the condition number of $\hat{A}_k$, and that of λ_V on the condition number of V in (6.59). In addition, the error in either estimate depends on the nonlinearity of the problem functions. The reader should refer to Example 5.3 in Section 5.1.5 for an illustration of the behaviour of these estimates.

6.6.2. Second-Order Multiplier Estimates

As in the linear-constraint case (Section 5.1.5.2), second-order multiplier estimates can be computed under certain conditions. Let q denote the (unknown) step from x_k to x^*. The optimality condition (6.54) implies that

$$g(x_k + q) = \hat{A}(x_k + q)^T \lambda^*.$$

Expanding g and $\hat{A}$ about x_k along q, we obtain

$$g_k + W_k(\lambda^*)p = \hat{A}_k^T \lambda^* + O(\|q\|^2), \tag{6.60}$$

where $W_k(\lambda)$ is defined as the Hessian of the Lagrangian function with multiplier λ, i.e.

$$W_k(\lambda) = G(x_k) - \sum_{i=1}^{t} \lambda_i \hat{G}_i(x_k).$$

It is not possible to estimate λ^* directly from (6.60), since q is unknown and the components of λ^* are included in the definition of W_k. However, (6.60) can be used as the basis for a multiplier estimate if we have a vector p that approximates q, and a vector λ that approximates λ^*. Substituting these approximations in (6.60) yields

$$g_k + W_k(\lambda)p = \hat{A}_k^T \lambda^* + O(\|q\|^2 + \|\lambda - \lambda^*\| \|q\| + \|p - q\|). \tag{6.61}$$

The relationship (6.61) suggests that a multiplier estimate η_L can be defined as the least-squares solution of the overdetermined system

$$\underset{\eta}{\text{minimize}} \, \|\hat{A}_k^T \eta - (g_k + W_k(\lambda)p)\|_2^2. \tag{6.62}$$

If $\hat{A}_k$ has full rank, η_L will be a second-order estimate of λ^* when the following three conditions hold: $\|q\| = O(\epsilon)$ for sufficiently small ϵ; $\|\lambda - \lambda^*\| = O(\epsilon)$; and $\|p - q\| = O(\epsilon^2)$. Thus, η_L will be an improved estimate only if x_k is sufficiently close to x^*, λ is at least a first-order estimate of λ^*, and p is a second-order estimate of q. We have seen in the linear-constraint case that second-order multiplier estimates can be completely inaccurate (refer to the example of second-order estimates in Section 5.1.5.2); this same observation obviously applies to second-order estimates in the nonlinear-constraint case.

The residual of (6.62) is zero when p is the solution of a QP subproblem

$$\begin{aligned}
&\underset{p \in \Re^n}{\text{minimize}} && g_k^T p + \frac{1}{2} p^T W_k(\lambda) p \\
&\text{subject to} && \hat{A}_k p = \hat{d}_k
\end{aligned} \tag{6.63}$$

for any choice of $\hat{d}_k$; in this case, η_L is the *exact multiplier* of (6.63). When the QP (6.63) is derived from the Newton equations for (6.48), $\|p - q\|$ will be of order ϵ^2 under certain conditions, (as observed in Section 6.5.3.2). In this case, η_L is a *second-order estimate* of λ^*. The quadratic convergence of η_L can be observed when Algorithm SQ is applied to Example 6.2, starting with $x_0 = (-1.5, -1.6)^T$ (see Figure 6m); for this example, the sequence $|\eta_L - \lambda^*|$, $k = 0, \ldots, 3$, is 2.22×10^{-1}, 2.12×10^{-2}, 2.23×10^{-4}, and 1.12×10^{-8}.

Another second-order multiplier estimate arises with the projected Lagrangian methods discussed in Section 6.5.1, in which the next iterate is the solution of a general linearly constrained subproblem, for example (6.37). In this case, the Lagrange multiplier λ_k^* of the k-th subproblem provides an estimate of the multipliers of the original problem. As noted in Section 6.5.2.2, under certain conditions the sequence λ_k^* converges *quadratically* to λ^*, and hence can serve as a higher-order multiplier estimate when x_k is sufficiently close to x^*. The quadratic convergence of this sequence can be seen when Algorithm SP is applied to Example 6.2, starting with x_0 as $(-1.5, -1.6)^T$ (see Figure 6l); the sequence $|\lambda_k^* - \lambda^*|$, $k = 0, \ldots, 3$, is 1.46×10^{-1}, 1.02×10^{-3}, 1.85×10^{-4}, and 7.25×10^{-9}. The quadratic convergence is obvious.

*6.6.3. Multiplier Estimates for Inequality Constraints

When the original problem contains inequalities, the components of λ^* associated with active inequality constraints must be non-negative (see Section 3.4.2). If the multipliers associated with inequality constraints are strictly positive, the corresponding components of either λ_L or λ_V will have the correct sign if x_k is sufficiently close to x^*. However, at a general point, these estimates may be positive, negative or zero. In some methods, it is undesirable for a negative multiplier estimate to be associated with an inequality constraint. For these methods, the question arises concerning what procedure should be followed to ensure non-negativity of the multiplier estimates associated with inequalities.

One strategy for obtaining non-negative estimates is to compute a first-order estimate as described in Section 6.6.1, and then simply to set to zero the negative components associated with any inequality constraints. Another possibility is to define the multiplier estimate as the solution of a more complicated problem than (6.57), namely a *bound-constrained least-squares problem*

$$\begin{array}{ll} \underset{\lambda}{\text{minimize}} & \|\hat{A}_k^T\lambda - g_k\|_2^2 \\ \text{subject to} & \lambda_i \geq 0, \quad i \in \mathcal{L}, \end{array} \tag{6.64}$$

where $\mathcal{L}$ denotes the set of indices of the inequality constraints; the problem (6.64) may be solved using a method of the type described in Section 5.3.3. In effect, this amounts to "deleting" inequality constraints with negative multiplier estimates from the working set.

Second-order multiplier estimates (Section 6.6.2) are typically defined as the Lagrange multipliers of a linearly constrained subproblem (either (6.37) or the QP (6.63)) that contains linear equality constraints corresponding to the active set of the original problem. In order to ensure that a second-order multiplier estimate for an inequality constraint is non-negative, it might seem reasonable to pose these subproblems with a *linear inequality constraint* representing each nonlinear inequality constraint, since the corresponding multiplier of the subproblem will necessarily be non-negative. Unfortunately, although this procedure would ensure non-negativity of second-order multiplier estimates for inequalities, it does *not* follow that the estimates are reliable. As shown in Section 6.6.2, a second-order estimate will be more accurate than a first-order estimate only under certain conditions. One of these conditions is that *the correct active set must be known before posing the subproblem* (otherwise, the objective function of the subproblem will not be an accurate reflection of the true Lagrangian function). If the correct active set *is* assumed to be known, the second-order multiplier estimate becomes one of those discussed in Section 6.6.2, and will automatically have the correct sign; if there is any doubt about the correct active set, the multiplier estimates obtained from the inequality-constrained subproblem cannot be relied upon.

6.6.4. Consistency Checks

Lagrange multiplier estimates are typically used to define the subproblem to be solved during a particular iteration. Because of the importance of accurate Lagrange multiplier estimates, it is advisable to include *consistency checks* if possible, especially when using second-order multiplier estimates. Such checks serve to confirm not only the estimates of Lagrange multipliers, but also any prediction of the active set that is based on multiplier estimates.

In general, a second-order multiplier estimate is only a *prediction* of the first-order multiplier estimate at an improved point. Therefore, a first-order estimate at x_{k+1} (computed with either a pre-assigned active set or the active set predicted by an inequality-constrained subproblem) can be used to assess the reliability of a second-order estimate computed at x_k. Eventually, both estimates should converge to λ^*. Consistency checks of this type are particularly useful when verifying the active set prediction of an inequality-constrained subproblem.

In many cases, there is no reason to use a second-order multiplier estimate from the previous iteration, since a first-order estimate at the new iterate should provide a better estimate, and can be computed with very little (if any) additional effort. For example, in a QP-based method in which the LQ factorization of $\hat{A}_k$ is used to compute the search direction, it is trivial to calculate λ_L at the beginning of each iteration.

We emphasize that if the sequence $\{x_k\}$ is converging superlinearly, the associated sequence of first-order estimates is also converging superlinearly. Furthermore, as we have seen, the use of a first-order multiplier estimate does not restrict the rate of convergence of projected Lagrangian methods. For example, under certain conditions, the iterates $\{x_k\}$ generated by Algorithm SQ (Section 6.5.3.2) will converge quadratically to x^*; therefore, *the sequence of first-order estimates evaluated at x_k also converges quadratically to λ^*.* When Algorithm SQ is applied to Example 6.2 with starting point $x_0 = (-1.5, -1.6)^T$, we have already observed in Section 6.5.3 that $\{x_k\}$ converges quadratically to x^*. The sequence $\{(\lambda_L)_k\}$ also converges quadratically; the values $\{|(\lambda_L)_k - \lambda^*|\}$, $k = 1, \ldots, 4$, are 8.57×10^{-2}, 7.67×10^{-3}, 4.02×10^{-5}, and 1.36×10^{-9} (note the improvement at each step over the sequence $\{\eta_L\}$ for the same problem given in Section 6.6.2).

Any algorithm that relies on Lagrange multiplier estimates should ensure that the estimates have *some meaningful interpretation*, even if the current iterate is not close to optimal. For example, the multiplier λ_L defined by (6.57) represents the set of coefficients in the expansion of the current gradient as a linear combination of the gradients of the constraints in the working set. Even when x_k is not close to x^*, this information is useful in analyzing the relationship of the objective function and the constraints — i.e., the effect on the objective function of perturbations in the constraints.

Notes and Selected Bibliography for Section 6.6

There have been very few papers concerned primarily with the topic of Lagrange multiplier estimates. In most instances, a procedure for estimating multipliers is included as part of an algorithm for constrained optimization. All the references concerning methods given in the Notes for Sections 6.3, 6.4, and 6.5 include some definition of Lagrange multiplier estimates.

A thorough treatment of Lagrange multiplier estimates, including computational procedures and consistency checks, is given by Gill and Murray (1979b).

*6.7. LARGE-SCALE NONLINEARLY CONSTRAINED OPTIMIZATION

In this section, we shall be concerned with the following problem:

$$
\begin{aligned}
&\underset{x \in \Re^n}{\text{minimize}} && F(x) \\
&\text{subject to} && Ax = b \\
& && c(x) \begin{Bmatrix} = \\ \geq \end{Bmatrix} 0 \\
& && l \leq x \leq u.
\end{aligned}
\tag{6.65}
$$

The matrix A is assumed to have m_1 rows; the vector $c(x)$ contains a set of twice-continuously differentiable nonlinear constraint functions $\{c_i(x)\}$, $i = 1, \ldots, m_2$. We assume that the number of variables and constraints is "large", and that A is sparse. Obviously, the definition of "large" depends on the available storage and computation time. It will generally be assumed that the number of nonlinear constraints is small relative to the number of linear constraints.

No general linear *inequality* constraints have been included in the form (6.65) because the methods to be discussed are based on implementations of the simplex method. In solving large linear programs, inequality constraints are converted to equalities by adding slack variables; the purpose of this transformation is to allow the simplex method to be implemented with only column operations on the constraint matrix (see Section 5.6.1).

The remarks at the beginning of Section 5.6.2 concerning the different factors that must be considered when solving large-scale linearly constrained problems also apply to methods for solving (6.65).

*6.7.1. The Use of a Linearly Constrained Subproblem

Given the sophisticated techniques available for large-scale linearly constrained optimization (see Section 5.6.2), it is logical to attempt to apply them to large-scale *nonlinearly* constrained problems. This can be done directly by use of a projected Lagrangian method based on a general linearly constrained subproblem (see Section 6.5.2).

In order to give the flavour of such an approach, we shall briefly describe an algorithm based on the method given in Section 5.6.2. Let x_k and λ_k denote the current iterate and the current estimate of the Lagrange multipliers; other quantities subscripted by k will denote those quantities evaluated at x_k. The next iterate, x_{k+1} is the solution of the linearly constrained subproblem

$$
\begin{aligned}
\underset{x \in \Re^n}{\text{minimize}} \quad & F(x) - \lambda_k{}^T \hat{c}(x) + \frac{\rho}{2} \hat{c}(x)^T \hat{c}(x) \\
\text{subject to} \quad & Ax = b \\
& A_k(x - x_k) \begin{Bmatrix} = \\ \geq \end{Bmatrix} - c_k \\
& l \leq x \leq u.
\end{aligned}
\tag{6.66}
$$

The objective function of the subproblem (6.66) is of the form of an augmented Lagrangian function (6.24). All the nonlinear constraint gradients are included in A_k, but only the active constraints appear in the augmented Lagrangian function. The penalty term is included to encourage progress from a poor starting point. When x_k is judged to be sufficiently close to x^*, the penalty parameter ρ is set to zero in order to achieve quadratic convergence.

Certain aspects of this method illustrate the compromises and decisions associated with solving large-scale problems. A method based on (6.66) generates the next "outer" iterate x_{k+1} through a subproblem whose solution also requires an iterative procedure (which generates "inner" iterates). A standard general-purpose method for large-scale linearly constrained optimization can be applied directly to solve (6.66). However, it seems essential to impose a limit on the number of "inner" iterations to be executed in solving (6.66); even if the solution is bounded, it is unlikely that the initial Jacobian approximation and multiplier estimates will remain appropriate if hundreds of iterations are required to reach optimality for the subproblem. A "good" choice for the penalty parameter ρ is crucial in the success and efficiency of the method on certain problems. The considerations in selecting ρ are similar to those in an augmented Lagrangian method (see Section 6.4.2.2).

The value of λ_k in (6.66) is taken as the multiplier vector of the previous subproblem. If the previous subproblem was solved to optimality, this ensures that the multipliers corresponding to inequality constraints have the correct sign. However, it means that multiplier estimates are not computed with the most recent information, but rather are based on the "old" Jacobian. In addition, the interpretation of the available Lagrange multiplier estimates is further complicated if an inner iteration is terminated before convergence.

*6.7.2. The Use of a QP Subproblem

We assume that QP-based methods will treat linear constraints by the techniques described in Chapter 5; this ensures that the nonlinear functions are evaluated only at points that satisfy the linear constraints. We shall consider only methods based on a QP subproblem of the form (6.41). In this case, the matrix $\hat{A}_k$ will include the general linear constraints and active bounds as well as the current gradients of the nonlinear constraints. Therefore it is essential to exploit the fact that only *part* of $\hat{A}_k$ changes from one iteration to the next, since the rows of $\hat{A}_k$ that correspond to linear constraints and simple bounds remain constant.

In the QP-based methods discussed in Section 6.5.3, the search direction was represented (and computed) in terms of matrices Y and Z obtained from the LQ factorization of $\hat{A}$ (see (6.42)). A similar representation of the solution of (6.41) must be developed that is suitable for large-scale problems. We shall outline one possible approach, which is similar to that described in Section 5.6.2 for the large-scale linear-constraint case (see (5.73) and (5.74)).

In this description, the subscript k associated with the current iteration will be omitted. The variables are partitioned into basic, superbasic, and nonbasic sets, with a corresponding partition of the columns of $\hat{A}$ and the components of p and g. Since the nonbasic variables are fixed on their bounds during a given iteration, the vector p_N must be zero (and can be ignored). To satisfy the linear constraints of (6.41), there must be an implicit ordering of the variables such that

$$(B \quad S)\begin{pmatrix} p_B \\ p_S \end{pmatrix} = d, \tag{6.67}$$

where B is a $t \times t$ square non-singular matrix, and d contains appropriate elements of $\hat{c}$. From (6.67) it follows that

$$Bp_B = d - Sp_S. \tag{6.68}$$

Note that the components of d will be zero in positions corresponding to linear constraints. Hence, for any p_S, the definition of p_B by (6.68) ensures that p will be feasible with respect to the linear constraints of both the subproblem and the original problem.

The vector p_S is determined by minimization of the quadratic objective function of the subproblem (6.41). Writing this objective in terms of the partitioned vector p, we obtain

$$\frac{1}{2}p_B^T H_B p_B + p_B^T H_{BS} p_S + \frac{1}{2}p_S^T H_S p_S + g_B^T p_B + g_S^T p_S, \tag{6.69}$$

where

$$H = \begin{pmatrix} H_B & H_{BS} \\ H_{BS}^T & H_S \end{pmatrix}.$$

Substituting for p_B using (6.68) makes (6.69) a quadratic function in p_S alone. The optimal p_S is the solution of a system of equations exactly analogous to (6.44) for the dense case:

$$Z^T H Z p_S = -Z^T g + Z^T H \begin{pmatrix} B^{-1}d \\ 0 \end{pmatrix}, \tag{6.70}$$

where Z is given by (5.75).

*6.7.2.1. Representing the basis inverse. At a typical iteration, B is given by

$$B = \begin{pmatrix} B_1 & B_2 \\ B_3 & B_4 \end{pmatrix} \begin{matrix} \}t_1 \\ \}t_2 \end{matrix}. \tag{6.71}$$

If we assume that the linear constraints are placed first, the first t_1 rows (the matrices B_1 and B_2) correspond to the linear constraints, and the last t_2 rows (the matrices B_3 and B_4) correspond to the nonlinear constraints. Both B_1 and B_4 are square.

In this section, we consider methods for representing B^{-1} as the iterations of a QP-based method proceed. The inverse is *never* represented explicitly. However, we use this terminology because the methods to be described solve the linear systems that involve B without a complete factorization of B.

Changes in the columns of B that result as variables move on and off bounds can be carried out exactly as in the linear-constraint case (see Section 5.6.2.2). The difficulty in a nonlinearly constrained problem is that the last t_2 *rows* of B will change at each iteration due to constraint nonlinearities. We assume that it is not computationally feasible to refactorize B at every iteration; however, periodic refactorization will be performed to condense storage and ensure accuracy in the factors.

If both t_2 and the number of non-zero elements in the last t_2 rows of B are small, the changes in B due to constraint nonlinearities represent only a small number of column changes. In this case, it would be practical to update the LU factors of B in a standard fashion. However, each iteration would involve several column updates, and hence the refactorizations necessary to condense the storage would be required at more frequent intervals.

Partitioning. Since B_1 includes only linear constraints, it is possible to recur a factorization of B_1 from iteration to iteration. This fact can be utilized to advantage because systems of equations involving B or B^T can be solved using factorizations of B_1 and a matrix the size of B_4.

For example, if the vector b is partitioned corresponding to (6.71) as $(\, b_1 \quad b_2 \,)$, the solution of $Bx = b$ can be represented as

$$x = \begin{pmatrix} u_1 + u_2 \\ v_1 \end{pmatrix},$$

where the vectors u_1, u_2 and v_1 are calculated from

$$\begin{aligned} B_1 u_1 &= b_1, \\ D v_1 &= b_2 - B_3 u_1, \\ B_1 u_2 &= -B_2 v_1, \end{aligned} \tag{6.72}$$

and

$$D = B_4 - B_3 B_1^{-1} B_2.$$

This procedure is sometimes described as a *partitioned inverse technique*. The steps of (6.72) are equivalent to block Gaussian elimination on B, with B_1 as the first block. Several strategies can be developed to exploit the structure of B in order to reduce the amount of work needed to perform the calculations of (6.72).

An approximate inverse; iterative improvement. One strategy for overcoming the difficulties of updating B^{-1} as its last rows change is simply not to update it. With the approach described in Section 6.7.1, the linear constraints remain constant until the general subproblem (6.66) has been solved.

This idea and its extensions can be applied to a QP-based method in several ways. Let $\bar{B}^{-1}$ denote an available representation of an *approximation* to the inverse of B (e.g., from the most recent factorization or some previous iteration). We shall mention two possible strategies for using $\bar{B}^{-1}$ to "solve" systems of equations such as $Bx = b$. Firstly, we can simply solve the system using $\bar{B}^{-1}$; in effect, this involves substituting $\bar{B}$ for B during some number of consecutive iterations. Secondly, $\bar{B}^{-1}$ could be used further in an *iterative improvement* procedure, assuming that B is also available.

Such approximations are acceptable in QP-based methods because the linear constraints of the QP subproblem are typically derived from optimality conditions, and the precise specification of the linear constraints is critical only near the solution (see Section 6.5.3.4). Consequently, there is substantial freedom to define the constraints (6.67) when x_k is not close to x^*, provided that a sufficient decrease in the merit function can be guaranteed.

When $\bar{B}$ is the basis matrix from a previous iteration, the error in the approximate inverse is of a special form because the first t_1 rows of B are constant. Because of the relationship between B and $\bar{B}$, the structure of the error in the approximate inverse is such that the equations (6.67) corresponding to *linear* constraints are always satisfied "exactly", even if $\bar{B}$ is used rather than B. In general, p_B should satisfy

$$Bp_B = d - Sp_S.$$

If $\bar{p}_B$ is defined instead from

$$\bar{B}\bar{p}_B = d - Sp_S,$$

and $\bar{p}_B$ is used instead of p_B, the equalities of the QP subproblem corresponding to the *linear* constraints remain satisfied (with exact arithmetic).

*6.7.2.2. The search direction for the superbasic variables. Given that we can obtain a representation of B^{-1} (and hence of Z), a second issue in implementing a QP-based method for large-scale problems is how to solve the equations (6.70) for p_S. The difficulties are similar to those described in Section 5.6.2.1 for the large-scale linear-constraint case — namely, the storage and computation associated with forming $Z^T H Z$ (or $Z^T H$) in order to solve (6.70) may be prohibitive. Since H is $n \times n$, there will in general be inadequate storage to retain a full version of H.

In many cases, the dimension of the projected Hessian matrix $Z^T H Z$ will be relatively small at every iteration, even when the problem dimension is large. If $Z^T H Z$ is small enough to be stored, standard approaches from the dense case may be used. For example, a quasi-Newton approximation of the projected Hessian of the Lagrangian function may be maintained using update procedures similar to those in the linear-constraint case. Any questions concerning such procedures apply generally to nonlinearly constrained optimization, and are not particular to large-scale problems (see the Notes for Section 6.5). However, the technique of computing finite-differences along the columns of Z, which is very successful for small problems, is too expensive in the large-scale case because of the effort required to form Z. Furthermore, even if W itself is available, it is probably too costly to form $Z^T W Z$.

When limitations of storage and/or computation preclude an explicit representation of $Z^T H Z$, one of the conjugate-gradient-type methods discussed in Section 5.1.2.5 can be used if the product of $Z^T H Z$ and a vector v can in some circumstances be computed efficiently even when $Z^T H Z$ is not available; the use of preconditioning (see Section 4.8.6) is essential if the solution is to be computed in a reasonable time.

A sparse matrix H can be obtained in several different ways. It may happen that the Hessian of the Lagrangian function (W) is sparse, with a known sparsity pattern. (This situation is less likely than in the unconstrained case, because the Hessians of all the active constraints as well as the objective function must be sparse.) If the Hessian of the Lagrangian function is not sparse, it is possible to estimate the vector WZv by a finite-difference along the vector Zv. Obviously, this computation requires additional evaluations of the problem functions.

Notes and Selected Bibliography for Section 6.7

The algorithm described in Section 6.7.1 is due to Murtagh and Saunders (1980), and has been implemented in the Fortran program MINOS/AUGMENTED, which can be obtained from the Systems Optimization Laboratory, Stanford University. Rosen (1978) also suggested the use of the program MINOS (see Section 5.6.2) to solve a linearly constrained subproblem of the form (6.37), following execution of a "phase 1" procedure to ensure a sufficiently close starting point.

The sequential linear programming (SLP) approach developed by Griffith and Stewart (1961) has been used extensively for large-scale problems. A great advantage of an LP-based approach is that an existing large-scale LP system can be applied directly. A difficulty is that the bounds on the variables must usually be adjusted by heuristic techniques, since the solution of an LP subproblem will lie at a vertex of the (linearized) constraint set. See Baker and Ventker (1980), Batchelor and Beale (1976), and Beale (1974, 1978) for a discussion of these methods.

Each iteration of a generalized reduced-gradient method (see Section 6.3.2) involves a subproblem that is closely related to a linearly constrained subproblem. Jain, Lasdon, and Saunders (1976) suggested extending the techniques used in the MINOS program to solve large-scale problems.

The application of QP-based methods to large-scale nonlinearly constrained problems has been considered by Escudero (1980), who suggested a method based on an inequality-constrained QP subproblem (which is solved using an existing system for large-scale quadratic programming). The ideas of Section 6.7.2 are discussed in more detail in Gill *et al.* (1981b). A QP subproblem (which can be adjusted at every iteration) is more flexible than a linearly constrained subproblem such as (6.66) in dealing with the unpredictable nature of nonlinear constraints. However, the price paid for this flexibility is a substantial increase in programming complexity, since methods based on general linearly constrained subproblems like (6.66) can use existing methods for large-scale linearly constrained optimization without significant modification.

The partitioned-inverse procedure (6.72) involves the *Schur complement*, which is discussed in detail by Cottle (1974). See the Notes for Section 5.6 for further references concerning techniques for solving and updating sparse linear systems. For a complete treatment of iterative improvement, see Wilkinson (1965).

6.8. SPECIAL PROBLEM CATEGORIES

Special methods have been developed for many categories of optimization problems. Some of these have been discussed in this book — for example, unconstrained nonlinear least-squares (Section 4.7), and linear and quadratic programming (Sections 5.3.1, 5.3.2 and 5.6.1). In other instances, we have briefly mentioned the special problem category — for example, the unconstrained minimization of certain non-differentiable functions (Section 4.2.3).

Obviously, it is not possible to treat (or even mention) the myriad of special problem categories in optimization, since each could form the subject of a complete volume. The purpose of this section is simply to mention some important special problems that have not been discussed, and to direct the reader to selected references for further information. Many of the special algorithms are modifications of the basic methods described in Chapters 4, 5 and 6 to exploit the structure or known properties of the problem.

6.8.1. Special Non-Differentiable Problems

Several important problems involve the unconstrained minimization of a non-differentiable function F that is a composite of smooth functions (the non-differentiability arises from the treatment of the subsidiary functions). We have discussed such problems in Section 4.2.3, and have indicated that the discontinuities have a very special structure. For example, the ℓ_∞ (minimax) problem is defined by

$$\underset{x \in \Re^n}{\text{minimize}} \ \underset{\{f_i\}}{\max} \ \{f_1(x), f_2(x), \ldots, f_m(x)\}, \tag{6.73}$$

where $\{f_i(x)\}$ are smooth functions. The discontinuities in the derivative of the objective function of (6.73) occur only at points where $f_i = f_j$, $i \neq j$. Similarly, the discontinuities in the derivative of the ℓ_1 objective function (4.4) occur only at points where $f_i = 0$.

An important special case of the minimax and ℓ_1 problems occurs when all the functions $\{f_i\}$ are *linear*. The transformed problems given in Section 4.3.2 for these problems are then very similar to *linear programming* problems (see Section 5.3.1), and can be solved by modifications of the simplex method (or other methods for LP). This approach can be extended to the nonlinear versions of the minimax and ℓ_1 problems, but it is less successful because the solution of the original problem does not necessarily lie at a vertex of the transformed constraint set.

Another approach to the minimax and ℓ_1 problems is based on the *a priori* characterizations noted above of points where the discontinuities will occur. The idea is to extend an efficient method for smooth problems to anticipate and predict the points of discontinuity. We have seen in Section 6.5 that a linear approximation of a smooth function can be used to predict a point where the function will become zero; this suggests defining a "working set" of the smooth functions that are likely to cause discontinuities at the next iterate. In this way, the computation of the search direction is similar to that in a QP-based projected Lagrangian method (Section 6.5.3). In addition, special step-length procedures and Lagrange multiplier estimates can be developed.

Although minimax and ℓ_1 algorithms involve the minimization of the *norm* of a vector-valued smooth function, the same solution techniques can be applied to other special non-differentiable problems. For example, the non-differentiable penalty function (6.8) is the sum of a smooth function and a special non-differentiable term (see the Notes for Section 6.2 and 6.5).

6.8.2. Special Constrained Problems

6.8.2.1. Convex programming. The properties of convexity and concavity can be defined in a very general way. We shall be concerned only with subclasses of convex (concave) functions, namely smooth functions whose Hessian matrices are everywhere positive (negative) semi-definite. (A linear function is both convex and concave.)

A *convex programming problem* (a *convex program*) is an optimization problem of the form NCP (see Section 1.1) in which $F(x)$ is convex, the equality constraints are linear, and the inequality constraints are concave. A fundamental property of a convex programming problem is that any local minimum x^* is a global minimum (see Section 3.1); furthermore, if $G(x^*)$ is positive definite, x^* is unique.

The fact that a problem is a convex program has many significant algorithmic implications. We shall mention three important special properties of convex programs that lead to the resolution of some of the difficulties that have been mentioned for general problems. Firstly, no strategies are necessary to treat an indefinite Hessian or an unbounded subproblem (see, for example, Sections 4.4.2 and 6.5.4.2). Similarly, it is not necessary to augment the Lagrangian function (see Section 6.4.1) in order to make x^* a local minimum.

Secondly, the linearization derived from the Taylor-series expansion of a concave function $c_i(x)$ (*cf.* (6.32)) has the following special property. If $\bar{x}$ is infeasible with respect to the constraint $c_i(x) \geq 0$, then the hyperplane $a_i(\bar{x})^T x \geq -c_i(\bar{x})$ *strictly separates* $\bar{x}$ from the feasible region. This property removes the difficulties of incompatible linearized constraints associated with the methods based on a linearly constrained subproblem.

Finally, there are numerous *duality results* concerning the optimal solution of a convex programming problem (see Sections 3.3.2.2 and 3.3.2.3 for a brief discussion of the dual problems for linear and quadratic programs). The existence of an equivalent dual problem means that algorithms can be developed based on solving the dual, or on combining the solution of the primal and dual problems.

Unfortunately, there is no general-purpose *practical* technique for determining whether a general function is convex. Thus, convex programming techniques should be applied only if convexity can be determined from some *a priori* analysis of the functions (see Section 6.8.2.3 for an example of a class of problems for which such an *a priori* analysis can be made).

6.8.2.2. Separable programming. The function $\Phi(x)$ is said to be *separable* if it can be written in the form

$$\Phi(x) = \sum_{i=1}^{n} \phi_i(x_i),$$

i.e., it is the sum of a set of univariate functions. An immediate consequence of this definition is that the Hessian matrix of a separable function is diagonal. Hence, the problem of minimizing a separable function (possibly subject to upper and lower bounds on the variables) can be decomposed immediately into n problems involving only one variable.

The term "separable program" usually refers to the problem

$$\begin{array}{ll} \underset{x \in \Re^n}{\text{minimize}} & F(x) \\ \text{subject to} & c_i(x) \geq 0, \end{array}$$

where F and $\{c_i\}$ are separable functions. Many of the methods developed for separable programs are designed for the special case in which the constraints are linear, and are usually extensions of linear programming techniques.

6.8.2.3. Geometric programming. The name "geometric programming" arises from the well-known inequality that relates the arithmetic and geometric means. A *posynomial* is a function Φ of a *positive vector* x ($x \in \Re^n$), of the form

$$\Phi(x) = \sum_{i=1}^{N} v_i(x),$$

where

$$v_i(x) = \alpha_i x_1^{a_{i1}} x_2^{a_{i2}} \cdots x_n^{a_{in}}, \quad i = 1, \ldots, N;$$

and

$$\alpha_i > 0, \quad i = 1, \ldots, N.$$

A posynomial program is usually cast in the form

$$\begin{aligned} \operatorname*{minimize}_{x \in \Re^n} \quad & F(x) \\ \text{subject to} \quad & c_i(x) \le 1, \quad i = 1, \ldots, m; \\ & x > 0, \end{aligned}$$

where F and $\{c_i\}$ are posynomials. It can be shown that a posynomial program is equivalent to a special case of a convex separable program.

A closely related, more general, problem class is *signomial programming*. A signomial is of the same form as a posynomial, but without the requirement that $\alpha_i > 0$. This difference means that a signomial program is not necessarily equivalent to a convex program.

Notes and Selected Bibliography for Section 6.8

There is a large literature on the minimax and ℓ_1 problems. One of the oldest algorithms for the minimax problem in the special case when all the functions are linear is that of Polya (1913). Most successful methods for the linear case are based on the exchange algorithm of Stiefel (1960).

For the nonlinear minimax problem, the most frequent approach has been to solve a sequence of linear minimax problems. For example, the algorithms of Zuhovickii, Polyak and Primak (1963) and Osborne and Watson (1969) are based on solving a linear program to obtain the search direction. Zangwill (1967b) and Charalambous and Conn (1978) compute the direction of search by solving several linear systems derived from the identification of a special subset of the problem functions (analogous to a "working set"). Generalizations of the Levenberg-Marquardt algorithm (Section 4.7.3) to the minimax problem have been suggested by Madsen (1975) and Anderson and Osborne (1977). Watson (1979) proposed a two-stage hybrid method. Several methods have been suggested that are closely related to QP-based projected Lagrangian methods (Section 6.5.3), in that the search direction can be interpreted as the solution of a quadratic program; see Han (1978a, b), Conn (1979), and Murray and Overton (1980a).

The linear ℓ_1 problem has a long history. Its connection with linear programming was observed by Charnes, Cooper and Ferguson (1955). Several algorithms have been based on exploiting this relationship; see, for example, Barrodale and Roberts (1973) and Armstrong and Godfrey (1979). For the nonlinear ℓ_1 problem, Osborne and Watson (1971) suggest solving a sequence of linear ℓ_1 problems. A method based on a penalty function is given by El-Attar, Vidyasagar and Dutta (1979). McLean and Watson (1979) give a method based on applying a Levenberg-Marquardt procedure. Murray and Overton (1980b) have suggested transforming the ℓ_1 problem into a nonlinearly constrained problem (see Section 4.2.3), which is then solved using a special QP-based projected Lagrangian method.

A comprehensive reference concerning all aspects of convexity is the book of Rockafellar (1970). The principal results used in convex programming are given by Mangasarian (1969).

The basic approach used in methods for separable programming with linear constraints (see Hartley, 1961, and Miller, 1963) is to approximate each nonlinear function of one variable by a piecewise linear function, and thereby to develop a constrained subproblem with piecewise linear objective function and constraints. This subproblem can be solved using a version of the simplex method (Section 5.3.1). Many commercial LP codes include the option of solving separable programs because of the close connection with the simplex method.

A comprehensive account of the early work in geometric programming is contained in the book of Duffin, Peterson and Zener (1967). The term "geometric programming" was originally restricted to posynomial programming, but now encompasses a much wider class of problems (see the review article of Peterson, 1976). Two excellent survey papers concerning the computational and practical aspects of geometric programming are Dembo (1978) and Ecker (1980). Both of these articles contain extensive bibliographies concerning geometric programming.

CHAPTER SEVEN

MODELLING

... who can tell
Which of her forms has shown her substance right?

—WILLIAM BUTLER YEATS, in *A Bronze Head* (1933)

7.1. INTRODUCTION

Mathematical models are frequently used to study real-world phenomena that are not susceptible to analytic techniques alone, and to investigate the relationships among the parameters that affect the functioning of complex processes. Models provide an effective — sometimes, the only — means of evaluating the results of alternative choices; for example, a model is essential in cases where experimentation with the real-world system is prohibitively expensive, dangerous, or even impossible.

Optimization methods play an important role in modelling, because a model is not usually developed as an end in itself. Rather, the model is formulated in order to determine values of free parameters that produce an optimum measure of "goodness" — for instance, the most stable structure, or the best performance on observed data.

The relationship between the formulation of a model and the associated optimization can take several forms. In many instances, virtually all the effort of model development is aimed toward constructing a model that reflects the real world as closely as possible. Only after the form of the model is essentially complete is some thought given to a method for finding optimal values of the parameters. However, selection of an off-the-peg algorithm without considering properties of the model often leads to unnecessary failure or gross inefficiency.

On the other hand, we do not advocate over-simplification or distortion in formulation simply in order to be able to solve the eventual optimization problem more easily. There has been a tendency, particularly in the large-scale area, to model even highly nonlinear processes as linear programs, because until recently no nonlinear methods were available for very large problems (see Sections 5.6.2 and 6.7 for a discussion of large-scale nonlinear problems). The effort to remove nonlinearities often leads to greatly increased problem size, and also significantly affects the nature of the optimal solution; for example, a linear programming solution (if it is unique) is always a vertex of the feasible region, but the solution of a nonlinear program is usually not (see Section 5.3.1).

A model to be optimized should be developed by striking a reasonable balance between the aims of improved accuracy in the model (which usually implies added complexity in the formulation) and increased ease of optimization. This might be achieved by invoking an optimization procedure on successively more complicated versions of the model, in a form of "stepwise" refinement. Thus, the effects of each refinement in the model on the optimization process can be monitored, and fundamental difficulties can be discovered much more quickly than if no optimization were applied until the model was essentially complete. This is especially important when dealing with models that contain many interconnected sub-systems, each requiring extensive calculation.

This chapter is not primarily concerned with how accurately models reflect the real world, but rather with aspects of modelling that influence the performance of optimization algorithms. In particular, we shall discuss considerations in formulating models that contribute to the success of optimization methods. Our observations of practical optimization problems have indicated that, even with the best available software, the efficient optimization of a model can be critically dependent on certain properties of the formulation. It is often the case that the formulator of the model must make numerous arbitrary decisions that do not affect the accuracy of the model, yet are crucial to whether the model is amenable to solution by an optimization algorithm.

In this chapter we shall describe some standard (and, for the most part, straightforward) techniques that can make optimization problems arising from modelling more amenable to solution by standard algorithms and software. Almost all of the methods given here have been used successfully in our own experiences with real problems. Certain cautionary guidelines will also be suggested in the hope of avoiding frequent pitfalls.

Of course, the nature of possible models varies so much that it is impossible to treat all relevant aspects of modelling. The main point of this chapter is that developers of models should consider in the initial stages the ultimate need to solve an optimization problem, since it is unlikely that optimization software will ever reach the state wherein a general routine can be used with impunity.

7.2. CLASSIFICATION OF OPTIMIZATION PROBLEMS

In Chapter 1 we introduced the following most general form of the optimization problem:

NCP	$\underset{x \in \Re^n}{\text{minimize}}$	$F(x)$
	subject to	$c_i(x) = 0, \quad i = 1, 2, \ldots, m';$
		$c_i(x) \geq 0, \quad i = m' + 1, \ldots, m.$

Constraints on the parameters may take other forms — e.g., some of the variables may be restricted to a finite set of values only. Problems of this type are generally much more difficult to solve than those of the form NCP; some possible approaches to models with such constraints are noted in Section 7.7.

An important point to be considered in modelling is whether the formulation has features that enhance ease of optimization, since a general algorithm for NCP will generally be inefficient if applied to a problem with special features. For purposes of choosing an algorithm, optimization problems are usually divided into categories defined by properties of the problem functions, where problems in each category are best solved by a different algorithm. A table depicting a typical classification scheme is given in Section 1.2.

Certain problem characteristics have a much greater impact on ease of optimization than others — for instance, consider problem size. Beyond univariate problems (which are invariably treated as a special case), the next dividing line occurs when the problem size becomes so large that: (a) the data cannot all be stored in the working memory of the computer; (b) exploiting the sparsity (proportion of zeros) in the problem data leads to a significant improvement in efficiency. Before that point, however, the effort required to solve a typical problem is, roughly speaking, bounded by a reasonably behaved polynomial function of problem size. Therefore, increasing the number of parameters in an unconstrained problem from, say, 9 to 12 is usually not significant.

By contrast, the form of the problem constraints can have an enormous effect on the ease of solution. In particular, there is generally a very small increase (or possibly even a reduction) in difficulty when moving from an unconstrained problem to one with simple bounds on the variables; in fact, the optimization library from the National Physical Laboratory, England, solves unconstrained problems by calling a bound-constraint subroutine (see Section 5.5.1). General linearly constrained problems are noticeably more difficult to solve than those with bound constraints only, and the presence of nonlinear constraints introduces a considerably larger increase in difficulty (see Chapter 6). For this reason, it is sometimes advisable to reformulate a model so as to eliminate nonlinear constraints; this topic will be discussed further in Section 7.4.

Probably the most fundamental property of the problem functions with respect to ease of optimization is *differentiability*, which is important because algorithms are based on using available information about a function at one point to deduce its behaviour at other points. If the problem functions are twice-continuously differentiable, say, the ability of an algorithm to locate the solution is greatly enhanced compared to the case when the problem functions are non-differentiable. Therefore, most optimization software is designed to solve smooth problems, and there is a great incentive to formulate differentiable model functions (see Section 7.3). For a smooth problem within a specific category, there still remains a great deal of choice in algorithm selection, depending, for example, on how much derivative information is available, the relative cost of computing certain quantities, and so on. As a general rule, algorithms tend to become more successful and robust as more information is provided (see Section 8.1.1).

7.3. AVOIDING UNNECESSARY DISCONTINUITIES

The word "unnecessary" appears in the title of this section because, strictly speaking, no function is continuous when evaluated with limited precision. Since only a finite set of numbers can be represented with a standard floating-point format, the usual mathematical definition of continuity, which involves arbitrarily small perturbations in the function and its arguments, is not applicable. In general, the *computed* version of any function is inherently discontinuous. Fortunately, for a well-scaled function, the discontinuities can be regarded as insignificant, in that they do not adversely affect the performance of optimization methods that assume smoothness; however, poor scaling can lead to difficulties. The topic of problem scaling will be briefly discussed in Section 7.5, and in more detail in Section 8.7.

Since optimization problems with general non-differentiable functions are difficult to solve, it is highly desirable for the user to formulate smooth mathematical models; some problems with structured discontinuities are discussed in Sections 4.2.3 and 6.8. Before discussing means of avoiding non-differentiability, we stress that there is a crucial distinction between a function that is non-differentiable and a function whose derivatives are (for some reason) not computable. If a function is truly non-differentiable, its derivatives simply do not exist mathematically at all points — e.g., the function $\max\{x_1, x_2\}$ is in general non-differentiable when $x_1 = x_2$. By contrast, a function may be smooth, but its derivatives are not available because, say, of the complexity or expense of computing them; nonetheless, an algorithm may rely on their existence.

Careful consideration of the underlying mathematical model can often indicate whether a given function should be differentiable. If there are critical points in the real-world process — for example, a reservoir overflows, or an activity shifts from one resource to another — there will probably be discontinuities in the derivatives. If the user is uncertain about differentiability, little will usually be lost by assuming that the derivatives are continuous. If the chosen optimization algorithm subsequently fails, the user may switch to an algorithm for non-smooth functions.

7.3.1. The Role of Accuracy in Model Functions

A common fallacy arises when only a limited accuracy is required in the optimal solution of a modelling problem (for example, when the model formulation is known to neglect significant elements in the real-world process, or the model function represents an observed phenomenon whose form is deduced from data of limited accuracy). In such an instance, the modeller may believe that the problem functions need to be evaluated to only slightly more than the required number of significant figures during optimization.

Because the real-world function, say $F_R(x)$, is only approximated by an ideal mathematical model function, say $F_M(x)$, the user is essentially assuming that an optimization method will tolerate convenient changes in the representation of $F_M(x)$ that are smaller in magnitude than the known accuracy to which F_M approximates F_R. However, this assumption is *not* warranted if the changes introduce serious discontinuities into the model function or its derivatives, or cause other substantive deviations in the nature of the model function. Let σ_M denote the percentage error in F_M due to fundamental deficiencies in the model. This error will not in general be known precisely, but often a lower bound can be estimated from, say, the accuracy of the data or the significance of neglected processes. If σ_M is very small, the modeller will tend to exercise the appropriate care in the computer implementation of F_M, in order to preserve the high accuracy. However, in a typical model, σ_M lies in the range 0.1%–5.0%. In this case, suppose that there are two possible computable approximations of F_M, say F_A and F_B, which can also be considered as functions that approximate F_R. The functions F_A and F_B differ from the idealized model function F_M in that, for convenience of implementation and computation, an additional error, say σ_C, has been introduced; however, σ_C is guaranteed to be much smaller than σ_M — say, $\sigma_C \approx .01\%$. Since this error does not significantly increase the existing error in approximating F_R, the three approximations F_M, F_A, and F_B could be considered of equal merit in one sense — their closeness in value to F_R.

Consider the specific univariate example illustrated in Figures 7a and 7b (the errors in the figures have been exaggerated to emphasize the aspect of interest). If the errors $|F_R(x) - F_A(x)|$ and $|F_R(x) - F_B(x)|$ were the sole concern, then the two approximations F_A and F_B would be equally good.

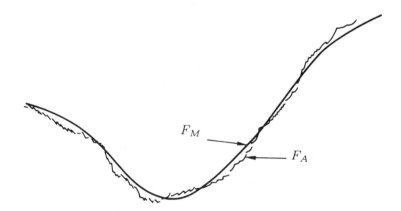

Figure 7a. This figure illustrates a problem that may occur when a discontinuous function F_A is used as an approximation to the mathematical model F_M. Although the differences between F_A and F_M may be less than the error with which the model reflects the real-world, an optimization method may locate a spurious minimum when minimizing F_A.

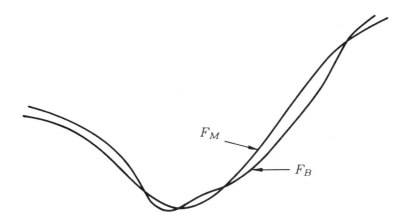

Figure 7b. In this case, the computed approximation to the mathematical model F_M is a continuous function. An optimization method designed for smooth problems should have no difficulty in minimizing F_B.

With respect to use by an optimization method, however, F_A and F_B are quite different. In particular, F_B has the same smoothness properties as the underlying (unknown) F_R, whereas F_A has discontinuities in both function and derivatives at many points. The derivative discontinuities alone would have several bad effects on an optimization method. First, the method might well converge to a spurious local minimum of F_A. Another harmful result of using F_A would occur within algorithms that approximate derivatives by finite differences. If the small step of the finite-difference interval happened to cross a discontinuity, the approximation of the gradient would be completely inaccurate, even if the gradient were well-defined at the current point.

It may seem that these cautionary remarks would apply only to a small number of uninformed people, since presumably no one would deliberately include significant discontinuities in the modelling function or its derivatives. Although discontinuities at the level of round-off error are inevitable in any model, unacceptably large discontinuities are sometimes introduced by modellers who assume that other "minor" changes are of no significance.

7.3.2. Approximation by Series or Table Look-Up

In our experience, one of the most common causes of lack of smoothness is the occurrence of a discontinuity in the evaluation of some subsidiary function, say $W(\gamma)$, upon which $F(x)$ depends. Since computers can perform only elementary arithmetic operations, more complicated functions are approximated in various ways — often by a truncated series expansion, the choice of which depends on the argument. Thus, it may happen that $W(\gamma)$ is evaluated using two formulae, one for small $|\gamma|$ and another for large $|\gamma|$. Although both should give the identical result at the crossover point, in general the truncation error will be different for the two series. Even if only a single series is used, discontinuities may occur because the evaluation process includes more terms of the series for certain values of the argument — e.g., at $\gamma = 4.7$ four terms of the series are used, whereas at $\gamma = 4.7 + 10^{-15}$ five terms are used.

To avoid such discontinuities (or at least minimize their effect), the user is advised to do the following:

(i) whenever possible, avoid switches in formulae (for example, by using a fixed number of terms in representing a function by an approximating series);

(ii) if there is a switch, ensure that the function values (and, *whenever possible*, the first derivatives) match at the cross-over point;

Unfortunately, switches in formulae sometimes occur without the user's knowledge. For example, a standard software library routine for evaluating the Bessel function $J_0(\gamma)$ uses two different methods, depending on whether γ is greater than 11. In such a case, the user may be required to utilize an alternative procedure for evaluating the subsidiary function.

A related way in which discontinuities are introduced is by including a "table look-up" during the computation of the model function. Suppose that $F(x)$ depends on the quantity $V(\gamma)$, and that $V(\gamma)$ is tabulated for the set of values $\gamma = 0(0.01)1$. If $V(0.6243)$ is required, the user may believe that $V(0.62)$ is an entirely adequate approximation. Although this might be true in some cases (as discussed in Section 7.3.1 with F_R and F_A), this treatment would make $V(\gamma)$ a piecewise constant function, with undesirable discontinuities if its properties are reflected in F. Linear interpolation within the table will produce continuity in $V(\gamma)$ (and hence, usually in F), but it will still produce discontinuities in the first derivatives. The best solution — which is always realizable — is to avoid tables completely, and to replace them by smooth approximating functions such as splines. Even two-way tables (those that require two parameters) can now be adequately represented by smooth surfaces.

7.3.3. Subproblems Based on Iteration

A more subtle source of discontinuities can be observed when evaluation of a function contains subproblems — for example, a system of differential equations or an integral. The solution of these subproblems to full machine precision (even if possible) generally requires considerable computational effort, and thus tends to be regarded as unwarranted by the modeller, since the integral, differential equation, or whatever, is only an approximation to some more complicated real-world phenomenon. A frequent example is the unconstrained minimization with respect to x of the integral

$$F(x) = \int_a^b f(x,t)\, dt. \tag{7.1}$$

Typically, the function $f(x,t)$ cannot be integrated analytically. Hence, a numerical quadrature scheme must be used, in which the integral is approximated by a weighted sum of function values at selected points:

$$\int_a^b f(x,t)\, dt \approx I(t,\omega) \equiv \sum_{j=1}^M \omega_j f(x,t_j), \tag{7.2}$$

where $\{\omega_j\}$ are the weights and $\{t_j\}$ are a set of abscissae such that $a \le t_1 \le \cdots \le t_M \le b$. The error in the approximation (7.2) depends on the higher derivatives of f, the number of abscissae $\{t_j\}$, and the position of $\{t_j\}$ within $[a,b]$.

Among the most efficient methods for numerical integration are the *adaptive* quadrature techniques, in which the abscissae in (7.2) are chosen dynamically, based on the sampled behaviour of f during an iterative procedure; the idea is to place more points in regions where f appears to be less well-behaved. Several good software packages are available for adaptive quadrature, and the user may well have chosen one of these state-of-the-art codes for evaluating the function

(7.1). However, unless the integrals are evaluated to full machine precision, the function (7.1) may not be "well-behaved" in all necessary senses. In the case of evaluating (7.1), use of an adaptive quadrature technique will tend to cause the same unfortunate consequences noted earlier with series representation. In particular, the inherently iterative nature of adaptive quadrature means that widely varying numbers of points may be placed in different parts of $[a, b]$ for very close values of x. Although a similar accuracy will generally be attained in the approximate integral for all values of x in $[a, b]$, the model function tends to contain undesirable discontinuities. Therefore, the curve of the approximate integral computed by an adaptive quadrature technique may well resemble that of F_A in Figure 7a.

It should be stressed that adaptive quadrature is inappropriate only because the subproblem that it enters is part of an outer problem in which smoothness is more important than accuracy, at least far from the solution.

An alternative way to proceed is to devise a fixed (smooth) quadrature formula I (as in (7.2)) to be used as input to the optimization routine, and thereby to determine $\bar{x}$, the point at which I achieves its minimum. It would be fortuitous indeed if $\bar{x}$ were an acceptable approximation to x^*, the minimum of (7.1), and therefore another step in the procedure is carried out. For instance, a more accurate quadrature formula (say, involving substantially more terms) can be devised, and the optimization process repeated, using $\bar{x}$ as the starting point; if $f(x, t)$ is well-behaved, a judicious choice of abscissae may allow a better estimate of the integral without unduly increasing the number of points. Since $\bar{x}$ should be a reasonably close approximation to x^*, only a relatively small number of evaluations of the more complex quadrature formula should be required. If a highly accurate integral at x^* is the ultimate aim, the final step could be application of an adaptive quadrature technique at the single point x^*. This example illustrates that it is often worthwhile to interleave modelling and optimization, since the creation of increasingly accurate quadrature formulae for smaller intervals is in fact a modelling process.

Notes and Selected Bibliography for Section 7.3

For more information concerning the error in the computed problem functions, see Section 8.5.

For details of methods of interpolation and approximation of data by smooth functions, see Hayes (1970) and Powell (1977c). For a state-of-the-art survey (*circa.* 1977) of methods and an extensive bibliography, see Cox (1977).

The basic quadrature formulae and their properties may be found in any elementary textbook on numerical analysis — see, for example, Dahlquist and Björck (1974). For a more extensive coverage, see Lyness (1977b). Lyness (1976) gives a full discussion of the difficulties associated with using adaptive quadrature schemes to define the objective function during an optimization.

7.4. PROBLEM TRANSFORMATIONS

7.4.1. Simplifying or Eliminating Constraints

In the past, algorithms for unconstrained optimization were more numerous and more effective than those for constrained problems. Today, however, algorithms for problems with only simple bounds or linear constraints are comparable in efficiency to unconstrained algorithms. Therefore, it is virtually never worthwhile to transform bound-constrained problems (in fact, it is often beneficial to *add* bounds on the variables), and it is rarely appropriate to alter linearly constrained problems.

7.4.1.1. Elimination of simple bounds. Any problem transformation should be undertaken only with extreme care. In particular, some "folklore" transformations may cause an increase in problem difficulty, and may not necessarily produce the desired result. As an example, we shall consider one of the earliest transformations concerned with the problem

$$\begin{array}{cc} \underset{x \in \Re^n}{\text{minimize}} & F(x) \\ \text{subject to} & x_i \geq 0, \quad i = 1, 2, \ldots, n. \end{array}$$

The idea is to solve a *squared-variable unconstrained problem*

$$\underset{w \in \Re^n}{\text{minimize}} \ \mathcal{F}(w)$$

where $\mathcal{F}(w)$ is the function $F(x)$ with each variable x_i replaced by $x_i = w_i^2$.

Example 7.1. As an illustration of this technique, consider the quadratic problem

$$\begin{array}{cc} \underset{x \in \Re^2}{\text{minimize}} & (x_1 + 1)^2 + (x_2 + 1)^2 \\ \text{subject to} & x_1 \geq 0. \end{array}$$

The gradient vector and Hessian matrix are given by

$$g(x) = \begin{pmatrix} 2(x_1 + 1) \\ 2(x_2 + 1) \end{pmatrix}, \qquad G(x) = \begin{pmatrix} 2 & 0 \\ 0 & 2 \end{pmatrix}.$$

The solution lies at the point $(0, -1)^T$. A bound-constraint method of the type discussed in Section 5.5.1 would use an active set strategy in which the variable x_1 is ultimately set at zero and the resulting objective function $(x_2 + 1)^2$ is minimized with respect to the single variable x_2.

The squared-variable transformation $x_1 = w_1^2$, $x_2 = w_2$ gives the unconstrained problem

$$\underset{w \in \Re^2}{\text{minimize}} \ (w_1^2 + 1)^2 + (w_2 + 1)^2$$

with derivatives

$$\nabla \mathcal{F}(w) = \begin{pmatrix} 4w_1(w_1^2 + 1) \\ 2(w_2 + 1) \end{pmatrix}, \qquad \nabla^2 \mathcal{F}(w) = \begin{pmatrix} 4(3w_1^2 + 1) & 0 \\ 0 & 2 \end{pmatrix}.$$

This transformed problem can be solved with an unconstrained minimization technique. The transformed function is "more nonlinear" than the original, but the Hessian matrix at the solution is positive definite and well conditioned.

Examples like this have encouraged the general use of the squared-variable transformation. Unfortunately, as the following example illustrates, the transformed unconstrained problem may not always be easy to solve.

Example 7.2. Consider the slightly modified form of Example 7.1:

$$\begin{array}{cc} \underset{x \in \Re^2}{\text{minimize}} & (x_1 - 1)^2 + (x_2 + 1)^2 \\ \text{subject to} & x_1 \geq 0. \end{array}$$

In this case, the solution lies at the point $(1, -1)^T$ and the restriction on x_1 is not binding. Note that the Hessian matrix is positive definite everywhere. The gradient and Hessian of the transformed function are now given

$$\nabla \mathcal{F}(w) = \begin{pmatrix} 4w_1(w_1^2 - 1) \\ 2(w_2 + 1) \end{pmatrix}, \qquad \nabla^2 \mathcal{F}(w) = \begin{pmatrix} 4(3w_1^2 - 1) & 0 \\ 0 & 2 \end{pmatrix}.$$

The transformation has had the effect of introducing a saddle point at $(0, -1)^T$. If the unconstrained algorithm used to minimize $\mathcal{F}$ is guaranteed to locate only *stationary points*, it is possible that the problem will not be solved (this will be true for all methods except Newton-type methods). On the other hand, suppose that a bound-constraint algorithm is used to solve the original problem. At the point $(0, -1)^T$, the variable x_1 will be released from its bound, and the minimization will proceed with respect to both variables. This form of transformation essentially replaces the combinatorial aspect of choosing the correct working set in an active set method by another, more difficult, combinatorial problem involving a set of stationary points of the transformed problem.

Unfortunately, this is not the only difficulty that may occur.

Example 7.3. Consider the problem

$$\underset{x \in \Re^2}{\text{minimize}} \quad x_1^{5/2} + (x_2 + 1)^2$$

$$\text{subject to} \quad x_1 \geq 0,$$

which has its solution at $(0, -1)^T$. The transformed problem becomes

$$\underset{w \in \Re^2}{\text{minimize}} \ (w_1^2)^{5/2} + (w_2 + 1)^2.$$

If the objective function of the transformed problem is computed as $w_1^5 + (w_2 + 1)^2$, then $\mathcal{F}$ does not have a minimum at $w_1 = 0$, $w_2 = -1$ (in fact, $\mathcal{F}$ is unbounded). However, if the term involving w_1 is computed by first squaring w_1 and then computing the positive root, $\mathcal{F}$ *will* have a minimum at $w_1 = 0$, $w_2 = -1$. In the latter case, we have effectively computed the transformation associated with a different original objective function in which the modulus of x_1 is taken when computing the square root. The modified objective function has a discontinuous derivative at $x_1 = 0$. In either case, the transformed function has a Hessian matrix

$$\nabla^2 \mathcal{F}(w) = \begin{pmatrix} 20w_1^3 & 0 \\ 0 & 2 \end{pmatrix},$$

that is singular at $(0, -1)^T$. This property will inhibit the rate of convergence of an unconstrained minimization method that is designed for problems with smooth derivatives (see Section 8.3.1). By contrast, when a bound-constraint method is used, the variable x_1 (which is fixed on its bound) is *eliminated* from the gradient and Hessian, and hence causes no difficulties.

7.4.1.2. Elimination of inequality constraints. It is now generally accepted that finding the solution of an *inequality*-constrained problem is more difficult than finding the solution of an *equality*-constrained problem. This has led to the widespread use of a constraint transformation in which inequality constraints of the form $c_i(x) \geq 0$ are replaced by equality constraints that include squared extra variables, i.e., $c_i(x) - y_i^2 = 0$. This transformation has disadvantages similar to those discussed in Section 7.4.1.1. If it is considered necessary to convert nonlinear inequality constraints into equalities, a preferable transformation is to add a slack variable whose non-negativity is imposed by means of a bound: $c_i(x) - y_i = 0$; with $y_i \geq 0$.

7.4.1.3. General difficulties with transformations. Transformation to an unconstrained problem or a problem with simple constraints can be an effective method of allowing the model to be solved more easily. This can sometimes be achieved simply by judicious choice of the model's independent variables. In any transformation, it is important to ensure that the new problem is not more difficult than the original one. Certain transformations of the variables may lead to the following difficulties:

(a) the desired minimum may be inadvertently excluded;

(b) the degree of nonlinearity may be significantly increased;

(c) the scaling may be adversely affected;

(d) the new function may contain singularities not present in the original problem;

(e) the new problem may have discontinuous derivatives;

(f) the Hessian matrix may become singular or ill-conditioned in the region of interest;

(g) the transformed problem may have additional local minima and stationary points;

(h) the function may be periodic in the new variables.

It is not easy to formulate general rules that will avoid these problems. In our experience, however, trigonometric and exponential transformations tend as a class to create more numerical difficulties than alternative approaches, especially as the number of variables increases.

The problem of periodicity can be offset to some extent in two ways. First, an unconstrained algorithm can be modified as follows. Suppose that the transformed variables are $\{y_i\}$, and that

$$F(y + j\alpha_i e_i) = F(y), \quad j = \pm 1, \pm 2, \ldots$$

where e_i is the i-th unit vector. If the step to be taken in y_i is p_i, p_i should be altered by adding or subtracting a multiple of α_i until $|p_i| < \alpha_i$. A second way to avoid difficulties with periodicity is to impose simple bounds on the appropriate variables — e.g., if x_1 represents an angle, add to the problem statement the requirement that $0 \le x_1 \le 2\pi$, and use a bound-constraint algorithm (see Section 5.5.1).

We shall use a simple example to illustrate other difficulties that may occur with problem transformations.

Example 7.4. Consider the nonlinearly constrained problem

$$\underset{x \in \Re^n}{\text{minimize}} \quad F(x) \tag{7.3a}$$

$$\text{subject to} \quad \sum_{i=1}^{n} x_i^2 = 1. \tag{7.3b}$$

If n is greater than one, we can make the transformation

$$x_1 = \sin y_1 \cdots \sin y_{n-1},$$

$$x_i = \cos y_{i-1} \sin y_i \cdots \sin y_{n-1}, \quad i = 2, \ldots, n-1,$$

and

$$x_n = \cos y_{n-1}.$$

The problem then becomes

$$\underset{y \in \Re^{n-1}}{\text{minimize}} \quad \mathcal{F}(y). \tag{7.4}$$

Some difficulties have been introduced into the transformed problem (7.4). In addition to the obvious periodicity, the new function is invariant to changes in any of the first $n-2$ variables if

y_{n-1} is zero. Furthermore, if any y_i $(i > 1)$ is close to zero, $\mathcal{F}(y)$ is almost invariant to changes in the other variables; clearly, the problem has become very badly scaled.

An alternative transformation to satisfy automatically the constraint (7.3b) is to define the new y variables such that:

$$a = \pm \left(1 + \sum_{i=1}^{n-1} y_i^2 \right)^{\frac{1}{2}}, \tag{7.5}$$

$$x_i = \frac{y_i}{a}, \quad i = 1, \ldots, n-1, \tag{7.6}$$

and

$$x_n = \frac{1}{a}. \tag{7.7}$$

The new problem is then

$$\underset{y \in \Re^{n-1}}{\text{minimize}} \ \min\{\bar{F}_P(y), \bar{F}_N(y)\},$$

where $\bar{F}_P(y)$ is the function obtained by choosing the plus sign in (7.5) and substituting for x in $F(x)$, with an analogous definition of $\bar{F}_N$.

In practice, if the sign of any of the optimal x_i is known, that variable could become the one whose value is fixed by (7.7), thereby removing the need to define two functions. It is preferable to choose an x_i whose value is not close to zero, since in this case some of the other transformed variables would become badly scaled. Note also that if x_i were subject to certain bounds, e.g. $0.1 \le x_i \le 0.2$, it would not be safe in general to eliminate that variable.

7.4.1.4. Trigonometric transformations. Despite their drawbacks, transformations involving trigonometric expressions are desirable in some situations. For example, consider a problem in which the variables are the co-ordinates in three dimensions of a set of k points, which are constrained to lie on the surface of a sphere. The problem in this form is then

$$\underset{x,y,z \in \Re^k}{\text{minimize}} \ \ F(x, y, z)$$

$$\text{subject to} \quad x_i^2 + y_i^2 + z_i^2 = r^2, \quad i = 1, 2, \ldots, k. \tag{7.8}$$

Note that there are $3k$ variables and k constraints.

In general, a proper elimination of t equality constraints from a problem with n unknowns leads to an unconstrained problem with $n - t$ variables (see Section 6.3). For this example, a trigonometric representation of the variables allows the constraints (7.8) to be satisfied automatically, by introducing a set of $2k$ angles $\{\theta_i, \psi_i\}$, which become the new variables, such that

$$x_i = r \sin\theta_i \cos\psi_i, \quad y_i = r\sin\theta_i\sin\psi_i \quad \text{and} \quad z_i = r\cos\theta_i.$$

To avoid difficulties with the periodic nature of the function, the simple bounds

$$0 \le \theta_i \le 2\pi \quad \text{and} \quad 0 \le \psi_i \le 2\pi,$$

can be imposed. In fact, it may be possible to make these bounds more restrictive to ensure some topological property of the set of points. In general, upper and lower bounds should be *as close as possible*.

Alternatively, the points might be restricted to lie *within* a sphere of radius r. In this case, k additional variables $\{d_i\}$ could be added, where d_i gives the distance of the i-th point from the origin, and satisfies the simple bound $d_i \leq r$. The constraints on the set of points would then become

$$x_i^2 + y_i^2 + z_i^2 + d_i^2 = r^2,$$

and the definitions of $\{\theta_i, \psi_i\}$ would be altered accordingly.

Although, in general, it is not worthwhile to eliminate only some, but not all, nonlinear constraints, this rule does not apply to sparse problems. When the sparsity of the constraints is significant, it is beneficial to replace constraints that involve a large number of variables by constraints in which only a small number of variables appear.

7.4.2. Problems Where the Variables are Continuous Functions

An important class of problems that are not immediately expressible in the finite-dimensional form NCP involves optimization with respect to a specified set of *functions*. For example, the problem may be to compute the minimum of the integral

$$\int_a^b f(x(t), t) \, dt \tag{7.11}$$

for a given f, over all smooth functions $x(t)$ defined on $[a, b]$. In many instances, this class of problem can be "solved" as a finite-dimensional problem; we shall illustrate the idea of such a transformation through a detailed treatment of (7.11).

Since the functional form of $x(t)$ cannot be obtained in general, it is necessary to represent $x(t)$ by a finite amount of information. Clearly it would be impractical to store the finite, but enormous, set of values of $x(t)$ at each machine-representable point in the interval. Instead, we must be content with storing a reasonable amount of information, from which a satisfactory approximation to $x(t)$ can be constructed. This usually involves little sacrifice because the desired result in most practical problems is simply a compact representation of the behaviour of $x(t)$ — typically, the values of $x(t)$ at a set of points in $[a, b]$. This set of information can be interpreted as an implicit definition of a new function $\tilde{x}(t)$, obtained by applying some form of interpolation to approximate the value of $x(t)$ at non-tabulated points. The accuracy of $\tilde{x}(t)$ depends on the smoothness of x and $\tilde{x}$, the number and placement of the interpolating points, etc.

A satisfactory solution to the original problem (7.11) is then a representation of $\tilde{x}(t)$. Let

$$\tilde{x}(t) = \sum_{j=1}^q c_j w_j(t), \tag{7.12}$$

where $\{c_j\}$ are a set of coefficients and $\{w_j(t)\}$ are a set of known basis functions. Examples of frequently used basis functions are: (i) polynomials: $w_j = t^{j-1}$; (ii) Chebyshev polynomials: $w_j = T_{j-1}(t)$; and (iii) B-splines: $w_j = M_j(t)$.

If the form (7.12) for $\tilde{x}$ is substituted for x in the objective function, the infinite-dimensional problem becomes a finite-dimensional problem with unknowns $\{c_j\}$. Depending on the nature of f, the integral (7.11) can then be computed analytically or from a quadrature rule; see Section 7.3.3 for further comments on the use of quadrature rules.

Notes and Selected Bibliography for Section 7.4

For a more detailed analysis of the squared-variable transformation, see Sisser (1981). An example of the solution of a time-dependent optimization problem that features many of the techniques discussed here can be found in Simpson (1969). The use of piecewise polynomial bases to solve a certain class of (quadratic) continuous optimization problems has been considered in great detail by Ciarlet, Schultz and Varga (1967); see also Gill and Murray (1973a). A definition of B-splines and Chebyshev polynomials can be found in Cox (1977).

7.5. SCALING

The term "scaling" is invariably used in a vague sense to discuss numerical difficulties whose existence is universally acknowledged, but cannot be described precisely in general terms. Therefore, it is not surprising that much confusion exists about scaling, and that authors tend to avoid all but its most elementary aspects.

The discussion of scaling in this section will be restricted to simple transformations of the variables, and special techniques in nonlinear least-squares problems. The reader is referred to Section 8.5 for a more detailed discussion, including suggestions for improving the scaling of constrained as well as unconstrained problems.

7.5.1. Scaling by Transformation of Variables

Scaling by variable transformation converts the variables from units that typically reflect the physical nature of the problem to units that display certain desirable properties during the minimization process.

There is an important distinction between transforming variables to improve the behaviour of an optimization method and transforming variables to change the problem category; the latter type of transformation is discussed in Section 7.4.1.

The first basic rule of scaling is that the variables of the scaled problem should be of similar magnitude and of order unity in the region of interest. Within optimization routines, convergence tolerances and other criteria are necessarily based upon an implicit definition of "small" and "large", and thus variables with widely varying orders of magnitude may cause difficulties for some algorithms. If typical values of all the variables are known, a problem can be transformed so that the variables are all of the same order of magnitude, as illustrated in the following example. Consider a problem that involves a gas/water heat exchanger. Table 7a gives some of the variables, their interpretation, and a typical value for each.

The magnitudes of the variables in Table 7a arise simply from the units in which they are expressed. Since most of the variables are measured in terms of different physical units, there

Table 7a

Typical values of unscaled variables

Variable	Interpretation	Units	Typical Value
x_1	Gas flow	lbs/hour	$11,000$
x_2	Water flow	lbs/hour	$1,675$
x_3	Steam thermal resistance	$(\text{BTU}/(\text{hour ft}^2\,{}^\circ\text{F}))^{-1}$	100
x_4	Waste build-up	$(\text{BTU}/(\text{hour ft}^2\,{}^\circ\text{F}))^{-1}$	6×10^{-4}
x_5	Gas-side radiation	$\text{BTU}/(\text{hour ft}^2\,{}^\circ\text{R}^4)$	5.4×10^{-10}

is no reason to suppose that they will be of similar size (in fact, the variables in the table are obviously of enormously different magnitudes). Even when the physical units of measure are the same, there may be a marked difference in typical values — for example, there is a difference between the physical properties of water and waste product.

Normally, only linear transformations of the variables should be used to re-scale (although occasionally nonlinear transformations are possible). The most commonly used transformation is of the form

$$x = Dy,$$

where $\{x_j\}$ are the original variables, $\{y_j\}$ are the transformed variables, and D is a constant diagonal matrix.

For the variables given in Table 7a, an adequate scaling procedure would be to set d_j, the j-th diagonal element of D, to a typical value of the j-th variable. For instance, d_1 could be set to 1.1×10^4.

Unfortunately, this simple type of transformation has the disadvantage that some accuracy may be lost, as the following example illustrates. Suppose that a variable x_j is known to lie in the range $[200.1242, 200.1806]$. If the variable is scaled by the "typical value" 200.1242, the scaled variables will lie in the range $[1.0, 1.000282]$ (to seven significant figures). On a computer with seven decimal digits of precision, only the three least significant figures are available to represent the variation in y_j, and consequently four figures of accuracy are lost whether the scaling is performed or not.

Another disadvantage of scaling by a diagonal matrix only is that the magnitude of a variable may vary substantially during the minimization. In this event, what might be a good scaling at one point may prove harmful at another.

Both of these disadvantages can be overcome if we know a realistic *range* of values that a variable is likely to assume during the minimization. For example, such a range may be provided by simple upper and lower bound constraints that have been imposed upon the variables. Suppose that the variable x_j will always lie in the range $a_j \leq x_j \leq b_j$. A new variable y_j can be defined as

$$y_j = \frac{2x_j}{b_j - a_j} - \frac{a_j + b_j}{b_j - a_j}. \tag{7.17}$$

The transformation (7.17) can be written in matrix form as

$$x = Dy + c, \tag{7.18}$$

where D is a diagonal matrix with j-th diagonal element $\frac{1}{2}(b_j - a_j)$, and c is a vector with j-th element $\frac{1}{2}(a_j + b_j)$. This transformation guarantees that $-1 \leq y_j \leq +1$ for all j, regardless of the value of x_j within the interval $[a_j, b_j]$. In the example noted above, the appropriate transformation (7.17) for the variable in the range $[200.1242, 200.1806]$ is

$$x_j = 0.0282y_j + 200.1524,$$

which allows y_j to be represented to full precision within the range $[-1, +1]$.

We emphasize that the interval specifying the range of values for a given variable must be a realistic one. Under no circumstances should this type of transformation be used when the value of a_j or b_j is simply a crude limit, possibly wrong by several orders of magnitude.

When the variables are scaled by a linear transformation of the form (7.18), the derivatives of the objective function are also scaled. Let g_y and G_y denote the gradient vector and Hessian matrix of the transformed problem; the derivatives of the original and transformed problems are then related by

$$g_y = Dg; \quad G_y = DGD.$$

Hence, even a "mild" scaling such as $x_j = 10y_j$ may have a substantial effect on the Hessian, and this in turn may significantly alter the convergence rate of an optimization algorithm.

7.5.2. Scaling Nonlinear Least-Squares Problems

Nonlinear least-squares problems most commonly arise when a model function, say $y(x, t)$, needs to be fitted as closely as possible to the set of observations $\{y_j\}$ at the points $\{t_j\}$. Methods for nonlinear least-squares problems are discussed in Section 4.7. The important feature of nonlinear data-fitting problems is that the variables to be estimated can sometimes be scaled automatically by scaling the independent variable t.

To see how this may happen, we consider the following example. The formulation is a simplified version of a real problem, but the original names of the variables have been retained. The function to be minimized is

$$\sum_{j=1}^{m} \left(\frac{y(p_j) - Y_j}{\Delta y_j} \right)^2,$$

where p is the independent variable, which lies in the range $[566, 576]$, and the data points $\{Y_j\}$ and their associated weights $\{1/\Delta y_j\}$ are given. The functional form assumed for $y(p)$ is

$$y(p) = \sum_{j=0}^{J} A_j p^j + \sum_{k=1}^{K} B_k \exp\left(-\frac{(p - \bar{p}_k)^2}{2\sigma_k^2} \right), \tag{7.19}$$

where the parameters to be estimated are $\{A_j\}, j = 1, \ldots, J$, and $\{B_k, \bar{p}_k, \sigma_k\}, \; k = 1, \ldots, K$. A typical data set and fitting function $y(p)$ are shown in Figure 7c.

The problem can be interpreted as fitting a Gaussian curve to each of the K peaks, together with a background function (the first term on the right-hand side of (7.19)). For the example depicted in Figure 7c, K is four, and $\{\bar{p}_k\}, k = 1, \ldots, 4$, are estimates of the corresponding peak positions; clearly each $\bar{p}_k$ lies in the range $[566, 576]$.

The major difficulty with solving this problem is that, even for moderate j, A_j must be very small because of the size of p_j. For example, if $j = 3$, A_3 is multiplied in (7.19) by at least $566^3 \approx 10^8$. Scaling the independent variable p so that each A_j lies in the range $[-1, +1]$ partially solves the problem. This can be done by defining a new independent variable z such that $p = 576z$. However, this transformation has the same disadvantages noted in the previous section for a purely diagonal scaling, namely, that relative precision in z is lost unnecessarily. However, since a meaningful range of y values is known exactly, the transformation

$$p = 5z + 571$$

may be used. With this transformation, both z and the transformed values of A_j are in the range $[-1, +1]$, and no relative precision is lost in the values of z. The transformed function is then

$$\Phi(z) \equiv \sum_{j=1}^{J} \bar{A}_j z^j + \sum_{k=1}^{K} B_k \exp\left(\frac{-(z - \bar{z}_k)^2}{2\bar{\sigma}_k^2} \right).$$

Often it is not necessary to recompute $\{A_j\}$, $\{\bar{p}_k\}$, and $\{\sigma_k\}$ from $\{\bar{A}_j\}$, $\{\bar{z}_k\}$, and $\{\bar{\sigma}_k\}$. For example, we may wish to compute the area under the $\Phi(z)$ curve, or to compute values of $y(p)$ at values other than $\{p_j\}$. In such cases the transformed function is just as useful as the original (and often much better).

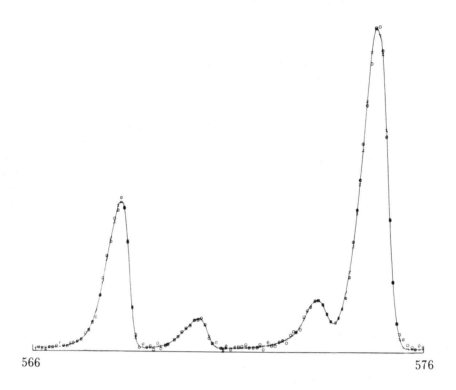

566   576

Figure 7c. This figure depicts the least-squares fit of a continuous function to a discrete data set. The parameters of the fitting function were determined using a nonlinear least-squares procedure. Each of the four peaks in the data are fitted by a Gaussian curve.

7.6. FORMULATION OF CONSTRAINTS

7.6.1. Indeterminacy in Constraint Formulation

A difficulty in formulating a model with constraints on the variables is the possibility of creating a poorly posed optimization problem, even though the underlying model has a well defined solution. This situation can exist for many reasons, which are too numerous to list here. For example, redundant constraints may be included that are simply linear combinations of other constraints, in order to provide a summary of certain activities. Such features may serve a useful purpose within the model, and the modeller knows that they "should" have no effect on the optimal values of the model parameters. Unfortunately, the performance of optimization algorithms may thereby be adversely affected.

A typical situation occurs when the variables in the optimization problem do not correspond directly to the model parameters. As an illustration of such a model, consider an optimization problem arising from a statistical model of data concerning the incidence of lung cancer. The model variables were three-way probabilities $\{p_{ijk}\}$, $i = 1, \ldots, I$; $j = 1, \ldots, J$; $k = 1, \ldots, K$, and the objective function depended only on $\{p_{ijk}\}$. In one of the models considered, it was assumed that these probabilities could be represented as the product of three two-way probabilities, i.e.

$$p_{ijk} = f_{ij}\, g_{jk}\, h_{ik}. \tag{7.20}$$

Further constraints were also imposed upon $\{f_{ij}\}$, $\{g_{jk}\}$, and $\{h_{ik}\}$, which then served as the variables in a nonlinearly constrained optimization problem.

The presence of indeterminacy in the optimization problem was revealed when the solution method consistently experienced great difficulty in converging, which was surprising in view of the quadratic convergence that was expected from the particular algorithm being used. In order to discover the source of the difficulty, several different starting points were tried. The method converged to completely different values of $\{f_{ij}\}$, $\{g_{jk}\}$, and $\{h_{ik}\}$, but always yielded the same values for all $\{p_{ijk}\}$ and the optimal objective function. This behaviour led to a re-examination of the model formulation, which showed that the problem variables were not uniquely defined by the optimization problem. Examination of (7.20) shows that the value of p_{ijk} is unaltered if f_{ij} and g_{jk}, say, are replaced by αf_{ij} and g_{jk}/α for any non-zero α. Such a change did not affect satisfaction of the remaining constraints, and hence the problem contained an inherent indeterminacy with respect to the chosen variables. In fact, if a less robust algorithm had been used to solve the nonlinearly constrained problem, it would have failed to converge, since the linear systems to be solved for the search direction were exactly singular. In this case, the lack of uniqueness was easily resolved by imposing some additional normalization constraints on one of the sets of variables — e.g., $\sum_k g_{jk} = 1$.

This example is not particularly unusual, and highlights the importance of applying modelling and optimization interactively. Although similar difficulties may be less simple to diagnose and correct, the general rule of thumb is to check that the solution of the optimization problem is as well defined as the underlying model.

7.6.2. The Use of Tolerance Constraints

Equality constraints occur in problem formulations for a variety of reasons. Often the very nature of the variables imposes an equality constraint — for example, if the variables $\{x_i\}$ represent proportions or probabilities, this gives rise to the constraint $\sum_i x_i = 1$ (as well as non-negativity restrictions). Constraints of this type are "genuine" equalities, in the sense that the computed solution must satisfy them exactly (where "exactly" means "within working precision"). However, it is not unusual in modelling that constraints that might seem initially to be firm equality constraints should be treated instead as constraints that need not be satisfied with maximum possible accuracy. For example, this situation occurs when the underlying model is known to contain inaccuracies. The term *tolerance constraint* refers to a range constraint with a very narrow range, which gives the effect of satisfying an equality constraint only to within a prescribed tolerance. Thus, the linear constraint

$$a^T x = b$$

would be replaced by

$$b - \epsilon_2 \le a^T x \le b + \epsilon_1, \tag{7.21}$$

where ϵ_1 and ϵ_2 are small, but not negligible, positive quantities (exactly the same transformation can be made for a nonlinear constraint).

Tolerance constraints of the type (7.21) differ from ordinary range constraints because the range of acceptable values, although non-zero, is very small. In some problems, treating constraints of this type as equalities may cause there to be no feasible solution, or may distort the properties of the solution if the corresponding constraints are ill-conditioned (e.g., points that satisfy the constraints exactly may be far removed from other points that lie within the given range).

The following detailed example illustrates this situation for both linear and nonlinear constraints, and also includes other forms of problem transformation. The statement of the problem has been simplified in order to highlight the features of interest. The model is to be used in the design of a platinum catalyst converter that controls the exhaust emission on car engines. The corresponding optimization problem was originally posed in terms of a set of equations modelling the chemical reactions within the converter. There are two types of equations: nonlinear equations that describe the reaction rates for various chemical processes, and linear relationships that arise from the conservation of the constituent elements. In total, there are eight variables and thirteen equations (eight nonlinear and five linear).

The variables $\{x_1, \ldots, x_8\}$ represent the partial pressures of the following species, in the order given: propane, carbon monoxide, nitrogen oxide, carbon dioxide, oxygen, water, nitrogen, and hydrogen. Clearly it is required that $x_i \geq 0$, $i = 1, \ldots, 8$, since negative values would have no physical meaning. The eight nonlinear reaction equations are as follows, where the constants $\{K_1, \ldots, K_8\}$ are the reaction constants whose logarithms are defined by logarithms in the temperature:

$$\frac{x_4^3 x_6^4 x_7^5}{x_1 x_3^{10}} - K_1 = f_1(x) = 0 \tag{7.22}$$

$$\frac{x_4^3 x_6^4}{x_1 x_5^5} - K_2 = f_2(x) = 0 \tag{7.23}$$

$$\frac{x_4^3 x_8^{10}}{x_1 x_6^6} - K_3 = f_3(x) = 0 \tag{7.24}$$

$$\frac{x_4 \sqrt{x_7}}{x_2 x_3} - K_4 = f_4(x) = 0 \tag{7.25}$$

$$\frac{x_4 x_8}{x_2 x_6} - K_5 = f_5(x) = 0 \tag{7.26}$$

$$\frac{x_6 \sqrt{x_7}}{x_3 x_8} - K_6 = f_6(x) = 0 \tag{7.27}$$

$$\frac{x_6}{x_8 \sqrt{x_5}} - K_7 = f_7(x) = 0 \tag{7.28}$$

$$\frac{x_4}{x_2 \sqrt{x_5}} - K_8 = f_8(x) = 0. \tag{7.29}$$

The linear equations derived from conservation of elements are the following, where the constants $\{a_1, \ldots, a_8\}$ represent the initial partial pressures of the various species:

Oxygen balance

$$x_2 + x_3 + 2x_4 + 2x_5 + x_6 - a_2 - a_3 - 2a_4 - 2a_5 - a_6 = f_9(x) = 0 \tag{7.30}$$

Carbon balance

$$3x_1 + x_2 + x_4 - 3a_1 - a_2 - a_4 = f_{10}(x) = 0 \tag{7.31}$$

Hydrogen balance

$$8x_1 + 2x_6 + x_8 - 8a_1 - 2a_6 - a_8 = f_{11}(x) = 0 \qquad (7.32)$$

Nitrogen balance

$$x_3 + 2x_7 - a_3 - 2a_7 = f_{12}(x) = 0 \qquad (7.33)$$

Total balance

$$\sum_{i=1}^{8} x_i - \sum_{i=1}^{8} a_i = f_{13}(x) = 0. \qquad (7.34)$$

The usual method for solving a set of overdetermined equations is to minimize the sum of squares of the residuals, i.e.

$$\text{minimize} \sum_{i=1}^{13} f_i^2(x). \qquad (7.35)$$

One difficulty with this approach is that the equations (7.22)–(7.34) are of two distinct types, and it is desirable to preserve the natural separation of linear and nonlinear equations during the process of solution. A means of allowing the latter is to include only the nonlinear equations in the sum of squares, and to solve the problem as

$$\text{minimize} \sum_{i=1}^{8} f_i^2(x) \qquad (7.36)$$

subject to the five linear equality constraints (7.30)–(7.34). If the problem is formulated as (7.36), the computed solution will satisfy the linear constraints exactly. In this situation, however, representation of the real-world phenomena by equality constraints may be undesirable, since it is known that not all possible chemical reactions have been included in the conservation equations (as many as thirty processes involving only minute quantities were omitted from the formulation). Therefore, forcing equality upon imprecise constraints may be imprudent. In fact, in some instances there would be no feasible solution for this model because of the additional non-negativity constraints. To avoid this difficulty, the equality constraints (7.30)–(7.34) might be replaced by tolerance constraints. The selection of each tolerance can be judged from the percentage of each reaction process that is dominated by the main reaction (the one represented in the constraints).

For some models, this adjustment of the constraints would suffice to allow the problem to be solved satisfactorily. However, in this instance poor scaling causes additional difficulties, since the reaction constants $\{K_i\}$ vary enormously (for instance, K_1 is of order 10^{250}). One method for overcoming poor scaling here is to replace each $f_i(x)$, $i = 1, \ldots, 8$ by the transformed function

$$F_i = \ln\left(f_i(x) + K_i\right) - \ln K_i,$$

and then to minimize the new objective function

$$\sum_{i=1}^{8} F_i^2(x).$$

Although such a transformation cures the difficulty due to the variation in magnitude of $\{K_i\}$, another indication of poor scaling is the extreme sensitivity of the solution to differences in parameter values that would ordinarily be considered negligible. For example, the functions vary dramatically depending on whether x_1 is 10^{-14} or 10^{-100}, whereas standard computer algorithms would undoubtedly treat these quantities as "equivalent". To overcome this difficulty, a *nonlinear* transformation of the variables is necessary, and therefore any suitable transformation destroys the linearity of the constraints (7.30)–(7.34). Fortunately, for this problem there is a nonlinear transformation that changes the nonlinear functions F_i into linear functions, namely

$$x_i = e^{y_i}. \tag{7.37}$$

Note that the transformation (7.37) also ensures that $x_i \geq 0$. To illustrate the effect on the nonlinear functions, consider F_6, which is transformed to

$$\hat{F}_6(y) = \frac{1}{2}y_7 + y_6 - y_3 - y_8 - \ln K_6.$$

The effect on the linear equations is illustrated by equation (7.32), which becomes

$$\hat{f}_{11}(y) = 8e^{y_1} + 2e^{y_6} + e^{y_8} - b_3 = 0.$$

With the transformation (7.37), we obtain the following linearly constrained problem:

$$\begin{aligned}
&\underset{y}{\text{minimize}} && \sum_{i=9}^{13} \hat{f}_i^2(y) \\
&\text{subject to} && \hat{F}_i(y) = 0, \quad i = 1, 2, \dots, 8.
\end{aligned} \tag{7.38}$$

Note that (7.38) would not be a satisfactory representation if the original linear constraints were expected to be satisfied exactly, since in general we would not expect $\hat{f}_i(y)$ to be zero at the solution of (7.38).

Even (7.38) is still not satisfactory because the linear constraints of (7.38) (the transformed nonlinear constraints of the original problem) will generally be incompatible, solely because the values of $\{K_i\}$ have been determined from inherently imprecise experimental data. Hence the equality constraints of (7.38) should be replaced by tolerance constraints, where the tolerance for each constraint is determined by the relative accuracy of the corresponding K_i. It is interesting to note that if the $\{K_i\}$ are only slightly in error (as they should be), the system of equations defined by the constraints of (7.38) is only slightly incompatible. In an initial solution of the catalyst converter problem, the incompatibility was much larger than expected, and this revealed an error in the original data.

Notes and Selected Bibliography for Section 7.6

The optimization problem arising from the lung-cancer model was brought to our attention by Professor A. Whittemore of Stanford University.

Publications on the application of optimization techniques to real-world problems are too numerous to be cited here. However, a good introduction to aspects of mathematical modelling related to engineering optimization is given by Wilde (1978). For a selection of applications, see Bracken and McCormick (1968) and the compendium of papers in Avriel, Rijckaert and Wilde (1973), Balinski and Lemaréchal (1978) and Avriel and Dembo (1979).

7.7. PROBLEMS WITH DISCRETE OR INTEGER VARIABLES

Many practical problems occur in which some of the variables are restricted to be members of a finite set of values. These variables are termed *discrete* or *integer*. Examples of such variables are: items that are obtainable or manufactured in certain sizes only, such as the output rating of pumps or girder sizes; or the number of journeys made by a travelling salesman. Such limitations mean that the standard definitions of differentiability and continuity are not applicable, and consequently numerical methods for differentiable problems must be used indirectly (except for a certain number of special cases where the solution of the continuous problem is known to satisfy the discrete/integer constraints automatically).

If the objective and constraint functions are linear, many special integer linear programming methods have been developed, notably variants of "branch and bound"; in some other special cases, dynamic programming methods can be applied. However, we shall be concerned with mixed integer-nonlinear problems, i.e. nonlinear problems with a mixture of discrete and continuous variables.

It is important to distinguish between two types of discrete variables, since different methods can be applied to help solve each problem category. We shall illustrate the distinction, and possible approaches for dealing with such variables, by considering two typical problems in some detail.

7.7.1. Pseudo-Discrete Variables

The first problem concerns the design of a network of urban sewer or drainage pipes. Within a given area, the position of a set of manholes is based on needs of access and the geography of the street layout. It is required to interconnect the manholes with straight pipes so that the liquid from a wide catchment area flows under gravity down the system and out of the area. Each manhole has several input pipes and a single output pipe. To facilitate the flow, the pipes are set at an angle to the horizontal.

The variables of the problem are the diameters of the pipes and the angles that the pipes make with the horizontal. The constraints are: the slope of a given pipe lies between upper and lower bounds (determined by the need to facilitate flow and comply with the topography of the street level); the pipe diameters are non-decreasing as flow moves down the system; and the flow in the pipes (a nonlinear function of the pipe diameters and slopes) lies between some maximum and minimum values when the system is subjected to a specific "steady state" load. The objective of the design is to minimize the cost of construction of the pipe network while still providing an adequate level of extraction. The major costs in constructing the system are the costs of digging the trenches for the pipes and the capital costs of the pipes themselves. These costs are complementary, since narrow pipes are cheap, but require the excavation of deep sloping trenches to carry the required load.

At first sight the problem appears to be a straightforward nonlinearly constrained problem with continuous variables. What makes it a mixed continuous-discrete problem is the fact that pipes are manufactured only in standard diameters. The variables corresponding to the diameters of the pipes are examples of the first type of discrete variable, which occurs when the solution to the continuous problem (in which the variables are not subject to the discrete restrictions) is perfectly meaningful, but cannot be accepted due to extraneous restrictions. Such variables will be termed *pseudo-discrete*, and occur frequently in practice.

As we shall now demonstrate by the sewer-network example, problems with pseudo-discrete variables can often be solved by utilizing the solution of the continuous problem. This suggestion relies on the well behaved nature of the functions in most practical models — i.e., if the optimal pipe diameter when treated as a continuous variable is, say, 2.5723 feet, the optimal discrete value is unlikely to be very different.

In a general problem with pseudo-discrete variables, suppose that x_1 must assume one of the values $d_1, d_2, \ldots, d_r$. Let x^c denote the value of x at the solution of the continuous problem, which is assumed to be unique. Suppose that x_1^c satisfies

$$d_s < x_1^c < d_{s+1}.$$

The value of the objective function F at x^c is a lower bound on the value of F at any solution of the discrete problem, since if x_1 is restricted to be any value other than x_1^c, the objective function for such a value must be larger than $F(x^c)$, irrespective of the values of $x_2, \ldots, x_n$.

The next stage of the solution process is to fix the variable x_1^c at either d_s or d_{s+1} (usually, the nearer value); similarly, any other discrete variable may be set in this manner. The problem is then solved again, minimizing with respect to the remaining continuous variables, using the old optimal values as the initial estimate of the new solution. Solving the restricted problem should require only a fraction of the effort needed to solve the continuous problem, since the number of variables is smaller, and the solution of the restricted problem should be close to the solution of the continuous problem. The solution of the restricted problem, say x^r, is not necessarily optimal since incorrect values may have been selected for the discrete variables. Since $F(x^c)$ is a lower bound on $F(x^r)$, a value of $F(x^r)$ close to $F(x^c)$ will be a satisfactory solution in most practical problems. If it is thought worthwhile to seek a lower value than $F(x^r)$, some of the discrete variables may be set at their alternative values. Usually, these trials are worthwhile for those variables whose "continuous" value lies close to the centre of the range. Typically, very few such trials are necessary in practice.

In some cases, the restriction of a discrete variable may have the beneficial effect of automatically narrowing the choice of others. For example, in the pipe network problem, pipes lower down the network cannot be smaller in diameter than those upstream. Consequently, setting x_1 to d_{s+1}, say, may fix the choice for x_2.

Discrete variables may also be chosen to achieve an overall balance in the value of $F(x)$. For example, if the problem concerns the selection of girder sizes for minimizing the weight of a bridge, some girder sizes could be increased to make the bridge take an increased load, but others could be simultaneously decreased to achieve only a small increase in overall weight.

It is important in such problems to note that the solution of the continuous problem can always be used as an initial estimate for the solution of a restricted problem. In many practical problems, only two or three solutions of a restricted problem are needed to determine an acceptable solution of a discrete problem. The extra computing cost in solving the additional restricted problems associated with the discrete variables is likely to be a fraction of the cost to solve the original full continuous problem; if not, this implies that the discrete solution differs substantially from the continuous solution. In such circumstances, it may be worthwhile to alter the extraneous conditions so that the two solutions are closer. For example, in problems of supply, such as the urban sewer problem, alternative supplies may be sought whose specifications are closer to the corresponding elements of the continuous problem, since by definition this change would yield a significant reduction in construction costs.

7.7.2. Integer Variables

The second type of discrete variable is one for which there is no sensible interpretation of a non-integer value — for example, when the variable represents a number of items, or a switch from one mutually exclusive strategy or resource to another (e.g., the change between coating materials in lens design). This type of discrete variable problem is much more difficult to solve than the first. If the number of such variables is small, say less than five, and the number of distinct values that each variable can take is also small, a combinatorial approach is possible. In this context a combinatorial technique is one in which the objective function is minimized for every possible combination of values that the discrete variables can assume. It may happen in practice that some combinations are considered unlikely, and so not all cases need to be tried. A combinatorial approach is often reasonable for constrained problems because many infeasible combinations can be eliminated before any computation takes place. In addition, with a combinatorial approach there is a useable solution no matter where the algorithm is terminated.

Unfortunately, for even a moderate number of variables the combinatorial approach becomes too expensive, as the number of possible cases grows extremely large very quickly. For some discrete-variable problems that arise in practice, it is possible to pose a related problem with only continuous variables, such that the solution of the new problem, although not identical to the solution of the original, serves as a guide to the likely values of the discrete variables.

We illustrate this approach by considering a method for the design of a distillation column. The left-most portion of Figure 7d depicts a simplified diagram of a distillation column. Vapour is introduced at the bottom of the column and flows upwards. Condensed vapour flows downwards and is recycled using a boiler. The column is divided into a number of stages, and at some of the stages additional liquid (known as *feed*) is introduced. At each stage the liquid and vapour mix and alter in composition. The liquid is then drawn off as a product, or is used as an input to the neighbouring stages. The optimization problem is to choose the level at which to place the feed input in order to achieve a specified performance at minimum cost.

At first sight it might be thought that the variable associated with the feed level has no continuous analogue. However, a set of new variables was introduced, where each new variable corresponds to a stage of the column, and represents the percentage of the total feed to be input at that particular level. The problem is then re-formulated and solved, treating the new variables as continuous, and its solution is taken to indicate properties of the solution of the original problem. For example, if the solution of the continuous problem indicates that 90 percent of the feed should go to a particular stage, this stage is likely to be the one at which to input the feed in the discrete model. Figure 7e shows some typical percentage feed levels for a 9-stage continuous model; stage four appears to be the most likely candidate for the value of the discrete variable.

It is interesting to note that the results for this problem suggested that a change in the design of some distillation columns should be considered. In some cases, the continuous solution indicated that the feed should be added at two separated stages — a design that had previously not been considered.

In conjunction with this and similar schemes, a term can be introduced into the objective function that has the effect of encouraging a single choice for a discrete variable. For example, in the continuous model of the distillation column, a term like $1/\sum x_i^2$ might be added to the objective function, where x_i is the fraction of the total input to the i-th stage.

Notes and Selected Bibliography for Section 7.7

Beale (1977) presents a survey of computational techniques for integer linear programming. Details of integer programming codes are given, amongst others, by Johnson (1978) and Johnson and Powell (1978).

The method for the optimal design of the plate distillation column was suggested by Sargent and Gaminibandara (1976).

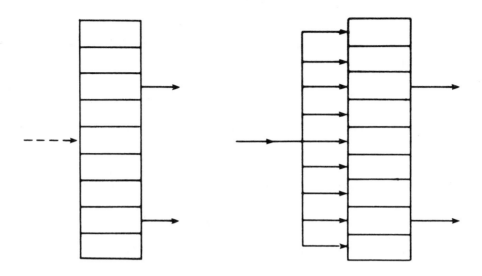

Figure 7d. The first figure depicts a schematic diagram of a traditional distillation column. Vapour is introduced at the bottom of the column and flows upwards. Condensed vapour flows downwards and is recycled using a boiler. The column is divided into a number of stages, and at some of the stages additional liquid (known as *feed*) is introduced. At each stage the liquid and vapour mix and alter in composition. The liquid is then drawn off as a product, or is used as an input to the neighbouring stages. The second figure depicts the simplified model in which a percentage of the total feed to be input is added at each level.

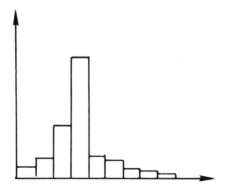

Figure 7e. This figure depicts some typical percentage feed levels for a 9-stage continuous model of a distillation column. Stage four appears to be the most likely candidate for the value of the discrete variable.

CHAPTER EIGHT

PRACTICALITIES

When one comes down to particular instances,
everything becomes more complicated.

—ALBERT CAMUS, in *The Notebooks: 1935–1942* (1963)

You can get it wrong and still you think that it's all right.

—JOHN LENNON AND PAUL McCARTNEY, from *We can work it out* (1965)

This chapter is intended as a guide to aspects of solving optimization problems that are useful in practice, but may not be treated in standard books or papers. Although there is no guaranteed strategy that will resolve every difficulty, some "conventional wisdom" about optimization can be applied to good effect in a surprisingly large number of cases. This chapter will play a role similar to that of a consultant whom the user might seek before, during, or after the numerical solution of an optimization problem. We wish to emphasize in advance that the advice that we shall offer is not foolproof, and many of the statements to be made should (and will often) be qualified.

8.1. USE OF SOFTWARE

In this section, we shall discuss selected aspects of the use of software to solve practical optimization problems. Many of our comments will refer to a "fictitious" software library that includes the methods recommended in Chapters 4, 5 and 6. Although no such library may exist at this time, our intention is to consider what we believe to be an "ideal" library, given the current state of the art of methods for optimization. Although many of the methods discussed in the earlier chapters are now available in the form of documented software, there will always be instances where a user will not have access to a particular code. For this reason we have included a discussion of how to make the best use of a limited amount of software (Section 8.1.5).

8.1.1. Selecting a Method

After an optimization problem has been formulated, a solution method must be selected. In terms of choosing a method, problems are usually classified based on properties of the objective and constraint functions, as indicated in Section 1.2. However, even within these categories, the solution method will vary depending on the information that the user can provide about the problem functions (e.g., first derivatives), the available storage, and the costs of evaluating the problem functions compared with the arithmetic operations performed by the algorithm.

The general rule in selecting a method is to make use of as much derivative information as can reasonably be provided. If it is impossible to calculate first derivatives, obviously a non-derivative method must be used. However, if computing first derivatives is merely *inconvenient*, the user should be aware of the increased complexity and decreased reliability that result when exact first derivatives are not available to an algorithm. Similarly, if the cost associated with computing the Hessian matrix is several orders of magnitude larger than that of calculating the gradient, it is probably not worthwhile to use a Newton-type method with exact second derivatives. The tradeoff between the cost of providing additional derivative information and the resulting gains in reliability and accuracy depends on the particular problem.

8.1.1.1. Selecting an unconstrained method. Methods for solving unconstrained problems are discussed in Chapter 4. When the objective function is smooth and n (the number of variables) is small enough so that an $n \times n$ symmetric matrix can be stored, the following ranking of methods indicates the confidence that can be placed in the method's success in finding an acceptable solution:

— Newton-type with second derivatives
— Newton-type without second derivatives
— Quasi-Newton with first derivatives
— Quasi-Newton without first derivatives
— Conjugate-gradient-type with first derivatives
— Conjugate-gradient-type without first derivatives
— Polytope

This hierarchy of algorithms is useful not only in selecting the initial solution method, but also in indicating possible alternatives if difficulties arise with the first choice.

The superiority of Newton-type methods (Sections 4.4 and 4.5.1) reflects the use of second-order properties of the function to improve algorithmic efficiency as well as to provide qualitative information about the computed solution. For example, the Hessian matrix at the solution can be used to estimate the conditioning and sensitivity of the solution (see Section 8.3.3); a quasi-Newton approximation (Section 4.5.2) cannot be guaranteed to be an accurate representation of the true Hessian. An additional benefit of using a Newton-type method is that there are certain problem classes (e.g., those with saddle points) where only a Newton-type method is assured of being able to proceed.

For most problems, the robustness and efficiency of Newton-type methods are not adversely affected if finite differences of gradients are used to approximate second derivatives. In addition, the logic of the method is essentially unaffected by the use of finite differences. Unfortunately, these same properties do not hold for quasi-Newton methods in which finite differences of function values are used to approximate first derivatives (see Section 4.6.2). The complications that can arise when it is necessary to approximate first derivatives by finite differences are discussed in detail in Section 8.6. Even with the best possible implementation, a finite-difference quasi-Newton method may fail to make any progress in a region where $\|g(x)\|$ is small but non-zero. Furthermore, a good finite-difference quasi-Newton algorithm will differ in several details from an exact-derivative method. In particular, a different step-length algorithm will usually be used in the non-derivative case, with a modified definition of "sufficient decrease" (see Section 4.3.2.1); the recommended termination criteria also change.

The polytope method (see Section 4.2.2) requires the least information about F, and hence is (by a wide margin) the least reliable. As mentioned in Section 4.2.1, *a method of this type should be used only when there is no alternative.*

The selection chart displayed in Figure 8a illustrates an aid that may be provided with a software library. Selection charts help the user to choose the most appropriate routine for a given problem, by determining the user's knowledge of the nature of F, and of the information that

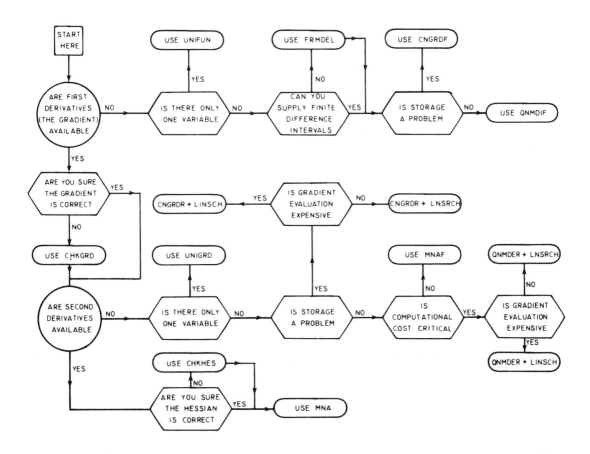

Figure 8a. This selection chart for dense unconstrained optimization provides a means for choosing the most suitable routine from a software library. The user starts at the square box, and continues to answer questions about the structure of the problem until termination occurs at the name of the required routine.

can be provided concerning derivatives. In addition to guiding the choice of a primary solution method, selection charts can also indicate which service routines (see Section 8.1.2.2) might be useful.

The state of the art for large-scale unconstrained optimization is less well-defined (see Section 4.8). Sparse discrete Newton methods tend to be extremely efficient when they can be used. The performance to date of sparse quasi-Newton methods has been disappointing relative to their success for dense problems. For some problems, a conjugate-gradient-type algorithm can be very efficient, particularly when combined with a careful preconditioning scheme.

8.1.1.2. Selecting a method for linear constraints. We have seen in Chapter 5 that the above methods for dense unconstrained minimization of a smooth function can be generalized for application to a linearly constrained problem. When the method has been properly adapted, the relative merits of these methods remain roughly the same as in the unconstrained case. The only change is a tendency for the scale of differences between methods to be "stretched", i.e. the variation is usually more marked. For example, some adverse effects might result from the imposition of bounds on the variables with the conjugate-gradient-type methods, whereas essentially no increase in difficulty would typically be caused for a quasi-Newton method.

Size of the problem. An important decision that must be made in the linear-constraint case is whether the problem should be categorized as "large" (sparse) or "small" (dense). To understand the fundamental differences between these two categories, it is illuminating to consider how codes for linearly constrained optimization have been developed. Historically, software for large-scale problems has been developed in the commercial field, whereas numerical methods for dense problems tend to be more the province of "academic" research. As a result, large-scale software (particularly linear programming codes) tends to be much more directed to the needs of the user, and to include features that are not part of the solution process proper. The effect of this difference can be seen by considering the format in which problems must be posed in order to use the software.

When solving a dense problem, the user is typically required to define explicit arrays that contain the values of the independent variables, the right-hand side of the constraints, and the coefficients of the constraints. The user is responsible for initializing all elements of the arrays. There are often numerous parameters associated with a given method that must be specified by the user (see Section 8.1.3 for further discussion of such parameters).

On the other hand, communication of information about a large-scale problem is invariably carried out by means of an *input file* created by the user and processed by the software. The usual format of the input file is the so-called *MPS standard form*. With this system, a *user-specified name* is associated with each variable and with each constraint. Since the coefficient matrices are usually sparse, only the non-zero elements need to be specified by the user. Furthermore, there are typically *default values* for parameters that characterize the method to be used, and hence the user needs to specify only the values that differ from the default choices. This feature is extremely convenient when there are a large number of parameters, many of which are of interest only to the expert user.

An additional difference between codes for small and large linearly constrained problems involves the form and structure of the software. The codes for dense problems are typically written in a portable high-level language (usually Fortran). They are most often *subroutines* that are inserted in the user's program. On the other hand, large-scale codes tend to be written in machine-specific languages, and hence will work only on particular machines; the actual code is usually proprietary, and is invisible to the user. Furthermore, a large-scale solution method is often only a part of a mathematical programming system that includes features besides the solution method (e.g., a report writer). The user thus communicates with the large-scale method by preparing an input file (and reading an output file) rather than calling a subroutine.

In general, methods for dense problems will have superior numerical properties because of the improved conditioning of the calculation of the search direction and the Lagrange multiplier estimates (see Sections 5.1.3 and 5.1.5). Hence, a dense code should be used (if possible) whenever there may be ill-conditioning in the problem. If the size of the problem is likely to increase substantially at some later date, it may prove to be advantageous to use a sparse code from the beginning. However, there is a price associated with changing from a dense to a sparse code — not only a loss of numerical stability, but also an increase in overhead costs (because of the complex logic associated with maintaining a sparse data structure).

The differences noted above make it non-trivial to switch between dense and sparse codes.

However, the ability to change from a dense to a sparse method should improve as more modern software is developed. Because of the popularity and convenience for the user of MPS-type input, future dense codes will tend to contain an input/output routine that can be used *separately from the solution method* to process the alphanumeric problem specification, transform it into the internal data structure of the subroutine, and then produce output in terms of the user-specified names. It is unfortunate that such input/output routines are invariably large and cumbersome pieces of software when they are written in a portable high-level language such as Fortran.

Representation of the problem. The discussion in Chapter 5 indicated that dense methods differ significantly in the representation of the matrix of constraint coefficients, and in the form assumed for the constraints. Given a set of assumptions about the most likely form of the problem, a method will tend to become inefficient (or inconvenient to use) when these assumptions are not satisfied.

In particular, in Sections 5.3.1 and 5.6.1, it was shown that some methods for linear programming treat all constraints (including bounds) in the same way, whereas other methods assume that the problem is given in *standard form*. Suppose that an n-variable LP problem contains m general constraints, and that all the variables are subject to simple bounds. Algorithms of the first type recur an explicit $n \times n$ matrix to represent the constraints in the working set. Algorithms of the second type convert all general constraints to equalities by adding slack variables, and recur an $m \times m$ column basis. Obviously, there are problems for which either of these algorithm types will be inefficient.

For example, consider a linear program with 10 general linear constraints and 100 variables; at a vertex, at least 90 variables must be on their bounds. For such a problem, methods of the first type will require storage for a 100×100 dense matrix, whereas the method based on the standard form will need to store only a 10×10 matrix. Suppose, on the other hand, that the problem has 10 variables and 100 general inequality constraints. Adding slack variables in order to convert the problem to standard form requires the storage of a 100×100 matrix, whereas the first method requires only the storage of a 10×10 matrix.

If the user must use a code that is unsuitable for the "shape" of the problem being solved, the only alternative is to apply the code to the dual linear program (see Section 3.3.2.2). However, the formation of the dual is a lengthy process that is prone to simple blunders.

Null-space or range-space method. For small problems, the cost of evaluating the objective function tends to dominate the work required to perform the linear algebraic operations associated with each iteration. Hence, in general the user should simply use the best available code for dense problems, which will probably be a null-space active-set method (see Section 5.2). However, as the size of the problem increases, the work and storage associated with such a method also increase to the extent that it may be worthwhile to use an alternative method for certain types of problems. The choice between a null-space and range-space method depends on the number of constraints active at the solution. See Section 5.4 for a discussion of the important tradeoffs in this decision.

Choosing a method that generates feasible iterates. Many practical problems are such that the objective function is defined only at points that lie inside the feasible region. For example, in Section 1.1 we discussed the application of optimization to the design of a nose cone. For this specific problem, the mathematical theory for the computation of the drag (the quantity to be minimized) does not apply if the volume of the nose cone (one of the constraints) is negative. In other problems, the objective function may be defined outside the feasible region, but the values there may have no physical significance. For example, suppose that a variable x enters the computation of the objective function by defining a subsidiary quantity that is the difference between a data point y_j and a Chebyshev expansion $\sum_i a_i T_i(x)$. Because the properties of the Chebyshev polynomials $\{T_i(x)\}$ hold only if x lies in the interval $[-1, 1]$, the value of the objective function will be defined but have no meaning for any x that lies outside this range.

If the user requires that the objective function be computed only at feasible points, there is some restriction on the choice of available algorithms. If exact arithmetic were used throughout, all the methods for linearly constrained minimization described in Section 5.2 that do not use finite-difference approximations to derivatives would compute the objective function only at feasible points. Rounding errors will inevitably cause a level of infeasibility of the order of machine precision; however, this need not be a problem if the right-hand sides of the constraints are perturbed by a quantity of similar magnitude before the solution commences. If finite differences are used to approximate either first or second derivatives, the function may be computed at a point where the constraint violation is much larger than the machine precision. If the working set contains only bound constraints, and there are no equality bound constraints, finite-difference formulae can be modified so that F is evaluated at only feasible points (see Section 5.5.1). However, *in general, the user must provide exact derivatives in order to be sure that F is always evaluated at feasible points.*

8.1.1.3. Selecting a method for nonlinear constraints. In contrast to the linear-constraint case, where most algorithms generate feasible points, many techniques for nonlinear constraints compute a sequence of iterates that are not guaranteed to be feasible (in fact, for some methods, it is known *a priori* that the iterates will be infeasible; see Section 6.2.1.1). Thus, the user must determine whether feasibility is important before selecting a method. In some instances, the choice will follow automatically from the nature of the problem. For example, if the objective function is undefined or meaningless for points in the infeasible region, a feasible-point method must be used.

A barrier transformation (see Section 6.2.1.2) cannot be applied to a nonlinear *equality* constraint. However, a feasible-point method based on a barrier transformation (such as the QP-based projected Lagrangian method described in Section 6.5.3.4) can be applied to a problem with *linear* equality constraints, if linear constraints are treated by one of the methods of Section 5.1.

Users sometimes attempt to "eliminate" nonlinear equality constraints from the problem by the following method. Suppose that there are t nonlinear equality constraints. The variables are partitioned into "independent" and "dependent" sets such that $x = (\ x_I \quad x_D\ )$, where x_D is an t-vector and x_I is an $(n-t)$-vector. The minimization is then performed only with respect to the independent variables x_I, and the dependent variables are determined by "solving" the equality constraints. (This method is the basis of the class of reduced-gradient-type methods considered in Section 6.3.) *We believe that this approach is rarely to be recommended.* We have indicated in Section 6.3 that attempting to satisfy a set of nonlinear constraints exactly at each iteration can lead to inefficiency. Furthermore, if the dependent variables are subject to simple bounds, it is not straightforward to incorporate these into such a method. The user would be better advised to use a method specifically designed for nonlinear equality constraints.

8.1.2. The User Interface

In the opinion of some, the purpose of software is to "... relieve [the user] ... of any need to think" (Davis and Rabinowitz, 1967). For some types of numerical problems, this aim may be achievable, within the limitations of finite-precision arithmetic and the inherent impossibility of "solving" every problem as the user might wish. Unfortunately, optimization problems in their full generality do not readily lend themselves to allowing the user to abandon the need to think, since certain critical aspects of the problem can in many instances be defined only by the user.

There is often a wide range of possible roles for the user in applying optimization software. At one extreme, the user is required to do nothing but specify the problem functions; at the other, the user is required to specify in detail certain features of the problem, qualities of the desired solution, and properties of the sequence of iterates. This variation in the role of the user with respect to the software is partially reflected by the number of parameters that the user must provide.

Optimization methods involve numerous decisions that are inherently problem-dependent, which are often reflected in certain parameters associated with the implementation of the algorithm. There are no *a priori* optimal choices of these parameters for all problems. However, in order to simplify life for the inexperienced or uninterested user, optimization software tends to have two forms of user interface that remove the need for the user to specify these parameters.

Firstly, when the optimization method is in the form of a subroutine to be called by the user, an "easy-to-use" version of the method may be provided. The "easy-to-use" routines have short calling sequences, so that only a minimal number of parameters need to be specified by the user. Secondly, when the user prepares an input file to be processed by the optimization routine, the user is required to specify only a minimal number of parameters.

With either of these approaches, the parameters not specified by the user must be chosen by the method. The parameters associated with optimization methods are generally considered to fall into two classes: those for which default values may be chosen, and those that can be computed automatically for each problem.

8.1.2.1. Default parameters. A reasonable *a priori* choice for some parameters is possible if the problem to be solved satisfies a specified set of assumptions. These *default values* are used if the user does not specify them. Default values are typically based on certain assumptions about the problem. For example, the choice of termination criteria depends on the problem's scaling. If the problem functions can be calculated to full machine precision, and if the problem is "well scaled" (see Section 8.7), then the termination criteria can be based on the machine precision (see Section 8.2.3 for a full discussion of termination criteria).

Default parameter values are considered to be "safe", in the sense that the idealized problem upon which their selection is based is expected to correspond fairly well to most practical problems. It is also expected that use of a default parameter value should not cause a severe decrease in efficiency for most problems.

For some problems, however, the default parameter values are not appropriate. In particular, if the user's problem does not conform to the assumed model, the algorithm's performance may be substantially influenced by the choice of these parameters, which may even be crucial to success for some problems. Furthermore, it may be inadvisable to use default values of parameters when the problem (or a set of similar problems) will be solved repeatedly, and is expensive to solve. Although the default parameters might produce an *adequate* solution, efficiency could probably be improved by "tuning" the chosen algorithm to the particular problem form. (Advice on the choice of optional parameters is given in Section 8.1.3.)

8.1.2.2. Service routines. Certain other parameters are too sensitive to the given problem for a fixed default value to be acceptable. Consequently, a set of *service routines* are sometimes provided to compute problem-dependent parameters. Although ideally these parameters would be specified based on the user's detailed knowledge of the problem, the process of selection can be automated to a significant extent. A carefully designed automatic routine is almost certain to give a better result than the random guess that might be made by an uninformed user.

For example, with a quasi-Newton method based on approximating first derivatives (see Section 4.6.2), a set of finite-difference intervals must be provided. The best choice of these intervals is quite complicated, and furthermore their values can have a major influence on whether the method is successful in solving the problem. Hence, a service routine may be used to compute a sensible set of finite-difference intervals (see Section 8.6.2). Although the technique cannot be guaranteed, the initial set of intervals typically remains a reasonable choice throughout the minimization, assuming that the scaling of the problem does not change too drastically.

Other service routines can provide a check on the consistency of the problem specification — for example, by testing whether the derivatives appear to be programmed correctly. An automatic procedure of this type is briefly described in Section 8.1.4.2.

When an easy-to-use routine is provided, service routines of this type are usually called automatically. When the user prepares an input file, the use of such service routines is sometimes one of the available options.

8.1.3. Provision of User-Defined Parameters

The user-specified parameters of an implemented algorithm allow the user to "tune" a method to work most effectively on a particular problem, and/or to provide additional information that may improve the algorithm's performance.

The significance of some user-supplied parameters is straightforward. For example, with a quasi-Newton method the usual initial approximation to the Hessian is taken as the identity matrix. However, a better estimate can be specified by the user if it is available.

Other, more complicated, parameters control decisions made within the algorithm. To select appropriate values for these parameters, it is necessary to understand how each enters the execution of the steps of the algorithm. The next eight sections contain a discussion of some typical parameters that occur in optimization software. Suggestions concerning the choice of finite-difference intervals are given in Section 8.6.

8.1.3.1. The precision of the problem functions. The user is often required to specify an estimate of the precision to which the problem functions are computed. Among those processes that depend critically upon the accuracy of the problem functions are: (i) the computation of the condition error in a finite-difference method (see Section 4.6.1.1); (ii) the determination of the correct scaling for x (see Section 8.7); (iii) the determination of whether or not an iterate satisfies the termination criteria (knowledge of the precision significantly reduces the probability of both premature termination and finding the solution to unnecessary accuracy; see Section 8.2); and (iv) the determination of the minimum spacing between points obtained during the computation of the step length (see Sections 4.3.2.1 and 8.4.2).

Note that in solving an optimization problem, *an estimate of the accuracy of the problem functions is needed at values of the variables that are unknown at the start of the minimization* (in contrast to other numerical problems, where the precision at only one point is required). Consequently, given an estimate of the precision at the starting point, an optimization algorithm must have the capability of estimating the accuracy at any point computed during the course of the solution. (Several techniques for estimating accuracy are described in Section 8.5.)

In general, the user will be required to give an estimate of a bound on the absolute error in the *computed function value* in a neighbourhood of the initial point x_0, i.e., the user must furnish a scalar ϵ_A such that

$$fl\big(F(x)\big) = F(x) + \sigma,$$

with $|\sigma| \leq \epsilon_A$ for all x "near" x_0 (see Section 2.1.6). Hence,

$$|fl\big(F(x)\big) - F(x)| \leq \epsilon_A.$$

Users often think of the accuracy of the computed function value in *relative* terms, i.e., that the computed value is correct to a certain number of significant figures. We shall show later (Section 8.5) that this form of error bound may be unsuitable when the function value is small. However, for some functions, it is possible to express the error in the computed value of F in terms of ϵ_R, *a bound on the relative precision*. The scalar ϵ_R is such that if

$$fl\big(F(x)\big) = F(x)(1 + \epsilon),$$

then $|\epsilon| \leq \epsilon_R$.

If the word length of the computer is large, the estimate of the absolute error can be incorrect by one or two orders of magnitude without unduly affecting the success of the minimization. In this case, if the user believes that approximately the first r significant digits of F are correct, then it is usually acceptable to specify the values $\epsilon_R = 10^{-r}$ and $\epsilon_A = 10^{-r}(1 + |F|)$. Accurate estimates of the precision are more important if the word length is small, because an over- or under-estimate of the error may cause difficulties.

We emphasize that ϵ_A and ϵ_R *should reflect only the error incurred during the floating-point computation of* F, *and not the accuracy with which* F *may represent some real-world problem*. (For a further discussion of this point, see Sections 7.3.1 and 8.5.1.1.)

If a user has no idea of the accuracy of a computed problem function, a service routine should be used that automatically estimates ϵ_A. For a further discussion of how ϵ_A may be estimated, see Section 8.5.

8.1.3.2. Choice of step-length algorithm. At a typical iteration of an unconstrained method, a step-length algorithm is executed to determine the step to be taken along the direction of search (see Section 4.3.2.1). Many step-length algorithms are based on safeguarded parabolic or cubic approximations to the behaviour of the given function along the search direction; the cubic procedure utilizes function and gradient values at every trial point, while the parabolic procedure uses only function values except possibly at the initial point of each iteration.

The choice of a parabolic or cubic step-length algorithm is sometimes straightforward. For example, within a Newton-type algorithm based on exact or approximated second derivatives, the slightly more robust cubic step-length algorithm is always used, since the cost of evaluating the gradient at every trial point is assumed to be reasonable. On the other hand, within an algorithm based on finite-difference approximations to first derivatives, the parabolic algorithm is chosen because of the excessive cost of the n function evaluations needed to approximate the gradient at each trial step.

With a quasi-Newton method that uses exact gradients, the best step-length algorithm ("best" in the sense of "with fastest execution time") depends on the expense involved in computation of the gradient vector relative to a calculation of the function value. In particular, the parabolic method will normally lead to faster execution time if calculation of the gradient is significantly more expensive than evaluation of the function. However, if difficulties in the line search occur, the cubic step-length procedure should be used because of its increased robustness.

8.1.3.3. Step-length accuracy. The parameter η, which satisfies $0 \le \eta < 1$, controls the accuracy to which the step-length procedure attempts to locate a local minimum of the function of interest along the search direction at each iteration (see Section 4.3.2.1). The smaller the value of η, the more accurately the univariate minimization is performed. Generally speaking, a smaller value of η causes an increase in the average number of function evaluations per iteration, and a decrease in the number of iterations required for convergence, with the reverse effect as η is increased. However, reduction of η eventually leads to a situation of diminishing returns, in the following sense. If η is reduced from, say, .9 to 10^{-3}, the number of iterations will typically be at least halved, and the total number of function evaluations required to find the solution will increase by about 50%. Decreasing η further, say to 10^{-6}, usually results in only a marginal reduction in the number of iterations, but still causes a significant increase in the total number of function evaluations. This tendency is particularly noticeable when the function-value-only step-length algorithm is used, since the polynomial approximation tends to break down close to a local minimum along the search direction.

The "best" value of η (that allows the solution to be found in the shortest time) depends on the problem. However, for many problems and algorithms the variation in the optimal η is small. Hence, there is little decrease in efficiency if η is set to an averaged value based on extensive computational experience; this value is the default option, and is selected for the easy-to-use routines.

Nonetheless, for some problem categories the optimal value of η may differ significantly from the default value. In particular, the default value is based on problems in which the evaluation of the problem functions dominates the computational cost. For cases where the problem functions are relatively cheap to evaluate, it is usually worthwhile to choose a value of η smaller than the default. By assumption, the increased number of function evaluations will then be offset by the decrease in computational effort associated with performing each iteration (e.g., solving a linear system, updating a matrix factorization, etc.).

Another instance in which a smaller value of η is justified occurs when gradient evaluations are significantly more costly than function evaluations. In this case, the user would select the parabolic step-length procedure as well as specify a smaller value of η in order to reduce the number of iterations.

The selection of η for conjugate-gradient-type algorithms (Section 4.8.3) is more complicated than for the other methods. With a conjugate-gradient method, the optimal η for many problems varies considerably from the default value. Therefore, if a user must solve a significant number of similar problems with a conjugate-gradient algorithm, it is worthwhile to conduct numerical experiments to determine a good value of η.

8.1.3.4. Maximum step length. In many algorithms, it is useful to limit the maximum change in x that can be made during any one iteration (see Section 4.3.2.1). Such a limit can be imposed with the user-supplied parameter Δ. At each iteration, the step length α must satisfy $\|\alpha p\| \le \Delta$.

There are several reasons for allowing this parameter to be set by the user:

(1) to prevent overflow in the user-supplied problem function routines;

(2) to increase efficiency by evaluating the problem functions only at "sensible" values of x;

(3) to prevent the step-length algorithm from returning an inordinately large value of α because the relevant convergence criteria are not satisfied at any smaller value (for instance, because the function is unbounded below along the given direction);

(4) to attempt to force convergence to the solution nearest to the initial estimate.

Similar objectives can also be achieved (more consistently) if reasonable upper and lower bound are always placed on all variables. Such bounds need not be highly accurate (although algorithms tend to perform more efficiently with better bounds).

8.1.3.5. A bound on the number of function evaluations. This parameter allows the user to terminate the computation after a fixed number of function evaluations. This "fail-safe" feature is useful in various circumstances when the user is uncertain that the method will succeed in solving the problem. For example, if an unconstrained problem has an unbounded solution, any descent method without such a safeguard would continue iterating until floating-point overflow occurs.

If the user is uncertain about a reasonable value for the maximum number of function evaluations, and there is little or no penalty for restarting, it is usually prudent to allow only a limited number of function evaluations at first. Examination of the results will usually indicate whether the method appears to be converging (see Section 8.3.1), and a conservative strategy minimizes the risk of wasting computer time by solving a problem that is posed incorrectly or that may need to be altered.

A lower bound on the number of function evaluations can be obtained for unconstrained problems by considering the number of evaluations that would be required to minimize a positive-definite quadratic function. For example, when a quasi-Newton method that uses exact gradients is applied to a quadratic function, the solution should be found in no more than $n + 1$ iterations. Each iteration requires at least one function/gradient evaluation, so that $n + 1$ evaluations would be required for the quadratic case. Hence, a useful approximate *lower bound* on the maximum number of function evaluations for a general problem using a quasi-Newton method would be $5n$.

8.1.3.6. Local search. When second derivative information is not available, it is impossible to confirm in general whether the sufficient conditions for optimality (see Chapter 3) are satisfied at the computed solution. In this situation, to avoid convergence to a saddle point, a different procedure — a *local search* — can be invoked, which attempts to find an improved point using a strategy radically different from the algorithm that produced the alleged solution. If no improved point can be found during a local search, it is considered that a satisfactory solution has been computed.

Local search procedures include "random" searching in the neighbourhood of the computed solution for a lower point, and hence require additional function evaluations. The decision as to whether a local search should be executed depends on the problem. If the nature of the problem makes it unlikely or even impossible that convergence to a saddle point will occur, the user may choose not to perform the local search, especially if function evaluations are very costly. However, for maximum robustness on general problems, the local search should be carried out.

8.1.3.7. The penalty parameter in an augmented Lagrangian method. An augmented Lagrangian method includes a penalty parameter ρ (see Section 6.4 for a discussion of augmented Lagrangian methods). The user is sometimes required to specify an initial value of ρ, which may subsequently be adjusted by the routine, depending on the results with the user's value. (Usually, the value of ρ is only increased.)

In general, the choice of the parameter ρ has a substantial effect on the performance of an augmented Lagrangian method. If ρ is too small or too large, an unconstrained subproblem within the method may be ill-conditioned (see Figure 6i). Furthermore, unless ρ exceeds an unknown threshold value, the unconstrained subproblem will have an unbounded solution.

It is difficult to offer good general-purpose advice concerning the selection of ρ, since the threshold value depends on the Hessian of the Lagrangian function at the solution. A reasonable choice in practice (assuming that the problem functions are well-scaled) is $\max\{10, 10|F(x_0)|\}$. If a good solution to the problem is known, but the user is unable to specify initial Lagrange multiplier estimates to the routine, ρ should be set to a large value in order to avoid diverging from the initial point.

If the user has previously solved a similar problem using an augmented Lagrangian method, the final value of the penalty parameter from a successful run is a good guide to the initial choice. If the previous initial value remained unaltered, it is probably worthwhile to try a smaller value (say, decreased by a factor of 10).

Some of the difficulties associated with a poor initial choice for ρ can be eased by placing sensible bounds on all the variables.

8.1.3.8. The penalty parameter for a non-smooth problem.

When using the algorithm of Section 6.2.2.2 to solve a non-smooth problem, the user is asked to specify an initial penalty parameter ρ. Exactly as in the augmented Lagrangian case discussed in Section 8.1.3.7, the user's value is likely to be adjusted by the algorithm; however, it may be either increased or decreased.

Fortunately, the choice of the initial ρ for a non-smooth problem tends to be less critical than in an augmented Lagrangian method. The safest strategy is to choose a relatively *large* value initially, in order to reduce the likelihood of an unbounded subproblem and the non-existence of the desired local minimum of the non-differentiable penalty function. Although the initial subproblem may be ill-conditioned because of an overly large value of ρ, the conditioning of subsequent subproblems should improve as the method automatically decreases the penalty parameter.

A "reasonable" choice under most circumstances is $\rho = \max\{100, 100|F(x_0)|\}$. In contrast to the situation with an augmented Lagrangian method, a *smaller* initial value should be specified if a good estimate of the solution is known — say, $\max\{1, |F(x_0)|\}$. Again, including sensible bounds on the variables can improve the efficiency of the method, and make its performance less sensitive to a poor choice of ρ. (In this case, the bound-constraint violations will be included in the penalty term.)

8.1.4. Solving the Correct Problem

We note the truism that even the best method is unlikely to find the correct solution to the wrong problem. Errors in the user's formulation are sufficiently common that they should always be checked for when unexplainable difficulties occur.

8.1.4.1. Errors in evaluating the function.

The first point to be verified is whether the user's code to evaluate the problem functions is correct. One obvious check is to compute the value of the function at a point where the correct answer is known. However, care should be taken that the chosen point fully tests the evaluation of the function. It is remarkable how often the values $x = 0$ or $x = 1$ are used to test function evaluation procedures, and how often the special properties of these numbers make the test meaningless.

Special care should be used in this test if the computation of the objective function involves some subsidiary data that is communicated to the function subroutine by means of an array parameter or (in Fortran) by COMMON storage. Although the first evaluation of the function may be correct, subsequent calculations may be in error because some of the subsidiary data has been accidentally overwritten. (This situation is one reason why the number of iterations should be severely restricted during the first solution attempt.)

Errors in programming the function may be quite subtle in that the function value is "almost" correct. For example, the function may be accurate to less than full precision because of the inaccurate calculation of a subsidiary quantity, or the limited accuracy of data upon which the function depends. A common error on machines where numerical calculations are usually performed in double precision is to include even one single-precision constant in the calculation of the function; since some compilers do not convert such constants to double precision, half the correct figures may be lost by such a seemingly trivial error.

Some optimization algorithms are affected in only a minor way by relative errors of small magnitude in the calculation of the function. With such methods, the correct solution will usually be found, but the routine will report an inability to verify the solution. However, other methods — those based on finite-difference approximations to first derivatives — are catastrophically affected by even small inaccuracies. They may perform many iterations in which essentially no progress is made, or may fail completely to move from the initial estimate (see Section 8.4). An indication that this type of error has been made is a dramatic alteration in the behaviour of the routine if the finite-difference interval is altered. One might also suspect this type of error if a switch is made to central differences even when the approximate gradient is large.

8.1.4.2. Errors in computing derivatives. Incorrect calculation of derivatives is by far the most common user error. Unfortunately, such errors are almost never small, and thus no algorithm can perform correctly in their presence. This is why we have recommended that some sort of consistency check on the derivatives should be performed.

The most straightforward means of checking for errors in the derivative involves comparing a finite-difference approximation with the supposedly exact value. Let x denote the point of interest. Choose a small scalar h and a random vector p of unit length, such that the elements of p are all similar in magnitude. It should hold that

$$F(x + hp) - F(x) \approx hg(x)^Tp. \tag{8.1}$$

If (8.1) does not hold, then (i) there is a programming error in the evaluation of g; (ii) there is a high level of rounding error in evaluating F; or (iii) F is badly scaled (see Section 8.7.1). This latter possibility can be tested by evaluating $F(x - hp)$, and then checking to see whether

$$F(x + hp) - F(x) \approx \frac{1}{2}\big(F(x + hp) - F(x - hp)\big). \tag{8.2}$$

If (8.2) holds, there is almost certainly a programming error in computing the gradient. When the right-hand sides of (8.1) and (8.2) are approximately the same, the gradient has probably been programmed correctly. If not, the whole sequence of calculations should be repeated with a larger value of h. An error in the gradient evaluation routine is overwhelmingly indicated if (8.1) does not hold for any reasonable value of h. A discussion of the behaviour of finite-difference approximations is given in Section 4.6.1 and 8.6. This finite-difference technique is quite effective in practice, although it cannot be guaranteed to find very small errors in g. Unfortunately, even small errors in g may cause the optimization to fail; the reader should refer to Section 8.4.2.4 for details of how to recognize such failures.

Second derivatives of F can be checked in a similar manner if the gradient is known to be programmed correctly. In this case, it should be true that

$$hp^TG(x)p \approx \big(g(x + hp) - g(x)\big)^Tp. \tag{8.3}$$

If (8.3) does not hold, the calculations should be repeated with a larger value of h. Let h_1 and h_2 be the two values of h; if (8.3) does nor hold for either value, but it is true that

$$h_2\big(g(x + h_1 p) - g(x)\big)^T p \approx h_1\big(g(x + h_2 p) - g(x)\big)^T p, \tag{8.4}$$

then it is likely that there is an error in the procedure that calculates the Hessian. If (8.4) does not hold, the testing procedure can be repeated with a different value of h, as in the case of first derivatives.

Small errors in the second derivatives are unlikely to prevent convergence. However, small errors may affect the asymptotic rate of convergence (see Section 8.3.1).

A good software library will include routines to check derivatives automatically, and to inform users of any inconsistencies or indications of bad scaling.

8.1.5. Making the Best of the Available Software

If a user does not have access to a comprehensive numerical software library or the library does not contain a routine for the particular problem category of interest, the user must make the best of what software he has. *The general rule is to select a solution method so that the maximum utilization is made of any structure in the problem.* In the remainder of this section, we shall discuss some instances in which available software can be used to solve problems for which it was not specifically designed.

8.1.5.1. Nonlinear least-squares problems. We have indicated in Section 4.7 that exploiting the special form of a least-squares problem can lead to substantial improvements in efficiency, compared to using a method for a general function. Suppose that the user wishes to solve a (possibly constrained) nonlinear least-squares problem, but has access only to methods for a general objective function.

Let $J(x)$ denote the Jacobian matrix of the set of functions (see Section 4.7 for a definition of the notation). If the objective function is small at the solution, the matrix $J(x)^T J(x)$ will be a good estimate of the Hessian matrix. Thus, it is preferable to use a Newton-type algorithm and "pretend" that the Hessian at x is given by $J(x)^T J(x)$, rather than use a general first-derivative method. A similar approach can be used in the non-derivative case by using a finite-difference approximation to $J(x)$ to compute the "Hessian".

If $J(x)$ may be rank-deficient, it is essential to use a positive-definite matrix as the Hessian (such as $J(x)^T J(x) + \sigma I$), rather than $J(x)^T J(x)$ alone. (The scalar σ should be a small positive quantity, e.g. $\sqrt{\epsilon_A}/(1 + |F(x)|)$ for a well-scaled problem.) This ensures that the Newton-type method will not encounter difficulties because the Hessian is not sufficiently positive definite.

8.1.5.2. Missing derivatives. We have shown that most software is designed to solve a class of problems where a specific level of derivative information is available. For certain problems, most (but not all) of the first and/or second derivatives may be available. In order to make use of the available derivatives, the user may implement a procedure to approximate only the (presumably small) number of missing elements.

In order to carry out this modification, the user must observe precautions similar to those noted in Section 4.6.2 in order to ensure that the finite-difference approximations are sufficiently accurate. For example, a central difference should be used if all the elements of the gradient (approximated and exact) become small. (For constrained problems, these same observations apply to the projected gradient.)

When the Hessian matrix is approximated by finite differences, it is usual for one column of the Hessian to be determined from a single gradient evaluation. Hence, any simple scheme by which a user plans to approximate only selected elements of the Hessian would result in savings only if all the elements in one column are known. In some cases it is possible to utilize known elements of the Hessian by taking finite differences along special vectors, rather than the ordinary co-ordinate directions. Methods of this type have been proposed to take advantage of a known pattern of zeros in the Hessian, but the strategies for constructing the vectors tend to be rather complex (see Section 4.8.1).

8.1.5.3. Solving constrained problems with an unconstrained routine.

If the user must solve a nonlinearly constrained problem, but has access only to a routine designed for smooth unconstrained problems, it is possible to transform the constrained problem into a sequence of smooth unconstrained problems by using an augmented Lagrangian method (see Section 6.4). (It is *undesirable* to effect such a transformation using a penalty function method; see Section 6.2.1.1.) However, certain precautions must be observed in this situation.

It is known that, for any value of the penalty parameter, the augmented Lagrangian function may be unbounded below. Hence, the user must take precautions to ensure that the unconstrained method does not iterate "forever" — for example, by choosing relatively small values for the maximum step allowed during an iteration and the maximum number of function evaluations (see Sections 8.1.3.4 and 8.1.3.5). It is even better to use a bound-constraint routine in this context, since bounds on all the variables should help to prevent difficulties with unboundedness.

8.1.5.4. Treatment of linear and nonlinear constraints.

In Chapter 5, we have presented many theoretical and practical reasons why linear and nonlinear constraints should be treated separately in a nonlinearly constrained routine. However, the user may have access only to one routine that treats *only linear constraints*, and to another routine that treats *all constraints as nonlinear*. If a significant number of the problem constraints are nonlinear, the user should apply the routine that treats all constraints as nonlinear (assuming that it is efficient). In this case, the efficient treatment of nonlinear constraints should compensate for the inefficiencies introduced by unnecessary re-evaluations of linear constraints. However, there may be an undesirable loss of feasibility with respect to the linear constraints during the solution process, depending on the method for treating the nonlinearities. On the other hand, if almost all the constraints are linear, or feasibility with respect to the linear constraints is essential, the user should treat the nonlinear constraints by means of an augmented Lagrangian transformation (subject to the precautions mentioned in Section 8.1.5.3).

8.1.5.5. Nonlinear equations.

When no routines for solving nonlinear equations are available, we have already observed in Section 4.7.6 that a least-squares method may be applied (usually without loss of efficiency). When the system of equations includes a mixture of linear and nonlinear functions, it should be possible to exploit the special structure of some of the equations by using a routine for linearly constrained least-squares. In addition, such a routine may improve efficiency if it is necessary to solve a zero-residual least-squares problem in which some of the functions are linear (see Section 7.6.2 for a discussion of a specific problem of this type).

Notes and Selected Bibliography for Section 8.1

The user-defined parameters listed here are very similar to those that need to be specified in the software libraries available from the Numerical Algorithms Group (NAG), and the National Physical Laboratory. These libraries also use service routines and easy-to-use routines.

For a general discussion of the design principles and structure of a portable Fortran library for optimization, see Gill, Murray, Picken and Wright (1979). Decision procedures similar in nature to selection charts occur in many program libraries for solving various sorts of problems; see Fletcher (1972c), the Eispack collection, Smith *et al.* (1974), and the Numerical Algorithms Group Reference Manual (1981).

A definition of the standard MPS input format is contained in IBM document number H20-0476-2, "Mathematical Programming System/360 Version 2, Linear and Separable Programming — User's Manual," pp. 141–151. For a more condensed description of the format and for a sample of user-supplied input in which default values are used, see Murtagh and Saunders (1977), and Murtagh (1981).

All of the major optimization software libraries have service routines to check that the derivatives are programmed correctly. For further references, see Murray (1972b), Wolfe (1976), and Moré (1979a).

8.2. PROPERTIES OF THE COMPUTED SOLUTION

8.2.1. What is a Correct Answer?

The issue of deciding when a computed answer is "correct" is extremely complicated, and a detailed treatment would fill an entire volume. Hence, only an overview of some of the major points will be presented here.

When a user decides to test an algorithm by solving a problem with a known solution, two disconcerting things can happen: the algorithm may compute the correct answer, but report a failure; even worse, the algorithm may claim to have converged successfully, but produce a solution completely different from the expected one. We shall explain briefly why such phenomena are inevitable, but are not necessarily cause for concern.

It has already been emphasized that computers will fail to find the "exact" solutions to even very simple problems (see Section 2.1). Therefore, it would be unreasonable to insist on the exact solution. Nonetheless, it is often possible to assess whether the computed answer is "correct", in the sense that it is a "good" solution of the given problem.

An absolutely essential point to bear in mind is that the qualities that define the mathematical solution to a problem do not necessarily carry over to assessing a computed solution. It is frequently the case that there exists a vector function Δ such that with exact arithmetic

$$\Delta(x^*) = 0.$$

However, let $\hat{x}$ be the closest representable version of x^*, and let $\hat{\Delta}$ denote the computed version of Δ. It will almost certainly be true that $\hat{\Delta}(\hat{x})$ is non-zero. The only reasonable definition of a computed "correct answer" $\hat{x}$ is thus any value that satisfies

$$\|\hat{\Delta}(\hat{x})\| \leq \delta,$$

where δ is a suitable "small" number. However, extreme care must be exercised in defining "small" (for some problems, δ may be ten!).

Furthermore, the measure $\hat{\Delta}$ need not be (and, in fact, generally will not be) *directly* related to the closeness of a given point to the solution. Ideally, it would hold that

$$\|x_1 - x^*\| < \|x_2 - x^*\| \;\Leftrightarrow\; \hat{\Delta}(x_1) < \hat{\Delta}(x_2). \tag{8.5}$$

If (8.5) were true, the smallness of $\hat{\Delta}(x)$ would be a consistent measure of the quality of x. However, in most instances (8.5) does not hold. This situation illustrates the fundamental difficulty that *computable criteria* are in no sense equivalent to exact properties. Therefore, it is inevitable that a computed "correct answer" may *not* be close to the mathematically exact solution. Furthermore, the more ill-conditioned the problem, the less likely it is that the computable criteria will directly reflect closeness to the solution. We shall illustrate this phenomenon for the case of an ill-conditioned quadratic function in Section 8.2.2.1.

The other possibility — of reporting a failure with the "correct" answer — can occur for various reasons. Every computation in finite precision has a limit on the maximum accuracy that can be achieved, which varies with the problem and with the available precision (see Section 8.2.2). Because the solution is, in general, unknown, the accuracy of any approximation can only be *estimated* (by means like $\hat{\Delta}$, mentioned above). A "failure" reported by an algorithm merely means that the computable criteria have not been satisfied. This happens most frequently because the estimated accuracy that has been achieved is insufficient to satisfy the user's specifications. Any criteria for assessing the accuracy of the solution are inadequate in some situations, and it may be that the routine fails to make adequate progress simply because no further measurable improvements can be made.

8.2.2. The Accuracy of the Solution

There is a fundamental limit on the accuracy that can be achieved in any computed estimate of x^*. A thorough knowledge of the accuracy that can be attained is useful for two reasons. Firstly, it is much easier for a user to select the parameter that terminates the algorithm if he knows precisely what accuracy can be expected. (Asking for unattainable accuracy is one of the most frequent reasons that a routine indicates that it has failed.) Secondly, a comparison of the final accuracy with that predicted from a theoretical analysis can provide a valuable insight into the scaling and conditioning of the problem.

The limiting accuracy in a computed solution $\bar{x}$ is critically dependent upon the accuracy to which the problem functions can be computed on the user's machine. Throughout this section we shall write ϵ_A for the bound on the absolute precision in F at the solution x^* (ϵ_A may be estimated by the algorithm from a value specified at the initial point by the user; see Section 8.5.3).

8.2.2.1. Unconstrained problems. For a smooth unconstrained problem, the computable criteria that are typically used to measure the quality of the solution are the values of the function F and the gradient vector g. Suppose that the computed value of F at the solution satisfies

$$fl\big(F(x^*)\big) = F(x^*) + \sigma, \tag{8.6}$$

where $|\sigma| \le \epsilon_A$ (see Section 2.1.6). Given a point $\bar{x}$ such that

$$|F(\bar{x}) - F(x^*)| \le \epsilon_A, \tag{8.7}$$

then no "better" point than $\bar{x}$ can be computed, since $F(\bar{x})$ is as close to $F(x^*)$ as the computed function value at the solution. Any point $\bar{x}$ that satisfies (8.7) will be called an *acceptable solution*.

General observations. Given the property (8.7) of $F(\bar{x})$, we wish to derive some bound on the accuracy of $\bar{x}$, i.e. a bound on $\|\bar{x} - x^*\|$. Assume that F is twice-continuously differentiable, and that x^* is a strong local minimum; then $g(x^*) = 0$ and $G(x^*)$ is positive definite. Expanding F in a Taylor-series expansion about x^* gives

$$F(\bar{x}) = F(x^* + hp) = F(x^*) + \frac{1}{2}h^2 p^T G(x^*)p + O(h^3), \qquad (8.8)$$

where $\|p\| = 1$ and $|h| = \|\bar{x} - x^*\|$. It follows from (8.7) and (8.8) that for any acceptable solution,

$$|F(\bar{x}) - F(x^*)| \approx \frac{1}{2}h^2 |p^T G(x^*)p| \approx \epsilon_A,$$

or, equivalently,

$$\|\bar{x} - x^*\|^2 \approx \frac{2\epsilon_A}{p^T G(x^*)p}. \qquad (8.9)$$

Without any further assumptions, (8.9) reveals that the conditioning of $G(x^*)$ can substantially affect the size of $\|\bar{x} - x^*\|$. If $G(x^*)$ is ill-conditioned, the error in $\bar{x}$ will vary with the direction of the perturbation p. In particular, if p is a linear combination of the eigenvectors of $G(x^*)$ corresponding to the *largest* eigenvalues, the size of $\|\bar{x} - x^*\|$ will be relatively small. On the other hand, if p is a linear combination of the eigenvectors of $G(x^*)$ corresponding to the *smallest* eigenvalues, the size of $\|\bar{x} - x^*\|$ will be relatively large. Thus, the error in an acceptable solution of an ill-conditioned problem may be very large along certain directions.

It is also useful to ascertain the limiting accuracy in the gradient vector, since the size of g is often used as an indication of the quality of the solution. If we expand $g(x)$ in its Taylor-series expansion about x^* we have

$$g(\bar{x}) = g(x^* + hp) = hG(x^*)p + O(h^2)$$

(recall that $g(x^*) = 0$). Substituting from (8.9), we obtain

$$\|g(\bar{x})\|^2 \approx 2\epsilon_A \frac{\|G(x^*)p\|^2}{p^T G(x^*)p}. \qquad (8.10)$$

Once again, we see the effect of the conditioning of $G(x^*)$. If $G(x^*)$ is badly conditioned, then $\|g(\bar{x})\|$ will tend to be *large* when p is a linear combination of the eigenvectors of $G(x^*)$ corresponding to *large* eigenvalues, and *small* when p is a linear combination of the eigenvectors of $G(x^*)$ corresponding to the *small* eigenvalues.

Example 8.1. We illustrate the effects of a badly conditioned Hessian with a two-dimensional quadratic function.

$$F(x_1, x_2) = (x_1 - 1)^2 + 10^{-6}(x_2 - 1)^2 + 1,$$

whose solution is obviously $x^* = (1,1)^T$. Assume that $\epsilon_A = 10^{-6}$. At the point $\bar{x} = (1,2)^T$, $F(\bar{x}) = 1 + 10^{-6}$, and hence $\bar{x}$ is an acceptable point, even though $\|\bar{x} - x^*\|_2 = 1$. At the point $\hat{x} = (1 + 10^{-3}, 1)^T$, $F(\hat{x}) = 1 + 10^{-6}$, and thus $\hat{x}$ is also acceptable (furthermore, $\hat{x}$ is closer to x^*). However, note that $\|g(\hat{x})\|_2 = 2 \times 10^{-3}$ and $\|g(\bar{x})\|_2 = 2 \times 10^{-6}$, so that using the size of $\|g\|_2$ to measure the quality of the solution would indicate that $\bar{x}$ was the "better" point. This emphasizes the observation of Section 8.2.1 that the available computable criteria are not directly related to the quality of the alleged solution.

Well-scaled problems. The results (8.9) and (8.10) indicate the effect of the conditioning of the Hessian at x^* on the accuracy of an acceptable solution. In order to give a general "feel" for the sizes of these errors, we shall consider the proverbial "well-scaled" problem (see Section 8.7). The properties of such a function include the following: $G(x^*)$ has a condition number that is not "too big"; $F(x^*)$ is of order unity; and $\|x^*\|$ is of order unity. Under these assumptions, the relationships (8.9) and (8.10) become

$$\|\bar{x} - x^*\| = O(\sqrt{\epsilon_A})$$

and

$$\|g(\bar{x})\| = O(\sqrt{\epsilon_A}).$$

These results lead to the common "folklore" observation that, if F is well-scaled at x^*, the *most* accuracy that can be expected in an acceptable solution is about *half* the correct figures obtained in F. (It is important to note that this is an estimate about the *norm* of $\bar{x}$ only, rather than a statement about the accuracy of each component.) Furthermore, the norm of the gradient vector at an acceptable solution will be similar to the accuracy in x.

Ordinarily, $\epsilon_A > \epsilon_M$; hence, the maximum number of correct figures that we can typically expect in $\|\bar{x}\|$ is half the number of figures in the mantissa. In some circumstances, more accuracy can be achieved in $\|\bar{x}\|$ because F can be calculated with greater-than-usual precision (see Section 8.5.1.3).

Non-smooth problems. All our comments regarding the accuracy of a solution have been with reference to twice-continuously differentiable functions. If x^* is a point of discontinuity of $F(x)$, there may exist points in an arbitrarily small neighbourhood for which the function is nothing like that at x^*. A possible configuration of points is depicted in Figure 8b. The accuracy of the final point will depend upon whether the algorithm terminated with the point $\bar{x}_1$ or the point $\bar{x}_2$. In general, the "larger" the neighbouring set of points for which $F(x) \approx F(x^*)$, the greater the probability that $\bar{x}$ is close to x^*; however, *it is difficult to make any statements that are guaranteed to hold under all circumstances.*

If x^* is a point discontinuity of the first derivatives it is more likely that $F(\bar{x}) \approx F(x^*)$ but it is still not possible to give any theoretical estimates except in very special cases.

8.2.2.2. Accuracy in constrained problems. When all the constraints are linear, the limiting accuracy of an iterate $\bar{x}$ may be determined by analyzing its precision in two orthogonal spaces — the range and null spaces of the matrix $\hat{A}$ of active constraints.

Accuracy in the range space – linear constraints. In contrast to the unconstrained case, we shall show that the accuracy of a computed solution in the range space depends not on the accuracy of F, but on the condition number of $\hat{A}$.

The exact solution x^* satisfies

$$\hat{A}x^* = \hat{b}.$$

We shall assume without loss of generality that the constraints have been scaled so $\|\hat{A}\|$ and $\|\hat{b}\|$ are of order unity. When a solution $\bar{x}$ is computed using any of the methods described in Chapter 5, it will hold that

$$\hat{A}\bar{x} = \hat{b} + O(\epsilon_M). \tag{8.11}$$

Let Y denote a matrix whose columns form a basis for the subspace of vectors spanned by the columns of $\hat{A}^T$, and let Z denote a matrix whose columns form a basis for the set of vectors

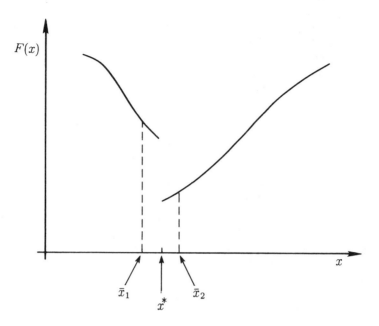

Figure 8b. Example of the accuracy obtained in a non-differentiable function. The accuracy of the final point will depend upon whether the algorithm terminated with the point $\bar{x}_1$ or the point $\bar{x}_2$.

orthogonal to the rows of $\hat{A}$. Then $\bar{x}$ may be written as

$$\bar{x} = x^* + Zp_z + Yp_Y.$$

It follows from substitution in (8.11) that

$$\hat{A}Yp_Y = O(\epsilon_M),$$

and hence that *the size of* $\|Yp_Y\|$ *depends on the condition number of* $\hat{A}$. If $\hat{A}$ is well-conditioned, the range-space perturbation in the computed solution will be of order ϵ_M. If $\hat{A}$ is ill-conditioned, the range-space perturbation can be arbitrarily large, even though $\bar{x}$ satisfies the active constraints almost exactly.

It is of some interest to consider the change in F that results from the perturbation Yp_Y. Expanding F in its Taylor series about x^*, we obtain

$$\begin{aligned} F(x^* + Yp_Y) &= F(x^*) + g(x^*)^T Yp_Y + O(\|Yp_Y\|^2) \\ &= F(x^*) + \lambda^{*T} \hat{A}Yp_Y + O(\|Yp_Y\|^2) \end{aligned} \tag{8.12}$$

(recall that $g(x^*) = \hat{A}^T \lambda^*$). If $\hat{A}$ is not ill-conditioned, then

$$F(x^* + Yp_Y) - F(x^*) \approx \lambda^{*T} \hat{A}Yp_Y = O(\epsilon_M \|\lambda^*\|).$$

If $\hat{A}$ is ill-conditioned, it is no longer possible in general to neglect the higher-order terms of (8.12).

Accuracy in the null space — linear constraints. The derivation of the limiting accuracy in the null space is similar to that in Section 8.2.2.1 for the unconstrained case. To simplify the discussion, we shall consider a vector $\hat{x}$ that is perturbed from x^* *only in the null space*, i.e. $\hat{x} = x^* + Zp_z$. An analysis similar to that in Section 8.2.2.1 gives formulae analogous to (8.9) and (8.10):

$$\|\hat{x} - x^*\|^2 \approx \frac{2\epsilon_A}{p_z^T Z^T G(x^*) Z p_z}$$

and

$$\|Z^T g(\hat{x})\|^2 \approx 2\epsilon_A \frac{\|Z^T G(x^*) Z p_z\|^2}{p_z^T Z^T G(x^*) Z p_z}.$$

Thus, the conditioning of the projected Hessian matrix affects the accuracy of the computed solution in the null space, exactly as with the full Hessian in the unconstrained case.

In summary, for a well-scaled, well-conditioned linearly constrained problem, the error in the computed solution will be dominated by the null-space component. When $\hat{A}$ is ill-conditioned, the analysis of accuracy is more complex and is beyond the scope of this text.

Nonlinear constraints. When the problem contains *nonlinear* constraints, for a well-scaled, well-conditioned problem, the estimates of the attainable accuracy are similar to those in the linear-constraint case (with $\hat{A}(x^*)$ replacing $\hat{A}$, and the Hessian of the Lagrangian function $W(x^*, \lambda^*)$ replacing $G(x^*)$). When the problem is ill-conditioned, the complications include not only those mentioned in the linear-constraint case, but also additional difficulties because of the need to consider nonlinear perturbations.

8.2.3. Termination Criteria

8.2.3.1. The need for termination criteria. Almost all optimization methods converge only in the limit, even those that would terminate in a finite number of steps with exact arithmetic. It is important for users who must specify termination criteria, or interpret the computed results, to understand the nature and significance of termination criteria.

Termination criteria are necessary for several reasons. First, there must be some means of deciding when the current estimate of the solution is acceptable. Furthermore, even when an acceptable solution has not been found, it may be desirable to stop the computation when it appears that progress is unreasonably slow, or when a specified amount of resources (e.g., computing time, function evaluations) has been consumed. It may also happen that an acceptable solution does not exist, or that the iterative process is cycling.

Termination criteria should therefore be designed to ensure that an adequate approximation to the solution has been found, minimize the risk of a false declaration of convergence, and avoid unwarranted effort.

In practice, the decision concerning whether the iterate x_k is an adequate approximation to the solution is based on two tests: whether x_k almost satisfies the sufficient conditions for optimality, and whether the sequence $\{x_k\}$ appears to have converged.

Unfortunately, for many problems only a few of the optimality conditions can be tested, and even these tests are approximate at best (see Section 8.2.1). Therefore, considerable reliance must be placed on a decision that the sequence of iterates is converging. Although certain computable characteristics can be observed when a sequence is converging, the decision as to convergence is necessarily speculative, in that convergence to a non-optimal point or an extended lack of significant progress over several iterations may be indistinguishable from convergence to the correct solution.

We emphasize that no set of termination criteria is suitable for all optimization problems and all methods. Termination criteria are necessarily based on a set of assumptions about the behaviour of the problem functions and the independent variables. Any criterion may be unsuitable under certain conditions — either by failure to indicate a successful termination when the solution has been found, or by terminating with an indication of success at a non-optimal point. Our intention is to highlight those quantities that can be used to terminate an algorithm and to discuss the complicated issues that are involved in deciding whether or not a point is a good estimate of the solution.

8.2.3.2. Termination criteria for unconstrained optimization. In this section we shall discuss conditions that can be used to test whether x_k should be regarded as a good approximation to the solution of an unconstrained problem. We shall suggest termination conditions that involve a *single* user-specified parameter that can be used to indicate the desired accuracy. The first question that arises is the form in which the user specifies the single parameter. Two natural choices are the desired accuracy in x_k and the desired accuracy in F_k. As we have shown in Section 8.2.2.1, there is a well-defined relationship between these two values when the problem is well-behaved. *However, if the problem is ill-conditioned, we can expect that F_k will be a good approximation to $F(x^*)$, but not that x_k will be a good approximation to x^*.* (See Section 8.2.2.1 for a discussion of this point.) Thus, we suggest that the user should specify a parameter τ_F that indicates the number of correct figures desired in F_k, where for this purpose, *leading zeros after the decimal point are to be considered as candidates for correct figures.* For example, if τ_F is given the value 10^{-6}, the user requires a solution with the objective function correct to six significant figures or six decimal places, depending on whether or not F is greater than or less than unity.

Smooth problems. We shall assume that τ_F can be converted into a measure θ_k of *absolute accuracy*, defined as

$$\theta_k = \tau_F\big(1 + |F_k|\big) \tag{8.13}$$

(the reasons for the definition (8.13) will be discussed below).

A sensible set of termination criteria within an algorithm for smooth unconstrained minimization would indicate a successful termination if the following three conditions are satisfied:

U1. $F_{k-1} - F_k < \theta_k$;

U2. $\|x_{k-1} - x_k\| < \sqrt{\tau_F}\,\big(1 + \|x_k\|\big)$;

U3. $\|g_k\| \leq \sqrt[3]{\tau_F}\,\big(1 + |F_k|\big)$.

Conditions U1 and U2 are designed to test whether the sequence $\{x_k\}$ is converging, while condition U3 is based on the necessary optimality condition $\|g(x^*)\| = 0$. Note that although the user-specified tolerance involves the change in the objective function, we have also included a condition on the convergence of the sequence $\{x_k\}$. For a well-scaled problem, the satisfaction of U1 will automatically imply the satisfaction of U2. However, for ill-conditioned problems, the condition U2 forces the algorithm to try harder to find a better point.

To illustrate the complementary effects of U1 and U2, consider the behaviour of an unconstrained algorithm on an ill-conditioned problem. If x_k satisfies U1, but is far from x^*, the previous search direction must have been nearly orthogonal to the steepest-descent direction. Thus, failure to satisfy U2 indicates that relatively large changes in x are being made, and there is some hope that better progress will be made at a later iteration. On the other hand, if x_k is close to the solution, failure to satisfy U2 is an indication that F has a "flat" minimum. Once again, the algorithm should continue, since it cannot be assumed that if $F_{k-1} - F_k < \theta_k$, then $F_k - F(x^*) < \theta_k$.

Although the general purpose of conditions U2 and U3 is clear, we shall explain in more detail the motivation for the specific form chosen. An alternative test in U1 would define θ_k as $\tau_F |F_k|$. However, if $|F_k|$ is small, the test U1 would then be unduly stringent. In our experience, a small value of F is usually the result of cancellation, and hence we prefer to use similar measures of error when $F_k = 0$ and when F_k is of order unity (for further comments on the expected accuracy in computed functions, see Section 8.5.1.3). The definition of θ_k (8.13) allows the criteria to reflect our requirements. From our previous analysis in Section 8.2.2.1, the right-hand side of U2 might have been taken as the square root of $\theta_k / \|G(x^*)\|$, but this test is impractical because $G(x^*)$ is not usually available. An alternative test in U2 would be the square root of $\theta_k / |F_k|$ or just τ_F if we use arguments similar to those used to justify (8.13). However, the Hessian is affected by the scaling of x as well as the scaling of F. For example, if x^* is $O(10)$ and the variables are changed so that $y_i = x_i / 10$, the objective function at the solution is unchanged by the transformation, yet the Hessian is scaled by a diagonal matrix of order 100. Clearly we would achieve the same relative accuracy in the y variables as in the x variables. This justifies the use of the norm of x_k on the right-hand side of U2. The term $1 + \|x_k\|$ is included because we are assuming that the attainable accuracy for a problem with a zero solution is the same as that with a unit solution.

From the justification for U2 and the results of Section 8.2.2.1, one might have expected the right-hand side of condition U3 to involve $\sqrt{\tau_F}$, and an argument can be made to support this choice. However, in our experience, such a value makes condition U3 overly stringent. Hence, provided that U1 and U2 are satisfied, we suggest imposing a slightly weakened condition on the gradient. The reason for weakening U3 rather than U2 is that a stringent test in U2 tends to be easier to satisfy. In fact, satisfaction of a stringent condition U3 almost always implies satisfaction of U2 (but not the converse).

Conditions U1–U3 are unsatisfactory if the initial point x_0 is in such a close neighbourhood of the solution that no further progress can be made, or if an iterate x_k lands, by chance, very close to the solution. To allow for such a possibility, the algorithm also terminates with "success" if the following condition is satisfied:

U4. $\|g_k\| < \epsilon_A$.

When using a Newton-type method, we impose the additional condition U5 that $G(x_k)$ (or a suitable finite-difference approximation) should be sufficiently positive definite. If the gradient vector g_k is approximated by finite differences, the approximate gradient can be used in conditions U3 and U4.

The specific norm used to measure the length of the vectors appearing in U1–U4 will depend upon n, the number of variables. If n is small, the two-norm is satisfactory. If n is large, the two-norm is unsatisfactory because the number of terms contributing to the magnitude of the norm may cause the conditions U3 or U4 to be overly stringent. Therefore, it is recommended that the infinity norm is used for large-scale problems. Alternatively, a suitable "pseudo-norm" may be defined that is scaled with respect to n.

Non-smooth problems. The termination criteria for an algorithm for non-differentiable problems (such as the polytope algorithm of Section 4.2.2) typically involve just a single test — on the accuracy of the objective function (see Section 8.2.2.1 for further comments on the non-differentiable case). Recall that at each stage of the polytope algorithm, $n + 1$ points $x_1, x_2, \ldots, x_n, x_{n+1}$ are retained, together with the function values at these points, which are ordered so that $F_1 \leq F_2 \leq \cdots \leq F_{n+1}$. An appropriate condition for termination is then

$$F_{n+1} - F_1 < \tau_F(1 + |F_1|). \tag{8.14}$$

The single criterion is appropriate because of possible discontinuities in the objective function (see Figure 8b), and because of the characteristic of the polytope method whereby a sequence of shrinking polytopes is generated (rather than the usual sequence of iterates).

8.2.3.3. Termination criteria for linearly constrained optimization. We shall state a sensible set of termination criteria for a linearly constrained problem that has been solved by an active set method that uses an orthogonal basis for the null space (Sections 5.1 and 5.2). The suggested tests are similar to the conditions for the unconstrained problem, with two major differences. Firstly, if x_k satisfies some of the constraints exactly, the tests involve the *projected* gradient $g_z(x_k)$ (and possibly the projected Hessian $G_z(x_k)$). Secondly, the signs of the Lagrange multipliers corresponding to active constraints are tested to ensure that the active set is correct.

As in the unconstrained case, θ_k is given by (8.13). Let t_k denote the *total* number of constraints in the working set at x_k; let λ_k denote the t_k-vector of Lagrange multiplier estimates; let $\sigma_{\min}$ denote the smallest Lagrange multiplier estimate corresponding to an *inequality* constraint; and let $\lambda_{\max}$ denote the Lagrange multiplier estimate of largest modulus (i.e., the largest multiplier for both the equality and inequality constraints). We shall assume that the working set contains at least one constraint (otherwise, the conditions for termination are given by U1–U4, subject to the requirement that x_k should be feasible). A successful termination at the point x_k is indicated if the following four conditions are satisfied:

LC1. $F_{k-1} - F_k < \theta_k$;

LC2. $\|x_{k-1} - x_k\| < \sqrt{\tau_F}\,(1 + \|x_k\|)$;

LC3. $\|g_z\| \leq \sqrt[3]{\tau_F}\,\|g_k\|$;

LC4. (to be used only if there is at least one inequality constraint in the working set) if $t_k > 1$, $\sigma_{\min} \geq \sqrt{\tau_F}\,|\lambda_{\max}|$; if $t_k = 1$, $\sigma_{\min} \geq \sqrt{\tau_F}\,(1 + |F_k|)$.

As in the unconstrained case, the algorithm should terminate if condition LC4 is satisfied and

LC5. $\|g_z\| < \epsilon_A$.

Note that we have not included any test involving the residuals of the constraints in the working set, since the residuals should be negligible for a carefully implemented active set method based on an orthogonal factorization of the matrix of active constraints. However, a test on the residuals is essential in large-scale linearly constrained problems (see Section 5.6).

8.2.3.4. Termination criteria for nonlinearly constrained optimization.

Smooth problems. The application of termination criteria for the nonlinear-constraint case depends to a large extent upon the solution strategy used by the algorithm. Many algorithms solve an adaptive subproblem to obtain each new iterate (see Section 6.1.2.1), and the termination criteria with respect to the original problem are invoked only after completion of a "major" iteration (the solution of a subproblem). Criteria for algorithms that utilize adaptive subproblems typically involve measures of the progress made during the last major iteration. Such measures may include tests upon the change in the Lagrange multiplier estimates, the amount of reduction of a merit function, etc. The termination criteria used by algorithms that solve a deterministic subproblem (Section 6.1.2.1) at each iteration may differ because of variations in the properties of the sequence $\{x_k\}$. For example, a reduced-gradient-type method that involves a strict feasibility requirement generates a sequence $\{x_k\}$ such that the step from x_{k-1} to x_k tends to lie in the null space of the matrix of active constraints $\hat{A}(x^*)$. In this case, if the problem is well-scaled, we expect

$$F_{k-1} - F_k = O\big(\|x_k - x_{k-1}\|^2\big)$$

in the neighbourhood of the solution (this is not the case for methods that approach the solution non-tangentially).

Let θ_k, t_k, λ_k, $\sigma_{\min}$ and $\lambda_{\max}$ denote the same quantities as in the linear-constraint case (Section 8.2.3.3); let $\hat{c}_k$ denote the set of constraints in the working set at x_k. Let τ_C denote a user-specified tolerance for the size of infeasibility that will be tolerated in a computed solution (based on some implicit scaling of the constraints; see Section 8.7). As in the linear-constraint case, we shall assume that the working set contains at least one constraint.

The following three general criteria are suggested for nonlinear-constraint problems:

NC1. $\|\hat{c}_k\| \leq \tau_C$;

NC2. $\|g_z\| \leq \sqrt{\tau_F}\,\|g_k\|$;

NC3. (to be used only if there is at least one *inequality* constraint in the working set) if $t_k > 1$, $\sigma_{\min} \geq \sqrt{\tau_F}\,|\lambda_{\max}|$; if $t_k = 1$, $\sigma_{\min} \geq \sqrt{\tau_F}\,(1 + |F_k|)$.

Note that the test NC2 on the projected gradient now involves $\sqrt{\tau_F}$ (in contrast to the unconstrained and linear-constraint cases). This is because a test analogous to U1 (or LC1) is not included (for reasons discussed above), and hence a more stringent test is imposed.

For problems that solve deterministic subproblems, we suggest the additional condition:

NC4. $\|x_{k-1} - x_k\| < \sqrt{\tau_F}\,(1 + \|x_k\|)$.

A single-point criterion similar to U4 and LC4 would be the satisfaction of conditions NC2 and NC3, in addition to the following condition:

NC5. $\|\hat{c}_k\| < \epsilon_M$.

Non-smooth problems. Two user-provided parameters are suggested to terminate the constrained polytope method (Section 6.2.2.2). The first, τ_F, has the same significance as in the unconstrained case. The second parameter, τ_C, defines an acceptable sum of the constraint violations for termination — i.e., the algorithm will terminate if (8.14) is satisfied and

$$\sum_{i=1}^{m} |c_i(x_1)| \leq \tau_C,$$

where x_1 is the vertex of the polytope with the best value of the non-differentiable penalty function.

8.2.3.5. Conditions for abnormal termination. In this section, we briefly consider criteria that might be used in deciding that an algorithm has "failed". Obviously, there should be an indication of abnormal termination if the specification of the problem contains errors — for example, if the parameter indicating the number of variables is negative. An abnormal termination should also occur when a user-specified maximum number of function evaluations or iterations has been exceeded (see Section 8.1.3.5).

Furthermore, it is advisable to terminate an algorithm if there appears to be a failure to make any "significant progress" (see Section 8.4.3). Unfortunately, there is no universal definition of progress, and hence the criteria will vary between methods.

One nearly-universal condition that should be satisfied at every iteration is a "sufficient decrease" in some suitable merit function. If a satisfactory sufficiently lower point cannot be found, this is a clear indication that there has been some form of failure (see Section 8.4.2). One common (but unsatisfactory) criterion is to fail if the directional derivative with respect to the merit function is not "sufficiently negative", i.e. exceeds some slightly negative scalar. A better criterion is to define a minimum acceptable distance δ_k between successive iterates (the value of δ_k should be an accurate estimate of the minimum value of $\|x_{k+1} - x_k\|$ that ensures distinct values of the objective function). A failure indication is then given if no sufficiently lower value of the merit function can be found for points farther than δ_k from the current iterate.

8.2.3.6. The selection of termination criteria.
The limiting accuracy that can be obtained in a solution approximately determines the *smallest* allowable values of the termination parameters. A sensible choice of termination criteria can reduce the amount of computation required by most optimization algorithms.

Initially we shall consider the selection of the tolerance τ_F (see Section 8.2.3.2) for the case of unconstrained optimization. It might appear that a "large" value of τ_F would achieve satisfactory results, since there would then be no danger from the complications caused by overly stringent convergence criteria. However, it is wrong to assume that a low accuracy requirement (a large value of τ_F) will always achieve the desired result. The user may reason that, since the data and/or model are accurate to, say, only 3%, there is little purpose in obtaining a solution to any greater accuracy. What this analysis fails to take into account is that, for most algorithms, the confidence in the accuracy of the solution increases with the accuracy obtained.

Suppose that the solution appears to be accurate to within, say, 3%, using criteria of the type described in the preceding section. Although this level of accuracy may in fact have been achieved, our confidence that this is so is likely to be low. By contrast, if the solution appears to be satisfy a more stringent accuracy requirement, say .0001%, in general one can be quite confident that the desired accuracy has been attained, and almost certain of having achieved a less stringent level of accuracy.

A serious danger of a low accuracy requirement is that premature termination can occur at a point well removed from the solution. For instance, consider the problem of minimizing the univariate function shown in Figure 8c with a termination criterion on the derivative such that $|g(x)| < \sigma(1 + |F(x)|)$. If σ were chosen as .05, all the points with function values in the circled regions would be considered an adequate approximation to x^*. By contrast, a more stringent accuracy requirement, say $\sigma = 10^{-8}$, would produce a good solution. Although it is possible to construct a similar, but badly scaled, function such that $|g(x)| < 10^{-8}(1 + |F(x)|)$ at points remote from the solution, such a function is much less likely to occur in practice. For most practical problems, the only points for which strict termination criteria are satisfied lie in an acceptably small neighbourhood of the solution. In addition, for all problems, the size of the region in which the criteria are satisfied clearly decreases as the conditions become more strict. If σ is small enough, there is a very low probability that a point will satisfy the criterion, yet will not be in a small neighbourhood of the solution.

To maximize the confidence that can be placed in the solution, τ_F should, in general, be the smallest value consistent with the maximum attainable accuracy. When a superlinearly convergent algorithm is used, the increase in accuracy during the last few iterations is usually

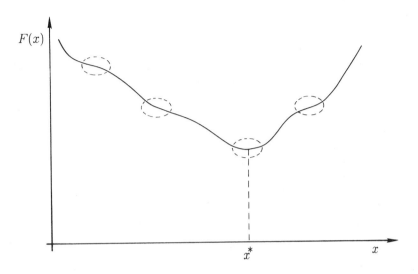

Figure 8c. This figure depicts the situation where a loose tolerance on the gradient may lead to false termination. With the termination criterion $|g(x)| < .05(1+|F(x)|)$, any point with function values in the circled regions would be considered an adequate approximation to x^*.

substantial (see Section 8.3.1). Moreover, the demonstration of superlinear convergence provides a further indication that an estimate is in the neighbourhood of the solution. However, it is not advisable to choose the smallest possible value of τ_F if the marginal effort required to obtain additional accuracy increases very rapidly beyond a certain value of τ_F. With algorithms that converge at an exceedingly slow rate, such as the polytope method, an effort to achieve other than a minimal accuracy is usually not justified by the resulting additional cost. Consequently, even if the polytope method is applied to a well-behaved smooth function, it is not worthwhile to specify a relative precision of less than 10^{-3} in the objective function.

Algorithms of the conjugate-gradient type lie between these two extremes. The rate of convergence for all the methods is effectively linear. On a machine with an s decimal digit mantissa, an appropriate value of τ_F would be approximately $10^{-s/2}$ if $s \geq 10$ and 10^{-4} otherwise. With a successful termination, this would give $F(x_k)$ to approximately $\min\{\frac{1}{2}s, 4\}$ significant digits of accuracy. For many applications this approximate solution is more than adequate, but *the low level of accuracy may imply that we are unable to determine that a correct solution has been found.*

A further factor that affects the choice of τ_F is the availability of first derivatives. When exact first derivatives are not available, the best accuracy that can be achieved is reduced because the error in the approximated gradient makes it impossible to detect errors of smaller magnitude in the solution. In a good software library, each algorithm will have an associated recommended value of τ_F. Furthermore, checks can be made as to whether the user-supplied value is too stringent or too slack. Obviously, if the value of τ_F specified by the user implies that θ_k is smaller than the absolute accuracy of F at a candidate solution, the absolute precision of $F(x_k)$ should be used in the tests in place of θ_k. (Since an algorithm may implicitly alter τ_F, it is inadvisable to use τ_F in order to force premature termination. There are often more appropriate parameters that can be used to limit the number of iterations or function evaluations (see Section 8.1.3.5).)

When there are nonlinear constraints, the value of τ_C can be much smaller than τ_F since we can usually expect to reduce the constraint violations much more than the projected gradient norm. If a smaller value of τ_C is used, the projected gradient norm will usually drop to its limiting accuracy and then hover around this value until the required accuracy is attained in the constraint violations.

Notes and Selected Bibliography for Section 8.2

The termination criteria for the special case of univariate minimization and zero-finding require the specification of the length of the final interval of uncertainty (see Section 4.1). For details of these criteria and a complete discussion of the attainable accuracy for univariate functions, see Brent (1973a).

The issues involved in the provision of *termination* criteria for nonlinearly constrained software are closely related to those associated with providing *assessment* criteria for performance evaluation; see Gill and Murray (1979c) for more details.

Termination criteria for nonlinear least-squares problems and nonlinear equations are explicitly given by Dennis (1977), Dennis, Gay and Welsch (1977), and Moré (1979b). The termination criteria for small-residual least-squares problems should reflect the fact that the sum of squares can be computed with unusually high accuracy (see Section 8.5.1.3).

A complete analysis of the relationship between the approximate solution of a set of linear algebraic equations and the corresponding residual vector is given by Wilkinson (1965).

8.3. ASSESSMENT OF RESULTS

8.3.1. Assessing the Validity of the Solution

It is an unfortunate fact of life that even a good optimization algorithm may find the solution and say that it has failed, or, what is worse, indicate that a point is a solution when it is not (see the discussion of termination criteria in Section 8.2.3). For this reason, *it is essential for a user to be suspicious about any pronouncements from an algorithm concerning the validity of a solution.* We believe that a good algorithm should err on the side of conservatism when judging whether an algorithm has converged successfully. If this philosophy is observed, then satisfaction of all the conditions for a successful termination means that a close approximation to a minimum has probably been found. *However, the prudent user will apply some additional checks to ensure that the routine has not terminated prematurely.* These additional checks are particularly pertinent when the user has requested a low-accuracy solution.

8.3.1.1. The unconstrained case. After an algorithm has indicated a successful termination with x_k, the user should check whether the following conditions are satisfied:

(i) $\|g(x_k)\| \ll \|g(x_0)\|$;

(ii) the iterates that precede x_k display a fast rate of convergence to x_k; and

(iii) the estimated condition number of the Hessian matrix (or its approximation) is small.

If all these conditions hold, x_k is likely to be a solution of the problem *regardless or not of whether the algorithm terminated successfully.*

Condition (i) states that the minimization is likely to have been successful if the norm of the final gradient is sufficiently reduced from that at the starting point. For example, this condition would imply that we should be happy with a point with gradient of order unity, if the gradient at the initial iterate had been of order 10^{10}. In general, the termination condition U3 of Section 8.2.3.2 would have accepted the iterate in this situation. However, the gradient at the starting point is included in condition (i) to ensure that a point is not rejected merely because poor scaling in the gradient is not reflected in the objective function. (It is not possible to include the magnitude of $\|g(x_0)\|$ in the termination criteria without running the risk of failure to terminate when x_0 is close to the solution.)

Many algorithms (e.g., Newton-type and quasi-Newton methods) will generally display a rate of convergence that is faster than linear in the vicinity of the solution. One of the most useful techniques for confirming that a solution has been found is to show that the final few computed iterates exhibit the expected rate of convergence. Unfortunately, it is much more difficult to assess the validity of the computed solution of an algorithm with a linear convergence rate, since one of the most common reasons for abnormal termination is the occurrence of an excessive number of iterations without significant progress (see Section 8.4.3).

To illustrate how condition (ii) can be verified, we consider solving an unconstrained optimization problem when first derivatives are available. In this case, a quick assessment of the rate of convergence can be made by observing $\|g_k\|$, where g_k denotes the gradient vector at x_k. Alternatively, a table can be constructed of

$$\xi_k \equiv F_{k-1} - F_k$$

for the last few (say, 5) values of k. If $|F_k|$ is large, then the table should include $\xi_k/|F_k|$. Superlinear convergence would be indicated if $\xi_{k+1} \approx \xi_k^r$, where $r > 1$. Fast linear convergence would be indicated if $\xi_{k+1} \approx \xi_k/M$, where $M > 2$.

An ideal sequence that displays superlinear (in this case, quadratic) convergence was obtained by minimizing Rosenbrock's function (see Example 4.2 and Figure 4k)

$$F(x_1, x_2) = (x_1^2 - x_2)^2 + 100(1 - x_1^2)$$

using a Newton-type method (see Section 4.4.1); the last few iterations are displayed in Table 8a. The quadratic convergence is easy to observe, and there is no necessity to compute the sequence $\{\xi_k\}$. (Note that this function is of the form for which extra accuracy can be expected in the neighbourhood of the solution; see Section 8.5.1.3.)

Table 8a

Final iterations of a Newton-type method on Rosenbrock's function.

k	F_k	$|(x^*)_1 - (x_k)_1|$	$|(g_k)_1|$
10	2.05×10^{-2}	1.1×10^{-1}	3.0
11	6.27×10^{-5}	3.0×10^{-3}	3.0×10^{-1}
12	1.74×10^{-7}	4.0×10^{-4}	7.0×10^{-3}
13	5.13×10^{-13}	4.0×10^{-7}	2.0×10^{-5}
14	1.68×10^{-24}	6.0×10^{-13}	3.0×10^{-11}

Table 8b

Iterations of the steepest-descent method on Rosenbrock's function.

k	F_k
34	1.87
35	1.83
36	1.79
37	1.71
38	1.65
39	9.29×10^{-1}
.	.
.	.
.	.
395	2.418×10^{-2}
396	2.411×10^{-2}
397	2.405×10^{-2}
398	2.397×10^{-2}

By contrast, the same problem was solved with the steepest-descent algorithm (see Section 4.3.2.2), which yielded the exceedingly slow linear convergence shown in Table 8b. Such a pattern of iterates in general is an indication (correct in this case) that the current iterate is not close to x^*.

In most instances, the behaviour of the iterates will lie between these two extremes, and the degree of confidence that can be placed in the result will depend on the position of the results within the range. For example, in Table 8c we show the results from solving a "real-world" problem. A cursory glance at the F_k column alone does not yield any significant differences between this example and that given in Table 8b. However, the $\{\xi_k\}$ sequence shows that the convergence rate, although linear, is fast.

In forming the sequence $\{\xi_k\}$, the user needs to be aware that eventually all algorithms either fail to make further progress or display slow linear convergence near the limiting accuracy of the solution. What may occur with a superlinearly convergent algorithm is that the sequence $\{\xi_k\}$ will demonstrate superlinear convergence for a few iterations only, and then lapse into slow linear convergence when the limiting accuracy has been achieved. Therefore, it is important, especially if a failure has been indicated, to examine the sequence $\{\xi_k\}$ at iterations that sufficiently precede the final stage. A similar phenomenon may occur with a Newton-type method if the routine takes

Table 8c

Final iterations of a quasi-Newton method on a real problem.

F_k	ξ_k	ξ_{k+1}/ξ_k
.986 333 666 2	—	—
.986 056 096 9	2.8×10^{-4}	2.0×10^{-2}
.986 050 500 0	5.6×10^{-6}	8.4×10^{-2}
.986 050 027 6	4.7×10^{-7}	4.3×10^{-3}
.986 050 027 4	2.0×10^{-9}	—

a step from a relatively poor point to one so close to the solution that rounding errors prevent further progress. In this case, the termination criteria given in Section 8.2.3.2 may not be satisfied (note that F_{k-1} will not be close to F_k, so that criterion U1 will not be satisfied, but the single point test U4 may not be satisfied either). It is usually easy to recognize this situation, since in general $\|g_k\| \ll \|g_{k-1}\|$ and the Hessian matrix at x_k is well conditioned.

If a Newton-type method does not exhibit a superlinear rate of convergence on a problem with a well-conditioned Hessian, the solution may be near a point of discontinuity of the second derivatives. Since small discontinuities are not likely to inhibit the rate of convergence significantly if the Hessian is well conditioned, the discontinuities are probably large enough to be detectable and (if possible) eliminated. Alternatively, a poor rate of convergence for a Newton-type method is often an indication of an error in calculating the Hessian matrix; hence, the user should verify that the second derivatives have been programmed correctly (see Section 8.1.4.2).

Nonlinear least-squares problems. A very small value of the objective function in a least-squares problem *should* be a strong indication that an acceptable solution has been found. However, since any objective function may be multiplied by a very small positive scalar without altering the solution, the user should be wary of badly-scaled problems.

Even when the optimal function value is not small, an acceptable solution may nonetheless have been found. In particular, it is possible for F to be large simply because there are a large number of terms in the objective function. The optimal F will also be large in the following situation. Suppose that the least-squares problem is derived from fitting a function $\phi(x, t)$ to a set of data $\{y_i\}$ at the points $\{t_i\}$. The term $(y_i - \phi(x, t_i))^2$ may be large because the derivative of ϕ with respect to t is large. An example of such a situation is depicted in Figure 8d. The enlargement shows that f_i is large in magnitude even though the fit to the model is quite satisfactory. If this difficulty occurs and such deviations are unacceptable, a user may wish to add more data points in the region of interest, or to give the offending points greater weight in the sum of squares.

8.3.1.2. The constrained case. The checklist of conditions (i)–(iii) for the unconstrained case given in Section 8.3.1.1 can be trivially extended to a constrained problem by performing the substitutions $Z^T g_k$ and $Z^T G_k Z$ for g_k and G_k (in the linear-constraint case), and $Z(x_k)^T g_k$ and $Z(x_k)^T W(x_k, \lambda_k) Z(x_k)$ for g_k and G_k (in the nonlinear-constraint case). When there are nonlinear constraints, the user should monitor the rate of convergence of the Lagrange multiplier estimates as well as of F_k and $\|Z(x_k)^T g_k\|$ (see Section 6.6).

Zero Lagrange multipliers. Complications arise in assessing the validity of the solution when there are zero, or near-zero, Lagrange multipliers. This is because the *sign* of the Lagrange multiplier corresponding to an active constraint is crucial not only in deciding whether a computed solution is optimal, but also in determining how to make further progress if the current estimate is non-optimal (see Section 5.8.3).

Computation in finite precision obviously causes difficulties in determining the correct sign of a very tiny multiplier, since a small perturbation (which could be due entirely to rounding error) may alter the sign of the multiplier. For example, if the computed estimate of a multiplier is 10^{-10}, its exact value could easily be -10^{-9}, in which case the corresponding constraint should not be active. Therefore, *near-zero multipliers create the danger that the computed solution could be in substantial error because the active set has been incorrectly identified.*

Accurate determination of the sign of a tiny multiplier is only part of the difficulty. If a multiplier is exactly zero, this implies that, to first order, a small change in the constraint produces no change in the objective function. (In order to ensure the optimality of a point with

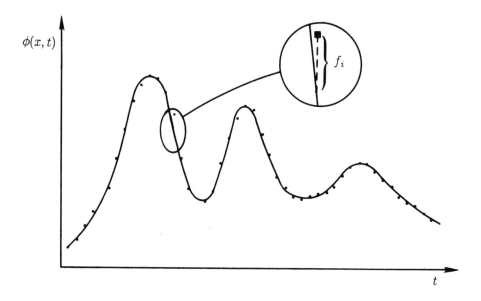

Figure 8d. The figure depicts an example in which the residual from a nonlinear parameter estimation is large even though there is a good fit to the data. The optimization problem is derived from fitting a function $\phi(x, t)$ to a set of data $\{y_i\}$ at the points $\{t_i\}$. The enlargement shows that the term $f_i^2 = (y_i - \phi(x, t_i))^2$ may be large because the derivative of ϕ with respect to t is large.

zero Lagrange multipliers, it is necessary to examine higher-derivatives that will not, in general, be available.) A zero multiplier sometimes means that the solution to the problem would be unaltered if the constraint were eliminated. Figure 8e displays such an example, where removal of $c_2(x) \geq 0$ would *not* change the solution. In other instances, however, the removal of a constraint with a zero multiplier *does* alter the solution. In Figure 8f, λ_1 is zero because the current point is a saddle point, and the solution would change if $c_1(x) \geq 0$ were deleted from the problem.

Even if the exact value of a multiplier is positive but very small, the given solution may be unsatisfactory, as illustrated in the example of Figure 8g. The point $x = a$ is, strictly speaking, a correct solution at which the constraint $x \geq a$ is active; however, x could also be regarded as the solution of an ill-conditioned problem, since a small perturbation of the constraint would lead to a significant change in the solution. In such an instance, it may be reasonable to expect an algorithm to attempt to determine a better solution.

The difficulty in treating zero Lagrange multipliers is in deciding which subset (if any) of the constraints with very small multipliers should be deleted from the working set. Any procedure for making this decision when there are a large number of zero multipliers is inherently combinatorial in nature, and hence has the potential of requiring a significant amount of computation.

The best way to proceed when there are near-zero multipliers at a computed solution will depend upon the algorithm that was used to solve the problem. It is safest to take a pessimistic view of any small positive multiplier, in that an attempt should be made to delete the corresponding constraint. Alternatively, the near-zero multipliers could be computed more accurately and the offending constraints could be perturbed in order to see whether the objective function is

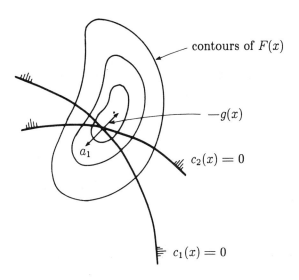

Figure 8e. Example where a zero Lagrange multiplier does not alter the solution. The constraint $c_2(x) \geq 0$ has a zero Lagrange multiplier and is redundant at the solution.

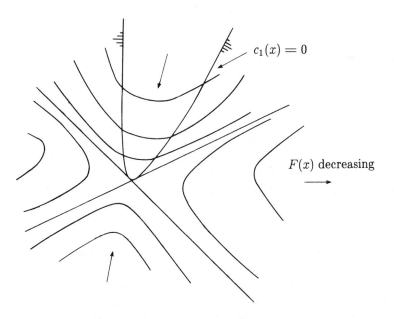

Figure 8f. Example where a zero Lagrange multiplier is significant. The Lagrange multiplier λ_1 corresponding to the constraint $c_1(x) \geq 0$ is zero at the saddle point. However, if the constraint $c_1(x) \geq 0$ were deleted from the problem, the solution would no longer lie at the saddle point.

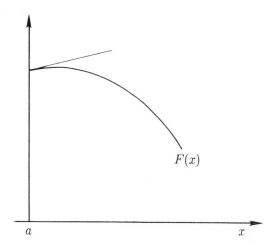

Figure 8g. The possible effect of a small positive Lagrange multiplier. The point $x = a$ is a legitimate solution at which the constraint $x \geq a$ is active. However, x could also be regarded as the solution of an ill-conditioned problem, since a small perturbation of the constraint would lead to a significant change in the solution.

increasing or decreasing on the constraint boundary. Note that good software should carry out such procedures automatically.

A consistent failure to verify optimality because of tiny Lagrange multipliers may require reformulation of the constraints — for example, to identify and eliminate near-redundancies.

Large Lagrange multipliers. In some nonlinearly constrained problems, the Lagrange multipliers do not exist (this can happen only when $\hat{A}(x^*)$, the Jacobian matrix of the active constraints, is rank-deficient; see Section 3.4.1). Fortunately, this situation is extremely rare in practice. It is more likely that, although the multipliers exist, in some sense the problem is "close" to one in which they do not exist, so that $\hat{A}(x^*)$ is "nearly" rank-deficient.

A poorly conditioned $\hat{A}(x^*)$ is indicated if *all* the multipliers are very large in magnitude relative to $\|g(x^*)\|$. However, poor scaling of the constraints can also cause the multipliers to be large, since multiplying a constraint by a positive scalar w causes the associated Lagrange multiplier to be divided by w (see Section 8.7.3). To distinguish near rank-deficiency from poor scaling, it is helpful to investigate the condition number of the *scaled* Jacobian matrix of the active constraints; the scaling should be chosen so that all rows of $\hat{A}(x^*)$ are of unit length. For example, if the LQ factorization (see Section 2.2.5.3) of this matrix is available, a simple estimate of its condition number is the ratio of the largest to the smallest (in magnitude) diagonal elements of L. If this quantity is large, the problem is probably close to one in which Lagrange multipliers do not exist. If the scaled Jacobian does not appear to be ill-conditioned, it is likely that the original constraints are not well-scaled. The user should then attempt to correct the scaling using one of the procedures suggested in Section 8.7.3, since the presence of large multipliers usually inhibits convergence and adversely affects efficiency.

8.3.2. Some Other Ways to Verify Optimality

If the output has been thoroughly analyzed and the user is still uncertain as to whether a true solution has been found, three options are open:

 (i) change the parameters of the algorithm;

(ii) use a different algorithm; or

(iii) alter the problem.

8.3.2.1. Varying the parameters of the algorithm. The simplest parameter to vary is the step-length accuracy (see (4.7) and (4.10) in Section 4.3.2.1). The default value of η for a particular algorithm (see Section 8.1.2.1) will typically correspond to a relatively inaccurate linear search (for example, $\eta = .9$ is often recommended for a Newton-type method). Consequently, finding a closer approximation to a univariate minimum at each iteration (i.e., choosing a smaller value of η) may produce a better result. This strategy is particularly recommended for conjugate-gradient-type methods, which are extremely sensitive to the choice of η (see Section 4.8.3).

If a local search option is available (see Section 8.1.3.6) and has not been tried, the local search should be executed at the final point, in order to determine whether a better point can be found.

One frequent reason that finite-difference quasi-Newton methods encounter difficulties in satisfying termination criteria is that the set of finite-difference intervals has become unsuitable. A new set of intervals can be computed at the best point found, and the algorithm restarted.

If a quasi-Newton method has difficulty in converging, this can be due to the fact that insufficient curvature information has been incorporated into the approximate Hessian matrix. In this case, it may be worthwhile to restart the algorithm with the initial Hessian approximation set to a finite-difference approximation of $G(x)$.

It is sometimes helpful to restart the algorithm with a new initial approximation to x^*. However, it is important to choose a point that is not too close to the old initial point or any of the previous iterates. Preferably, the new initial estimate should be on the "opposite side" of the final value from the previous starting point.

If convergence to the same point does occur after one of the indicated alterations, the user's confidence in the solution should be marginally improved. Unfortunately, there is always the danger — particularly when second derivatives are not available — of repeatedly converging to a point that has a very small gradient, yet is not a solution.

8.3.2.2. Using a different method. If the results of altering the parameters have proved inconclusive, the user should next contemplate using an alternative method — preferably one which is considered to be more robust than the old. However, it may also be useful to apply a less robust algorithm in some circumstances; for example, if the Newton direction is nearly orthogonal to the gradient, a quasi-Newton method may be able to make better progress.

Consider an attempt to verify the result of a minimization by a quasi-Newton method that uses first derivatives. The user has at least three options: programming the second derivatives and using a Newton-type method, using a finite-difference Newton-type method, or using a quasi-Newton method without derivatives. A further option is to replace the subroutine that calculates exact derivatives by one that always computes a central-difference approximation to the gradient. If a Newton-type method is selected to try to verify optimality, the starting point should be the last iterate of the previous run, in order to determine the properties of the Hessian at that point. With any alternative method, the run should be started from a completely different initial point, to allow a better chance of avoiding the region of difficulty.

8.3.2.3. Changing the problem. If the user is consistently unable to verify optimality using the above strategies, it is likely that the problem formulation is defective in some way — for example, the scaling may be poor (see Sections 7.5 and 8.7), or there may be dependencies among the variables (see Section 7.6.1). In this case, the only hope is to reformulate the problem in a more amenable form.

8.3.3. Sensitivity Analysis

It is often desirable to have information about the "sensitivity" of the solution to various aspects of the problem. For example, in a constrained problem the user may wish to estimate the effect on $F(x^*)$ and x^* of perturbing the constraints. In data-fitting problems, it may be important to know the effect on the solution of errors in the data. In addition, one might wish to determine changes in x^* that produce the largest or smallest changes in the optimal function value.

8.3.3.1. The role of the Hessian. In analyzing the sensitivity of the optimal value of an unconstrained function to changes in x, it is enlightening to consider the unconstrained minimization of the quadratic function

$$\Phi(x) = c^T x + \frac{1}{2} x^T G x,$$

where G is positive definite. The behaviour of this function in the neighbourhood of a local minimum is determined by the eigensystem of the matrix G (see Section 3.2.3). Let the eigenvalues and eigenvectors of G be denoted, respectively, by $\{\lambda_i\}$ and $\{u_i\}$, $i = 1, 2, \ldots, n$, with $\lambda_1 \geq \lambda_2 \geq \cdots \geq \lambda_n$. The condition number of G (see Section 2.2.4.3) is defined as

$$\text{cond}(G) = \frac{\lambda_1}{\lambda_n}.$$

When $\text{cond}(G) = 1$, the contours of $\Phi(x)$ are circular, as shown in Figure 8h; as $\text{cond}(G)$ increases, the contours become more elongated. Figure 8h shows that if $\text{cond}(G)$ is large, the relative change in $F(x)$ due to a perturbation of constant norm in x will vary radically depending on the direction of the perturbation. The directions that produce the largest and smallest changes in the function value are u_1 and u_n.

For most nonlinear functions that occur in practice, the Hessian matrix at the solution thus provides a measure of the sensitivity, or conditioning, of the function value with respect to changes in x (see Section 8.2.2). The exceptions are functions that do not have continuous second derivatives, or functions such as $F(x) = x_1^4 + x_2^4$, where the Hessian is the null matrix at the solution. When the Hessian matrix of a function is null at the solution, the local behaviour can be deduced only from (unavailable) higher derivatives — for example, to distinguish between the functions $x_1^4 + x_2^4$ and $x_1^4 + 10^6 x_2^4$.

8.3.3.2. Estimating the condition number of the Hessian. To obtain full information about a Hessian matrix, strictly speaking it is necessary to compute its eigensystem. Fortunately, it is rarely necessary to resort to a complete eigensystem analysis. Many algorithms that evaluate or approximate the Hessian compute the Cholesky factorization of the matrix (see Section 2.2.5.2), which can be used to estimate λ_1 and λ_n and their corresponding eigenvectors.

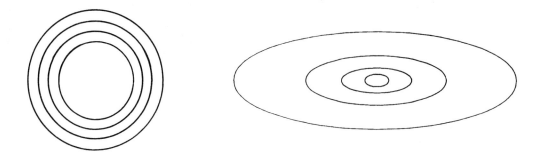

Figure 8h. Contours of quadratic functions with $\text{cond}(G) = 1$ and $\text{cond}(G) = 10$. The first figure illustrates the circular contours of a quadratic function when $\text{cond}(G) = 1$. As $\text{cond}(G)$ increases, the contours become more elongated, as shown in the second figure. If $\text{cond}(G)$ is large, the relative change in the objective function due to a perturbation of constant norm in the variables will vary radically depending on the direction of the perturbation.

Let the Cholesky factorization of G be denoted as LDL^T, and let s and r denote the indices of the largest and smallest elements of D respectively, i.e.

$$d_i \leq d_s \quad \text{and} \quad d_r \leq d_i, \quad \text{for} \quad i = 1, 2, \ldots, n.$$

(In many cases, $s = 1$ and $r = n$.) A lower bound on the condition number is given by

$$\kappa = \frac{d_s}{d_r}.$$

More importantly, κ is usually a good estimate of $\text{cond}(G)$, not in the sense that the two values agree to a certain number of figures, but rather that they are similar in magnitude.

Example 8.2. Consider the matrix

$$G = \begin{pmatrix} 22.3034 & 18.4384 & 7.6082 & 13.7463 \\ 18.4384 & 15.2666 & 6.2848 & 11.3694 \\ 7.6082 & 6.2848 & 2.5964 & 4.6881 \\ 13.7463 & 11.3694 & 4.6881 & 8.4735 \end{pmatrix}. \tag{8.15}$$

which has condition number $\text{cond}(G) = 1.1305997 \times 10^9$. To four significant figures, the factors of G are

$$L = \begin{pmatrix} 1 & & & \\ .8267 & 1 & & \\ .3411 & -.2105 & 1 & \\ .6163 & .2217 & -.1305 & 1 \end{pmatrix},$$

and $D = \text{diag}(2.230 \times 10^1, 2.344 \times 10^{-2}, 7.785 \times 10^{-5}, 5.613 \times 10^{-8})$, giving the condition estimator $\kappa = 3.9738 \times 10^8$.

By doing a little extra computation, it is possible to obtain a much better estimate of the condition number. Define the vectors

$$w_r = (L^T)^{-1} e_r \quad \text{and} \quad w_s = L e_s.$$

It can be shown that

$$\lambda_n \le \frac{d_r}{\|w_r\|_2} \quad \text{and} \quad \|w_s\|_2 d_s \le \lambda_1,$$

and consequently

$$\frac{\lambda_1}{\lambda_n} \ge \frac{\|w_r\|_2 \|w_s\|_2 d_s}{d_r}.$$

Define the vector $\bar{w}_r$ as

$$\bar{w}_r = \frac{1}{\|w_r\|_2} w_r.$$

It can be shown that, if $d_r / \|w_r\|_2 \approx \lambda_n$ and $\lambda_n \ll \lambda_{n-1}$, then

$$\bar{w}_r \approx u_n.$$

If λ_{n-1} is similar to λ_n and $\lambda_{n-1} \ll \lambda_{n-2}$, $\bar{w}_r$ will be a vector that approximately lies in the subspace spanned by u_n and u_{n-1}. In either case, $\bar{w}_r$ will be a vector along which the function is slowly varying.

A similar argument can be used to show that $\bar{w}_s \approx u_1$, where $\bar{w}_s = w_s / \|w_s\|_2$.

For the matrix (8.15), $\lambda_1 = 4.8622797 \times 10^1$ and $\lambda_4 = 4.3006198 \times 10^{-8}$, with corresponding eigenvectors

$$u_1 = \begin{pmatrix} .677217 \\ .560151 \\ .230952 \\ .417455 \end{pmatrix} \quad \text{and} \quad u_4 = \begin{pmatrix} -.437963 \\ -.170007 \\ .114370 \\ .875332 \end{pmatrix}.$$

The vectors w_4 and $\bar{w}_4$ are given by

$$w_4 = \begin{pmatrix} -.500240 \\ -.194261 \\ .130467 \\ 1.000000 \end{pmatrix} \quad \text{and} \quad \bar{w}_4 = \begin{pmatrix} -.437898 \\ -.170052 \\ .114208 \\ .875377 \end{pmatrix}.$$

Note that $\bar{w}_4$ is a good approximation to u_4. Similarly, the vectors w_1 and $\bar{w}_1$ are given by

$$w_1 = \begin{pmatrix} 1.000000 \\ .826707 \\ .341121 \\ .616332 \end{pmatrix} \quad \text{and} \quad \bar{w}_1 = \begin{pmatrix} .677336 \\ .559959 \\ .231054 \\ .417464 \end{pmatrix}.$$

Finally,

$$\frac{\|w_r\|_2 \|w_s\|_2 d_s}{d_r} = 1.1303443 \times 10^9$$

which is a better estimate of the condition number of Example 8.2 than κ.

8.3.3.3. Sensitivity of the constraints. The Lagrange multipliers associated with a constrained problem provide useful quantitative information about the sensitivity of the optimum with respect to perturbations in the constraints. Assume that the set of active constraints at x^* is $\hat{c}$, and that the full-rank matrix $\hat{A}$ contains the gradients of the active constraints.

Consider a perturbation q that lies entirely in the range space of $\hat{A}^T$, and that, to first order, alters only one constraint, such that

$$\hat{a}_j^T q = 1; \quad \hat{a}_i^T q = 0, \quad i \neq j. \tag{8.16}$$

Let the perturbed point $\bar{x}$ be defined as $x^* + hq$.

Substituting q for $Y p_Y$ in equation (8.12) of Section 8.2.2.2, we obtain

$$F(\bar{x}) = F(x^*) + h\lambda_j^* + O(h^2\|q\|_2^2). \tag{8.17}$$

The relationship (8.17) can be used to obtain an estimate of the change in the function resulting from the loss of feasibility with respect to one constraint. Suppose that the point $\bar{x}$ satisfies all the active constraints exactly except the j-th, which has residual δ_j, i.e.

$$\hat{c}_j(\bar{x}) = \delta_j.$$

Then, to first order, $\bar{x} = x^* + \delta_j q$, where q is defined by (8.16). From (8.17), the first-order change in F when moving from x^* to $\bar{x}$ is $\delta_j \lambda_j^*$.

Lagrange multipliers thus provide a relative measure of the sensitivity of $F(x^*)$ to changes in the constraints. For example, if $\lambda_1 = 1000$ and $\lambda_2 = 10^{-3}$, we conclude that changes in $\hat{c}_1(x^*)$ will tend to have a much larger effect on the value of F than changes in $\hat{c}_2(x^*)$. We caution, however, that in order to interpret Lagrange multipliers in this way, the constraint functions must have been suitably scaled with respect to each other, in that similar perturbations in x should lead to similar perturbations in the values of the constraints (see Section 8.7.3).

Information about the sensitivity of the constraints is often useful because the constraints may not be defined rigidly within the context of the outer problem. For example, in a model the constraints often represent desirable rather than essential properties of the solution. Hence, loosening a restriction may not be critical if the objective function will thereby improve significantly; alternatively, it may be beneficial to impose more stringent requirements, provided that the change in the optimal objective is minor.

Notes and Selected Bibliography for Section 8.3

For further information on the interpretation of results associated with linear least-squares problems, see Lawson and Hanson (1974). There is an extensive literature on the statistical interpretation of results from nonlinear least-squares problems; for more details and references, see Bard (1976). Aspects of the analysis of results for the unconstrained case are discussed by Murray (1972b).

With an increased amount of computation, condition estimators more accurate than that of Section 8.3.3.2 are available; see Cline *et al.* (1979) and O'Leary (1980b).

Methods for computing the sensitivity of a nonlinearly constrained problem when using penalty function and augmented Lagrangian methods are discussed by Fiacco and McCormick (1968), Fiacco (1976) and Buys and Gonin (1977).

8.4. WHAT CAN GO WRONG (AND WHAT TO DO ABOUT IT)

The user who is faced with the failure of an algorithm to solve an optimization problem should be aware of some of the things that can go wrong — even with a well-designed implementation of the best available algorithm. It has already been noted that an algorithm may fail because the wrong problem was posed. Unfortunately, this type of error may not be revealed directly, but rather through the failure of some portion of the algorithm. Ideally, the user should not need to perform a detailed investigation to determine why a routine has failed, since a good routine should provide some reason for an unsuccessful termination. In this section, we shall describe the failures that occur most frequently in practice, outline the likely causes, and suggest some possible cures.

8.4.1. Overflow in the User-Defined Problem Functions

This is not the most common cause of failure, but it can cause the most distress amongst users, because an overflow exit transfers control from the user to the computer operating system.

Overflow may occur when the optimization problem has an unbounded solution, since in this case the function value or the iterates will necessarily become increasingly large. An unbounded problem is indicated when there is a consistent decrease in the function being minimized, with no sign of convergence. It is generally assumed that a user would not knowingly specify an *unconstrained* problem with an unbounded solution, but it may happen nonetheless through inadvertent errors in formulation. Another instance in which a user-specified unconstrained subproblem may be unbounded is that the user wants to find a particular *local minimum* of a function that is elsewhere unbounded, but the iterates wander away from the region of the local minimum. Unboundedness is quite common when solving subproblems that arise in methods for constrained optimization (e.g., the methods of Sections 6.2, 6.4, and 6.5).

The appropriate remedy depends on the reason for the unboundedness. Obviously, the user must redefine the problem if the original unconstrained function was incorrectly formulated. If a function is indeed unbounded, but the user wishes to find a particular local minimum, this can sometimes be achieved by imposing a small value of the maximum step allowed during each iteration (Section 8.1.3.4), or by adding bounds on the variables to keep the iterates within some desired region. When an unbounded subproblem occurs within a method for nonlinearly constrained optimization that involves penalty functions, the unboundedness can sometimes be eliminated by increasing the penalty parameter. However, it may also be necessary to undertake the suggestions made above to encourage convergence to a particular local minimum, since for some problems the unconstrained subproblem will be unbounded for any value of the penalty parameter (see, for example, Figure 6h).

Overflow can also occur when minimizing a function with a bounded solution, especially in the early stages of the solution process. In particular, the components of the search direction are often very large for the first few iterations of a quasi-Newton method, and thus a trial step length may produce a large change in the variables during the linear search. If the objective function contains any terms that grow rapidly, such as exponentials or large powers of x, overflow will occur even if the trial step length is not particularly large. For example, consider a function that includes the term $\exp(10x_1)$, and suppose that x_1 is altered by 10^2. This difficulty can be avoided by a judicious choice of the maximum step length (see Section 8.1.3.4).

8.4.2. Insufficient Decrease in the Merit Function

In many optimization methods, a "sufficient decrease" must be achieved in some (merit) function

Φ_M at each iteration; the search for a "lower point" is performed by a step-length algorithm (Section 4.3.2.1). The most common cause of failure of such an algorithm is that the step-length procedure has failed to find a point that yields an adequate decrease.

We cannot treat all possible failures of this type, since the criteria that define an adequate decrease vary from one implementation to another. For example, *some software will declare a failure when the directional derivative of the merit function along the direction of search is not sufficiently negative*; however, this symptom indicates a failure to find a descent direction, and will be discussed in Section 8.4.6. We shall discuss failures in a step-length procedure that requires the determination of a step α_k such that the following three conditions hold

$$\Phi_M(x_k + \alpha_k p_k) \leq \Phi_M(x_k) - \mu_k; \tag{8.18}$$

$$\delta_k \leq \|\alpha_k p_k\|; \tag{8.19}$$

and

$$\|\alpha_k p_k\| \leq \Delta_k, \tag{8.20}$$

where μ_k ($\mu_k > 0$) denotes the "sufficient decrease" in Φ_M, δ_k ($\delta_k > 0$) denotes the minimum acceptable separation of iterates, and Δ_k ($\Delta_k > 0$) denotes the maximum allowed step (see Section 8.1.3.4). The choice of δ_k is based on the assumption that changes in x_k that are less than $\|\delta_k p_k\|$ will produce changes in Φ_M that are less than ϵ_A, the absolute precision of Φ_M at x_k. One possible value for δ_k is

$$\delta_k = \max\left(\frac{\epsilon_A}{|p_k^T \nabla \Phi_M| + \sqrt{\epsilon_A}}, \ \epsilon_A \frac{1 + \|x_k\|}{\epsilon_M + \|p_k\|} \right). \tag{8.21}$$

For further details of these and other conditions that might be imposed on α_k, see Sections 4.3.2.1 and 8.1.3.4.

8.4.2.1. Errors in programming.

One common reason for failure in the step-length procedure is an error in the calculation of Φ_M and its gradient. In general, step-length algorithms fit a simple polynomial to the function along the search direction, and use the minimum of the fitted polynomial as an estimate of the minimum of the original function. Errors in computing the function or gradient can cause the step-length procedure to be faced with nonsensical values. For example, if the iterates were as shown in Figure 8i, the algorithm would take ever-smaller trial values of α as candidates for α_k, but would never find a lower value of Φ_M.

Errors in programming can often be detected by plotting the trial step lengths and the associated values of Φ_M and its directional derivative. If Φ_M or its gradient have been programmed incorrectly, the treatment is obvious (see Section 8.1.4 for suggestions on how to detect such errors).

8.4.2.2. Poor scaling.

Another cause of a failure of a step-length algorithm to find an acceptable point involves what can loosely be described as "poor scaling" of the merit function at the current point along p_k. The effect of this poor scaling is to invalidate the criteria (8.18), (8.19) and (8.20) used to accept the step length.

Probably the most common form of such poor scaling is an imbalance between the values of the function and changes in x. Firstly, the function values may be changing very little even though x is changing significantly. This happens when the derivatives of Φ_M are not "balanced" (see Section 8.7.1.3). In this case, no value of α_k exists that will satisfy (8.18) and (8.20).

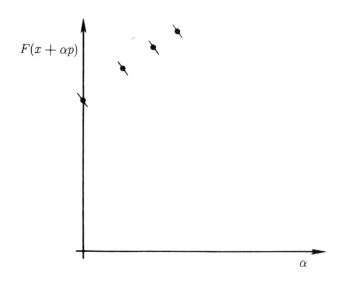

Figure 8i. Configuration of step-length iterates that are symptomatic of an error in the way the function value and derivatives are programmed. The figure depicts the graph of $F(x + \alpha p)$ as a univariate function of the step-length parameter α. The dots indicate the values of the function and the lines through the dots indicate the sign and magnitude of the directional derivative (i.e., the derivative of F with respect to α).

On the other hand, Φ_M may be changing extremely rapidly even though x is changing hardly at all. This will cause difficulties because the value of μ_k in (8.18) is typically based on the size of the gradient of Φ_M (see (4.8) in Section 4.3.2.1), and hence a large directional derivative of Φ_M along p_k will lead to a large value of μ_k. However, if the directional derivative of Φ_M is so large that the second term in (8.21) applies, it may be impossible to satisfy (8.18) and (8.19).

A failure in the step-length procedure can also be caused by poor scaling among the variables and the search direction. For example, suppose that δ_k is computed using (8.21) on a five decimal digit machine with parameters given by $\epsilon_A = 10^{-5}$, $x_k = (10^6, 1)^T$, and $p_k = (0, 1)^T$. Although the first variable would not enter the linear search (since it would remain unaltered), its large value would cause the value of δ_k (8.21) to be approximately 10^{-1}. This value of δ_k in (8.19) would impose an unreasonably large "minimum separation" requirement, and might well lead to a failure of the step-length algorithm.

When the poor scaling is caused by an unsuitable search direction, the difficulty can sometimes be removed by restarting. The Newton search direction tends to be "badly scaled" when the Hessian of Φ_M at x_k is close to singularity, yet the gradient is not necessarily small (i.e., on the "wall" of a long narrow valley in n dimensions). Since a restart at the same point would produce exactly the same failure, the algorithm should be restarted at some nearby point. In contrast, quasi-Newton methods tend to fail when x_k is on the "valley floor", but the approximate Hessian is a poor approximation to the true Hessian. Thus, a quasi-Newton method should be restarted with a different approximate Hessian (possibly a finite-difference approximation to the Hessian) at the point where the failure occurred. If the user is uncertain whether the difficulty lies with a locally poor search direction or with general poor scaling of the problem, one test is to attempt a linear search from x_k along the steepest-descent direction. If no progress can be made along this direction, a fundamental scaling problem is indicated, and one of the re-scaling procedures given in Section 8.7 should be used.

8.4.2.3. Overly stringent termination criteria. Overly stringent accuracy requirements (see Section 8.2) may also cause a failure of the step-length procedure. If x_k is close enough to x^* so that the limiting precision has been reached in Φ_M, local perturbation will lead only to "noise-level" (i.e., meaningless) changes in Φ_M. Hence, although a sufficiently lower point could be found with exact arithmetic, no further "significant" improvement in Φ_M is possible. This is why in some instances an algorithm may indicate that it has been unable to terminate successfully, whereas in fact the solution has been found.

When the step-length procedure fails at a point in a close neighbourhood of the solution, no corrective action usually needs to be taken. In this case, it is advisable for the user to re-define the termination criteria if a similar problem is to be solved, in order to avoid unnecessary iterations seeking unattainable accuracy (and further "failures" by the algorithm).

8.4.2.4. Inaccuracy in a finite-difference approximation. An inaccurate finite-difference approximation to the gradient or Hessian of Φ_M can lead to a failure to find a sufficiently lower point. There are several causes for a poor relative accuracy in the gradient approximation.

Firstly, when x_k is nearly optimal, no simple finite-difference formula can produce high relative accuracy in the gradient approximation (see Section 8.6.1), and thus a failure in the step-length procedure may indicate simply that the limiting accuracy has been reached.

When x_k is not close to optimal, a failure to find a lower point may indicate substantial inaccuracy in the finite-difference approximation to the gradient. Let $\hat{g}_k$ denote an inaccurate approximation to the gradient of Φ_M at x_k. With a well-implemented descent method, the computed value of $\hat{g}_k^T p_k$ will always be negative (see Section 8.4.6); however, if $\hat{g}_k$ is not the true gradient, p_k may not be a descent direction (and hence the step-length procedure may fail to find a lower point).

If the set of finite-difference intervals is unsuitable, the corresponding finite-difference approximation may have very little accuracy. In this case, a new set of intervals should be computed (see Section 8.6.2 for a suitable method).

Another reason for inaccuracies in the gradient is the presence of *small discontinuities* in Φ_M (large discontinuities usually reveal themselves in a rather obvious fashion). Discontinuities in the function should be suspected if there are disproportionately large components in the approximate gradient. One technique of testing for discontinuities is to increase the appropriate finite-difference interval and observe the effect on the approximated gradient. For example, suppose that a finite-difference interval is of order 10^{-8}. If this step crosses a discontinuity in Φ_M of order 10^{-3}, the corresponding element of the gradient will be of order 10^5. However, changing the finite-difference interval to 10^{-6} would lead to an approximate gradient of order 10^3 — a much larger change than would be expected for a well-scaled problem.

By far the best solution to the problem of small discontinuities is to avoid them in the first place (see Section 7.3). If this is impossible, a purely local difficulty with small discontinuities can sometimes be cured by using a finite-difference formula that does not cross the discontinuity.

8.4.3. Consistent Lack of Progress

8.4.3.1. Unconstrained optimization. Depending on the termination criteria used by an algorithm for unconstrained minimization, an algorithm may indicate an unsuccessful termination if no significant decrease in the objective function has been achieved for some number of iterations. The most frequent cause for this failure is that the search direction is almost orthogonal to the negative gradient. In this case, one of the restart procedures suggested in Section 8.4.2.2 for this same difficulty may cure the problem. If the algorithm continues to display inadequate progress, it is sometimes useful to change algorithms, if possible. For example, if a quasi-Newton method displays this behaviour, a Newton-type method could be tried (or vice versa).

Unfortunately, there are problems where the local poor scaling causes all standard methods to cease to make progress. In these instances, it may be necessary to reformulate or rescale the problem (as in Section 8.7).

8.4.3.2. Linearly constrained optimization. A special form of lack of progress associated with linearly constrained problems occurs when constraints are repeatedly added to and deleted from the working set, yet x_k does not change. A special case of this situation is that of *cycling* (see Section 5.8.2), in which the same working set recurs after some number of additions and deletions. To avoid cycling, some algorithms terminate with an error condition if a specified number of changes to the working set are performed without changing the objective function.

The first situation to display this form of lack of progress occurs when an algorithm is attempting to determine a suitable set of n linearly independent constraints at a *vertex* of the feasible region (i.e., to resolve degeneracy); see Section 5.8.2. If the number of constraints is not too large, the user may be able to identify the dependent constraints and remove them from the problem. However, if there are many constraints, and many degeneracies, a technique of combinatorial complexity must be used. If this is unacceptable, one possible course of action is to restart the algorithm and hope that the degeneracy will be resolved during a later iteration. Alternatively, it may be advisable to reformulate the problem, or to perturb the right-hand sides of the offending constraints.

A failure to progress will also happen when zero Lagrange multipliers are associated with some of the constraints in the working set. In this case, the algorithm may be trying to identify a linearly independent subset of the constraints that define a positive-definite projected Hessian (see Section 5.8.3). A guaranteed resolution of this situation also requires a combinatorial procedure. However, it is to be hoped that the algorithm "stopped" at a point that the user can recognize as the solution, in which case there is no need to perform further computation (recall from Section 5.8.3 that zero Lagrange multipliers cause difficulties only at points where the projected gradient vanishes). Otherwise, the user may be able to identify a suitable set of linearly independent constraints. If not, progress can sometimes be made by perturbing the right-hand sides of the constraints and restarting the algorithm.

8.4.3.3. Nonlinearly constrained optimization. There are several reasons why algorithms for nonlinearly constrained minimization may display a consistent lack of progress. In algorithms based on an unconstrained or linearly constrained subproblem, the observations of Sections 8.4.3.1 and 8.4.3.2 apply to a failure to progress within the subproblem. However, there are other situations in which no meaningful progress can be made when solving nonlinearly constrained problems.

Failure to identify the correct active set. A particular difficulty with certain nonlinear-constraint problems occurs in a Lagrangian-based method when the correct set of active constraints cannot be identified. Suppose that, for some set $\bar{c}$ of inactive constraints,

$$\bar{c}(x^*) = \delta,$$

where $\|\delta\|$ is very small. Until $\|\bar{c}(x_k)\| \ll \delta$, even a good algorithm is likely to be unable to detect that the constraints in $\bar{c}$ are not active at the solution. This situation tends to cause a lack of progress because the algorithm's rate of convergence will be inhibited until the correct active set has been identified. Furthermore, no accuracy can be expected in the computed solution until the correct active set has been identified.

The difficulty is related in a general sense to those associated with zero Lagrange multipliers (see Section 8.3.1.2), which sometimes indicate that the solution would be unaltered if the corresponding constraints were not present. Unfortunately, we cannot expect the Lagrange multiplier estimates associated with the constraints in $\bar{c}$ to be small.

A lack of progress because of a failure to identify the correct active set is indicated if the working set changes significantly even though very little change in the iterates has occurred. The Lagrange multiplier estimates may also change dramatically, often in a cyclic fashion — i.e., the multiplier estimates at a given iteration are very similar to those from an earlier iteration with the same working set.

Difficulty in identifying the active set frequently occurs when the minimax and ℓ_1 problems are solved using the transformations given in Section 4.2.3. The most critical data points define the active constraints; however, the remaining data points are "almost" critical.

8.4.4. Maximum Number of Function Evaluations or Iterations

Good software for optimization usually allows the user to impose an upper bound upon the number of times the problem functions are evaluated, and/or an upper bound on the number of iterations. Such bounds are useful in several situations, and serve as a protection against errors in formulation that would otherwise not be revealed.

In particular, the upper bound on the number of function evaluations may be reached when an optimization problem has an unbounded solution (see also Section 8.4.1). This failure will also occur if a large number of iterations are performed without any significant improvement in the objective function. In this case, the analysis and remedies of Section 8.4.3 should be applied.

8.4.5. Failure to Achieve the Expected Convergence Rate

A user who expects a certain algorithm to converge rapidly may consider that a "failure" has occurred if the expected rate of convergence is not displayed. Before discussing the reasons for this situation, we emphasize that the user must have a realistic expectation of the performance of the algorithm; techniques for estimating the rate of convergence are noted briefly in Section 8.3.1.

This failure occurs most often with respect to Newton-type methods, where most users are aware that quadratic convergence to the solution is typical. There are three major reasons that a Newton-type method may succeed in converging, but will not converge quadratically.

Firstly, the most common cause is an error in programming the calculation of the Hessian matrix. Even one incorrect element can lead to a loss of quadratic convergence. The user should suspect an error of this type when the "Hessian matrix" is well-conditioned, or the step lengths consistently differ substantially from unity.

A similar effect occurs when the Hessian matrix is approximated by finite-differences of the gradient, and there are small discontinuities in the computed gradient. The symptoms of this problem are similar to those mentioned in Section 8.4.2.4 concerning small discontinuities in the function — in particular, very large elements in the Hessian matrix, or an extreme change in the Hessian approximation when a small change is made to the finite-difference interval.

The third reason for failure to converge quadratically is ill-conditioning (or possibly singularity) in the Hessian matrix. The proof of quadratic convergence of Newton's method depends on the non-singularity of the Hessian at the solution, and the region in which quadratic convergence will occur decreases in size as the condition number of the Hessian at the solution increases. Hence, the convergence of a Newton-type method will be unavoidably degraded if the Hessian at the solution is very ill-conditioned. If the ill-conditioning arises from poor scaling of the problem, it may sometimes be corrected by applying the techniques of Section 8.7.

When the Hessian at the solution is *singular*, this usually indicates that the minimum is not uniquely determined (see the discussion of quadratic functions in Section 3.2.3 for an explanation of this point). This situation may be an indication of indeterminacy in the problem formulation with respect to the chosen variables (see Section 7.6.1 for an example of indeterminacy and a suggested correction).

8.4.6. Failure to Obtain a Descent Direction

When a "sufficient decrease" in some merit function Φ_M must be achieved at each iteration, it is customary to require that the search direction p_k should be a *descent direction* with respect to Φ_M, i.e. that

$$p_k^T \nabla \Phi_M(x_k) < 0 \qquad (8.22)$$

(when x_k is not a stationary point). If (8.22) is not satisfied, the algorithm will fail.

The search direction p_k is generally computed from a linear system of the form

$$M p_k = -\nabla \Phi_M(x_k). \qquad (8.23)$$

If M is positive definite, in theory (8.22) must hold, regardless of errors in computing $\nabla \Phi_M$. In a properly implemented method, M should always be "sufficiently positive definite", and this error should never happen; note that such methods may need to *alter* some matrix in the problem (see Section 4.4.2).

When M is indefinite, obviously the solution of (8.23) is not necessarily a descent direction. However, if M is positive definite, but extremely ill-conditioned, the computed p_k may fail to satisfy (8.22) because of numerical error in solving (8.23).

A similar difficulty can occur in quasi-Newton methods that rely only on mathematical conditions like (4.41) to ensure that an approximation to the Hessian (or inverse Hessian) will remain positive definite; see Example 4.8 for an illustration of why this is inadequate.

Instead of the test (8.22), an algorithm may require that p_k should satisfy

$$p_k^T \nabla \Phi_M(x_k) < -\varsigma_k, \qquad (8.24)$$

for some positive scalar ς_k. Depending on how ς_k is defined, (8.24) may not hold even if p_k is a descent direction. This difficulty is usually related to overly-strict termination criteria.

8.5. ESTIMATING THE ACCURACY OF THE PROBLEM FUNCTIONS

8.5.1. The Role of Accuracy

8.5.1.1. A definition of accuracy. It is important in many optimization algorithms to have some idea of the accuracy with which a *computed* function value represents the *exact* function value. Since optimization algorithms utilize the values of the function at many points, our concern is not with an exact value of the "accuracy" at only one point. Rather, we are interested in obtaining a *good bound* on the error in the computed function value for *any point* at which the function must be evaluated during the optimization.

As we shall see, it is inadvisable to make *a priori* assumptions about the accuracy of the computed function values. On the other hand, a significant amount of computation is required in order to obtain a reasonably reliable estimate of the accuracy at a single point. Therefore, in order to provide a good estimate of the accuracy without excessive computation, we suggest the following approach, which is usually quite successful on practical problems.

For a given function Φ, a good estimate should be obtained of the accuracy at a "typical" point (usually, the initial point of the minimization); procedures for estimating the accuracy will be described in Section 8.5.2. We then assume a certain model of the behaviour of the error, which allows an estimate of the accuracy of Φ at any point to be computed from the initial estimate; this procedure will be described in Section 8.5.3.

Before describing how to estimate the accuracy, it is important to understand the definition and role of the estimated accuracy. Given a procedure for computing a function Φ, we wish to know a positive scalar ϵ_A such that, at the point x,

$$|fl\big(\Phi(\bar{x})\big) - \Phi(x)| \leq \epsilon_A, \tag{8.25}$$

where $\bar{x}$ is the representable number corresponding to x. Thus, we include in ϵ_A the errors that occur in the computation of Φ and the rounding of x. (We shall assume that x and Φ are well removed from the overflow and underflow limits of the given machine.)

Note that the relationship (8.25) does not define ϵ_A uniquely (if the inequality held for one value of ϵ_A, it would also hold for all larger values). Fortunately, for our purposes it is not necessary for ϵ_A to be unique. Even in a small neighbourhood of a given point, the *actual error* may vary significantly; for example, there may be points for which there is no error in representing x or computing Φ. The desired property of ϵ_A is thus that it should represent a good estimate of the error in computing Φ at all points in some small neighbourhood of x.

We emphasize that the definition (8.25) involves *only the error that occurs in computing the function*. The value of ϵ_A is *not* intended to include the accuracy to which the mathematical function reflects some process in the real world. For example, if the function involves observed real-world data values that are accurate to only three decimal places, this does *not* mean that ϵ_A should be of order 10^{-3}. Of course, this type of inaccuracy in the model will affect the interpretation of the solution of the optimization problem, since the mathematical problem is a perturbation of the "real" problem; thus, the user must analyze the relationship between the solution of the perturbed problem, and the desired solution of the real-world problem. This point is discussed further in Section 7.3.1.

8.5.1.2. How accuracy estimates affect optimization algorithms. The estimated accuracy of the problem functions is important in many aspects of an optimization algorithm. In particular, the accuracy of the problem functions affects: (i) the specification of termination criteria (Section

8.2.3); (ii) the minimum separation between points imposed during the step-length procedure (Sections 4.3.2.1 and 8.4.2); (iii) the estimated condition error in a finite-difference method (Sections 4.6.1.1 and 8.6); and (iv) the estimation of the correct scaling (see Section 8.7).

In cases (i) and (ii) above, the estimated accuracy indicates the limiting accuracy that is obtainable in the problem functions (see Section 8.2.2), and is used to indicate when further calculation is unnecessary because the variation in the function is at "noise level" (less than ϵ_A). Thus, a too-small value of ϵ_A may cause unnecessary iterations without any real progress, and may lead to an indication by the algorithm that it has "failed" (see Section 8.4.3). On the other hand, if ϵ_A is unduly large, an algorithm may terminate prematurely (see Section 8.2.3.6).

Fortunately, if the available computer precision is sufficiently high, the estimate of the accuracy can be incorrect by one or two orders of magnitude without unduly affecting the success of the minimization. The estimate of ϵ_A is more crucial in termination criteria when the working precision is relatively small.

When gradients are estimated by finite-differences, the estimated accuracy is used to determine when a forward-difference approximation to the gradient should be abandoned in favour of a central-difference approximation. A pessimistic estimate of the accuracy may cause central differences to be used earlier than necessary, and thereby lead to unnecessary function evaluations. On the other hand, an overly optimistic estimate of the accuracy may cause a deterioration in the rate of convergence because of inaccurate approximate gradients.

8.5.1.3. The expected accuracy. We have indicated in Section 2.1.4 that in the usual model of floating-point arithmetic, the result of every elementary arithmetic operation between machine-representable values a and b can be expressed as

$$fl(a \text{ op } b) = (a \text{ op } b)(1 + \epsilon),$$

where $|\epsilon|$ is bounded by $\gamma\epsilon_M$ for some constant γ of order unity. Since most functions encountered in practice involve numerous elementary operations, it is not practicable to perform a detailed rounding error analysis in order to obtain ϵ_A.

The standard lower bound on ϵ_A. If $\Phi(x)$ is not zero, the value of ϵ_A can be expressed as a *relative error*, i.e.

$$\epsilon_A \sim \epsilon_R |\Phi(x)|. \tag{8.26}$$

However, a relationship like (8.26) is useful *only when ϵ_R is "small"*. When Φ is a *standard function*, such as sine, cosine or exponential, (8.26) usually holds with ϵ_R of order ϵ_M.

Unfortunately, ϵ_R in (8.26) will generally *not* be small when $|\Phi|$ is small relative to unity. Even for very simple functions, we have seen that the *relative error* in the calculated value can be extremely *large*. For example, consider the case when $\Phi(x) = 1 - x$ and x is close to unity; see also Section 2.1.5 and Example 8.3. Therefore, as a general rule in practice, the *smallest* value that can be expected of ϵ_A is

$$\epsilon_A^* \sim \epsilon_M(1 + |\Phi|), \tag{8.27}$$

and we shall refer to ϵ_A^* as the *standard lower bound* on ϵ_A. Note that (8.27) implies that ϵ_A^* is bounded below by ϵ_M, but corresponds to a relative error of ϵ_M when $|\Phi|$ is *large*.

Functions with unusually small values of ϵ_A. We have already observed that ϵ_A may be smaller than ϵ_A^* when $\Phi(x)$ is a standard function. Another important class of functions for which the value of ϵ_A may be smaller than ϵ_A^* is that of *small-residual least-squares problems*. To see why this is so, assume that Φ is of the form

$$\Phi(x) = \sum_{i=1}^{m} f_i(x)^2,$$

and that, for every i,

$$fl(f_i) = f_i + \delta_i,$$

where $|\delta_i| \leq \epsilon_i$, and $|f_i|$ is "small" for each i. Using the definition of Φ and the assumed form for $fl(f_i)$, we can obtain an expression for $fl(\Phi)$ (neglecting the errors that arise from the arithmetic operations of squaring and adding):

$$fl(\Phi) - \Phi \sim \sum_{i=1}^{m} 2f_i \delta_i + \delta_i^2. \tag{8.28}$$

Examination of (8.28) shows that, when $|\epsilon_i|$ and $|f_i|$ are "small", the value of ϵ_A for this particular function can be *much smaller than* ϵ_M. We have emphasized this point because the frequent use of *zero-residual sums of squares* as test problems — for example, Rosenbrock's function (Example 4.2) — sometimes leads to an unrealistically high expectation of the achievable accuracy for functions that do not happen to be of this special form.

8.5.2. Estimating the Accuracy

Given the enormous variety of functions that occur in optimization calculations, *it is impossible to give an a priori estimate of their accuracy*. In particular, it is inadvisable to assume that all functions can be calculated to full machine precision. Therefore, in this section we shall suggest some techniques that may be used to provide reasonable estimates of the accuracy for many functions that occur in practice. However, we emphasize that these estimates are based on the *behaviour of the particular function at selected points*; the quality of the estimate depends on whether the associated assumptions are satisfied.

For simplicity of description, we shall consider estimating the accuracy in the calculation of a twice-continuously differentiable *univariate* function $f(x)$. (The techniques may be applied in the n-dimensional case by considering the behaviour of the function along a direction p such that $\|p\| = 1$.)

In some cases, the error in computing f may be attributable almost entirely to a subcalculation that is carried out to a fixed, known accuracy. Ordinarily, the user would be aware of such calculations and their expected accuracy, and can use this information to specify ϵ_A.

For general functions, we shall give one method that may be used when extra precision is available, two methods that may be used when accurate derivatives of f are available, and one method for use when only function values are available.

8.5.2.1. Estimating the accuracy when higher precision is available. A very simple and effective method for estimating ϵ_A is available if it is easy to compute f in multiple precision. For example, this is possible with Fortran compilers that allow the specification of variable types using the REAL*n statement.

Suppose that the minimization is to be performed in "short" precision. If fl_s and fl_L denote the results of floating-point arithmetic in "short" and "long" modes, an estimate of ϵ_A can be obtained by calculating f at x and several neighbouring points in both long and short precision, and taking the maximum magnitude of the difference as the estimate of ϵ_A. Thus,

$$\epsilon_A \sim \max_i |fl_L(f(x_i)) - fl_S(f(x_i))|.$$

The function must be evaluated at several neighbouring points (say, 3 or 4) in order to avoid the possibility that, by chance, f may be evaluated with exceptionally high accuracy at one particular point (e.g., when f is a relatively simple function, and the point can be represented with no rounding error).

8.5.2.2. Estimating the accuracy when derivatives are available. The procedures to be described in this section require the availability of the first derivative of f.

An estimate derived from the Taylor series. The estimate of ϵ_A to be described in this section is based on the assumption that the error in computing f at x is similar in magnitude to the change in the computed value of f when the *smallest meaningful perturbation* is made in x.

   Let $h_{\min}$ denote the smallest possible meaningful perturbation in x. If $|x|$ is of order unity or greater, $h_{\min}$ is typically given by $h_{\min} = \epsilon_M |x|$. However, if $|x|$ is small, the size of $h_{\min}$ involves the problem-dependent determination of whether *relative* perturbations in small values of x are meaningful. For example, if $\epsilon_M = 10^{-8}$, and $x = 10^{-16}$, a perturbation of 10^{-24} in x will be meaningful for some problems, but will certainly not be for others. In many practical problems, the smallest meaningful perturbation in x is of order ϵ_M.

   The exact value of f at the perturbed point $x + h_{\min}$ is given by the Taylor-series expansion of f about x:

$$f(x + h_{\min}) = f(x) + h_{\min} f'(\xi),$$

where ξ satisfies $x \le \xi \le x + h_{\min}$. If $f'(\xi) \approx f'(x)$, then, under the assumption mentioned above, two possible estimates of ϵ_A are

$$\epsilon_A \sim |f(x + h_{\min}) - f(x)|; \tag{8.29}$$

and

$$\epsilon_A \sim h_{\min} |f'(x)|. \tag{8.30}$$

Using the error in finite-difference approximations. For a *very small* value of the finite-difference interval, the error in a finite-difference approximation to f' is dominated by a term (the *condition error*) that reflects the accuracy with which f can be computed (see Sections 4.6.1.1 and 8.6). This suggests that a *known error* in a finite-difference approximation to f' may be used to estimate ϵ_A.

   Consider the value of a finite-difference estimate φ of f' computed with an interval h so small that the truncation error can be ignored, and for which the condition error is of the form

$$|\varphi(h) - f'(x)| \sim \frac{\epsilon_A}{h}. \tag{8.31}$$

An estimate of ϵ_A is given by

$$\epsilon_A \sim h|\varphi(h) - f'(x)|. \tag{8.32}$$

   The estimate (8.32) will be accurate only when the error in the finite-difference approximation to f' satisfies (8.31). This may not be the case when the errors in the computed values of f at nearby points *cancel* when these quantities are subtracted. Therefore, we suggest that the estimate (8.32) should be computed for three choices of φ (the forward-, backward-, and central-difference approximations φ_F, φ_B, and φ_C; see Section 8.6.1), and that ϵ_A should be defined as

$$\epsilon_A \sim \max_{\varphi_F, \varphi_B, \varphi_C} h|\varphi(h) - f'(x)|. \tag{8.33}$$

As a check on the estimated accuracy, it is advisable to compute the value of (8.33) for at least two (small) finite-difference intervals. We emphasize that the finite-difference interval h must be small enough so that condition error dominates the error in the derivatives. Note that the estimate (8.32) becomes essentially the right-hand side of (8.30) when φ is zero.

We can illustrate the effect of the estimate (8.32) with a simplified example. Suppose that all computation is carried out on a 12-digit decimal machine. We assume that $|x|$, $|f(x)|$, $|f'(x)|$, and $|f''(x)|$ are of order unity, and that $\epsilon_A \sim 10^{-10}$. Figure 8j depicts the mantissas of the computed values of $f(x)$, $f(x+h)$, and $\Delta f = f(x+h) - f(x)$ when $h = 10^{-8}$. The digits of the mantissas of $f(x)$ and $f(x+h)$ that are unreliable are marked by "$\times$". Because of our assumption about the sizes of f and its derivatives at x, the computed function value at $x+h$ will agree with that at x in the first eight significant digits, which are marked with a "$\bullet$". When $f(x)$ and $f(x+h)$ are subtracted, only the four least significant digits will be retained. Since two of those retained digits are unreliable because of the limited accuracy of f, Δf will have only *two* correct significant digits. (The ten least-significant digits of Δf, marked with a "$\times$", are unreliable.)

Thus, since division by h introduces an error only in the least significant digit, the computed value of φ_F will differ from $f'(x)$ by an error of order 10^{-2}. Since $h = 10^{-8}$, this gives the (correct) estimate

$$\epsilon_A \sim h|\varphi_F - f'(x)| = 10^{-8} 10^{-2} = 10^{-10}.$$

8.5.2.3. Estimating the accuracy when only function values are available.

When only (inaccurate) function values are available, strictly speaking it is *impossible* to estimate the accuracy of the function values, since there is no known exact quantity that can be compared to the inaccurate values. However, if we are prepared to make some assumptions about the behaviour of the higher

mantissa of $f(x)$	$\bullet$	$\bullet$	$\bullet$	$\bullet$	$\bullet$	$\bullet$	$\bullet$	$\bullet$	m_9	m_{10}	$\times$	$\times$
mantissa of $f(x+h)$	$\bullet$	$\bullet$	$\bullet$	$\bullet$	$\bullet$	$\bullet$	$\bullet$	$\bullet$	$\bar{m}_9$	$\bar{m}_{10}$	$\times$	$\times$
mantissa of Δf	m_1	m_2	$\times$	$\times$	$\times$	$\times$	$\times$	$\times$	$\times$	$\times$	$\times$	$\times$

Figure 8j. Schematic diagram of 12-digit mantissas during a first-order differencing procedure to estimate the accuracy of f. This figure depicts the mantissas of the computed values of $f(x)$, $f(x+h)$, and $\Delta f = f(x+h) - f(x)$ when $h = 10^{-8}$. The digits of the mantissas of $f(x)$ and $f(x+h)$ that are unreliable are marked by "$\times$". The computed function value at $x+h$ agrees with that at x in the first eight significant digits, which are marked with a "$\bullet$". When $f(x)$ and $f(x+h)$ are subtracted, only the four least significant digits will be retained. Since two of those retained digits are unreliable because of the limited accuracy of f, Δf will have only *two* correct significant digits.

higher derivatives of f and the statistical distribution of the errors in the computed function values, it is possible to obtain an estimate of the accuracy.

Suppose that f has been computed at a set of values $\{x_i\}$, where the point x_i is defined by $x_i = x + ih$, and $|h|$ is small. We assume that *each computed value* $\bar{f}_i$ is of the form

$$\bar{f}_i = f(x_i) + \delta_i \equiv f(x_i) + \theta_i \epsilon_A, \tag{8.34}$$

where $|\theta_i| \leq 1$. We can obtain a *difference table* (see Section 2.3.5) by considering the set of values $\bar{f}_i$ as the first column of a table, and defining each successive column as the difference of the values in the previous column. By the linearity of the difference operator, after k differences we will have

$$\Delta^k \bar{f}_i = \Delta^k f_i + \Delta^k \delta_i.$$

As discussed in Section 2.3.5, $\Delta^k f = h^k f^{(k)}$. Under mild conditions on f, the value $|h^k f^{(k)}|$ will become very small for moderate values of k (say, $k \geq 4$) if h is small enough. Thus, the higher differences of the computed values should reflect almost entirely the *differences of the errors* δ_i.

Under certain assumptions about the distribution of $\{\theta_i\}$ as random variables, the later differences $\Delta^k \bar{f}_i$ tend to be *similar* in magnitude and to *alternate in sign*. An estimate of ϵ_A can thus be obtained when this general pattern has been observed in one column of the difference table. One formula that has been suggested for the estimate of ϵ_A from the k-th column is

$$\epsilon_A^{(k)} \sim \frac{\max_i |\Delta^k \bar{f}_i|}{\beta_k}, \tag{8.35}$$

where

$$\beta_k = \sqrt{\frac{(2k)!}{(k!)^2}}.$$

In practice, the desired pattern of behaviour typically begins when k is 4 or 5, and the largest value of k that would usually be required is 10. The constants β_k for this small set of values of k can be stored in the routine that estimates the value of ϵ_A.

8.5.2.4. Numerical examples. In this section, we shall consider applying the techniques described in this section to a specific numerical example.

Example 8.3. Consider the function

$$f(x) = e^x + x^3 - 3x - 1.1.$$

The calculations were performed in short precision on an IBM 370 ($\epsilon_M = 16^{-5} \approx 9.537 \times 10^{-7}$). Two points were used in the tests: $x_1 = 10$, for which $f(x_1) = 2.29954 \times 10^4$ and $f'(x_1) = 2.23235 \times 10^4$; and $x_2 = 1.115$, for which $f(x_2) = -9.23681 \times 10^{-3}$ and $f'(x_2) = 3.77924$ (all rounded to six figures). The point x_2 was chosen because it is close to a point where the function is zero, and shows how the procedures work when the value of ϵ_R in (8.26) is *not* of order ϵ_M.

Firstly, the two methods of Section 8.5.2.2 were applied (with the derivative of f computed in short precision). For x_1, the exact error in the computed value of f was 2.5×10^{-3}, which is less than the "standard" lower bound ϵ_A^* (8.27) ($\epsilon_A^* = 2.2 \times 10^{-2}$). Note that this is perfectly reasonable, since we know that the *actual error* may vary substantially at very close points. When the finite-difference interval was taken as 1.049×10^{-5}, the estimate of ϵ_A defined by (8.33) was

3.7×10^{-3}; when the interval was taken as 1.049×10^{-4}, the estimate of ϵ_A from (8.33) was 6.1×10^{-3}. The estimates (8.29) and (8.30) were 2.13×10^{-1} and 2.11×10^{-1}; these large values indicate the magnitude of $f'(x_1)$.

For x_2, the exact value of ϵ_A was 1.4×10^{-6}, which is larger than the "standard" lower bound ϵ_A^* ($\epsilon_A^* = 9.6 \times 10^{-7}$). Note that the *exact value* of ϵ_R at x_2 is 1.5×10^{-4}. When the finite-difference interval was taken as 2.017×10^{-6}, the estimate of ϵ_A from (8.33) was 2.9×10^{-6}; however, the estimate corresponding to only φ_F is of order 10^{-9}, which shows why it is advisable to take the maximum of the three estimates. When the finite-difference interval was 2.017×10^{-5}, the estimate of ϵ_A from (8.33) was 2.7×10^{-6}. The estimates (8.29) and (8.30) were 4.02×10^{-6} and 3.81×10^{-6}, which are comparable to the other estimates at this point.

When the method of Section 8.5.2.3 is applied to Example 8.3 at the point x_1, Table 8d gives the columns of differences, computed with the finite-difference interval taken as 10^{-2}. We observe that the fourth and fifth differences (and, in addition, columns six through ten, which are not shown) generally display the expected alternation in sign. Using formula (8.35), the estimates of ϵ_A corresponding to $k = 4,\ldots,7$ are 1.02×10^{-2}, 6.40×10^{-3}, 6.30×10^{-3}, and 5.93×10^{-3}. The *maximum* exact error at the points involved in the difference table was 2.5×10^{-2}, and the *average* exact error was 8.1×10^{-3}, and hence the estimate of ϵ_A is quite good. It is interesting to observe that the values of θ_i in (8.34) are *not* uniformly distributed in $[-1, 1]$ for this example, since the exact error in evaluating f was *positive* at every point in the neighbourhood of x_1. However, the expected behaviour of the error depends only on the assumptions that the errors are uncorrelated and have the same variance.

When the method of Section 8.5.2.3 is applied to Example 8.3 at the point x_2, Table 8e gives the columns of differences, computed with the finite-difference interval taken as 10^{-3}. In Table 8e, we observe that, although the elements in columns 4 and 5 are of similar magnitude, the pattern of sign alternation occurs in groups of four or five rather than throughout the entire column; in our experience, this is typical. Using formula (8.35), the estimates of $\epsilon_A^{(k)}$ corresponding to $k = 4,\ldots,7$ are 1.03×10^{-6}, 1.08×10^{-6}, 1.10×10^{-6}, and 1.07×10^{-6}. The maximum *exact* error at the points involved in the difference table was 5.1×10^{-6}, and the *average* exact error was 4.1×10^{-6}; thus, again the estimate of ϵ_A is quite good. As at the previous point, the exact error in evaluating f was *positive* at every point in the neighbourhood of x_2.

Finally, the function f was calculated in *short precision* within a routine in which all other calculations were performed in *long precision*. (On the IBM 370, long precision arithmetic corresponds to $\epsilon_M = 16^{-13} \approx 2.22 \times 10^{-16}$.) This test was made in order to demonstrate how

Table 8d

Difference table for Example 8.3 at x_1, with $h = 10^{-2}$.

$\bar{f}_i$	Δ^1	Δ^2	Δ^3	Δ^4	Δ^5
2.300×10^4	2.24×10^2	2.62×10^0	-1.56×10^{-2}	$+5.08 \times 10^{-2}$	-9.38×10^{-2}
2.322×10^4	2.27×10^2	2.25×10^0	$+3.52 \times 10^{-2}$	-4.30×10^{-2}	$+1.21 \times 10^{-1}$
2.345×10^4	2.29×10^2	2.28×10^0	-7.81×10^{-3}	$+7.81 \times 10^{-2}$	-1.45×10^{-1}
2.368×10^4	2.31×10^2	2.27×10^0	$+7.03 \times 10^{-2}$	-6.64×10^{-2}	$+8.20 \times 10^{-2}$
2.391×10^4	2.33×10^2	2.34×10^0	$+3.91 \times 10^{-3}$	$+1.56 \times 10^{-2}$	-3.52×10^{-2}
2.414×10^4	2.36×10^2	2.35×10^0	$+1.95 \times 10^{-2}$	-1.95×10^{-2}	$+9.77 \times 10^{-2}$
2.438×10^4	2.38×10^2	2.37×10^0	0.0	$+7.81 \times 10^{-2}$	-1.68×10^{-1}

Table 8e

Difference table for Example 8.3 at x_2, with $h = 10^{-3}$.

$\bar{f}_i$	$\Delta^1 \bar{f}_i$	$\Delta^2 \bar{f}_i$	$\Delta^3 \bar{f}_i$	$\Delta^4 \bar{f}_i$	$\Delta^5 \bar{f}_i$
-9.238×10^{-3}	3.78×10^{-3}	1.24×10^{-5}	-4.77×10^{-6}	$+9.54 \times 10^{-6}$	-2.00×10^{-5}
-5.456×10^{-3}	3.80×10^{-3}	7.63×10^{-6}	$+4.77 \times 10^{-6}$	-1.05×10^{-5}	$+2.19 \times 10^{-5}$
-1.661×10^{-3}	3.80×10^{-3}	1.24×10^{-5}	-5.72×10^{-6}	$+1.14 \times 10^{-5}$	-1.91×10^{-5}
2.141×10^{-3}	3.82×10^{-3}	6.68×10^{-6}	$+5.72 \times 10^{-6}$	-7.63×10^{-6}	$+5.72 \times 10^{-6}$
5.956×10^{-3}	3.82×10^{-3}	1.24×10^{-5}	-1.91×10^{-6}	-1.91×10^{-6}	$+1.14 \times 10^{-5}$
9.777×10^{-3}	3.83×10^{-3}	1.05×10^{-5}	-3.82×10^{-6}	$+9.54 \times 10^{-6}$	-2.00×10^{-5}
1.361×10^{-2}	3.84×10^{-3}	6.68×10^{-6}	$+5.72 \times 10^{-6}$	-1.05×10^{-5}	$+2.00 \times 10^{-5}$

the procedure of Section 8.5.2.3 works when ϵ_A is much larger than the standard lower bound ϵ_A^* (8.27). The added precision in the arithmetic should *not* significantly alter the estimate of ϵ_A. The added precision allowed a smaller value ($h = 10^{-5}$) to be used to calculate the difference table; in this case, the higher differences did display a consistent pattern of sign alternation beyond column three. Using formula (8.35), the estimates of $\epsilon_A^{(k)}$ corresponding to $k = 4, \ldots, 7$ are 1.14×10^{-6}, 1.20×10^{-6}, 1.25×10^{-6}, and 1.29×10^{-6}, which are essentially the same as in the single-precision case.

8.5.3. Re-Estimation of the Accuracy

In most optimization algorithms, an estimate of ϵ_A is required at more than one point. It would be inadvisable to use the same value of ϵ_A at all iterates, since the magnitudes of x and of f may vary dramatically from one point to another. However, it would be too costly in terms of function evaluations to use any of the procedures of Section 8.5.2. at every point. Hence, an effective strategy is to use the initial estimate of the accuracy to estimate the accuracy at other points.

It is generally *wrong* to assume that the value of $\epsilon_A(x_0)/|f(x_0)|$ can be used as a *relative error*, i.e. that

$$\epsilon_A(x_k) = |f(x_k)| \frac{\epsilon_A(x_0)}{|f(x_0)|},$$

since the resulting estimate of ϵ_A can be much too small when $|f|$ becomes very small. As discussed in Section 8.5.1, for many functions the lower bound of ϵ_A^* (8.27) applies.

We suggest the following model of the behaviour of ϵ_A. The value of ϵ_A at x_0 is interpreted as the "number of correct figures" in the computed value of f at all points of interest. *When $|f|$ is small, the "correct figures" include the leading zeros after the decimal point.* However, we impose the restriction that the estimated number of correct figures may not exceed the number of figures corresponding to full machine precision. Under this assumption, the value of ϵ_A at any point x is given by

$$\epsilon_A(x) = (1 + |f(x)|) \max\left(\epsilon_M, \frac{\epsilon_A(x_0)}{1 + |f(x_0)|}\right). \tag{8.36}$$

We illustrate the rule (8.36) on Example 8.3. When x_0 is taken as 10, the lower bound $\overset{*}{\epsilon_A}$ exceeds any of the computed estimates of ϵ_A, and hence the value of ϵ_A at x_2 computed from (8.36) would be $\epsilon_M(1 + |f(x_2)|)$ ($= 9.6 \times 10^{-7}$). When x_0 is taken as 1.115 and the value $\epsilon_A(x_0)$ is taken as the estimated value 1.2×10^{-6}, the value of ϵ_A at x_1 from (8.36) is 2.7×10^{-2}.

Notes and Selected Bibliography for Section 8.5

The topic of estimating accuracy is discussed in Stewart (1967), Brent (1973a), and Curtis and Reid (1974). The technique discussed in Section 8.5.2.3 follows a method suggested by Hamming (1962, 1973). The method is discussed further by Lyness (1977a).

8.6. COMPUTING FINITE DIFFERENCES

Quasi-Newton methods with finite-difference approximations to the gradient vector are considered to be among the most efficient algorithms for a smooth unconstrained or linearly constrained problem of reasonable size in which the derivatives of $F(x)$ are too difficult or expensive to evaluate (see Sections 4.6.2 and 5.1.2.4). The success of such algorithms critically depends on obtaining "good" approximations to the necessary first derivatives — much more so than with Newton-type methods that use finite differences of gradients to approximate second derivatives.

We shall see that certain "standard" choices for the finite-difference interval sometimes produce acceptable approximations. However, they may lead to poor results in the presence of bad scaling or a non-typical starting point. Although no procedure is guaranteed for every case, the methods discussed in this section are designed to overcome the most common difficulties in choosing the finite-difference interval, and can lead to substantial improvements in performance of the associated minimization algorithms.

For simplicity of description, the discussion in this section will concern the choice of interval for estimating derivatives of the *univariate* function $f(x)$; the extension to the multivariate case was discussed in Section 4.6.1.3. We assume throughout that the accuracy of the computed value of f is known (see Section 8.5).

8.6.1. Errors in Finite-Difference Approximations; The Well-Scaled Case

In general, the errors in finite-difference approximations depend on quantities that are *unknown*, such as higher derivatives of the function. Therefore, the derivation of a representation of the error is useful mainly in indicating how the error varies with the finite-difference interval. However, under certain assumptions about the function f and its derivatives, a good a *priori* finite-difference interval can be specified.

8.6.1.1. The forward-difference formula.

In this section we shall summarize the results of Section 4.6.1.1 concerning the errors in approximating $f'(x)$ using the forward-difference formula

$$\varphi_F(f, h) = \frac{f(x + h) - f(x)}{h},$$

where $h > 0$ (*cf.* (4.50)). The most important errors in the *computed* value of φ_F are *truncation error* (caused by the neglected terms of the Taylor series) and *condition (cancellation) error* (caused by inaccuracies in the computed function values). The error from these two sources is bounded by the expression

$$\frac{h}{2}|f''(\xi)| + \frac{2}{h}\epsilon_A, \tag{8.37}$$

where $\xi \in [x, x+h]$ and ϵ_A is a bound on the error in the computed function values at x and $x + h$ (see Section 8.5.1). The value h_F that minimizes the sum (8.37) is

$$h_F = 2\sqrt{\frac{\epsilon_A}{|f''(\xi)|}}, \tag{8.38}$$

and the minimal sum (8.37) is $2\sqrt{\epsilon_A|f''(\xi)|}$. Unfortunately, since f'' and ξ are unknown, further assumptions are necessary in order to use (8.38) to estimate h_F.

In particular, suppose that f is non-zero, and that ϵ_A can be expressed in terms of a known bound ϵ_R on the *relative error*, i.e.

$$\epsilon_A = |f|\epsilon_R. \tag{8.39}$$

If it also holds that

$$O(|f|) = O(|f''|)$$

for all points in $[x, x+h]$, then it follows from (8.38) that

$$h_F \sim \sqrt{\epsilon_R}.$$

Furthermore, if $O(|f'|) = O(|f|)$, the bound on the *relative error* in the approximation of $f'(x)$ is of order $\sqrt{\epsilon_R}$. This result leads to the "folklore" observation that, in general, the number of correct figures in φ_F with the best finite-difference interval is *half* the number of correct figures in f.

We emphasize that the preceding analysis applies only under the stated assumptions about the values of f and its derivatives. Note also that when $|f'|$ is *small* relative to $|f|$ and $|f''|$ (which will usually be the case when x is near a local minimum of f), the *relative* error in φ_F may become large, even for the best possible choice of finite-difference interval.

Exactly the same analysis applies to the errors in the *backward-difference formula*

$$\varphi_B(f, h) = \frac{f(x) - f(x-h)}{h},$$

with the obvious difference that $\xi \in [x-h, x]$.

8.6.1.2. The central-difference formula. The first derivative can also be approximated using the central-difference formula

$$\varphi_C(f, h) = \frac{f(x+h) - f(x-h)}{2h}, \tag{8.40}$$

where $h > 0$. In this case, the truncation error is bounded by $\frac{1}{6}h^2|f'''(\mu)|$, where $\mu \in [x-h, x+h]$, and the condition error is bounded by ϵ_A/h. Thus, the overall bound on the error in the computed value of φ_C is $\frac{1}{6}h^2|f'''(\mu)| + \epsilon_A/h$. The value h_C that minimizes this bound is given by

$$h_C = \sqrt[3]{\frac{3\epsilon_A}{|f'''(\mu)|}}.$$

If (8.39) holds and if $O(|f|) = O(|f'''|)$ for all points in $[x-h, x+h]$, then h_C satisfies

$$h_C \sim \sqrt[3]{\epsilon_R}. \tag{8.41}$$

As in the forward-difference case, we emphasize that (8.41) is not valid except under the stated conditions.

8.6.1.3. Second-order differences. The second-order difference formula

$$\Phi(f, h) = \frac{f(x + h) - 2f(x) + f(x - h)}{h^2} \tag{8.42}$$

will be used to compute an approximation to f''. The truncation error caused by neglecting higher-order terms of the Taylor series is given by $\frac{1}{12}h^2 f^{(4)}(\eta)$, where $\eta \in [x - h, x + h]$, and the condition error arising from inaccuracies in the computed values of $f(x)$, $f(x + h)$ and $f(x - h)$ is bounded by $4\epsilon_A/h^2$, where ϵ_A is a bound on the error in the computed function value at x, $x - h$ and $x + h$. Thus, the total error when the computed value of Φ is used to approximate $f''(x)$ is bounded by $4\epsilon_A/h^2 + \frac{1}{12}h^2 |f^{(4)}(\eta)|$. The value h_ϕ that minimizes this bound is

$$h_\phi = 2 \sqrt[4]{\frac{3\epsilon_A}{|f^{(4)}(\eta)|}}. \tag{8.43}$$

If f satisfies (8.39) and $O(|f|) = O(|f^{(4)}|)$ for all points in $[x - h, x + h]$, then h_ϕ satisfies

$$h_\phi \sim \sqrt[4]{\epsilon_R}.$$

8.6.2. A Procedure for Automatic Estimation of Finite-Difference Intervals

In this section we outline the main features of an automatic procedure for computing a "good" finite-difference interval for an optimization method that primarily uses forward-difference approximations to f'. *The intention is not to compute the most accurate possible estimate of the first derivative at a single point. Rather, we wish to obtain a "sensible" interval that will produce reasonably accurate approximations of the derivative throughout the course of the minimization.* The technique may be applied to an n-variable function by computing an interval for each variable.

8.6.2.1. Motivation for the procedure. Since the bound (8.37) involves the unknown second derivative f'' at the unknown point ξ, the procedure is based on approximating $f''(\xi)$ by the second-order formula (8.42). After an acceptable non-zero value of (8.42), say Φ, has been computed, the estimate of the "optimal" interval h_F is given by

$$h_F = 2 \sqrt{\frac{\epsilon_A}{|\Phi|}}. \tag{8.44}$$

Thus, the procedure involves primarily the determination of an interval h_ϕ that may be used to compute an order-of-magnitude estimate of Φ from (8.42).

Two assumptions underlie this choice of Φ. Firstly, it is necessary for $f''(x)$ to be an adequate approximation of $f''(\xi)$, which will be true only if the second derivative is not changing too rapidly near x. Secondly, the value of Φ must be a sufficiently accurate estimate of $f''(x)$. The procedure is directed toward finding an interval h_ϕ whose associated value of Φ is a correct order-of-magnitude estimate of $f''(x)$ (i.e., has at least one correct decimal figure).

The procedure is based on the fact that the bound on the truncation error tends to be an increasing function of h, while the condition error bound is generally a decreasing function of h.

In particular, the value of h_ϕ is selected from a sequence of trial values $\{h_i\}$. The decision as to whether a given value of Φ is sufficiently accurate involves $\hat{C}(\Phi)$, the following bound on the *relative* condition error in Φ:

$$\hat{C}(\Phi) \equiv \frac{4\epsilon_A}{h^2|\Phi|}. \tag{8.45}$$

(When Φ is zero, $\hat{C}(\Phi)$ is taken as an arbitrarily large number.) No attempt is made to compute an explicit estimate of the truncation error.

If the value of $\hat{C}(\Phi)$ for the trial value h_i is "acceptable" (i.e., lies in the interval $[.001, .1]$), h_ϕ is taken as h_i, and the current value of Φ is used to compute h_F from (8.44). There is a clear need for an upper bound on the value of $\hat{C}(\Phi)$. A *lower bound* on the acceptable value of $\hat{C}(\Phi)$ is imposed because a small value of $\hat{C}(\Phi)$ may indicate that the truncation error is "too large" (recall that condition error and truncation error are inversely related). Thus, a smaller interval will reduce the truncation-error bound and may also yield a sufficiently small value of $\hat{C}(\Phi)$.

If $\hat{C}(\Phi)$ is unacceptable, the next trial interval is chosen so that the relative condition error bound will either decrease or increase, as required. If the bound on the relative condition error is "too large", a *larger* interval is used as the next trial value in an attempt to reduce the condition error bound. On the other hand, if the relative condition error bound is "too small", h_i is reduced. We have chosen simply to multiply h_i by a fixed factor to obtain the next trial value. More complicated iterative schemes can be devised, based on the special form of (8.45), but they tend to require further assumptions that do not necessarily hold in practice (e.g., that Φ remains of a similar magnitude as h changes).

The procedure will fail to produce an acceptable value of $\hat{C}(\Phi)$ in two situations. Firstly, if Φ is extremely small, then $\hat{C}(\Phi)$ may never become small, even for a very large value of the interval. This will happen if f is nearly constant, linear, or $f(y - x)$ is odd in y. Secondly, $\hat{C}(\Phi)$ may never exceed $.001$, even for a very small value of the interval. This implies that Φ is extremely large, which usually occurs when x is near a singularity.

As a check on the validity of the estimated derivative, the procedure provides a comparison of the forward-difference approximation computed with h_F and the central-difference approximation computed with h_ϕ. If these values are not in some agreement, neither can be considered reliable.

In order to obtain h_0 (the initial trial value of h_ϕ), we make the following assumption about the relationship between the magnitudes of x, $f(x)$, and $f''(x)$:

$$O(|f''|) = O\left(\frac{\omega + |f(x)|}{(\eta + |x|)^2}\right), \tag{8.46}$$

where $\omega > 0$ and $\eta > 0$. This assumption derives from several considerations. The quantity ω is included so that no special assumptions are required when $|f|$ is very small; the value of $|f|$ is included because a large value of $|f|$ often indicates a large value of $|f''|$ (if the user has eliminated superfluous constants from the definition of f). Similarly, the denominator of (8.46) is intended to avoid special assumptions when $|x|$ is very small, and to reflect to some degree the influence on the second derivative of changes in the scale of x. The assumption (8.46) is used merely to obtain an initial interval, and is not critical in the final value of h_ϕ; however, the values of ω and η do affect the efficiency of the procedure. In our experience, the values $\omega = 1$ and $\eta = 1$ have always been satisfactory.

If (8.46) holds, then, from (8.38), the interval $\bar{h} = 2(\eta + |x|)\sqrt{\epsilon_A/(\omega + |f|)}$ will produce a condition error in φ_F comparable to its "optimal" value, but the relative condition error $\hat{C}(\Phi)$ will be of order one (i.e., there will be *no* correct figures in Φ). Therefore, we choose the initial h_0 as $10\bar{h}$, which should tend to produce a value for $\hat{C}(\Phi)$ of order $.01$ (since $\hat{C}(\Phi)$ is inversely proportional to h^2). If the function is well scaled in the sense of Section 8.6.1.1, taking h_0 as $10\bar{h}$ will produce an acceptable value of $\hat{C}(\Phi)$, so that no further intervals need be computed.

8.6.2.2. Statement of the algorithm. Algorithm FD requires an initial point x, the value of $f(x)$, and the value of ϵ_A. The algorithm computes an interval h_F that should produce an adequate forward-difference approximation of f'. The algorithm also produces φ (an estimate of $f'(x)$), Φ (an estimate of $f''(x)$), and E_F (an estimate of the error bound in φ).

The positive integer K is an upper bound on the number of trial intervals. In situations where the maximum number of trials is made without success, it is useful to store h_s, the smallest value (if any) of h_i for which the relative condition errors in the forward- and backward-difference approximations are acceptable. For this reason, we define the following bounds on the *relative* condition errors in φ_F and φ_B:

$$\hat{C}(\varphi_F) \equiv \frac{2\epsilon_A}{h|\varphi_F|} \quad \text{and} \quad \hat{C}(\varphi_B) \equiv \frac{2\epsilon_A}{h|\varphi_B|}.$$

(When φ_F or φ_B is zero, the corresponding error bound is taken as an arbitrarily large number.) The value h_s is the smallest interval in the sequence (if any) for which these bounds are "acceptable" (i.e., less than .1).

The formal statement of the algorithm is the following.

Algorithm FD (*Automatic estimation of* h_F, f' *and* f'' *using finite differences*).
FD1. [Initialization.] Choose ω, η and K. Evaluate $f(x)$. Define $h_0 = 10\bar{h}$, where

$$\bar{h} = 2(\eta + |x|)\sqrt{\frac{\epsilon_A}{\omega + |f(x)|}}, \tag{8.47}$$

and set $k \leftarrow 0$. Evaluate $f(x+h_0)$, $f(x-h_0)$, $\varphi_F(h_0)$, $\varphi_B(h_0)$, $\varphi_C(h_0)$, $\Phi(h_0)$, $\hat{C}(\varphi_F)$, $\hat{C}(\varphi_B)$ and $\hat{C}(\Phi)$. Set $h_s \leftarrow -1$.
FD2. [Decide whether to accept the initial interval.] If $\max\{\hat{C}(\varphi_F), \hat{C}(\varphi_B)\} \leq .1$, set $h_s \leftarrow h_0$. If $.001 \leq \hat{C}(\Phi) \leq .1$, set $h_\phi \leftarrow h_0$ and go to Step FD5. If $\hat{C}(\Phi) < .001$, go to Step FD4. Otherwise, continue at Step FD3.
FD3. [Increase h.] Set $k \leftarrow k + 1$ and $h_k \leftarrow 10h_{k-1}$. Compute the associated finite-difference estimates and their relative condition errors. If $h_s < 0$ and $\max\{\hat{C}(\varphi_F), \hat{C}(\varphi_B)\} \leq .1$, set $h_s \leftarrow h_k$. If $\hat{C}(\Phi) \leq .1$, set $h_\phi \leftarrow h_k$ and go to Step FD5. If $k = K$, go to Step FD6. Otherwise, repeat Step FD3.
FD4. [Decrease h.] Set $k \leftarrow k + 1$ and $h_k \leftarrow h_{k-1}/10$. Compute the associated finite-difference estimates and their relative condition errors. If $\hat{C}(\Phi) > .1$, set $h_\phi \leftarrow h_{k-1}$ and go to step FD5. If $\max\{\hat{C}(\varphi_F), \hat{C}(\varphi_B)\} \leq .1$, set $h_s \leftarrow h_k$. If $.001 \leq \hat{C}(\Phi) \leq .1$, set $h_\phi \leftarrow h_k$ and go to step FD5. If $k = K$, go to Step FD6. Otherwise, repeat Step FD4.
FD5. [Compute the estimate of the optimal interval.] Define h_F from (8.44), and set φ to $\varphi_F(h_F)$. Set the estimated error bound to

$$E_F = \frac{h_F|\Phi|}{2} + \frac{2\epsilon_A}{h_F}. \tag{8.48}$$

Compute the difference between φ and $\varphi_C(h_\phi)$ as

$$\bar{E} = |\varphi - \varphi_C(h_\phi)|.$$

If $\max\{E_F, \bar{E}\} \leq .5|\varphi|$, terminate successfully. Otherwise, terminate with an error condition.

FD6. [Check unsatisfactory cases.] If $h_s < 0$ (i.e., $\max\{\hat{C}(\varphi_F), \hat{C}(\varphi_F)\} > .1$ for all values of h_k), then f appears to be nearly constant; set $h_F \leftarrow \bar{h}$ (8.47), and set φ, Φ and E_F to zero. If $\hat{C}(\Phi) > .1$ and $h_s > 0$, then f appears to be odd or nearly linear; set h_F to h_s, set φ to $\varphi_F(h_F)$, set Φ to zero, and compute E_F from (8.48). Otherwise, f'' appears to be increasing rapidly as h decreases (since $\hat{C}(\Phi) < .001$ for all values of h_k); set $h_F \leftarrow h_K$, set φ to $\varphi_F(h_F)$, set Φ to $\Phi(h_F)$, and compute E_F from (8.48). In all these cases, terminate with an error condition. ∎

8.6.2.3. Numerical examples.

All calculations were performed in short precision on an IBM 370 ($\epsilon_M = .9375 \times 10^{-6}$). In all the examples, K was taken as 6, and ω and η in (8.47) were both taken as 1. The value of ϵ_A was computed using the procedure of Section 8.5.2.3.

Example 8.4. When Algorithm FD is applied to the function of Example 4.9, Section 4.6, at the point $x = 1$, with $\epsilon_A = .4 \times 10^{-5}$, the relative condition error with the interval h_0 ($.398 \times 10^{-1}$) is $\hat{C}(\Phi) = .416 \times 10^{-3}$, which is less than the desired lower bound; the value of $\hat{C}(\Phi)$ corresponding to h_1 is $.424 \times 10^{-1}$, and thus $h_\phi = h_1$. The value of Φ is $.238294 \times 10^2$, which is a good order-of-magnitude estimate of the exact value of $f''(x)$ ($.242661 \times 10^2$). The value of h_F is $.819 \times 10^{-3}$, and $\varphi_F(h_F) = .955636 \times 10^1$, giving a relative error in f' of $.807 \times 10^{-3}$. Note that the estimate h_F agrees with the results given in Table 4a, and that the estimates of f' and f'' are as accurate as we might expect from six-decimal arithmetic.

Example 8.5. Consider the function

$$f(x) = (x - 100)^2 + 10^{-6}(x - 300)^3$$

at the point $x = 0$, with $\epsilon_A = .9 \times 10^{-2}$. The exact values of $f(x)$ and $f''(x)$ are $.997300 \times 10^4$ and $.199820 \times 10^1$, so that condition (8.46) is not satisfied. Since h_0 is too small, the interval is increased twice before the relative condition error is acceptable (the final value of Φ is $.199567 \times 10^1$). The value of h_F is $.134$; the corresponding φ_F is $-.199632 \times 10^3$, and the exact value of $f'(x)$ is $-.199730 \times 10^3$, which gives a relative accuracy in f' of $.49 \times 10^{-3}$.

Example 8.6. It was observed in Section 8.6.1.1 that a forward-difference approximation will tend to have poor relative accuracy when $|f'|$ is small, even with the best choice of interval. As an example of this situation, consider the function $f(x) = x^4 + 3x^2 - 10x$ at the point $x = .99999$, with $\epsilon_A = .7 \times 10^{-5}$. The exact value of $f'(x)$ is $-.180244 \times 10^{-3}$. The algorithm computes a good estimate of Φ with h_0 ($\Phi = .180008 \times 10^2$ compared to the exact value $f'' = .179999 \times 10^2$), and h_F is given by $.125 \times 10^{-2}$. However, $\varphi_F(h_F)$ is $.913 \times 10^{-2}$, so that the approximate derivative has poor relative accuracy, even though h_F is a good estimate of the optimal forward-difference interval. The central-difference approximation computed with h_ϕ ($-.357 \times 10^{-2}$) also has poor relative accuracy. Note that the difference between the forward- and central-difference estimates would cause Algorithm FD to produce an error message.

8.6.2.4. Estimating the finite-difference interval at an arbitrary point.

The procedure of Section 8.6.2.2 requires at least three evaluations of the function for each component of the gradient, even for a well-scaled problem. An optimization algorithm that required this number of function evaluations at every iteration would be uncompetitive with alternative non-derivative methods. Ideally, an algorithm should be able to compute an approximation to the gradient with only one function evaluation for each component.

Fortunately, this aim can often be achieved, for several reasons. First, and most important, it is our experience that, for many functions encountered in practice, the finite-difference intervals generated by a procedure such as that of Section 8.6.2.2 do not vary significantly from one iteration to the next. Second, the intervals produced by the procedure of Section 8.6.2.2 do not usually differ widely from the "optimal" intervals. Finally, finite-difference gradient methods can generally make satisfactory progress as long as the overall gradient vector has a "reasonable" level of accuracy; it is not essential for each component of the gradient to have close-to-maximal accuracy at every iterate.

Based on these observations, we suggest that the procedure of Section 8.6.2.2 should be executed at a "typical" point (usually, x_0). The set of intervals obtained should then be used to compute forward-difference approximations at subsequent iterates.

We illustrate the effects of this strategy with the function from Example 8.4 when $x_0 = 1$. In Section 8.6.2.3, it was shown that Algorithm FD produces the interval $h_F = .819 \times 10^{-3}$. When this interval is used to compute a forward-difference approximation at the point $x = 10$, we obtain a relative accuracy of $.409 \times 10^{-3}$ in f' (the values of φ and f' are $.971 \times 10^9$ and $.970286 \times 10^9$). When Algorithm FD is executed at the point $x_0 = 10$ (with $\epsilon_A = .5 \times 10^3$), the value of h_F is $.101 \times 10^{-2}$; if this interval is used at $x = 1$, the relative error in the forward-difference approximation is $.102 \times 10^{-2}$. *These results are typical of those in many numerical experiments.* Although it is not difficult to construct examples for which the optimal finite-difference intervals vary significantly from point to point, our experience is that the procedure works well in practice.

An additional benefit of the procedure of Section 8.6.2.2 is that the quantity Φ associated with the i-th variable is usually a good estimate of the i-th diagonal element of the Hessian matrix at x_0, and thus we can obtain an initial diagonal approximation of the Hessian. An improved initial estimate of the Hessian often reduces the number of iterations required for convergence with a quasi-Newton method.

8.6.2.5. Finite-difference approximations in constrained problems.

In theory, the analysis of Section 8.6.2.1 can be applied when the vector to be approximated is the projected gradient $Z(x)^T g(x)$ rather than $g(x)$. One would simply consider the behaviour of the function when perturbations are made along the columns of $Z(x)$ rather than along the co-ordinate directions. Unfortunately, since Z may change completely when the working set changes, there is no guarantee that the set of "optimal " intervals computed at one point have any straightforward relationship to the set of "optimal" intervals at another point. One strategy that often gives good results for linearly constrained problems is to obtain a set of intervals $\{h_i\}$ for the full gradient (as in Section 8.6.2.2); the interval to be used along the direction z is then taken as $\sum_i |h_i z_i|$.

It might be considered that the difficulties associated with a loss of relative accuracy in the approximate gradient near the solution could be avoided by approximating the *full* gradient vector (as in an unconstrained problem), since in general $g(x^*)$ does not vanish for a constrained problem. However, it can be shown that the approximate gradient will retain high relative accuracy at a point x in the neighbourhood of x^* only in the range space of $\hat{A}(x)^T$; thus, premultiplying the approximate gradient by $Z(x)^T$ simply annihilates the accurate part of the approximation. In order to achieve the maximum accuracy in the solution, a method must ensure that there is sufficiently high relative accuracy in the projected gradient (even if the projected gradient is not formed explicitly).

Notes and Selected Bibliography for Section 8.6

The references concerning finite-difference approximations are given in the Notes for Section 4.6.

8.7. MORE ABOUT SCALING

We have already discussed in Section 7.5.1 an elementary form of problem scaling, in which a diagonal linear transformation is applied. The purpose of this scaling is to make *all the variables of a similar order of magnitude* in the region of interest, with the aim of causing each variable to be of similar "weight" during the optimization. In this section, we shall consider the idea of "scaling" in more general terms, and shall describe scaling techniques for both unconstrained and constrained problems. All the scaling procedures to be described require the availability of the derivatives of the problem functions.

8.7.1. Scaling the Variables

In this section, we shall consider the effect on problem scaling of replacing the original variables x of the problem by a set of *transformed* variables y. In theory, such a transformation does not affect the optimal solution; however, it should be clear from all the previous sections of this chapter that the scaling of the variables may have an enormous effect on the behaviour of finite-precision calculations.

We shall consider only *linear* transformations of the variables. In this case, it holds that

$$x = Ly, \tag{8.49}$$

where L is a fixed non-singular matrix. (Nonlinear transformations are briefly discussed in Section 7.4.1.) Let $\mathcal{F}(y)$ denote the function of the transformed variables

$$\mathcal{F}(y) = F(Ly). \tag{8.50}$$

The derivatives of the function $\mathcal{F}$ with respect to the transformed variables are $\nabla_y \mathcal{F} = Lg(x)$, $\nabla_y^2 \mathcal{F} = L^T G(x)L$.

8.7.1.1. Scale-invariance of an algorithm. Suppose that a problem is originally formulated with variables x, and then is transformed to a problem with variables y defined by (8.49). If an algorithm is applied to both the original problem and the transformed problem, and the resulting sequences of iterates $\{x_k\}$ and $\{y_k\}$ satisfy

$$x_k = Ly_k \tag{8.51}$$

for all k, the algorithm is said to be *invariant with respect to linear transformations of the variables* (or simply *scale-invariant*).

Some algorithms are scale-invariant under suitable conditions *if the iterates are calculated with exact arithmetic*. For example, Newton methods that use the exact Hessian and certain quasi-Newton methods that use the exact gradient can be shown to satisfy (8.51) when the Hessian (or Hessian approximation) is positive definite at every iteration. Unfortunately, *the property of scale-invariance cannot be achieved in a practical implementation of an algorithm*.

Firstly, computer arithmetic is not scale-invariant. The rounding error associated with forming, say, $x_1 + x_2$ is not related in a straightforward way to the rounding error in computing the sum $\alpha x_1 + \beta x_2$ (unless $\alpha = \beta$).

Secondly, it is essential for numerical stability that an algorithm should control the conditioning of the associated numerical processes, and hence good algorithms treat quantities that are "sufficiently small" in magnitude as "negligible" (in effect, as zero). However, since there is no universal definition of "small", it is impossible to formulate an ironclad procedure for distinguishing between quantities that *should* be treated as zero, and scale-dependent quantities that *should not* be neglected. For this reason, some algorithms may be nearly invariant to scaling when all the variables are multiplied by a *large* number, but not when the variables are multiplied by a *small* number.

8.7.1.2. The conditioning of the Hessian matrix. We have seen (Section 8.2.2.1) that the conditioning of the Hessian matrix at the solution of an unconstrained problem determines the accuracy with which the solution can be computed in different directions. In particular, when $G(x^*)$ is ill-conditioned, the function will vary much more rapidly along some directions than along others. An ill-conditioned Hessian at the solution is thus a form of bad scaling, in the sense that similar changes in $\|x\|$ do not lead to similar changes in F.

This type of bad scaling near the solution may have several harmful effects on a minimization algorithm. In particular, F may vary so slowly along an eigenvector associated with a near-zero eigenvalue that changes in F that "should" be significant may be lost amongst the rounding error. Ill-conditioning of $G(x^*)$ also causes a degradation in performance in methods that need to solve a system of linear equations that involves the Hessian matrix (or an approximation to the Hessian).

One way to ensure that a problem is "well-scaled" in the above sense is *to devise a linear transformation that minimizes the condition number of the Hessian at the solution.* If $G(x^*)$ were known (and positive-definite), L in (8.51) could be taken as

$$L = G(x^*)^{-1/2}, \qquad (8.52)$$

and the Hessian of the transformed problem at the solution would be the identity matrix. (We note that some optimization methods implicitly compute a scaling of the form (8.52); in particular, quasi-Newton and conjugate-gradient-type methods effectively construct the matrix $G^{-1/2}$ when applied with exact arithmetic to a quadratic function with a positive-definite Hessian matrix G.)

Since $G(x^*)$ is generally unknown, the choice (8.52) is not practical. If second derivatives are available and we are prepared to assume that the Hessian does not vary too much over the region of interest, the matrix L may be taken as $G(x_0)^{-1/2}$. If exact first derivatives are available, L may be based on a finite-difference approximation of the Hessian. If only function values are available, L is sometimes defined as a *diagonal* scaling based on the approximated diagonal elements of the Hessian matrix at x_0 (estimates of these elements are produced by the procedure of Section 8.6.2.2). *However, any of these procedures may produce a poor scaling if the Hessian at x^* differs significantly from that at x_0.*

Despite these difficulties, simple diagonal scaling procedures often help to reduce the condition number of the Hessian at the solution.

Example 8.7. Consider the following function of two variables:

$$F(x) = x_1^b x_2^c,$$

where b and c are known parameters. It is readily verified that the Hessian matrix of $F(x)$ is

$$G(x) = \frac{F}{x_1^2 x_2^2} \begin{pmatrix} b(b-1)x_2^2 & bcx_1x_2 \\ bcx_1x_2 & c(c-1)x_1^2 \end{pmatrix}. \qquad (8.53)$$

The matrix (8.53) may be factorized as

$$G(x) = \frac{F}{x_1^2 x_2^2} LL^T,$$

where

$$L = \begin{pmatrix} \beta x_2 & 0 \\ \Gamma x_1 & \delta x_1 \end{pmatrix},$$

with $\beta^2 = b(b-1)$, $\Gamma = bc/\beta$, and $\delta^2 = c(1-b-c)/(b-1)$. Since the squared ratio of the largest and smallest diagonals of L often provides an order-of-magnitude estimate of the condition number of $G(x)$ (see Section 8.3.3.2), we have that

$$\text{cond}(G) = \text{cond}(L)^2 \sim \left(\frac{\beta x_2}{\delta x_1}\right)^2$$

(if $|\beta x_2| \geq |\delta x_1|$). Assuming that β and δ are of similar magnitude, the ratio of x_2^2/x_1^2 at the solution can be used as an indicator of the condition number of $G(x^*)$. To make the condition number close to unity, x_1 and x_2 should be scaled to have approximately the same values at the solution. (Exactly the same scaling results if $|\beta x_2| \leq |\delta x_1|$.)

8.7.1.3. Obtaining well-scaled derivatives.

The notion of bad scaling discussed in Section 7.5.1 involved variables of widely differing magnitudes. Another form of bad scaling (which is sometimes related to the first) occurs when the *partial derivatives* of a function with respect to a particular variable are not "balanced". All the scaling schemes to be described in this section are most effective when used after the variables have been scaled in magnitude using the method described in Section 7.5.1.

Scaling based on the first derivative. Difficulties can arise when the first partial derivative of a function is "small" (or "large") relative to the accuracy with which the function can be evaluated (ϵ_A).

Example 8.8. To illustrate this situation, we consider the seven-variable function:

$$
\begin{aligned}
F(x) = {} & (x_1 - 100)^2 + (x_2 - 1000)^2 + (x_3 + x_4 - 1000)^2 \\
& + (x_5 - 1000)^2 - (x_6 + x_7 - 200)^2 \\
& + 10^{-6}(x_1 + 2x_2 + 3x_3 + 4x_4 + 5x_5 + 6x_6 + 7x_7 - 1900)^3.
\end{aligned}
$$

The exact value of this function at the point $x = (0, 0, 400, 100, 0, 0, 0)^T$ is $F(x) = .2219973 \times 10^7$.

When F is evaluated in short precision on an IBM 370 ($\epsilon_M = 16^{-5} \approx .9537 \times 10^{-6}$), the value of ϵ_A (the accuracy in the computed value of F) in the region of interest of the order of $\epsilon_M|F|$, so that the function can be evaluated to full machine precision.

The bad scaling of F with respect to the first variable can be seen by evaluating F at the perturbed point $\bar{x} = x + \bar{h}e_1$, where $\bar{h}$ is defined by (8.47). If F were well-scaled, a perturbation of this size in x should change about *half* the figures of F (see Section 8.6.1). Instead, only the *last* (hexadecimal) figure of the mantissa varies (by one unit). The effect of the bad scaling is that the change in F is much smaller than "expected".

Bad scaling of this type can be critical when finite differences are used to approximate derivatives, *particularly when a "standard" finite-difference interval is used*. If a finite-difference approximation is computed with an interval that produces a too-small change in F, the approximate derivatives will be swamped by condition error.

Example 8.9. Consider the calculation of a forward-difference approximation to $g_1(x)$ for the badly scaled function of Example 8.8, in short precision on an IBM 370 ($\epsilon_M = 16^{-5} \approx .9537 \times 10^{-6}$). Based on the reasoning of Section 8.6.1 and (8.47), the finite-difference interval h_1 is taken as

$$h_1 = \bar{h} = 2\sqrt{\frac{\epsilon_A}{1 + |F(x)|}}. \tag{8.54}$$

The resulting forward-difference approximation to $g_1(x)$ is $-.512000 \times 10^3$, whereas the true first partial derivative at x is $-.199730 \times 10^3$. *The poor scaling of F has caused a "standard" choice of finite-difference interval to produce a completely inaccurate estimate of the derivative.*

We emphasize that the seriousness of this form of "bad scaling" is related to the available machine precision. If $F(\bar{x})$ is evaluated in long precision on an IBM 370 (where the relative precision is $\epsilon_M = 16^{-13} \approx .222 \times 10^{-15}$), nine of the fourteen (hexadecimal) digits of the mantissa of $F(\bar{x})$ differ from those of $F(x)$. However, when a forward-difference approximation to $g_1(x)$ is computed with a finite-difference interval analogous to (8.54) with the appropriate value of ϵ_A, the derivative approximation has **five correct** (decimal) figures (rather than the eight that would be expected for a well-scaled problem), and the bad scaling of F can still be observed.

The root of the difficulty in Example 8.9 is that the magnitude of the first partial derivative is of order $\sqrt{\epsilon_A(1 + |F(x)|)}$ (or smaller). From the Taylor-series expansion of F about x along e_j, it follows that

$$F(x + h_j e_j) - F(x) = h_j g_j(x) + \frac{1}{2} h_j^2 G_{jj}(x) + O(h_j^3). \tag{8.55}$$

If the first-order term of (8.55) dominates the right-hand side, the change in F is of order $|h_j g_j(x)|$. Substituting the values of h_1 and $g_1(x)$ from Example 8.9, we see that the change in F will be of order ϵ_A or less (i.e., at "noise level").

This type of poor scaling of derivatives may be remedied under certain conditions by using a diagonal scaling of the form

$$x = Dy. \tag{8.56}$$

Consider the Taylor-series expansion of the function $\mathcal{F}(y)$ (8.50) about the point y ($y = D^{-1}x$) with perturbation $\gamma_j e_j$, we obtain

$$\mathcal{F}(y + \gamma_j e_j) - \mathcal{F}(y) = \gamma_j e_j^T \nabla_y \mathcal{F}(y) + \frac{1}{2} \gamma_j^2 e_j^T \nabla_y^2 \mathcal{F}(y) e_j + O(\gamma_j^3)$$
$$= \gamma_j d_j g_j(x) + \frac{1}{2} \gamma_j^2 d_j^2 G_{jj}(x) + O(\gamma_j^3). \tag{8.57}$$

If second-order terms can be neglected on the right-hand side of (8.57), it follows that a change of γ_j in y_j produces a change in $\mathcal{F}$ of magnitude $|\gamma_j d_j g_j|$.

The choice of d_j is based on the idea that perturbation by the "optimal" finite-difference interval $\bar{h}$ (8.54) should lead to a perturbation of "half" the significant figures of F. This leads to the equation

$$2|d_j g_j| = 1 + |F(x)|,$$

so that

$$d_j = \frac{1 + |F(x)|}{2|g_j(x)|}. \tag{8.58}$$

Scaling based on the second derivative. The second case of interest occurs when we can neglect the *first-order* terms of (8.57). In particular, when $g_j(x)$ is zero, the change in $\mathcal{F}$ resulting from a perturbation $\gamma_j e_j$ is

$$\mathcal{F}(y + \gamma_j e_j) - \mathcal{F}(y) = \frac{1}{2}\gamma_j^2 d_j^2 G_{jj}(x) + O(\gamma_j^3). \qquad (8.59)$$

If the third-order terms in (8.59) can be neglected, for a well-scaled problem we expect that, when γ_j is taken as $\bar{h}$ (8.54), the change in function value should be of order ϵ_A. To achieve this objective, the value of d_j should be defined as

$$d_j = \sqrt{\frac{1 + |F(x)|}{2|G_{jj}(x)|}}. \qquad (8.60)$$

The scale factor defined by (8.60) is still appropriate when $|g_j|$ is small enough so that the second-order terms are dominant in (8.57).

When this form of scaling procedure is used, it is necessary to decide whether to use (8.58) or (8.60) to compute the scale factors $\{d_j\}$. Unfortunately, a wrong decision concerning the choice of scaling formula may lead to serious difficulties during the subsequent minimization. If the scale factors are computed using (8.58), but the second-order terms of the Taylor-series expansion should not have been neglected, d_j may be a drastic overestimate of a useful scaling factor. *A consequence of an overly large value of d_j is that the scaled objective function will be more "nonlinear" than necessary.* If the magnitude of the second derivative term dominates the Taylor-series expansion of the scaled problem, a linear approximation to $\mathcal{F}(y)$ will be adequate only in a very small neighbourhood of y. This may cause difficulties because many optimization algorithms assume that the objective function can be adequately approximated by the first-order terms of the Taylor-series expansion.

Scaling based on the estimated finite-difference interval. It is possible to avoid the need to choose between (8.58) and (8.60) by using a scaling scheme based upon the estimate of the finite-difference interval that minimizes the sum of condition error and truncation error in a forward-difference estimate of $g_j(x)$ (see Section 8.6).

Let $\hat{h}_j$ denote the computed estimate of the "optimal" interval corresponding to the j-th original variable (see Sections 4.6.1.3 and 8.6.2). From (8.56), the perturbation $\hat{h}_j$ in x_j is equivalent to a perturbation

$$\gamma_j = \frac{\hat{h}_j}{d_j}$$

in y_j. For a well-scaled problem, the optimal finite-difference interval is simply $\bar{h}$ (8.54). This implies that d_j should be taken as

$$d_j = \frac{\hat{h}_j}{2}\sqrt{\frac{1 + |F(x)|}{\epsilon_A}}. \qquad (8.61)$$

An example of scaling the derivatives. Consider the calculation of the first partial derivative when the first variable in Example 8.8 is scaled according to (8.58) and (8.61). The value of d_1 computed from (8.58) for Example 8.8 is $.56 \times 10^4$. The perturbed value of $\mathcal{F}(y + \bar{h}e_1)$ is

.221792 $\times$ 10^7; three of the six hexadecimal digits of the mantissa of $\mathcal{F}$ are changed, so that the scaling has had the desired effect on the relative change in $\mathcal{F}$. However, the forward-difference approximation to the first partial derivative of $\mathcal{F}$ using the interval $\bar{h}$ is $-.105062 \times 10^7$, compared to the exact value $-.110999 \times 10^7$. This reflects the increase in truncation error in the forward-difference approximation because of the larger second partial derivatives (which have been inflated by a factor of 3.1×10^7).

The value of d_1 given by (8.61) for Example 8.8 is $.106 \times 10^4$. The perturbed value $\mathcal{F}(y + \bar{h}e_1)$ for this scale factor is $.221956 \times 10^7$, and, as with the first scaling, three of the six hexadecimal digits of the mantissa are changed by the perturbation $\bar{h}$. The forward-difference approximation to the first partial derivative using the interval $\bar{h}$ is $-.209920 \times 10^6$, compared to the exact value $-.211682 \times 10^6$. As we might expect, the relative error in the finite-difference approximation with the scaling (8.61) is better than with (8.58).

8.7.2. Scaling the Objective Function

Certain properties of the objective function have already been mentioned in the discussion of scaling the variables. It is sometimes believed that scaling of the objective function is not important, since in theory the solution of a given problem is unaltered if $F(x)$ is multiplied by a positive constant, or if a constant value is added to $F(x)$. (Note that multiplication of F by a constant also causes all the derivatives to be multiplied by the constant, whereas adding a constant does not change any of the derivatives, but only the value of F.)

It is generally considered desirable for the magnitude of the objective function to be of order no larger than unity in the region of interest. This can be achieved easily when the magnitude of F is known to be *large* at all points of interest, by simply choosing an appropriate large constant to be divided into F and all its derivatives. For example, if in the original formulation $F(x)$ is of the order of 10^5 (say), then the value of $F(x)$ (and appropriate derivatives) should be multiplied by 10^{-5} when evaluating the function within the optimization routines.

Obvious difficulties arise if F is *very small everywhere*. As noted in several previous sections, optimization methods designed for practical computation invariably include an *absolute* test that defines a "small" quantity, in addition to relative tests that apply when quantities are large. Hence, although in theory the solution is unaltered if F is multiplied by, say, 10^{-20}, the effect of this scaling is likely to be that the algorithm declares convergence at the starting point, because the norm of the gradient is less than some pre-assigned absolute tolerance.

Including the addition of a constant to F can cause difficulties because the error associated with forming the sum in floating point may reflect the size of the constant rather than F. Hence, it is preferable whenever possible to omit such constants. For example, $F(x)$ should be formulated as $x_1^2 + x_2^2$, rather than $x_1^2 + x_2^2 + 1000$ (or even $x_1^2 + x_2^2 + 1$).

8.7.3. Scaling the Constraints

A "well-scaled" set of constraint functions has two main properties. Firstly, each constraint function should be well conditioned with respect to perturbations of the variables. Secondly, the constraint functions should be *balanced* with respect to each other, i.e. all the constraints should have "equal weight" in the solution process.

8.7.3.1. Some effects of constraint scaling. The scaling of the constraints has several effects on the computation of the solution and in the interpretation of the results. For example, recall that the accuracy of exact and computed Lagrange multiplier estimates is critically dependent upon the condition number of the Jacobian matrix of the constraints that are in the working set (see Section 5.1.5). Furthermore, this same condition number affects the limiting accuracy of the solution in the range-space of the active set (see Section 8.2.2.2).

A less obvious instance in which scaling is significant concerns the choice of constraint to be deleted in an active set method for linear constraints, when an algorithm has reached a constrained stationary point x_k. The Lagrange multipliers λ_k with respect to the working set are defined by the compatible system

$$\hat{A}_k^T \lambda_k = g_k.$$

The effect of multiplying each constraint by a different scale factor is to alter the rows of $\hat{A}_k$ (and, hence, the values λ_k). In particular, if the j-th constraint is multiplied by a constant ω_j, the j-th component of λ_k is divided by ω_j.

In most algorithms, the constraint chosen for deletion is taken as the one with the most negative Lagrange multiplier estimate, so that the constraint to be deleted will generally *change* if the constraints are re-scaled in this simple way. This means that a different sequence of iterates will be generated by the algorithm.

Example 8.10. Consider a two-variable problem. Suppose that the value of the inner product $\hat{a}_1^T x$ represents the amount of dissolved oxygen in water (with a typical value of 1 part per million), and that the inner product $\hat{a}_2^T x$ represents the velocity of flow of the water (with a typical value of 5 miles per hour). Assume that these inner products define two range constraints

$$0 \leq \hat{a}_1^T x \leq 8 \times 10^{-6}$$
$$3 \leq \hat{a}_2^T x \leq 10,$$

and that the 2×2 matrix $\hat{A}$ with rows $\hat{a}_1^T$ and $\hat{a}_2^T$ is given by

$$\hat{A} = \begin{pmatrix} 10^{-6} & -10^{-6} \\ 1 & 1 \end{pmatrix}. \tag{8.62}$$

If both these constraints were equal to their lower bounds and had negative Lagrange multiplier estimates, the multiplier estimate corresponding to the first constraint would tend to be much larger in magnitude than the multiplier corresponding to the second constraint. Therefore, the first constraint would almost always be deleted first.

In this simple example, the constraints can be balanced simply by multiplying the first constraint by 10^6. With this re-scaling, the coefficient matrix becomes

$$\begin{pmatrix} 1 & -1 \\ 1 & 1 \end{pmatrix}. \tag{8.63}$$

which is very well-conditioned (in fact, the columns are orthogonal).

A further lesson to be drawn from Example 8.10 is that an *automatic* procedure for balancing the constraints should not be used without some reference to the origins of the problem. Consider a second problem with exactly the same coefficient matrix as (8.62), but in which the first constraint is defined by the sum of several other constraints, all of whose coefficients lie in the range $[-1, 1]$. For the second problem, the small coefficients in (8.62) occur because of cancellation in forming the aggregated constraint; thus, it would be inappropriate to multiply this constraint by a large factor in order to give it the same weight as the "real" constraints in the problem.

8.7.3.2. Methods for scaling linear constraints. Several options are open to the user who wishes to scale linear constraints. For example, the variables can be scaled using one of the methods outlined in Section 8.7.1, and the matrix of constraint coefficients can then be balanced by multiplying the rows by appropriate weights (as discussed in Section 8.7.3.1). Such a scheme would be based on the assumption that a good scaling of the variables relative to the objective function will generally be a good scaling relative to the constraints also. However, the user should be careful of applying this type of scaling procedure when the solution of the problem is largely determined by the constraints. In this case, re-scaling the variables with respect to the (relatively unimportant) objective function may have an undesirable effect on the original relationship among the variables and constraints.

A second scaling procedure is to scale *both the rows and columns* of the coefficient matrix, and to ignore the scaling of the objective function. Schemes of this type are often used in linear and quadratic programming. The scaled coefficient matrix is of the form

$$D_1 A D_2,$$

where D_1 and D_2 are diagonal matrices with positive entries. We shall describe one method for computing D_1 and D_2 that has been used extensively in large-scale linear programming (details of other methods may be found in the references cited in the Notes). The basis of the method is to repeatedly scale the rows (columns) of the matrix by the geometric mean of the largest and smallest elements of the row (column). The formal statement of the algorithm is the following.

Algorithm SC (*Row and column scaling of linear constraints*).

SC1. [Compute the greatest ratio of two elements in the same column.] Compute ρ_0, which is defined as

$$\rho_0 = \max_j \max_{r,s}\{|a_{rj}/a_{sj}|\},$$

where $a_{sj} \neq 0$.

SC2. [Perform row scaling.] Divide each row i of A and its corresponding right-hand side value by $((\min_j |a_{ij}|)(\max_j |a_{ij}|))^{1/2}$, where the minimum is taken over all $a_{ij} \neq 0$.

SC3. [Perform column scaling.] Divide each column j of A by $((\min_i |a_{ij}|)(\max_i |a_{ij}|))^{1/2}$, where the minimum is taken over all $a_{ij} \neq 0$.

SC4. [Compute the greatest ratio of two elements in the same column.] Compute the value ρ, defined as

$$\rho = \max_j \max_{r,s}\{|a_{rj}/a_{sj}|\},$$

where $a_{sj} \neq 0$.

SC5. [Check for termination.] If $|\rho - \rho_0| \geq 10^{-1}|\rho_0|$, go back to Step SC1. Otherwise, the procedure terminates. ∎

Criteria other than the test used in step SC5 may be used to terminate the algorithm. For example, a fixed number of iterations (say, 5) could be performed.

If the Algorithm SC is applied to the matrix of Example 8.10, the well-conditioned matrix (8.63) is obtained after one iteration.

8.7.3.3. Methods for scaling nonlinear constraints.

When a problem contains *linear* constraints, results from numerical linear algebra can be applied to specify the accuracy with which the computed solution should (and will) satisfy the active constraints. In general, if the constraints have been properly scaled, the residual in an active linear constraint will be of the order of machine precision. For nonlinear constraint functions, however, a correct decision can be made concerning the acceptable size of the constraint residual only if a value of the attainable accuracy (ϵ_A) is provided *for every constraint*. Lacking this information, most algorithms will attempt to ensure that all the active constraints satisfy some "reasonable" definition of "small". However, the values of the active constraints tend to vary in magnitude at the best computed solution, even if all are small (for example, $\hat{c}_1 = 10^{-10}$, $\hat{c}_2 = -10^{-6}$, and so on).

Assuming that all the constraints are of equal significance, it is desirable for a unit change in x to produce a similar change in *each* constraint. A set of weights that will achieve this objective are often suggested by the form of the problem. For example, in the nose-cone problem of Section 1.1, the nonlinear constraints associated with the volume and the length of the nose cone can be weighted so that their scaled values are both of order one. Similarly, the simple bound constraints may be scaled by making the final radius equal to one and by expressing the angle of each conical section in radians.

The techniques of Section 8.7.1 may be applied to scale each nonlinear constraint function so that it is well-conditioned with respect to perturbations in x. Unfortunately, this procedure is time-consuming and difficult, since a different transformation of the variables might be required for each constraint. When using Lagrangian-based methods such as those of Sections 8.4, 8.5 and 8.6, ideally it is necessary to scale only the Lagrangian function. However, this is not usually possible, since only unreliable multiplier estimates may be available.

Many methods for nonlinearly constrained optimization perform computations that involve the Jacobian matrix of the constraint functions in the working set. In order to improve the numerical stability of these calculations, the Jacobian matrix could be equilibrated at every iteration using an algorithm similar to Algorithm SC; note, however, that such a scaling must be performed as part of the optimization method rather than by the user.

Other scalings of the Jacobian matrix may involve normalizing each constraint gradient to have unit norm. Since this process breaks down if the gradient of a constraint is zero, there may be difficulties when the gradient of a constraint is very small. The perennial question then arises as to whether the gradient is small because it is negligible, or whether it is small merely because of poor scaling. As we have observed several times, there is no all-purpose resolution of this dilemma. In practice, it is probably desirable to err on the side of conservatism, and to consider very small numbers as negligible. If this strategy produces an unacceptable solution, the gradients that were neglected may be re-scaled appropriately.

Notes and Selected Bibliography for Section 8.7

Example 8.7 was suggested by Murtagh and Saunders (1977). The geometric-mean scaling scheme has been used successfully in large-scale linear programming for many years; see, for example, Benichou *et al.* (1977). The form of the method given here is that used by Fourer (1979). For other scaling schemes for the linear-constraint case, see Hamming (1971), Fulkerson and Wolfe (1962) and Curtis and Reid (1972). A numerical comparison of several schemes is given by Tomlin (1975b).

QUESTIONS AND ANSWERS

Amid the pressure of great events, a general principle gives no help.

—G. W. F. HEGEL (1832)

During our involvement in the development of software for numerical optimization, we have noticed that certain questions regarding the use of optimization methods are asked repeatedly by software users. The following section includes twenty of the most frequently asked questions together with some possible answers. Note that the order of the questions is not significant.

Q1. *How do I know that the answer from my computer run is the global minimum?*

A1. There is no guarantee that the computed solution will be a global minimum unless your problem is quite special (see Section 6.8). General methods for locating a global minimum are not guaranteed to work except under very stringent conditions on the problem, and usually require so much computation that they are impractical unless the number of variables is very small.

Q2. *I am solving an optimization problem in which the objective function involves an iterative subproblem that may be solved to a user-specified accuracy. The optimization code to be used assumes that the accuracy of the objective function is always at the level of machine precision. Is it possible to avoid the need to solve the iterative subproblem to full machine precision?*

A2. In most optimization codes, a parameter is set to the machine precision. This parameter can be set to a larger value during an initial run, so that the subproblem needs to be solved less accurately at points far from the solution of the optimization problem. The machine precision can then be changed back to its correct value (and the subproblem solved more accurately) in order to refine the solution.

Q3. *I have solved my unconstrained optimization problem by applying a non-derivative quasi-Newton method. If I increase the number of variables, what change can I expect in the number of function evaluations and iterations?*

A3. Since quasi-Newton methods are iterative, it is not possible to estimate in advance the number of iterations that will be required to find the solution for a general problem. The number of iterations will depend on the distance from the starting point to the solution, and on the nonlinearity and conditioning of the problem. Each iteration of a non-derivative quasi-Newton method requires at least as many function evaluations as there are variables, and hence each iteration will become more costly as the number of variables increases. In certain situations, a larger number of variables causes only a minor increase in the number of iterations — for example, when the variables are the coefficients in the expansion of a function, and the number of variables represents the accuracy of the mathematical model.

357

Q4. *I am solving a nonlinearly constrained problem in which I know that some of the constraints will be active at the solution. Should I include these constraints as inequalities or as equalities?*

A4. The best strategy will depend upon the software that is available. If you have access to a good program that always computes feasible estimates of the solution, and you know a point that is strictly feasible, you should treat the constraints as inequalities and use a feasible-point method. However, if the available software is based on a method that may compute the problem functions at infeasible points, the constraints should be formulated as equalities, in order to reduce the difficulties in identifying the active set.

Q5. *I can compute the first derivatives of my problem analytically. Should I use a discrete Newton method or quasi-Newton method?*

A5. The answer to this question will depend on the desired quality in the solution and on the number of variables. The approximate Hessian matrix computed by a discrete Newton method at the solution can be used to estimate the conditioning and sensitivity of the solution (see Section 8.3.3.1); a quasi-Newton approximation (Section 4.5.2) cannot be guaranteed to be an accurate representation of the true Hessian. Moreover, there are certain problem classes (e.g., those with saddle points) where only a discrete Newton method is assured of being able to proceed. However, a discrete Newton method will require at least n gradient evaluations every iteration in order to approximate the Hessian. Consequently, unless the Hessian matrix has some special structure (see Section 4.8.1), discrete Newton methods tend to require more gradient evaluations than a quasi-Newton method as the number of variables increases.

Q6. *I am using some library software to check the gradients by finite differences. The checking routine indicates that one gradient element is wrong, although I know that it is correct. Is there an error in the library routine?*

A6. Not necessarily. Your problem could be so badly scaled that the changes in the objective function computed by the gradient-checking routine are lost below the level of rounding error. In this case, it is desirable to re-scale the problem along the lines of Section 8.7. However, if you are absolutely certain that the bad scaling is only local, you may simply ignore the error indication.

Q7. *I am estimating some parameters using a nonlinear least-squares routine. The data values to which I am fitting the function are accurate to only two significant figures. Is it reasonable to ask for only two correct figures from the optimization program?*

A7. Not necessarily! Despite the inaccurate data, it is likely that the least-squares objective problem can be computed to much higher relative precision than 10^{-2}. The effect of the errors in the data is simply to make the exact solution of the optimization problem different from the real-world solution. Furthermore, the specification of a "large" value for the termination criterion does not necessarily imply that the computed solution will be a low-accuracy approximation to the solution (see Section 8.2.3.6).

Q8. *After using an unconstrained routine, I found a minimum of my problem at nonsensical values of the variables. How should I proceed?*

A8. The fact that you were able to ascertain that the values of the variables were nonsensical implies that you must know values of the variables that are reasonable. It would be better, therefore, if the function were minimized subject to simple upper and lower bounds that restrict the variables to lie in acceptable regions (see Section 5.5.1).

Q9. *I am using a function comparison method for multivariate unconstrained minimization. How can I be sure that I have found the correct answer?*

A9. A serious disadvantage of function comparison methods for the multivariate case is that it is impossible to give a general guarantee that the method has found a solution. (Since the method does not compute any derivatives of the objective function, the first-order necessary conditions for optimality cannot be checked.) Your confidence that the correct answer has been found may be increased if the algorithm converges to the same point after starting the algorithm at a different point.

Q10. *My unconstrained code says that the correct answer has been found, yet the elements of the gradient vector are quite large. Doesn't this indicate an error in the program?*

A10. If the problem is badly scaled, the gradient vector may be large at a perfectly legitimate solution. In most cases, the norm of the gradient should be scaled by the magnitude of the objective function in performing the tests for convergence (see Section 8.2.3).

Q11. *My objective function and its gradient are defined by a very long and complicated subroutine. When I use a quasi-Newton method, the convergence rate is very slow, and the objective function changes only imperceptibly. What can be wrong?*

A11. The two most likely possibilities are an error in the code that computes the derivatives, or discontinuities in the derivatives. You should attempt to check the derivatives along the lines of Section 8.1.4.2. Small discontinuities in the function or the derivatives can often be detected by comparing forward- and backward- difference approximations at the point where convergence is poor (see Section 8.4.2.4).

Q12. *A nonlinearly constrained problem includes a set of nonlinear equality constraints such that a fast special method can be used to solve for one ("dependent") subset of the variables in terms of the remaining ("independent") variables. Is it worth using this special procedure to eliminate some of the constraints?*

A12. We do not recommend this approach if the constraints have any significant degree of nonlinearity. Firstly, it is difficult to impose any simple bounds upon the dependent variables. Secondly, the resulting optimization method is of the "constraint-following" type described in Section 6.3, and thus will tend to be less efficient than other methods (unless the constraints are nearly linear). Finally, in our experience the computational effort required to solve for the dependent variables at every trial point is not usually worthwhile, compared to expending a similar amount of effort in the optimization without eliminating the variables.

Q13. *All the functions in my problem are smooth, but highly nonlinear. Is it true that Newton or quasi-Newton methods will not work well on highly nonlinear problems, so that I should use a function comparison method?*

A13. Emphatically no! The idea that Newton and quasi-Newton methods are prone to failure on highly nonlinear problems arose from the days when these methods were poorly implemented (e.g., did not include a line search). A function comparison method is the *worst* possible choice for a difficult problem, since it is likely to be extremely inefficient, and to provide no assurances about convergence.

Q14. *My output from a Newton-type method indicated that convergence to the solution was very* *rapid, yet the program indicated a failure because a sufficiently lower point could not be found by a step-length procedure. How can this happen?*

A14. Sometimes, Newton-type methods reach a close neighbourhood of the solution before the algorithm can ascertain with any certainty that successful convergence has been achieved. The algorithm is then unable to make any further progress because there is no significant difference in the function values at points in the region. See Section 8.2.3 for a full discussion of termination criteria.

Q15. *When looking over the output from a linearly constrained minimization, I noticed that several Lagrange multipliers were close to zero at the final point. Does this imply that the relevant constraints are redundant and can be removed from the problem?*

A15. Not necessarily! See Figures 8e and 8f in Chapter 8.

Q16. *I have access to a double precision optimization code but my objective function and constraints are all coded in single precision. Can I go ahead and use the program?*

A16. It depends upon whether or not the code allows the user to specify the precision of the problem functions. If the accuracy of the problem functions is a parameter, the program may be used with this parameter given a value that reflects the single precision accuracy. If no such parameter is available, the routine probably assumes that the problem functions can be computed to full precision. In this case, the routine should be used with caution, since the computed solution may not correspond exactly to the problem that you wished to solve.

Q17. *I plan to estimate some parameters by fitting certain model functions to some data. Should I minimize the difference between the data and the model in terms of the two-norm, the one-norm or the infinity norm?*

A17. Unless the norm has some special statistical significance, we would recommend using the two-norm. The use of the one-norm and infinity-norm leads to an unconstrained optimization problem with discontinuities in the derivatives. Although special methods exist for these problems, the complexity of the problem is significantly increased compared to the least-squares problem (see Sections 4.2.3 and 6.8.1).

Q18. *Some constraints in my problem are not expected to be satisfied exactly at the solution. Should I use an algorithm for unconstrained or constrained optimization?*

A18. The answer to this question depends upon a number of factors. If the effort required to program the constraints (and possibly their derivatives) is significant, or the quality of the available software for unconstrained minimization is better than that for constrained problems, an unconstrained method should be tried first. If the function may be unbounded for some values of the variables, a constrained routine should be used. (Note that the imposition of simple-bound constraints tends to increase the efficiency of the minimization, regardless of whether or not they are active at the solution.)

Q19. *I believe that my objective function is smooth at the solution but may have discontinuities at other points. Must I use an algorithm for non-smooth functions?*

A19. It is usually worth trying an algorithm for smooth functions first. However, it is worth bearing in mind that some algorithms for smooth optimization are more susceptible to discontinuities than others. Newton-type methods are likely to be the most effective in this situation because they adapt instantaneously to discrete changes in the Hessian matrix. Any routine that computes difference approximations to derivatives should be used with caution if there is the chance that a finite-difference will be made across a discontinuity.

Q20. *What general features of an optimization problem affect the choice of algorithm?*

A20. The following list presents some features of optimization problems that have an important effect on the choice of method. Where appropriate, references are given to the sections in which the relevant method is discussed.

(a) the number of variables;

(b) whether there is a range of values within which the variables must lie (see Section 5.5.1);

(c) whether the problem functions and their derivatives are smooth (see Sections 4.2.1 and 6.8);

(d) the highest level of derivatives that can be efficiently coded and evaluated (see Section 8.1.1);

(e) if the problem is large, the proportion of zero elements in the Hessian matrix and Jacobian matrix of the constraints (see Sections 4.8, 5.6 and 6.7);

(f) the number of general linear constraints compared to the number of variables, and the number of constraints that are likely to be active at the solution (see Sections 5.4 and 5.5.2); and

(h) whether the problem functions can be evaluated or are meaningful outside the feasible region (see Sections 6.2.1.2 and 8.1.1.3).

BIBLIOGRAPHY

Knowledge is of two kinds. We know a subject ourselves,
or we know where we can find information upon it.

—SAMUEL JOHNSON (1775)

Aasen, J. O. (1971). On the reduction of a symmetric matrix to tridiagonal form, *Nordisk Tidskr. Informationsbehandling (BIT)* **11**, pp. 233–242.

Abadie, J. and Carpentier, J. (1965). Généralisation de la méthode du gradient réduit de Wolfe au cas de contraintes non-linéaires, Note HR6678, Électricité de France, Paris.

Abadie, J. and Carpentier, J. (1969). "Generalization of the Wolfe reduced-gradient method to the case of nonlinear constraints", in *Optimization* (R. Fletcher, ed.), pp. 37–49, Academic Press, London and New York.

Abadie, J. and Guigou, J. (1970). "Numerical experiments with the GRG method", in *Integer and Nonlinear Programming* (J. Abadie, ed.), pp. 529–536, North-Holland, Amsterdam.

Abadie, J. (1978). "The GRG method for nonlinear programming", in *Design and Implementation of Optimization Software* (H. J. Greenberg, ed.), pp. 335–362, Sijthoff and Noordhoff, Netherlands.

Ablow, C. M. and Brigham, G. (1955). An analog solution of programming problems, *Operations Research* **3**, pp. 388–394.

Anderson, D. H. and Osborne, M. R. (1977). Discrete, nonlinear approximations in polyhedral norms: a Levenberg-like algorithm, *Num. Math.* **28**, pp. 157–170.

Anderson, N. and Björck, Å. (1973). A new high-order method of the regula falsi type for computing a root of an equation, *Nordisk Tidskr. Informationsbehandling (BIT)* **13**, pp. 253–264.

Anderssen, R. S. and Bloomfield, P. (1974). Numerical differentiation procedures for non-exact data, *Num. Math.* **22**, pp. 157–182.

Apostol, T. M. (1957). *Mathematical Analysis*, Addison-Wesley, Massachusetts and London.

Armstrong, R. D. and Godfrey, J. P. (1979). Two linear programming algorithms for the discrete ℓ_1 norm problem, *Mathematics of Computation* **33**, pp. 289–300.

Aspvall, B. and Stone, R. E. (1980). Khachiyan's linear programming algorithm, *Journal of Algorithms* **1**, 1–13.

Avila, J. H. and Concus, P. (1979). Update methods for highly structured systems of nonlinear equations, *SIAM J. Numer. Anal.* **16**, pp. 260–269.

Avriel, M. (1976). *Nonlinear Programming: Analysis and Methods*, Prentice-Hall, Inc., Englewood Cliffs, New Jersey.

Avriel, M. and Dembo, R. S. (eds.) (1979). Engineering Optimization, *Math. Prog. Study* **11**.

Avriel, M., Rijckaert, M. J. and Wilde, D. J. (eds.) (1973). *Optimization and Design*, Prentice-Hall, Inc., Englewood Cliffs, New Jersey.

Axelsson, O. (1974). On preconditioning and convergence acceleration in sparse matrix problems, Report 74-10, CERN European Organization for Nuclear Research, Geneva.

Baker, T. E. and Ventker, R. (1980). "Successive linear programming in refinery logistic models", presented at ORSA/TIMS Joint National Meeting, Colorado Springs, Colorado.

Balinski, M. L. and Lemaréchal, C. (eds.) (1978). Mathematical Programming in Use, *Math. Prog. Study* **9**.

Bard, Y. (1976). *Nonlinear Parameter Estimation*, Academic Press, London and New York.

Bard, Y. and Greenstadt, J. L. (1969). "A modified Newton method for optimization with equality constraints", in *Optimization* (R. Fletcher, ed.), pp. 299–306, Academic Press, London and New York.

Barrodale, I. and Roberts, F. D. K. (1973). An improved algorithm for discrete ℓ_1 linear approximation, *SIAM J. Numer. Anal.* **10**, pp. 839–848.

Bartels, R. H. (1971). A stabilization of the simplex method, *Num. Math.* **16**, pp. 414–434.

Bartels, R. H. (1980). A penalty linear programming method using reduced-gradient basis-exchange techniques, *Linear Algebra and its Applics.* **29**, pp. 17–32.

Bartels, R. H. and Conn, A. R. (1980). Linearly constrained discrete ℓ_1 problems, *ACM Trans. Math. Software* **6**, pp. 594–608.

Bartels, R. H. and Golub, G. H. (1969). The simplex method of linear programming using the *LU* decomposition, *Comm. ACM* **12**, pp. 266–268.

Bartels, R. H., Golub, G. H. and Saunders, M. A. (1970). "Numerical techniques in mathematical programming", in *Nonlinear Programming* (J. B. Rosen, O. L. Mangasarian and K. Ritter, eds.), pp. 123–176, Academic Press, London and New York.

Batchelor, A. S. J. and Beale, E. M. L. (1976). "A revised method of conjugate-gradient approximation programming", presented at the Ninth International Symposium on Mathematical Programming, Budapest.

Beale, E. M. L. (1959). On quadratic programming, *Naval Res. Logistics Quarterly* **6**, pp. 227–243.

Beale, E. M. L. (1967a). "An introduction to Beale's method of quadratic programming", in *Nonlinear Programming* (J. Abadie, ed.), pp. 143–153, North-Holland, Amsterdam.

Beale, E. M. L. (1967b). "Numerical methods", in *Nonlinear Programming* (J. Abadie, ed.), pp. 132–205, North-Holland, Amsterdam.

Beale, E. M. L. (1972). "A derivation of conjugate gradients", in *Numerical Methods for Nonlinear Optimization* (F. A. Lootsma, ed.), pp. 39–43, Academic Press, London and New York.

Beale, E. M. L. (1974). "A conjugate-gradient method of approximation programming", in *Optimization Methods for Resource Allocation* (R. W. Cottle and J. Krarup, eds.), pp. 261–277, English Universities Press.

Beale, E. M. L. (1975). The current algorithmic scope of mathematical programming systems, *Math. Prog. Study* **4**, pp. 1–11.

Beale, E. M. L. (1977). "Integer Programming", in *The State of the Art in Numerical Analysis* (D. Jacobs, ed.), pp. 409–448, Academic Press, London and New York.

Beale, E. M. L. (1978). "Nonlinear programming using a general mathematical programming system", in *Design and Implementation of Optimization Software* (H. J. Greenberg, ed.), pp. 259–279, Sijthoff and Noordhoff, Netherlands.

Benichou, M., Gauthier, J. M., Hentges, G. and Ribière, G. (1977). The efficient solution of large-scale linear programming problems — some algorithmic techniques and computational results, *Math. Prog.* **13**, pp. 280–322.

Ben-Israel, A. (1967). On iterative methods for solving nonlinear least-squares problems over convex sets, *Israel J. of Maths.* **5**, pp. 211–224.

Bertsekas, D. P. (1975a). Necessary and sufficient conditions for a penalty function to be exact, *Math. Prog.* **9**, pp. 87–99.

Bertsekas, D. P. (1975b). Combined primal-dual and penalty methods for constrained minimization, *SIAM J. Control and Optimization* **13**, pp. 521–544.

Bertsekas, D. P. (1976a). Multiplier methods: a survey, *Automatica* **12**, pp. 133–145.

Bertsekas, D. P. (1976b). On penalty and multiplier methods for constrained minimization, *SIAM J. Control and Optimization* **14**, pp. 216–235.

Bertsekas, D. P. (1979). "Convergence analysis of augmented Lagrangian methods", *presented at the IIASA Task Force Meeting on "Generalized Lagrangians in Systems and Economic Theory"*, IIASA, Laxenburg, Austria (proceedings to be published in 1981).

Best, M. J., Bräuninger, J., Ritter, K. and Robinson, S. M. (1981). A globally and quadratically convergent algorithm for general nonlinear programming problems, *Computing* **26**, pp. 141–153.

Betts, J. T. (1976). Solving the nonlinear least-square problem, *J. Opt. Th. Applics.* **18**, pp. 469–483.

Biggs, M. C. (1972). "Constrained minimization using recursive equality quadratic programming", in *Numerical Methods for Non-Linear Optimization* (F. A. Lootsma, ed.), pp. 411–428, Academic Press, London and New York.

Biggs, M. C. (1974). *The Development of a Class of Constrained Optimization Algorithms and Their Application to the Problem of Electric Power Scheduling*, Ph.D. Thesis, University of London.

Biggs, M. C. (1975). "Constrained minimization using recursive quadratic programming: some alternative subproblem formulations", in *Towards Global Optimization* (L. C. W. Dixon and G. P. Szegö, eds.), pp. 341–349, North-Holland, Amsterdam.

Bland, R. G. (1977). New finite pivoting rules for the simplex method, *Math. of Oper. Res.* **2**, pp. 103–107.

Boggs, P. T. (1975). The solution of nonlinear operator equations by A-stable integration techniques, *SIAM J. Numer. Anal.* **8**, pp. 767–785.

Boggs, P. T. and Tolle, J. W. (1980). Augmented Lagrangians which are quadratic in the multiplier, *J. Opt. Th. Applics.* **31**, pp. 17–26.

Bracken, J. and McCormick, G. P. (1968). *Selected Applications of Nonlinear Programming*, John Wiley and Sons, New York and Toronto.

Brayton, R. K. and Cullum, J. (1977). "Optimization with the parameters constrained to a box", in *Proceedings of the IMACS International Symposium on Simulation Software and Numerical Methods for Differential Equations*, IMACS.

Brayton, R. K. and Cullum, J. (1979). An algorithm for minimizing a differentiable function subject to box constraints and errors, *J. Opt. Th. Applics.* **29**, pp. 521–558.

Brent, R. P. (1973a). *Algorithms for Minimization without Derivatives*, Prentice-Hall, Inc., Englewood Cliffs, New Jersey.

Brent, R. P. (1973b). Some efficient algorithms for solving systems of nonlinear equations, *SIAM J. Numer. Anal.* **10**, pp. 327–344.

Brodlie, K. W. (1977a). "Unconstrained optimization", in *The State of the Art in Numerical Analysis* (D. Jacobs, ed.), pp. 229–268, Academic Press, London and New York.

Brodlie, K. W. (1977b). An assessment of two approaches to variable metric methods, *Math. Prog.* **12**, pp. 344–355.

Broyden, C. G. (1965). A class of methods for solving nonlinear simultaneous equations, *Mathematics of Computation* **19**, pp. 577–593.

Broyden, C. G. (1967). Quasi-Newton methods and their application to function minimization, *Mathematics of Computation* **21**, pp. 368–381.

Broyden, C. G. (1970). The convergence of a class of double-rank minimization algorithms, *J. Inst. Maths. Applics.* **6**, pp. 76–90.

Broyden, C. G., Dennis, J. E., Jr. and Moré, J. J. (1973). On the local and superlinear convergence of quasi-Newton methods, *J. Inst. Maths. Applics.* **12**, pp. 223–245.

Buckley, A. G. (1975). An alternative implementation of Goldfarb's minimization algorithm, *Math. Prog.* **8**, pp. 207–231.

Buckley, A. G. (1978). A combined conjugate-gradient quasi-Newton minimization algorithm, *Math. Prog.* **15**, pp. 200–210.

Bunch, J. R. and Kaufman, L. C. (1977). Some stable methods for calculating inertia and solving symmetric linear equations, *Mathematics of Computation* **31**, pp. 163–179.

Bunch, J. R. and Kaufman, L. C. (1980). A computational method for the indefinite quadratic programming problem, *Linear Algebra and its Applics.* **34**, pp. 341–370.

Bunch, J. R. and Parlett, B. N. (1971). Direct methods for solving symmetric indefinite systems of linear equations, *SIAM J. Numer. Anal.* **8**, pp. 639–655.

Bus, J. C. P. and Dekker, T. J. (1975). Two efficient algorithms with guaranteed convergence for finding a zero of a function, *ACM Trans. Math. Software* **1**, pp. 330–345.

Businger, P. and Golub, G. H. (1965). Linear least-squares solutions by Householder transformations, *Num. Math.* **7**, pp. 269–276.

Buys, J. D. (1972). *Dual Algorithms for Constrained Optimization Problems*, Ph.D. Thesis, University of Leiden, Netherlands.

Buys, J. D. and Gonin, R. (1977). The use of augmented Lagrangian functions for sensitivity analysis in nonlinear programming, *Math. Prog.* **12**, pp. 281–284.

Byrd, R. H. (1976). *Local convergence of the diagonalized method of multipliers*, Ph.D. Thesis, Rice University, Texas.

Byrd, R. H. (1978). Local convergence of the diagonalized method of multipliers, *J. Opt. Th. Applics.* **26**, pp. 485–500.

Carroll, C. W. (1959). *An Operations Research Approach to the Economic Optimization of a Kraft Pulping Process*, Ph.D. Thesis, Institute of Paper Chemistry, Appleton, Wisconsin.

Carroll, C. W. (1961). The created response surface technique for optimizing nonlinear restrained systems, *Operations Research* **9**, pp. 169–184.

Cauchy, A. (1847). Méthode Générale pour la Résolution des Systéms d'Équations Simultanées, *Comp. Rend. Acad. Sci. Paris*, pp. 536–538.

Chamberlain, R. M. (1979). Some examples of cycling in variable metric methods for constrained minimization, *Math. Prog.* **16**, pp. 378–383.

Chamberlain, R. M., Lemaréchal, C., Pederson, H. C. and Powell, M. J. D. (1980). The watchdog technique for forcing convergence in algorithms for constrained optimization, Report DAMTP 80/NA 1, University of Cambridge.

Charalambous, C. (1978). A lower bound for the controlling parameter of the exact penalty function, *Math. Prog.* **15**, pp. 278–290.

Charalambous, C. and Conn, A. R. (1978). An efficient method to solve the minimax problem directly, *SIAM J. Numer. Anal.* **15**, pp. 162–187.

Charnes, A. (1952). Optimality and degeneracy in linear programming, *Econometrica* **20**, pp. 160–170.

Charnes, A., Cooper, W. W. and Ferguson, R. (1955). Optimal estimation of executive compensation by linear programming, *Management Science* **2**, pp. 138–151.

Ciarlet, P. G., Schultz, M. H. and Varga, R. S. (1967). Nonlinear boundary value problems I, *Num. Math.* **9**, pp. 394–430.

Cline, A. K., Moler, C. B., Stewart, G. W. and Wilkinson, J. H. (1979). An estimate for the condition number of a matrix, *SIAM J. Numer. Anal.* **16**, pp. 368–375.

Coleman, T. F. (1979). *A Superlinear Penalty Function Method to Solve the Nonlinear Programming Problem*, Ph.D. Thesis, University of Waterloo, Ontario, Canada.

Coleman, T. F. and Conn, A. R. (1980a). Second-order conditions for an exact penalty function, *Math. Prog.* **19**, pp. 178–185.

Coleman, T. F. and Conn, A. R. (1980b). Nonlinear programming via an exact penalty function method: asymptotic analysis, Report CS–80–30, Department of Computer Science, University of Waterloo, Ontario, Canada.

Coleman, T. F. and Conn, A. R. (1980c). Nonlinear programming via an exact penalty function method: global analysis, Report CS–80–31, Department of Computer Science, University of Waterloo, Ontario, Canada.

Coleman, T. F. and Moré, J. J. (1980). "Coloring large sparse Jacobians and Hessians", presented at the SIAM 1980 Fall Meeting, Houston. November 1980.

Colville, A. R. (1968). A comparative study on nonlinear programming codes, Report No. 320-2949, IBM New York Scientific Center.

Concus, P., Golub, G. H. and O'Leary, D. P. (1976). "A generalized conjugate-gradient method for the numerical solution of elliptic partial differential equations", in *Sparse Matrix Computations* (J. R. Bunch and D. J. Rose, eds.), pp. 309–332, Academic Press, London and New York.

Conn, A. R. (1973). Constrained optimization using a non-differentiable penalty function, *SIAM J. Numer. Anal.* **10**, pp. 760–779.

Conn, A. R. (1976). Linear programming via a non-differentiable penalty function, *SIAM J. Numer. Anal.* **13**, pp. 145–154.

Conn, A. R. (1979). An efficient second-order method to solve the (constrained) minimax problem, Report CORR-79-5, University of Waterloo, Canada.

Conn, A. R. and Pietrzykowski, T. (1977). A penalty function method converging directly to a constrained optimum, *SIAM J. Numer. Anal.* **14**, pp. 348–375.

Conn, A. R. and Sinclair, J. W. (1975). Quadratic programming via a non-differentiable penalty function, Report 75/15, Department of Combinatorics and Optimization, University of Waterloo, Canada.

Cottle, R. W. (1974). Manifestations of the Schur complement, *Linear Algebra and its Applics.* **8**, pp. 189–211.

Courant, R. (1936). *Differential and Integral Calculus* (two volumes), Blackie, London and Glasgow.

Courant, R. (1943). Variational methods for the solution of problems of equilibrium and vibrations, *Bull. Amer. Math. Soc.* **49**, pp. 1–23.

Cox, M. G. (1977). "A survey of numerical methods for data and function approximation", in *The State of the Art in Numerical Analysis* (D. Jacobs, ed.), pp. 627–668, Academic Press, London and New York.

Curtis, A. R., Powell, M. J. D. and Reid, J. K. (1974). On the estimation of sparse Jacobian matrices, *J. Inst. Maths. Applics.* **13**, pp. 117–119.

Curtis, A. R. and Reid, J. K. (1972). On the automatic scaling of matrices for Gaussian elimination, *J. Inst. Maths. Applics.* **10**, pp. 118–124.

Curtis, A. R. and Reid, J. K. (1974). The choice of step lengths when using differences to approximate Jacobian matrices, *J. Inst. Maths. Applics.* **13**, pp. 121–126.

Dahlquist, G. and Björck, Å. (1974). *Numerical Methods*, Prentice-Hall Inc., Englewood Cliffs, New Jersey.

Daniel, J. W., Gragg, W. B., Kaufman, L. C. and Stewart, G. W. (1976). Reorthogonalization and stable algorithms for updating the Gram-Schmidt QR factorization, *Mathematics of Computation* **30**, pp. 772–795.

Dantzig, G. B. (1963). *Linear Programming and Extensions*, Princeton University Press, Princeton, New Jersey.

Dantzig, G. B., Orden, A. and Wolfe, P. (1955). Generalized simplex method for minimizing a linear form under linear inequality restraints, *Pacific J. Math.* **5**, pp. 183–195.

Dantzig, G. B., Dempster, M. A. H. and Kallio, M. J. (eds.) (1981). *Large-Scale Linear Programming (Volume 1)*, IIASA Collaborative Proceedings Series, CP-81-51, IIASA, Laxenburg, Austria.

Davidon, W. C. (1959). Variable metric methods for minimization, A. E. C. Res. and Develop. Report ANL-5990, Argonne National Laboratory, Argonne, Illinois.

Davidon, W. C. (1975). Optimally conditioned optimization algorithms without line searches, *Math. Prog.* **9**, pp. 1–30.

Davidon, W. C. (1979). Conic approximations and collinear scalings for optimizers, *SIAM J. Numer. Anal.* **17**, pp. 268–281.

Davis, P. J. and Rabinowitz, P. (1967). *Numerical Integration*, Blaisdell, London.

Dekker, T. J. (1969). "Finding a zero by means of successive linear interpolation", in *Constructive Aspects of the Fundamental Theorem of Algebra* (B. Dejon and P. Henrici, eds.), pp. 37–48, Wiley Interscience, London.

Dembo, R. S. (1978). Current state of the art of algorithms and computer software for geometric programming, *J. Opt. Th. Applics.* **26**, pp. 149–184.

Dembo, R. S., Eisenstat, S. C. and Steihaug T. (1980). Inexact Newton methods, Working Paper #47, School of Organization and Management, Yale University.

Dembo, R. S. and Steihaug T. (1980). Truncated-Newton algorithms for large-scale unconstrained optimization, Working Paper #48, School of Organization and Management, Yale University.

Dennis, J. E., Jr. (1973). "Some computational techniques for the nonlinear least-squares problem", in *Numerical Solution of Systems of Nonlinear Algebraic Equations* (G. D. Byrne and C. A. Hall, eds.), pp. 157–183, Academic Press, London and New York.

Dennis, J. E., Jr. (1977). "Nonlinear Least Squares", in *The State of the Art in Numerical Analysis* (D. Jacobs, ed.), pp. 269–312, Academic Press, London and New York.

Dennis, J. E., Jr. and Moré, J. J. (1974). A characterization of superlinear convergence and its application to quasi-Newton methods, *Mathematics of Computation* **28**, pp. 549–560.

Dennis, J. E., Jr. and Moré, J. J. (1977). Quasi-Newton methods, motivation and theory, *SIAM Review* **19**, pp. 46–89.

Dennis, J. E., Jr., Gay, D. M. and Welsch, R. E. (1977). An adaptive nonlinear least-squares algorithm, Report TR 77-321, Department of Computer Sciences, Cornell University.

Dennis, J. E., Jr. and Schnabel, R. E. (1979). Least change secant updates for quasi-Newton methods, *SIAM Review* **21**. pp. 443–469.

Dennis, J. E., Jr. and Schnabel, R. E. (1980). A new derivation of symmetric positive definite secant updates, Report CU-CS-185-80, Department of Mathematical Sciences, Rice University.

Dixon, L. C. W. (1972a). Quasi-Newton algorithms generate identical points, *Math. Prog.* **2**, pp. 383–387.

Dixon, L. C. W. (1972b). Quasi-Newton algorithms generate identical points. II. The proof of four new theorems, *Math. Prog.* **3**, pp. 345–358.

Dixon, L. C. W. (1975). Conjugate-gradient algorithms: quadratic termination without linear searches, *J. Inst. Maths. Applics.* **15**, pp. 9–18.

Djang, A. (1980). *Algorithmic Equivalence in Quadratic Programming*, Ph.D. Thesis, Stanford University, California.

Dongarra, J. J., Bunch, J. R., Moler, C. B. and Stewart, G. W. (1979). *LINPACK Users Guide*, SIAM Publications, Philadelphia.

Duff, I. S. and Reid, J. K. (1978). An implementation of Tarjan's algorithm for the block triangularization of a matrix, *ACM Trans. Math. Software* **4**, pp. 137–147.

Duffin, R. J., Peterson, E. L. and Zener, C. (1967). *Geometric Programming — Theory and Applications*, John Wiley and Sons, New York and Toronto.

Dumontet, J. and Vignes, J. (1977). Determination du pas optimal dans le calcul des dérivées sur ordineur, *Revue française d'automatique, d'information et de recherche opérationelle, Analyse numérique (RAIRO)* **11**, pp. 13–25.

Ecker, J. G. (1980). Geometric programming: methods, computations and applications, *SIAM Review* **22**, pp. 338–362.

El-Attar, R. A., Vidyasagar, M. and Dutta, S. R. K. (1979). An algorithm for ℓ_1-norm minimization with application to nonlinear ℓ_1 approximation, *SIAM J. Numer. Anal.* **16**, pp. 70–86.

Escudero, L. (1980). A projected Lagrangian method for nonlinear programming, Report No. G320-3401, IBM Palo Alto Scientific Center.

Evans, J. P., Gould, F. J. and Tolle, J. W. (1973). Exact penalty functions in nonlinear programming, *Math. Prog.* **4**, pp. 72–97.

Fiacco, A. V. and McCormick, G. P. (1968). *Nonlinear Programming: Sequential Unconstrained Minimization Techniques*, John Wiley and Sons, New York and Toronto.

Fiacco, A. V. (1976). Sensitivity analysis for mathematical programming using penalty functions, *Math. Prog.* **10**, pp. 287–311.

Fletcher, R. (1968). Generalized inverse methods for the best least-squares solution of systems of nonlinear equations, *Computer Journal* **10**, pp. 392–399.

Fletcher, R. (1970a). A new approach to variable metric algorithms, *Computer Journal* **13**, pp. 317–322.

Fletcher, R. (1970b). "A class of methods for nonlinear programming with termination and convergence properties", in *Integer and Nonlinear Programming* (J. Abadie, ed.), pp. 157–175, North-Holland, Amsterdam.

Fletcher, R. (1971a). A modified Marquardt subroutine for nonlinear least squares, Report R6799, Atomic Energy Research Establishment, Harwell, England.

Fletcher, R. (1971b). A general quadratic programming algorithm, *J. Inst. Maths. Applics.* **7**, pp. 76–91.

Fletcher, R. (1972a). An algorithm for solving linearly constrained optimization problems, *Math. Prog.* **2**, pp. 133–165.

Fletcher, R. (1972b). "Minimizing general functions subject to linear constraints", in *Numerical Methods for Non-linear Optimization* (F. A. Lootsma, ed.), pp. 279–296, Academic Press, London and New York.

Fletcher, R. (1972c). Methods for the solution of optimization problems, *Comput. Phys. Comm.* **3**, pp. 159–172.

Fletcher, R. (1973). An exact penalty function for nonlinear programming with inequalities, *Math. Prog.* **5**, pp. 129–150.

Fletcher, R. (1974). "Methods related to Lagrangian functions", in *Numerical Methods for Constrained Optimization* (P. E. Gill and W. Murray, eds.), pp. 219–240, Academic Press, London and New York.

Fletcher, R. (1976). Factorizing symmetric indefinite matrices, *Linear Algebra and its Applics.* **14**, pp. 257–272.

Fletcher, R. (1977). "Methods for solving nonlinearly constrained optimization problems", in *The State of the Art in Numerical Analysis* (D. Jacobs, ed.), pp. 365–448, Academic Press, London and New York.

Fletcher, R. (1980). *Practical Methods of Optimization, Volume 1, Unconstrained Optimization*, John Wiley and Sons, New York and Toronto.

Fletcher, R. and Freeman, T. L. (1977). A modified Newton method for minimization, *J. Opt. Th. Applics.* **23**, pp. 357–372.

Fletcher, R. and Jackson, M. P. (1974). Minimization of a quadratic function of many variables subject only to upper and lower bounds, *J. Inst. Maths. Applics.* **14**, pp. 159–174.

Fletcher, R. and Lill, S. A. (1970). "A class of methods for non-linear programming: II. computational experience", in *Nonlinear Programming* (J. B. Rosen, O. L. Mangasarian and K. Ritter, eds.), pp. 67–92, Academic Press, London and New York.

Fletcher, R. and McCann, A. P. (1969). Acceleration techniques for nonlinear programming, in *Optimization* (R. Fletcher, ed.), pp. 37–49, Academic Press, London and New York.

Fletcher, R. and Reeves, C. M. (1964). Function minimization by conjugate gradients, *Computer Journal* **7**, pp. 149–154.

Fletcher, R. and Powell, M. J. D. (1963). A rapidly convergent descent method for minimization, *Computer Journal* **6**, pp. 163–168.

Fletcher, R. and Powell, M. J. D. (1974). On the modification of LDL^T factorizations, *Mathematics of Computation* **28**, pp. 1067–1087.

Forrest, J. J. H. and Tomlin, J. A. (1972). Updating triangular factors of the basis to maintain sparsity in the product form simplex method, *Math. Prog.* **2**, pp. 263–278.

Forsythe, G. E. and Moler, C. B. (1967). *Computer Solution of Linear Algebraic Systems*, Prentice-Hall, Inc., Englewood Cliffs, New Jersey.

Fourer, R. (1979). Sparse Gaussian elimination of staircase linear systems, Report SOL 79-17, Department of Operations Research, Stanford University, California.

Frisch, K. R. (1955). The logarithmic potential method of convex programming, Memorandum of May 13, 1955, University Institute of Economics, Oslo, Norway.

Fulkerson, D. R. and Wolfe, P. (1962). An algorithm for scaling matrices, *SIAM Review* **4**, pp. 142–146.

Gács, P. and Lovász, L. (1981). Khachiyan's algorithm for linear programming, *Math. Prog. Study* **14**, pp. 61–68.

Garcia Palomares, U. M. and Mangasarian, O. L. (1976). Superlinearly convergent quasi-Newton algorithms for nonlinearly constrained optimization problems, *Math. Prog.* **11**, pp. 1–13.

Gay, D. M. (1979a). On robust and generalized linear regression problems, Report 2000, Mathematics Research Center, University of Wisconsin, Madison, Wisconsin.

Gay, D. M. (1979b). Computing optimal locally constrained steps, Report 2013, Mathematics Research Center, University of Wisconsin, Madison, Wisconsin.

Gay, D. M. and Schnabel, R. B. (1978). "Solving systems of nonlinear equations by Broyden's method with projected updates", in *Nonlinear Programming 3* (O. L. Mangasarian, R. R. Meyer and S. M. Robinson, eds.), pp. 245–281, Academic Press, London and New York.

Gill, P. E. (1975). *Numerical Methods for Large-Scale Linearly Constrained Optimization Problems*, Ph.D. Thesis, University of London.

Gill, P. E., Golub, G. H., Murray, W. and Saunders, M. A. (1974). Methods for modifying matrix factorizations, *Mathematics of Computation* **28**, pp. 505–535.

Gill, P. E. and Murray, W. (1972). Quasi-Newton methods for unconstrained optimization, *J. Inst. Maths. Applics.* **9**, pp. 91–108.

Gill, P. E. and Murray, W. (1973a). The numerical solution of a problem in the calculus of variations, in *Recent Mathematical Developments in Control* (D. J. Bell, ed.), pp. 97–122, Academic Press, London and New York.

Gill, P. E. and Murray, W. (1973b). Quasi-Newton methods for linearly constrained optimization, Report NAC 32, National Physical Laboratory, England.

Gill, P. E. and Murray, W. (1973c). A numerically stable form of the simplex method, *Linear Algebra and its Applics.* **7**, pp. 99–138.

Gill, P. E. and Murray, W. (1974a). Newton-type methods for unconstrained and linearly constrained optimization, *Math. Prog.* **28**, pp. 311–350.

Gill, P. E. and Murray, W. (eds.) (1974b). *Numerical Methods for Constrained Optimization*, Academic Press, London and New York.

Gill, P. E. and Murray, W. (1974c). "Newton-type methods for linearly constrained optimization", in *Numerical Methods for Constrained Optimization* (P. E. Gill and W. Murray, eds.), pp. 29–66, Academic Press, London and New York.

Gill, P. E. and Murray, W. (1974d). "Quasi-Newton methods for linearly constrained optimization", in *Numerical Methods for Constrained Optimization* (P. E. Gill and W. Murray, eds.), pp. 67–92, Academic Press, London and New York.

Gill, P. E. and Murray, W. (1974e). Safeguarded steplength algorithms for optimization using descent methods, Report NAC 37, National Physical Laboratory, England.

Gill, P. E. and Murray, W. (1976a). "Nonlinear least squares and nonlinearly constrained optimization", in *Numerical Analysis, Dundee 1975* (G. A. Watson, ed.), pp. 135–147, Springer-Verlag Lecture Notes in Mathematics 506, Berlin, Heidelberg and New York.

Gill, P. E. and Murray, W. (1976b). Minimization subject to bounds on the variables, Report NAC 71, National Physical Laboratory, England.

Gill, P. E. and Murray, W. (1977a). "Linearly constrained problems including linear and quadratic programming", in *The State of the Art in Numerical Analysis* (D. Jacobs, ed.), pp. 313–363, Academic Press, London and New York.

Gill, P. E. and Murray, W. (1977b). The computation of Lagrange multiplier estimates for constrained minimization, Report NAC 77, National Physical Laboratory, England.

Gill, P. E. and Murray, W. (1978a). Algorithms for the solution of the nonlinear least-squares problem, *SIAM J. Numer. Anal.* **15**, pp. 977–992.

Gill, P. E. and Murray, W. (1978b). Numerically stable methods for quadratic programming, *Math. Prog.* **14**, pp. 349–372.

Gill, P. E. and Murray, W. (1978c). The design and implementation of software for unconstrained optimization, in *Design and Implementation of Optimization Software* (H. Greenberg, ed.), pp. 221–234, Sijthoff and Noordhoff, Netherlands.

Gill, P. E. and Murray, W. (1979a). Conjugate-gradient methods for large-scale nonlinear optimization, Report SOL 79-15, Department of Operations Research, Stanford University, California.

Gill, P. E. and Murray, W. (1979b). The computation of Lagrange multiplier estimates for constrained minimization, *Math. Prog.* **17**, pp. 32–60.

Gill, P. E. and Murray, W. (1979c). "Performance evaluation for optimization software", in *Performance Evaluation of Numerical Software* (L. D. Fosdick, ed.), pp. 221–234, North-Holland, Amsterdam.

Gill, P. E., Murray, W. and Nash, S. G. (1981). Newton-type minimization methods using the linear conjugate-gradient method, Report (to appear), Department of Operations Research, Stanford University, California.

Gill, P. E., Murray, W., Picken, S. M. and Wright, M. H. (1979). The design and structure of a Fortran program library for optimization, *ACM Trans. Math. Software* **5**, pp. 259–283.

Gill, P. E., Murray, W. and Saunders, M. A. (1975). Methods for computing and modifying the *LDV* factors of a matrix, *Mathematics of Computation* **29**, pp. 1051–1077.

Gill, P. E., Murray, W., Saunders, M. A. and Wright, M. H. (1979). Two step-length algorithms for numerical optimization, Report SOL 79-25, Department of Operations Research, Stanford University, California.

Gill, P. E., Murray, W., Saunders, M. A. and Wright, M. H. (1980). "A projected Lagrangian method for problems with both linear and nonlinear constraints", presented at the SIAM 1980 Fall Meeting, Houston, Texas.

Gill, P. E., Murray, W., Saunders, M. A. and Wright, M. H. (1981a). "A numerical investigation of ellipsoid algorithms for large-scale linear programming", in *Large-Scale Linear Programming (Volume 1)* (G. B. Dantzig, M. A. H. Dempster and M. J. Kallio, eds.), pp. 487–509, IIASA Collaborative Proceedings Series, CP-81-51, IIASA, Laxenburg, Austria.

Gill, P. E., Murray, W., Saunders, M. A. and Wright, M. H. (1981b). QP-based methods for large-scale nonlinearly constrained optimization, Report SOL 81-1, Department of Operations Research, Stanford University, California. To appear in *Nonlinear Programming 4*, (O. L. Mangasarian, R. R. Meyer and S. M. Robinson, eds.), Academic Press, London and New York.

Glad, S. T. (1979). Properties of updating methods for the multipliers in augmented Lagrangians, *J. Opt. Th. Applics.* **28**, pp. 135–156.

Glad, S. T. and Polak, E. (1979). A multiplier method with automatic limitation of penalty growth, *Math. Prog.* **17**, pp. 140–155.

Goffin, J. L. (1980). Convergence results in a class of variable metric subgradient methods, Working Paper 80-08, Faculty of Management, McGill University, Montreal, Canada. To appear in *Nonlinear Programming 4*, (O. L. Mangasarian, R. R. Meyer and S. M. Robinson, eds.), Academic Press, London and New York.

Goldfarb, D. (1969). Extension of Davidon's variable metric method to maximization under linear inequality and equality constraints, *SIAM J. Appl. Math.* **17**, pp. 739–764.

Goldfarb, D. (1970). A family of variable metric methods derived by variational means, *Mathematics of Computation* **24**, pp. 23–26.

Goldfarb, D. (1980). Curvilinear path step length algorithms for minimization which use directions of negative curvature, *Math. Prog.* **18**, pp. 31–40.

Goldfarb, D. and Reid, J. K. (1977). A practicable steepest-edge simplex algorithm, *Math. Prog.* **12**, pp. 361–371.

Goldfeld, S. M., Quandt, R. E. and Trotter, H. F. (1966). Maximization by quadratic hill-climbing, *Econometrica* **34**, pp. 541–551.

Goldstein, A. and Price, J. (1967). An effective algorithm for minimization, *Numer. Math.* **10**, pp. 184–189.

Golub, G. H. and Pereyra, V. (1973). The differentiation of pseudo-inverses and nonlinear least-squares problems whose variables separate, *SIAM J. Numer. Anal.* **10**, pp. 413–432.

Golub, G. H. and Reinsch, C. (1971). "Singular value decomposition and least-squares solutions", in *Handbook for Automatic Computation, Vol. II* (J. H. Wilkinson and C. Reinsch, eds.), pp. 134–151, Springer-Verlag, Berlin, Heidelberg and New York.

Graham, S. R. (1976). *A matrix factorization and its application to unconstrained minimization*, Project thesis for BSc. (Hons) in Mathematics for Business, Middlesex Polytechnic, Enfield, England.

Greenberg, H. J. (1978a). "A tutorial on matricial packing", in *Design and Implementation of Optimization Software* (H. J. Greenberg, ed.), pp. 109–142, Sijthoff and Noordhoff, Netherlands.

Greenberg, H. J. (1978b). "Pivot selection tactics", in *Design and Implementation of Optimization Software* (H. J. Greenberg, ed.), pp. 143–174, Sijthoff and Noordhoff, Netherlands.

Greenberg, H. J. and Kalan, J. E. (1975). An exact update for Harris' TREAD, *Math. Prog. Study* **4**, pp. 26–29.

Greenstadt, J. L. (1967). On the relative efficiencies of gradient methods, *Mathematics of Computation* **21**, pp. 360–367.

Greenstadt, J. L. (1970). Variations on variable-metric methods, *Mathematics of Computation* **24**, pp. 1–22.

Greenstadt, J. L. (1972). A quasi-Newton method with no derivatives, *Mathematics of Computation* **26**, pp. 145–166.

Griffith, R. E. and Stewart, R. A. (1961). A nonlinear programming technique for the optimization of continuous processing systems, *Management Science* **7**, pp. 379–392.

Gue, R. L. and Thomas, M. E. (1968). *Mathematical Methods in Operations Research*, The Macmillan Company, New York.

Haarhoff, P. C. and Buys, J. D. (1970). A new method for the optimization of a nonlinear function subject to nonlinear constraints, *Computer Journal* **13**, pp. 178–184.

Hamming, R. W. (1962). *Numerical Methods for Scientists and Engineers*, McGraw-Hill Book Co., New York.

Hamming, R. W. (1971). *Introduction to Applied Numerical Analysis*, McGraw-Hill Book Co., New York.

Hamming, R. W. (1973). *Numerical Methods for Scientists and Engineers* (2nd Edition), McGraw-Hill Book Co., New York.

Han, S.-P. (1976). Superlinearly convergent variable metric algorithms for general nonlinear programming problems, *Math. Prog.* **11**, pp. 263–282.

Han, S.-P. (1977a). Dual variable metric algorithms for constrained optimization, *SIAM J. Control and Optimization* **15**, pp. 546–565.

Han, S.-P. (1977b). A globally convergent method for nonlinear programming, *J. Opt. Th. Applics.* **22**, pp. 297–310.

Han, S.-P. (1978a). Superlinear convergence of a minimax method, Computer Science Department, Cornell University, Ithaca, New York.

Han, S.-P. (1978b). On the validity of a nonlinear programming method for solving minimax problems, Report 1891, Mathematics Research Center, University of Wisconsin, Madison, Wisconsin.

Han, S.-P. and Mangasarian, O. L. (1979). Exact penalty functions in nonlinear programming, *Math. Prog.* **17**, pp. 251–269.

Harris, P. M. J. (1973). Pivot selection methods of the Devex LP code, *Math. Prog.* **5**, pp. 1–28. [Reprinted in *Math. Prog. Study* **4** (1975), pp. 30–57.]

Hartley, H. O. (1961). Nonlinear programming by the simplex method, *Econometrica* **29**, pp. 223–237.

Hayes, J. G. (ed.) (1970). *Numerical Approximation to Functions and Data*, Academic Press, London and New York.

Heath, M. T. (1978). *Numerical Algorithms for Nonlinearly Constrained Optimization*, Ph.D. Thesis, Stanford University, California.

Hebden, M. D. (1973). An algorithm for minimization using exact second derivatives, Report TP515, Atomic Energy Research Establishment, Harwell, England.

Hellerman, E. and Rarick, D. (1971). Reinversion with the preassigned pivot procedure, *Math. Prog.* **1**, pp. 195–216.

Hellerman, E. and Rarick, D. (1972). "The partitioned preassigned pivot procedure (P^4)", in *Sparse Matrices and their Applications* (D. J. Rose and R. A. Willoughby eds.), pp. 67–76, Plenum Press, New York.

Hestenes, M. R. (1946). Sufficient conditions for the isoperimetric problem of Bolza in the calculus of variations, *Trans. Amer. Math. Soc.* **60**, pp. 93–118.

Hestenes, M. R. (1947). An alternative sufficiency proof for the normal problem of Bolza, *Trans. Amer. Math. Soc.* **61**, pp. 256–264.

Hestenes, M. R. (1969). Multiplier and gradient methods, *J. Opt. Th. Applics.* **4**, pp. 303–320.

Hestenes, M. R. (1979). "Historical overview of generalized Lagrangians and augmentability", presented at the IIASA Task Force Meeting on "Generalized Lagrangians in Systems and Economic Theory", IIASA, Laxenburg, Austria (proceedings to be published in 1981).

Hestenes, M. R. (1980a). *Conjugate-Direction Methods in Optimization*, Springer-Verlag, Berlin, Heidelberg and New York.

Hestenes, M. R. (1980b). Augmentability in optimization theory, *J. Opt. Th. Applics.* **32**, pp. 427–440.

Hestenes, M. R. and Stiefel, E. (1952). Methods of conjugate gradients for solving linear systems, *J. Res. Nat. Bur. Standards* **49**, pp. 409–436.

Howe, S. (1973). New conditions for exactness of a simple penalty function, *SIAM J. Control* **11**, pp. 378–381.

Jain, A., Lasdon, L. S. and Saunders, M. A. (1976). "An in-core nonlinear mathematical programming system for large sparse nonlinear programs", presented at ORSA/TIMS Joint National Meeting, Miami, Florida.

Jarratt, P. and Nudds, D. (1965). The use of rational functions in the iterative solution of equations on a computer, *Computer Journal* **8**, pp. 62–65.

Johnson, E. L. (1978). "Some considerations in using branch-and-bound codes", in *Design and Implementation of Optimization Software* (H. J. Greenberg, ed.), pp. 241–248, Sijthoff and Noordhoff, Netherlands,

Johnson, E. L. and Powell, S. (1978). "Integer programming codes", in *Design and Implementation of Optimization Software* (H. J. Greenberg, ed.), pp. 225–240, Sijthoff and Noordhoff, Netherlands,

Kahan, W. (1973). The implementation of algorithms: Part 1, Technical Report 20, Department of Computer Science, University of California, Berkeley.

Kaniel, S. and Dax, A. (1979). A modified Newton's method for unconstrained minimization, *SIAM J. Numer. Anal.* **16**, pp. 324–331.

Kantorovich, L. V. and Akilov, G. P. (1964). *Functional Analysis in Normed Spaces*, MacMillan, New York.

Kaufman, L. C. and Pereyra, V. (1978). A method for separable nonlinear least-squares problems with separable nonlinear equality constraints, *SIAM J. Numer. Anal.* **15**, pp. 12–20.

Kelley, J. E. (1960). The cutting plane method for solving convex programs, *J. Soc. Indust. Appl. Math.* **8**, pp. 703–712.

Khachiyan, L. G. (1979). A polynomial algorithm in linear programming, *Doklady Akademiia Nauk SSSR Novaia Seriia* **244**, pp. 1093–1096. [English translation in *Soviet Mathematics Doklady* **20**, (1979), pp. 191–194.]

Knuth, D. E. (1979). *TEX and METAFONT, New Directions in Typesetting*, American Mathematical Society and Digital Press, Bedford, Massachusetts.

Kort, B. W. (1975). "Rate of convergence of the method of multipliers with inexact minimization", in *Nonlinear Programming 2* (O. L. Mangasarian, R. R. Meyer and S. M. Robinson, eds.), pp. 193–214, Academic Press, London and New York.

Kort, B. W. and Bertsekas, D. P. (1976). Combined primal dual and penalty methods for convex programming, *SIAM J. Control and Optimization* **14**, pp. 268–294.

Kuhn, H. W. (1976). "Nonlinear programming: a historical view", in *SIAM-AMS Proceedings, Volume IX, Mathematical Programming* (R. C. Cottle and C. E. Lemke, eds.), pp. 1–26, American Mathematical Society, Providence, Rhode Island.

Kuhn, H. W. and Tucker, A. W. (1951). "Nonlinear Programming", in *Proceedings of the Second Berkeley Symposium on Mathematical Statistics and Probability* (J. Neyman, ed.), pp. 481–492, Berkeley, University of California Press.

Lasdon, L. S., Fox, R. L. and Ratner, M. (1973). An efficient one-dimensional search procedure for barrier functions, *Math. Prog.* **4**, pp. 279–296.

Lasdon, L. S. and Waren, A. D. (1978). "Generalized reduced gradient software for linearly and nonlinearly constrained problems", in *Design and Implementation of Optimization Software* (H. J. Greenberg, ed.), pp. 335–362, Sijthoff and Noordhoff, Netherlands.

Lasdon, L. S., Waren, A. D., Jain, A. and Ratner, M. (1978). Design and testing of a GRG code for nonlinear optimization, *ACM Trans. Math. Software* **4**, pp. 34–50.

Lawler, E. L. (1980). The great mathematical sputnik of 1979, University of California, Berkeley, California (February 1980).

Lawson, C. L. and Hanson, R. J. (1974). *Solving Least-Squares Problems*, Prentice-Hall, Inc., Englewood Cliffs, New Jersey.

Lemaréchal, C. (1975). An extension of Davidon methods to non-differentiable problems, *Math. Prog. Study* **3**, pp. 95–109.

Lemke, C. E. (1965). Bimatrix equilibrium points and mathematical programming, *Management Science* **11**, pp. 681–689.

Lenard, M. L. (1979). A computational study of active set strategies in nonlinear programming with linear constraints, *Math. Prog.* **16**, pp. 81–97.

Levenberg, K. (1944). A method for the solution of certain problems in least squares, *Quart. Appl. Math.* **2**, pp. 164–168.

Lill, S. A. (1972). "Generalization of an exact method for solving equality constrained problems to deal with inequality constraints", in *Numerical Methods for Non-linear Optimization* (F. A. Lootsma, ed.), pp. 383–394, Academic Press, London and New York.

Lootsma, F. A. (1969). Hessian matrices of penalty functions for solving constrained optimization problems, *Philips Res. Repts* **24**, pp. 322–331.

Lootsma, F. A. (1970). Boundary properties of penalty functions for constrained optimization problems, *Philips Res. Repts Suppl.* **3**.

Lootsma, F. A. (1972). "A survey of methods for solving constrained optimization problems via unconstrained minimization", in *Numerical Methods for Non-linear Optimization* (F. A. Lootsma, ed.), pp. 313–347, Academic Press, London and New York.

Luenberger, D. G. (1973). *Introduction to Linear and Nonlinear Programming*, Addison-Wesley, Menlo Park, California.

Luenberger, D. G. (1974). A combined penalty function and gradient projection method for nonlinear programming, *J. Opt. Th. Applics.* **14**, pp. 477–495.

Lyness, J. N. (1976). "An interface problem in numerical software", *Proceedings of the 6th Manitoba Conference on Numerical Mathematics*, pp. 251–263.

Lyness, J. N. (1977a). "Has numerical differentiation a future?" *Proceedings of the 7th Manitoba Conference on Numerical Mathematics*, pp. 107–129.

Lyness, J. N. (1977b). "Quid, quo, quadrature?" in *The State of the Art in Numerical Analysis* (D. Jacobs, ed.), pp. 535–560, Academic Press, London and New York.

Lyness, J. N. and Moler, C. B. (1967). Numerical differentiation of analytic functions, *SIAM J. Numer. Anal.* **4**, pp. 202–210.

Lyness J. N. and Sande, G. (1971). ENTCAF and ENTCRE: Evaluation of normalized Taylor coefficients of an analytic function, *Comm. ACM* **14**, pp. 669–675.

Madsen, K. (1975). An algorithm for the minimax solution of overdetermined systems of linear equations, *J. Inst. Maths. Applics.* **16**, pp. 321–328.

Mangasarian, O. L. (1969). *Nonlinear Programming*, McGraw-Hill Book Co., New York.

Mangasarian, O. L. (1975). Unconstrained Lagrangians in nonlinear programming, *SIAM J. Control and Optimization* **13**, pp. 772–791.

Maratos, N. (1978). *Exact Penalty Function Algorithms for Finite-Dimensional and Control Optimization Problems*, Ph. D. Thesis, University of London.

Markowitz, H. M. (1957). The elimination form of the inverse and its applications to linear programming, *Management Science* **3**, pp. 255–269.

Marquardt, D. (1963). An algorithm for least-squares estimation of nonlinear parameters, *SIAM J. Appl. Math.* **11**, pp. 431–441.

Marwil, E. (1978). *Exploiting Sparsity in Newton-Type Methods*, Ph. D. Thesis, Cornell University, Ithaca, New York.

Marsten, R. E. (1978). XMP: A structured library of subroutines for experimental mathematical programming, Report 351, Department of Management Information Systems, University of Arizona, Tucson, Arizona.

Marsten, R. E. and Shanno, D. F. (1979). Conjugate-gradient methods for linearly constrained nonlinear programming, Report 79-13, Department of Management Information Systems, University of Arizona, Tucson, Arizona.

Mayne, D. Q. and Maratos, N. (1979). A first-order, exact penalty function algorithm for equality constrained optimization problems, *Math. Prog.* **16**, pp. 303–324.

Mayne, D. Q. and Polak, E. (1976). Feasible direction algorithms for optimization problems with equality and inequality constraints, *Math. Prog.* **11**, pp. 67–80.

McCormick, G. P. (1969). Anti-zigzagging by bending, *Management Science* **15**, pp. 315–320.

McCormick, G. P. (1970a). The variable reduction method for nonlinear programming, *Management Science* **17**, pp. 146–160.

McCormick, G. P. (1970b). "A second-order method for the linearly constrained nonlinear programming problem", in *Nonlinear Programming* (J. B. Rosen, O. L. Mangasarian and K. Ritter, eds.), pp. 207–243, Academic Press, London and New York.

McCormick, G. P. (1977). A modification of Armijo's step-size rule for negative curvature, *Math. Prog.* **13**, pp. 111–115.

McLean, R. A. and Watson, G. A. (1979). Numerical methods for nonlinear discrete ℓ_1 approximation problems, proceedings of the Oberwolfach Conference on Approximation Theory (to appear).

Miele, A., Cragg, E. E., Iyer, R. R. and Levy, A. V. (1971). Use of the augmented penalty function in mathematical programming, Part I, *J. Opt. Th. Applics.* **8**, pp. 115–130.

Miele, A., Cragg, E. E. and Levy, A. V. (1971). Use of the augmented penalty function in mathematical programming, Part II, *J. Opt. Th. Applics.* **8**, pp. 131–153.

Mifflin, R. (1975). A superlinearly convergent algorithm for minimization without evaluating derivatives, *Math. Prog.* **9**, pp. 100–117.

Mifflin, R. (1977). Semismooth and semiconvex functions in constrained optimization, *SIAM J. Control and Optimization* **15**, pp. 959—972.

Miller, C. E. (1963). "The simplex method for local separable programming", in *Recent Advances in Mathematical Programming* (R. L. Graves and P. Wolfe, eds.), pp. 89–100, McGraw-Hill Book Co., New York.

Moré, J. J. (1977). "The Levenberg-Marquardt algorithm: implementation and theory", in *Numerical Analysis* (G. A. Watson, ed.), pp. 105–116, Lecture Notes in Mathematics 630, Springer-Verlag, Berlin, Heidelberg and New York.

Moré, J. J. (1979a). On the design of optimization software, Report DAMTP 79/NA 8, University of Cambridge.

Moré, J. J. (1979b). "Implementation and testing of optimization software", in *Performance Evaluation of Numerical Software* (L. D. Fosdick, ed.), pp. 253–266, North-Holland, Amsterdam.

Moré, J. J. and Sorensen, D. C. (1979). On the use of directions of negative curvature in a modified Newton method, *Math. Prog.* **15**, pp. 1–20.

Murray, W. (1967). "Ill-conditioning in barrier and penalty functions arising in constrained nonlinear programming", presented at the Princeton Mathematical Programming Symposium, August 14–18, 1967.

Murray, W. (1969a). *Constrained Optimization*, Ph.D. Thesis, University of London.

Murray, W. (1969b). "An algorithm for constrained minimization", in *Optimization* (R. Fletcher, ed.), pp. 247–258, Academic Press, London and New York.

Murray, W. (1971a). An algorithm for finding a local minimum of an indefinite quadratic program, Report NAC 1, National Physical Laboratory, England.

Murray, W. (1971b). Analytical expressions for the eigenvalues and eigenvectors of the Hessian matrices of barrier and penalty functions, *J. Opt. Th. Applics.* **7**, pp. 189–196.

Murray, W. (1972a). "Second derivative methods", in *Numerical Methods for Unconstrained Optimization* (W. Murray, ed.), pp. 57–71, Academic Press, London and New York.

Murray, W. (1972b). "Failure, the causes and cures", in *Numerical Methods for Unconstrained Optimization* (W. Murray, ed.), pp. 107–122, Academic Press, London and New York.

Murray, W. (1976). "Constrained Optimization", in *Optimization In Action* (L. C. W. Dixon, ed.), pp. 217–251, Academic Press, London and New York.

Murray, W. and Overton, M. L. (1980a). A projected Lagrangian algorithm for nonlinear minimax optimization, *SIAM J. Sci. Stat. Comput.* **1**, pp. 345–370.

Murray, W. and Overton, M. L. (1980b). A projected Lagrangian algorithm for nonlinear ℓ_1 optimization, Report SOL 80-4, Department of Operations Research, Stanford University, California.

Murray, W. and Wright, M. H. (1976). Efficient linear search algorithms for the logarithmic barrier function, Report SOL 76-18, Department of Operations Research, Stanford University, California.

Murray, W. and Wright, M. H. (1978). Projected Lagrangian methods based on the trajectories of penalty and barrier functions, Report SOL 78-23, Department of Operations Research, Stanford University, California.

Murray, W. and Wright, M. H. (1980). Computation of the search direction in constrained optimization algorithms, Report SOL 80-2, Department of Operations Research, Stanford University, to appear in *Math. Prog. Study on Constrained Optimization*.

Murtagh, B. A. (1981). *Advanced Linear Programming*, McGraw-Hill Book Co., New York.

Murtagh, B. A. and Sargent, R. H. W. (1969). "A constrained minimization method with quadratic convergence," in *Optimization* (R. Fletcher, ed.), pp. 215–246, Academic Press, London and New York.

Murtagh, B. A. and Saunders, M. A. (1977). MINOS User's Guide, Report SOL 77-9, Department of Operations Research, Stanford University, California.

Murtagh, B. A. and Saunders, M. A. (1978). Large-scale linearly constrained optimization, *Math. Prog.* **14**, pp. 41–72.

Murtagh, B. A. and Saunders, M. A. (1980). A projected Lagrangian algorithm and its implementation for sparse nonlinear constraints, Report SOL 80-1R, Department of Operations Research, Stanford University, California, to appear in *Math. Prog. Study on Constrained Optimization.*

NAG Fortran Library Reference Manual (Mark 8) (1981). Numerical Algorithms Group Limited, Oxford, England.

Nazareth, L. (1977). A conjugate-direction algorithm without line searches, *J. Opt. Th. Applics.* **23**, pp. 373–388.

Nazareth, L. (1979). A relationship between the BFGS and conjugate-gradient algorithms and its implications for new algorithms, *SIAM J. Numer. Anal.* **16**, pp. 794–800.

Nazareth, L. and Nocedal, J. (1978). A study of conjugate-gradient methods, Report SOL 78-29, Department of Operations Research, Stanford University, California.

Nelder, J. A. and Mead, R. (1965). A simplex method for function minimization, *Computer Journal* **7**, pp. 308–313.

Nemirovsky, A. S. and Yudin, D. B. (1979). Effective methods for solving convex programming problems of large size, *Ékonomika i Matematičeskie Metody* **15**, pp. 135–152.

Nocedal, J. (1980). Updating quasi-Newton matrices with limited storage, *Mathematics of Computation* **35**, pp. 773–782.

Numerical Optimization Software Library Reference Manual (1978). Division of Numerical Analysis and Computing, National Physical Laboratory, England.

O'Leary, D. P. (1980a). A discrete Newton algorithm for minimizing a function of many variables, Report 910, Computer Science Center, University of Maryland, College Park, Maryland.

O'Leary, D. P. (1980b). Estimating matrix condition numbers, *SIAM J. Sci. Stat. Comput.* **1**, pp. 205–209.

Oliver, J. (1980). An algorithm for numerical differentiation of a function of one real variable, *J. Comp. Appl. Math.* **6**, pp. 145–160.

Oliver, J. and Ruffhead, A. (1975). The selection of interpolation points in numerical differentiation, *Nordisk Tidskr. Informationsbehandling (BIT)* **15**, pp. 283–295.

Orchard-Hays, W. (1968). *Advanced Linear Programming Computing Techniques*, McGraw-Hill, New York.

Orchard-Hays, W. (1978a). "History of mathematical programming systems", in *Design and Implementation of Optimization Software* (H. J. Greenberg, ed.), pp. 1–26, Sijthoff and Noordhoff, Netherlands,

Orchard-Hays, W. (1978b). "Scope of mathematical programming software", in *Design and Implementation of Optimization Software* (H. J. Greenberg, ed.), pp. 27–40, Sijthoff and Noordhoff, Netherlands,

Orchard-Hays, W. (1978c). "Anatomy of a mathematical programming system", in *Design and Implementation of Optimization Software* (H. J. Greenberg, ed.), pp. 41–102, Sijthoff and Noordhoff, Netherlands,

Oren, S. S. (1974a). Self-scaling variable metric (SSVM) algorithms, Part II: implementation and experiments, *Management Science* **20**, pp. 863–874.

Oren, S. S. (1974b). On the selection of parameters in self-scaling variable metric algorithms, *Math. Prog.* **7**, pp. 351–367.

Oren, S. S. and Luenberger, D. G. (1974). Self-scaling variable metric (SSVM) algorithms, Part I: criteria and sufficient conditions for scaling a class of algorithms, *Management Science* **20**, pp. 845–862.

Oren, S. S. and Spedicato, E. (1976). Optimal conditioning of self-scaling and variable metric algorithms, *Math. Prog.* **10**, pp. 70–90.

Ortega, J. M. and Rheinboldt, W. C. (1970). *Iterative Solution of Nonlinear Equations in Several Variables*, Academic Press, London and New York.

Osborne, M. R. and Ryan, D. M. (1972). "A hybrid algorithm for nonlinear programming", in *Numerical Methods for Non-linear Optimization* (F. A. Lootsma, ed.), pp. 395–410, Academic Press, London and New York.

Osborne, M. R. and Watson, G. A. (1969). An algorithm for minimax approximation in the nonlinear case, *Computer Journal* **12**, pp. 63–68.

Osborne, M. R. and Watson, G. A. (1971). An algorithm for discrete nonlinear ℓ_1 approximation, *Computer Journal* **10**, pp. 172–177.

Paige, C. C. (1980). Error analysis of some techniques for updating orthogonal decompositions, *Mathematics of Computation* **34**, pp. 465–471.

Parkinson, J. M. and Hutchinson, D. (1972). "An investigation into the efficiency of variants of the simplex method", in *Numerical Methods for Non-linear Optimization* (F. A. Lootsma, ed.), pp. 115–135, Academic Press, London and New York.

Perold, A. F. (1981a). "Exploiting degeneracy in the simplex method", in *Large-Scale Linear Programming (Volume 1)* (G. B. Dantzig, M. A. H. Dempster and M. J. Kallio, eds.), pp. 55–66, IIASA Collaborative Proceedings Series, CP-81-51, IIASA, Laxenburg, Austria.

Perold, A. F. (1981b). "A degeneracy-exploiting *LU* factorization for the simplex method", in *Large-Scale Linear Programming (Volume 1)* (G. B. Dantzig, M. A. H. Dempster and M. J. Kallio, eds.), pp. 67–96, IIASA Collaborative Proceedings Series, CP-81-51, IIASA, Laxenburg, Austria.

Perry, A. (1977). A class of conjugate-gradient algorithms with a two-step variable-metric memory, Discussion paper 269, Center for Mathematical Studies in Economics and Management Science, Northwestern University.

Peters, G. and Wilkinson, J. H. (1970). The least-squares problem and pseudo-inverses, *Computer Journal* **13**, pp. 309–316.

Peterson, E. L. (1976). Geometric programming, *SIAM Review* **18**, pp. 1–51.

Polya, G. (1913). Sur un algorithme toujours convergent pour obtenir les polynomes de meilleure approximation de Tchebycheff pour une fonction continue quelconque, Comptes Rendus Hébdomadaires, Scéances de l'Académie des Sciences, Paris.

Pietrzykowski, T. (1962). "Application of the steepest-ascent method to concave programming", in *Proceedings of the IFIPS Congress, Munich, 1962*, pp. 185–189, North-Holland, Amsterdam.

Pietrzykowski, T. (1969). An exact potential method for constrained maxima, *SIAM J. Numer. Anal.* **6**, pp. 299–304.

Powell, M. J. D. (1964). An efficient method for finding the minimum of a function of several variables without calculating derivatives, *Computer Journal* **7**, pp. 155–162.

Powell, M. J. D. (1969). "A method for nonlinear constraints in minimization problems", in *Optimization* (R. Fletcher, ed.), pp. 283–298, Academic Press, London and New York.

Powell, M. J. D. (1970a). "A new algorithm for unconstrained optimization", in *Nonlinear Programming* (J. B. Rosen, O. L. Mangasarian and K. Ritter, eds.), pp. 31–65, Academic Press, London and New York.

Powell, M. J. D. (1970b). "A hybrid method for nonlinear equations", in *Numerical Methods for Nonlinear Algebraic Equations* (P. Rabinowitz, ed.), pp. 87–114, Gordon and Breach, London.

Powell, M. J. D. (1971). On the convergence of the variable metric algorithm, *J. Inst. Maths. Applics.* **7**, pp. 21–36.

Powell, M. J. D. (1972). "Problems relating to unconstrained optimization", in *Numerical Methods for Unconstrained Optimization* (W. Murray, ed.), pp. 29–55, Academic Press, London and New York.

Powell, M. J. D. (1974). "Introduction to constrained optimization", in *Numerical Methods for Constrained Optimization* (P. E. Gill and W. Murray, eds.), pp. 1–28, Academic Press, London and New York.

Powell, M. J. D. (1975). "Convergence properties of a class of minimization algorithms", in *Nonlinear Programming 2* (O. L. Mangasarian, R. R. Meyer and S. M. Robinson, eds.), pp. 1–27, Academic Press, London and New York.

Powell, M. J. D. (1976a). "A view of unconstrained optimization", in *Optimization In Action* (L. C. W. Dixon, ed.), pp. 117–152, Academic Press, London and New York.

Powell, M. J. D. (1976b). Some convergence properties of the conjugate-gradient method, *Math. Prog.* **11**, pp. 42–49.

Powell, M. J. D. (1976c). "Some global convergence properties of a variable metric algorithm without exact line searches", in *SIAM-AMS Proceedings, Volume IX, Mathematical Programming* (R. C. Cottle and C. E. Lemke, eds.), pp. 53–72, American Mathematical Society, Providence, Rhode Island.

Powell, M. J. D. (1977a). Restart procedures for the conjugate-gradient method, *Math. Prog.* **12**, pp. 241–254.

Powell, M. J. D. (1977b). A fast algorithm for nonlinearly constrained optimization calculations, Report DAMTP 77/NA 2, University of Cambridge, England.

Powell, M. J. D. (1977c). "Numerical methods for fitting functions of two variables", in *The State of the Art in Numerical Analysis* (D. Jacobs, ed.), pp. 563–604, Academic Press, London and New York.

Powell, M. J. D. (1978). "The convergence of variable metric methods for nonlinearly constrained optimization calculations", in *Nonlinear Programming 3* (O. L. Mangasarian, R. R. Meyer, and S. M. Robinson, eds.), pp. 27–63, Academic Press, London and New York.

Powell, M. J. D. (1980). "An upper triangular matrix method for quadratic programming", presented at the symposium: Nonlinear Programming 4, Madison, Wisconsin, July 1980.

Powell, M. J. D. (1981). A note on quasi-Newton formulae for sparse second derivative matrices, *Math. Prog.* **20**, pp. 144–151.

Powell, M. J. D. and Toint, P. L. (1979). On the estimation of sparse Hessian matrices, *SIAM J. Numer. Anal.* **16**, pp. 1060–1074.

Ramsin, H. and Wedin, P. Å. (1977). A comparison of some algorithms for the nonlinear least-squares problem, *Nordisk Tidskr. Informationsbehandling (BIT)* **17**, pp. 72–90.

Reid, J. K. (1975). A sparsity-exploiting variant of the Bartels-Golub decomposition for linear programming bases, Report CSS 20, Atomic Energy Research Establishment, Harwell, England.

Reid, J. K. (1976). Fortran subroutines for handling sparse linear programming bases, Report R8269, Atomic Energy Research Establishment, Harwell, England.

Robinson, S. M. (1972). A quadratically convergent algorithm for general nonlinear programming problems, *Math. Prog.* **3**, pp. 145–156.

Robinson, S. M. (1974). Perturbed Kuhn-Tucker points and rates of convergence for a class of nonlinear programming algorithms, *Math. Prog.* **7**, pp. 1–16.

Rockafellar, R. T. (1970). *Convex Analysis*, Princeton University Press, Princeton, New Jersey.

Rockafellar, R. T. (1973a). A dual approach to solving nonlinear programming problems by unconstrained optimization, *Math. Prog.* **5**, pp. 354–373.

Rockafellar, R. T. (1973b). The multiplier method of Hestenes and Powell applied to convex programming, *J. Opt. Th. Applics.* **12**, pp. 555–562.

Rockafellar, R. T. (1974). Augmented Lagrange multiplier functions and duality in nonconvex programming, *SIAM J. Control and Optimization* **12**, pp. 268–285.

Rosen, J. B. (1960). The gradient projection method for nonlinear programming, Part I — linear constraints, *SIAM J. Appl. Math.* **8**, pp. 181–217.

Rosen, J. B. (1961). The gradient projection method for nonlinear programming, Part II — nonlinear constraints, *SIAM J. Appl. Math.* **9**, pp. 514–532.

Rosen, J. B. (1978). "Two-phase algorithm for nonlinear constraint problems", in *Nonlinear Programming 3* (O. L. Mangasarian, R. R. Meyer and S. M. Robinson, eds.), pp. 97–124, Academic Press, London and New York.

Rosen, J. B. and Kreuser, J. (1972). "A gradient projection algorithm for nonlinear constraints", in *Numerical Methods for Non-Linear Optimization* (F. A. Lootsma, ed.), pp. 297–300, Academic Press, London and New York.

Rosenbrock, H. H. (1960). An automatic method for finding the greatest or least value of a function, *Computer Journal* **3**, pp. 175–184.

Ruhe, A. (1979). Accelerated Gauss-Newton algorithms for nonlinear least-squares problems, *Nordisk Tidskr. Informationsbehandling (BIT)* **19**, pp. 356–367.

Ruhe, A. and Wedin, P. Å. (1980). Algorithms for separable nonlinear least-squares problems, *SIAM Review* **22**, pp. 318–337.

Ryan, D. M. (1971). *Transformation Methods in Nonlinear Programming*, Ph.D. Thesis, Australian National University.

Ryan, D. M. (1974). "Penalty and barrier functions", in *Numerical Methods for Constrained Optimization* (P. E. Gill and W. Murray, eds.), pp. 175–190, Academic Press, London and New York.

Sargent, R. W. H. (1974). "Reduced-gradient and projection methods for nonlinear programming", in *Numerical Methods for Constrained Optimization* (P. E. Gill and W. Murray, eds.), pp. 149–174, Academic Press, London and New York.

Sargent, R. W. H. and Gaminibandara, K. (1976). "Optimal design of plate distillation columns", in *Optimization In Action* (L. C. W. Dixon, ed.), pp. 267–314, Academic Press, London and New York.

Sargent, R. W. H. and Murtagh, B. A. (1973). Projection methods for nonlinear programming, *Math. Prog.* **4**, pp. 245–268.

Saunders, M. A. (1976). "A fast, stable implementation of the simplex method using Bartels-Golub updating", in *Sparse Matrix Computations* (J. R. Bunch and D. J. Rose, eds.), pp. 213–226, Academic Press, New York.

Saunders, M. A. (1980). Private communication.

Schittkowski, K. (1980). *Nonlinear Programming Codes*, Springer-Verlag Lecture Notes in Economics and Mathematical Systems, Volume 183, Berlin, Heidelberg and New York.

Schittkowski, K. and Stoer, J. (1979). A factorization method for the solution of constrained linear least-squares problems allowing data changes, *Num. Math.* **31**, pp. 431–463.

Shanno, D. F. (1970). Conditioning of quasi-Newton methods for function minimization, *Mathematics of Computation* **24**, pp. 647–657.

Shanno, D. F. (1978). Conjugate-gradient methods with inexact searches, *Math. of Oper. Res.* **3**, pp. 244–256.

Shanno, D. F. (1980). On variable metric methods for sparse Hessians, *Mathematics of Computation* **34**, pp. 499–514.

Shanno, D. F. and Phua, K. H. (1976). Algorithm 500 — Minimization of unconstrained multivariate functions, *ACM Trans. Math. Software* **2**, pp. 87–94.

Shor, N. Z. (1970). Convergence rate of the gradient descent method with dilation of the space, *Kibernetika* **2**, pp. 102–108.

Shor, N. Z. (1977). The cut-off method with space dilation for solving convex programming problems, *Kibernetika* **13**, pp. 94–95. [English translation in *Cybernetics* **13** (1978), 94–96.]

Shor, N. Z. and Gershovich, V. I. (1979). A family of algorithms for solving convex programming problems, *Kibernetika* **15**, pp. 62–67. [English translation in *Cybernetics* **15** (1980), 502–508.]

Schubert, L. K. (1970). Modification of a quasi-Newton method for nonlinear equations with a sparse Jacobian, *Mathematics of Computation* **24**, pp. 27–30.

Sisser, F. S. (1981). Elimination of bounds in optimization problems by transforming variables, *Math. Prog.* **20**, pp. 110–121.

Smith, B. T., Boyle, J. M., Garbow, B. S., Ikebe, Y., Klema, V. C. and Moler, C. B. (1974). *Matrix Eigensystem Routines — EISPACK Guide*, Lecture Notes in Computer Science 6, Springer-Verlag, Berlin, Heidelberg and New York.

Sorensen, D. (1980a). Newton's method with a model trust region modification, Report ANL-80-106, Argonne National Laboratory, Argonne, Illinois.

Sorensen, D. (1980b). The Q-superlinear convergence of a collinear scaling algorithm for unconstrained minimization, *SIAM J. Numer. Anal.* **17**, pp. 84–114.

Spedicato, E. (1975). "On condition numbers of matrices in rank two minimization algorithms", in *Towards Global Optimization* (L. C. W. Dixon and G. P. Szegö, eds.), pp. 196–210, North-Holland, Amsterdam.

Spendley, W., Hext, G. R. and Himsworth F. R. (1962). Sequential application of simplex designs in optimization and evolutionary design, *Technometrics* **4**, pp. 441–461.

Stepleman, R. S. and Winarsky, N. D. (1979). Adaptive numerical differentiation, *Mathematics of Computation* **33**, pp. 1257–1264.

Stewart, G. W. (1967). A modification of Davidon's method to accept difference approximations of derivatives, *J. ACM* **14**, pp. 72–83.

Stewart, G. W. (1973). *Introduction to Matrix Computations*, Academic Press, London and New York.

Stiefel, E. (1960). Note on Jordan elimination, linear programming and Tschebyscheff approximation, *Num. Math.* **2**, pp. 1–17.

Stoer, J. (1971). On the numerical solution of constrained least-squares problems, *SIAM J. Numer. Anal.* **8**, pp. 382–411.

Stoer, J. (1975). On the convergence rate of imperfect minimization algorithms in Broyden's β-class, *Math. Prog.* **9**, pp. 313–335.

Stoer, J. (1977). On the relation between quadratic termination and convergence properties of minimization algorithms, Part I, theory, *Num. Math.* **28**, pp. 343–366.

Strang, G. (1976). *Linear Algebra and its Applications*, Academic Press, London and New York.

Swann, W. H. (1972). "Direct search methods", in *Numerical Methods for Unconstrained Optimization* (W. Murray, ed.), pp. 13–28, Academic Press, London and New York.

Swann, W. H. (1974). "Constrained optimization by direct search", in *Numerical Methods for Constrained Optimization* (P. E. Gill and W. Murray, eds.), pp. 191–217, Academic Press, London and New York.

Tapia, R. A. (1974a). Newton's method for problems with equality constraints, *SIAM J. Numer. Anal.* **11**, pp. 174–196.

Tapia, R. A. (1974b). Newton's method for optimization problems with equality constraints, *SIAM J. Numer. Anal.* **11**, pp. 874–886.

Tapia, R. A. (1977). Diagonalized multiplier methods and quasi-Newton methods for constrained optimization, *J. Opt. Th. Applics.* **22**, pp. 135–194.

Tapia, R. A. (1978). "Quasi-Newton methods for equality constrained optimization: equivalence of existing methods and a new implementation", in *Nonlinear Programming 3* (O. L. Mangasarian, R. R. Meyer and S. M. Robinson, eds.), pp. 125–164, Academic Press, London and New York.

Tarjan, R. (1972). Depth-first search and linear graph algorithms, *SIAM J. Comput.* **1**, pp. 146-160.

Thapa, M. N. (1979). A note on sparse quasi-Newton methods, Report SOL 79-13, Department of Operations Research, Stanford University, California.

Thapa, M. N. (1980). *Optimization of Unconstrained Functions with Sparse Hessian Matrices*, Ph.D. Thesis, Stanford University, California.

Toint, P. L. (1977). On sparse and symmetric matrix updating subject to a linear equation, *Mathematics of Computation* **31**, pp. 954–961.

Toint, P. L. (1978). Some numerical results using a sparse matrix updating formula in unconstrained optimization, *Mathematics of Computation* **32**, pp. 839–851.

Toint, P. L. (1979). On the superlinear convergence of an algorithm for solving a sparse minimization problem, *SIAM J. Numer. Anal.* **16**, pp. 1036–1045.

Tomlin, J. A. (1975a). An accuracy test for updating triangular factors, *Math. Prog. Study* **4**, pp. 142–145.

Tomlin, J. A. (1975b). On scaling linear programming problems, *Math. Prog. Study* **4**, pp. 146–166.

Tomlin, J. A. (1976). Robust implementation of Lemke's method for the linear complementarity problem, Report SOL 76-24, Department of Operations Research, Stanford University.

Topkis, D. M. and Veinott, A. F., Jr. (1967). On the convergence of some feasible direction algorithms for nonlinear programming, *SIAM J. Control* **5**, pp. 268–279.

Van der Hoek, G. (1979). Asymptotic properties of reduction methods applying linearly equality constrained reduced problems, Report 7933, Econometric Institute, Erasmus University, Rotterdam.

Watson, G. A. (1979). The minimax solution of an overdetermined system of nonlinear equations, *J. Inst. Maths. Applics.* **23**, pp. 167–180.

Wedin, P. Å. (1974). On the Gauss-Newton method for the nonlinear least-squares problem, Report 23, Swedish Institute for Applied Mathematics (ITM), Stockholm.

Wilde, D. J. (1978). *Globally Optimal Design*, John Wiley and Sons, New York and Toronto.

Wilkinson, J. H. (1963). *Rounding Errors in Algebraic Processes*, Notes on Applied Sciences 32, Her Majesty's Stationery Office, London; Prentice-Hall, Inc. [also published by Englewood Cliffs, New Jersey].

Wilkinson, J. H. (1965). *The Algebraic Eigenvalue Problem*, Oxford University Press.

Wilkinson, J. H. and Reinsch, C. (1971). *Handbook for Automatic Computation, Vol. II*, Springer-Verlag, Berlin, Heidelberg and New York.

Wilson, R. B. (1963). *A Simplicial Algorithm for Concave Programming*, Ph.D. Thesis, Harvard University.

Wolfe, P. (1959). The simplex method for quadratic programming, *Econometrica* **27**, pp. 382–398.

Wolfe, P. (1962). The reduced-gradient method, unpublished manuscript, the RAND Corporation.

Wolfe, P. (1963a). A technique for resolving degeneracy in linear programming, *SIAM J. Appl. Math.* **11**, pp. 205–211.

Wolfe, P. (1963b). "Methods of nonlinear programming", in *Recent Advances in Mathematical Programming* (J. Abadie, ed.), pp. 67–86, North-Holland, Amsterdam.

Wolfe, P. (1966). On the convergence of gradient methods under constraints, IBM Research report, Zurich Laboratory.

Wolfe, P. (1967). "Methods of nonlinear programming", in *Nonlinear programming* (J. Abadie, ed.), pp. 97–131, North-Holland, Amsterdam.

Wolfe, P. (1969). Convergence conditions for ascent methods, *SIAM Review,* **11**, pp. 226–235.

Wolfe, P. (1976). Checking the calculation of gradients, Report RC 6007, IBM Yorktown Heights Research Center (May 1976).

Wolfe, P. (1980a). A bibliography for the ellipsoid algorithm, Report RC 8237, IBM Yorktown Heights Research Center (April 1980).

Wolfe, P. (1980b). The ellipsoid algorithm, in *Optima,* **1** (Newsletter of the Mathematical Programming Society, June 1980).

Wright, M. H. (1976). *Numerical Methods for Nonlinearly Constrained Optimization,* Ph.D. Thesis, Stanford University, California.

Zangwill, W. I. (1965). Nonlinear programming by sequential unconstrained maximization, Working Paper 131, Center for Research in Management Science, University of California, Berkeley.

Zangwill, W. I. (1967a). Nonlinear programming via penalty functions, *Management Science* **13**, pp. 344–358.

Zangwill, W. I. (1967b). Algorithm for the Chebyshev problem, *Management Science* **14**, pp. 58–78.

Zoutendjik, G. (1970). "Nonlinear programming, computational methods", in *Integer and Nonlinear Programming* (J. Abadie, ed.), pp. 37–86, North-Holland, Amsterdam.

Zuhovickii, S. I., Polyak, R. A. and Primak, M. E. (1963). An algorithm for the solution of convex Cebysev approximations, *Soviet Math.* **4**, pp. 901–904.

INDEX

*I leaned back and took down the great index to which he referred...
the record of old cases, mixed with the accumulated information of a lifetime.*

—A. CONAN DOYLE, *in "The Adventure of the Sussex Vampire"* (1920)